WITHDRAW

Methods in Microbiology
Volume 38

Recent titles in the series

Volume 24 *Techniques for the Study of Mycorrhiza*
JR Norris, DJ Reed and AK Varma

Volume 25 *Immunology of Infection*
SHE Kaufmann and D Kabelitz

Volume 26 *Yeast Gene Analysis*
AJP Brown and MF Tuite

Volume 27 *Bacterial Pathogenesis*
P Williams, J Ketley and GPC Salmond

Volume 28 *Automation*
AG Craig and JD Hoheisel

Volume 29 *Genetic Methods for Diverse Prokaryotes*
MCM Smith and RE Sockett

Volume 30 *Marine Microbiology*
JH Paul

Volume 31 *Molecular Cellular Microbiology*
P Sansonetti and A Zychlinsky

Volume 32 *Immunology of Infection, 2nd edition*
SHE Kaufmann and D Kabelitz

Volume 33 *Functional Microbial Genomics*
B Wren and N Dorrell

Volume 34 *Microbial Imaging*
T Savidge and C Pothoulakis

Volume 35 *Extremophiles*
FA Rainey and A Oren

Volume 36 *Yeast Gene Analysis, 2nd edition*
I Stansfield and MJR Stark

Volume 37 *Immunology of Infection*
D Kabelitz and SHE Kaufmann

Methods in Microbiology

Volume 38
Taxonomy of Prokaryotes

Edited by

Fred Rainey

Department of Biological Sciences,
Louisiana State University,
Baton Rouge,
LA, USA

and

Aharon Oren

Department of Plant & Environmental Sciences,
The Alexander Silberman Institute of Life Sciences,
The Hebrew University of Jerusalem,
Edmond Safra Campus Givat Ram,
Jerusalem, Israel

Amsterdam • Boston • Heidelberg • London • New York • Oxford
Paris • San Diego • San Francisco • Singapore • Sydney • Tokyo
Academic Press is an imprint of Elsevier

Academic Press is an imprint of Elsevier
32 Jamestown Road, London, NW1 7BY, UK
525 B Street, Suite 1900, San Diego, CA 92101-4495, USA
225 Wyman Street, Waltham, MA 02451, USA
Radarweg 29, PO Box 211, 1000 AE Amsterdam, The Netherlands

First edition 2011

ISBN: 978-0-12-387730-7
ISSN: 0580-9517 (Series)

For information on all Academic Press publications
visit our website at www.elsevierdirect.com

Printed and bound in UK

11 12 13 14 10 9 8 7 6 5 4 3 2 1

Cover image: **"Organized Variation" by Jason Paul Fristensky**

Dedication

The editors dedicate this volume to the memory of Peter H.A. Sneath (1923–2011) and George A. Zavarzin (1933–2011) in recognition of their great contributions to bacterial taxonomy.

Contents

Colour Plate Section can be found at the back of this book

Series Advisors

Gordon Dougan The Wellcome Trust Sanger Institute, Wellcome Trust Genome Campus, Hinxton, Cambridge CBIO ISA, UK

Graham J Boulnois Schroder Ventures Life Science Advisers (UK) Limited, 71 Kingsway, London WC2B 6ST, UK

Jim Prosser School of Medical Sciences, University of Aberdeen, Cruickshank Building, St Machar Drive, Aberdeen, AB24 3UU, UK

Ian R Booth School of Medical Sciences, University of Aberdeen, Institute of Medical Sciences, Foresterhill, Aberdeen AB25 2ZD, UK

David A Hodgson Department of Biological Sciences, University of Warwick, Conventry CV4 7AL, UK

David H Boxer University of Dundee, Dundee DD1 4HN, UK

List of Contributors

Luciana Albuquerque, Center for Neuroscience and Cell Biology, University of Coimbra, Coimbra, Portugal

Oliver Bezuidt, University of Pretoria, Department of Biochemistry, Bioinformatics and Computational Biology Unit, Pretoria, South Africa

Jochen Blom, Center for Biotechnology (CeBiTec), Bielefeld University, Bielefeld, Germany

Rainer Borriss, ABiTEP GmbH, Berlin, Germany

Hans-Jürgen Busse, Institut für Bakteriologie, Mykologie und Hygiene, Abteilung für Pathobiologie, Veterinärmedizinische Universität Wien, Wien, Austria

Milton S. da Costa, Department of Life Sciences, University of Coimbra, Coimbra, Portugal

Paul De Vos, Laboratory of Microbiology, Ghent University, Gent, Belgium

Michel Drancourt, Unité de Recherche sur les Maladies Infectieuses et Tropicales Emergentes UMR CNRS 6236 IRD 198, IFR48, Institut Hospitalier Universitaire POLMIT, Université de la Méditerranée, Marseille, France

B. Bédis Dridi, Unité de Recherche sur les Maladies Infectieuses et Tropicales Emergentes UMR CNRS 6236 IRD 198, IFR48, Institut Hospitalier Universitaire POLMIT, Université de la Méditerranée, Marseille, France

Frank Oliver Glöckner, Max Planck Institute for Marine Microbiology, Celsiusstrasse 1, 28359 Bremen, Germany; Jacobs University Bremen GmbH, Campus Ring 1, 28759 Bremen, Germany

Hans-Peter Klenk, DSMZ - German Collection of Microorganisms and Cell Cultures, Braunschweig, Germany

Noel R. Krieg, Department of Biological Sciences, Virginia Tech, Blacksburg, Virginia 24061, USA

Arantxa López-López, Marine Microbiology Group, Department of Ecology and Marine Resources, IMEDEA (CSIC-UIB), C/Miquel Marqués 21, E-07190 Esporles, Illes Balears, Spain

Wolfgang Ludwig, Lehrstuhl für Mikrobiologie, Technische Universität München, Am Hochanger 4, 85350 Freising, Germany

Mostafa Mesbah, Department of Pharmacognosy, Faculty of Pharmacy, Suez Canal University, Ismailia 41522, Egypt

Noha M. Mesbah, Department of Biochemistry, Faculty of Pharmacy, Suez Canal University, Ismailia 41522, Egypt

M. Fernanda Nobre, Department of Life Sciences, University of Coimbra, Coimbra, Portugal

Aharon Oren, Department of Plant and Environmental Sciences, Institute of Life Sciences, and the Moshe Shilo Minerva Center for Marine Biogeochemistry, The Hebrew University of Jerusalem, Jerusalem 91904, Israel

Penelope J. Padgett, Environment, Health & Safety, University of North Carolina, North Carolina 27599-1650, USA

Natal'ya V. Potekhina, Department of Biology, M. V. Lomonosov Moscow State University, Moscow, Russia

Fred A. Rainey, Department of Biological Sciences, Louisiana State University, Baton Rouge, Louisiana 70803, USA

Oleg Reva, University of Pretoria, Department of Biochemistry, Bioinformatics and Computational Biology Unit, Pretoria, South Africa

Manfred Rohde, Department of Medical Microbiology, Helmholtz Centre for Infection Research (HZI), Braunschweig, Germany

Ramon Rosselló-Móra, Marine Microbiology Group, Department of Ecology and Marine Resources, IMEDEA (CSIC-UIB), C/Miquel Marqués 21, E-07190 Esporles, Illes Balears, Spain

Christian Rueckert, Center for Biotechnology (CeBiTec), Bielefeld University, Bielefeld, Germany

Peter Schumann, Leibniz-Institut DSMZ-Deutsche Sammlung von Mikroorganismen und Zellkulturen GmbH, Braunschweig, Germany

Alexander S. Shashkov, N. D. Zelinsky Institute of Organic Chemistry, Russian Academy of Sciences, Moscow, Russia

Galina M. Streshinskaya, Department of Biology, M. V. Lomonosov Moscow State University, Moscow, Russia

Elena M. Tul'skaya, Department of Biology, M. V. Lomonosov Moscow State University, Moscow, Russia

Mercedes Urdiain, Marine Microbiology Group, Department of Ecology and Marine Resources, IMEDEA (CSIC-UIB), C/Miquel Marqués 21, E-07190 Esporles, Illes Balears, Spain

Robin Wait, Kennedy Institute of Rheumatology, Imperial College, London, United Kingdom

William B. Whitman, Department of Microbiology, University of Georgia, Athens, GA 30602, USA

Atteyet-Alla Fetouh Yassin, Institut für Medizinische Mikrobiologie und Immunologie der Universität Bonn, Bonn, Germany

Pelin Yilmaz, Max Planck Institute for Marine Microbiology, Celsiusstrasse 1, 28359 Bremen, Germany; Jacobs University Bremen GmbH, Campus Ring 1, 28759 Bremen, Germany

1 Taxonomy of Prokaryotes – Introduction

Fred A. Rainey[1] **and Aharon Oren**[2]
[1] Department of Biological Sciences, Louisiana State University, Baton Rouge, Louisiana 70803, USA; [2] Department of Plant and Environmental Sciences, Institute of Life Sciences, and the Moshe Shilo Minerva Center for Marine Biogeochemistry, The Hebrew University of Jerusalem, Jerusalem 91904, Israel

Prokaryote diversity has been the subject of increased attention in the last two decades. With a total estimated number of $4-6 \times 10^{30}$ individuals, prokaryotes are the most numerous organisms on Earth (Whitman *et al.*, 1998). The application of molecular-based, culture-independent approaches has shown that we know only a small fraction of the diversity as existing species in culture. The result has been an intensification of culturing efforts by the community of microbiologists, aiming to discover the role of the yet-uncultured diversity in the biosphere and the biotechnological potential within. Such culturing efforts have yielded numerous bacterial and archaeal isolates, many of which have been fully characterized and described as novel taxa. The number of descriptions of new species of prokaryotes is growing at an ever-increasing rate: The numbers of validly published names of new species in the years 1990, 1995, 2000, 2005 and 2010 were 140, 217, 275, 528 and 611, respectively (information derived from www.bacterio.cict.fr; see also Euzéby, 1997). Still, the total numbers of prokaryote taxa described are small: As of 10 May 2011, the number of prokaryotic species names with standing in the nomenclature was 10,706 (of which 31 are later homotypic synonyms, 1237 are new combinations, 13 are nomina nova, about 288 are considered as later heterotypic synonyms and 67 are illegitimate). These were classified in 2010 genera (of which about 105 are considered as synonyms and 20 are illegitimate) and 291 families (of which 6 are illegitimate). The majority of descriptions of new prokaryote taxa are published in the *International Journal of Systematic and Evolutionary Microbiology*. However, a number of descriptions are published in other journals such as *Systematic and Applied Microbiology*, *Extremophiles*, *Applied and Environmental Microbiology*, *Journal of Bacteriology*, *Journal of Clinical Microbiology*, etc., and procedures have been established for the valid publication of names earlier published in different frameworks (Tindall *et al.*, 2006).

The species concept for prokaryotes is a long-debated issue; an in-depth discussion on this topic is outside the scope of this book, but it has been discussed

METHODS IN MICROBIOLOGY, VOLUME 38
0580-9517 DOI 10.1016/B978-0-12-387730-7.00001-2

and reviewed in many publications. The most accepted concept, considered to be pragmatic and useful for species definition, is the so-called phylo-phenetic species concept – a monophyletic and genomically coherent cluster of individual organisms (strains) that show a high degree of overall similarity in many independent characteristics – and is diagnosable by one or more discriminative phenotypic properties (Rosselló-Mora and Amann, 2001). Based on such a definition, it is clear that a distinct genospecies (i.e. a species discernible only by nucleic acid comparisons) that cannot be differentiated from another genospecies on the basis of any known phenotypic property should not be named as a new species until some phenotypic differentiating property is found.

Based on the experience of the last 25 years, a pragmatic definition of the prokaryote species has emerged, on the basis of the recommendations published in 1987 by a committee of experts (Wayne *et al.*, 1987) and updated in 2002 (Stackebrandt *et al.*, 2002). The species concept is based on a 'polyphasic' approach, which includes description of diagnostic phenotypic features combined with genomic properties (Vandamme *et al.*, 1996; Gillis *et al.*, 2001). Individually, many of the phenotypic and chemotaxonomic characters used as diagnostic properties are insufficient to delineate a species, but together they provide sufficient descriptive information to allow the definition of a species. The more properties are included in the descriptions, the more robust and stable the resulting classification will be. Different properties have different resolving power; some are species specific, while others are valuable for discriminating genera, families and orders. Properties tested generally include morphological characters (cell shape and size, the Gram reaction, cell inclusions and presence and nature of the surface layers, including extracellular capsules), information on motility (presence of flagella, their number and the way they are inserted into the cell, and gliding movement), the mode of nutrition (assimilatory metabolism) and energy generation (dissimilatory metabolism), the cells' relationship to molecular oxygen, temperature, pH, tolerance towards and requirement for salt and many others. Genotypic information includes 16S rRNA gene sequence information, which is a very valuable tool for rapidly placing any isolate within the current classification scheme, at least down to the family and genus level. For final identification of the species, DNA–DNA hybridization tests are the ultimate tool to decide whether two isolates should be classified in the same species. More specific tests such as serotyping or phage typing may be necessary for certain groups of microorganisms to obtain a reliable identification.

In a recent paper titled 'Notes on the characterization of prokaryote strains for taxonomic purposes' Tindall *et al.* (2010) presented an outline of the key elements required for the characterization of new species of prokaryotes, 'with a view to providing an overview of some of the pitfalls commonly encountered in taxonomic papers'. When a strain or set of strains are shown to be novel taxa, they should be characterized as comprehensively as possible, based on a polyphasic approach. The goal of this characterization is to place them within the hierarchical framework laid down by the Bacteriological Code (1990 revision) (Lapage et al., 1992) and provide a description of the taxa. Strains should be allocated to a genus and a species (and/or subspecies). For many groups of prokaryotes, taxonomic subcommittees of the International Committee on

Systematics of Prokaryotes (ICSP) have established recommended minimal standards for the description of new isolates. Groups of prokaryotes for which such minimal standards have been published are the genera *Brucella* (Corbel and Brinley-Morgan, 1975), *Moraxella* and *Acinetobacter* (Bøvre and Hendriksen, 1976), *Mycobacterium* (Levy-Frebault and Portales, 1992), *Helicobacter* (Dewhirst *et al.*, 2000) and *Staphylococcus* (Freney *et al.*, 1999); the families *Campylobacteraceae* (Ursing *et al.*, 1994), *Flavobacteriaceae* (Bernardet *et al.*, 2002), *Halomonadaceae* (Arahal *et al.*, 2007) and *Pasteurellaceae* (Christensen *et al.*, 2007); the order *Halobacteriales* (Oren *et al.*, 1997); the suborder *Micrococcineae* (Schumann *et al.*, 2009); the class *Mollicutes* (Brown *et al.*, 2007); the methanogenic Archaea (Boone and Whitman, 1988); the root- and stem-nodulating bacteria (Graham *et al.*, 1991); the anoxygenic phototrophic bacteria (Imhoff and Caumette, 2004); and the aerobic, endospore-forming bacteria (Logan *et al.*, 2009). Thus, recommended minimal standards are not available for the majority of groups of prokaryotes, and those available may not necessarily be current.

Obtaining and analysing the data to be collected for a full characterization of isolates towards their description as new species involve a wide range of experimental and analytical techniques, many of which are not at all practised in an individual laboratory. Earlier volumes in the series *Methods in Microbiology*, published from 1969 onwards, have been devoted to different aspects of the taxonomic characterization of isolates, most recently in the 1980s (volumes 18 and 19; Gottschalk (1985) and Colwell and Grigorova (1987)). Since the publication of these volumes, the field of prokaryote taxonomy has been revolutionized by the application of 16S rRNA gene sequence–based phylogenetic analysis. Many other established methodologies have been modified with time, and new equipment and materials have become available in recent years. The aim of this volume is therefore to bring together the techniques needed to characterize and describe new prokaryote isolates and to present the underlying scientific theory as well as detailed protocols that anyone with laboratory experience could carry out independently. It starts with a chapter describing the steps and process of characterizing a new strain and describing it as a novel taxon (Rainey). Subsequent chapters are dedicated to phenotypic and physiological characterization methods (Krieg and Padgett), electron and light microscopy (Rohde), peptidoglycan structure (Schumann), teichoic acids (Potekhina *et al.*), polar lipids, fatty acids and quinones (Da Costa *et al.*), mycolic acids (Yassin), polyamines (Busse), pigments (Oren) and characterization of cellular components by matrix-assisted laser desorption ionization time of flight (MALDI-TOF) mass spectrometry (Dridi and Drancourt). Other chapters are devoted to different aspects of the genotypic characterization of prokaryotes: the determination of the G+C content of the DNA (Mesbah *et al.*), DNA–DNA hybridization (Rosselló-Móra *et al.*), analysis of 16S rRNA gene sequence data (Ludwig *et al.*), multilocus sequence analysis (De Vos) and whole genome sequence comparisons in taxonomy (Borriss *et al.*). The final chapter gives advice on how to name new genera and species of prokaryotes (Oren).

The editors hope that the information collected in this volume will provide useful practical guidelines for the characterization and description of novel types of prokaryotes in the years to come, by experienced taxonomists as well as by newcomers in the field of prokaryote diversity and taxonomy.

References

Arahal, D. R., Vreeland, R. H., Litchfield, C. D., Mormile, M. R., Tindall, B. J., Oren, A., Bejar, V., Quesada, E. and Ventosa, A. (2007). Recommended minimal standards for describing new taxa of the family *Halomonadaceae*. *Int. J. Syst. Evol. Microbiol.* **57**, 2436–2446.

Bernardet, J.-F., Nakagawa, Y. and Holmes, B. (2002). Proposed minimal standards for describing new taxa of the family *Flavobacteriaceae* and emended description of the family. *Int. J. Syst. Evol. Microbiol.* **52**, 1049–1070.

Boone, D. R. and Whitman, W. B. (1988). Proposal of minimal standards for describing new taxa of methanogenic bacteria. *Int. J. Syst. Bacteriol.* **38**, 212–219.

Bøvre, K. and Hendriksen, S. D. (1976). Minimal standards for description of new taxa within the genera *Moraxella* and *Acinetobacter*: proposal by the subcommittee on *Moraxella* and *Acinetobacter*. *Int. J. Syst. Bacteriol.* **26**, 92–96.

Brown, D. R., Whitcomb, R. F. and Bradbury, J. M. (2007). Revised minimal description of new species of the class *Mollicutes* (division *Tenericutes*). *Int. J. Syst. Evol. Microbiol.* **57**, 2703–2719.

Christensen, H., Kuhnert, P., Busse, H.-J., Frederiksen, W. C. and Bisgaard, M. (2007). Proposed minimal standards for the description of genera, species and subspecies of the *Pasteurellaceae*. *Int. J. Syst. Evol. Microbiol.* **57**, 166–178.

Colwell, R. R. and Grigorova, R. (eds.), (1987). Current methods for classification and identification of microorganisms. *Methods in Microbiology,* Vol. 19, Academic Press, London.

Corbel, M. J. and Brinley-Morgan, W. J. (1975). Proposal for minimal standards for description of new species and biotypes of the genus. *Brucella. Int. J. Syst. Bacteriol.* **25**, 83–89.

Dewhirst, F. E., Fox, J. G. and On, S. L. W. (2000). Recommended minimal standards for describing new species of the genus *Helicobacter*. *Int. J. Syst. Evol. Microbiol.* **50**, 2231–2237.

Euzéby, J. (1997). List of bacterial names with standing in nomenclature: a folder available on the internet. *Int. J. Syst. Bacteriol.* **47**, 590–592.

Freney, J., Kloos, W. E., Hajek, V., Webster, J. A., Bes, M., Brun, Y. and Vernozy-Rozand, C. (1999). Recommended minimal standards for description of new staphylococcal species. *Int. J. Syst. Bacteriol.* **49**, 489–502.

Gillis, M., Vandamme, P., de Vos, P., Swings, J., and Kersters, K. (2001). Polyphasic taxonomy, In *Bergey's Manual of Systematic Bacteriology, 2nd ed., Vol. 1, The Archaea and the Deeply Branching and Phototrophic Bacteria* (D. R. Boone, and R. W. Castenholz, Eds., G. M. Garrity, Editor-in-Chief), pp. 43-48. Springer-Verlag, New York.

Gottschalk, G. (1985). *Methods in Microbiology*. Vol. 18, Academic Press, London.

Graham, P. H., Sadowsky, M. J., Keyser, H. H., Barnet, Y. M., Bradley, R. S, Cooper, J. E., de Ley, D. J., Jarvis, B. D. W., Roslycky, E. B., Strijdom, B. W. and Young, J. P. W. (1991). Proposed minimal standards for the description of new genera and species of root- and stem-nodulating bacteria. *Int. J. Syst. Bacteriol.* **41**, 582–587.

Imhoff, J. F. and Caumette, P. (2004). Recommended standards for the description of new species of anoxygenic phototrophic bacteria. *Int. J. Syst. Evol. Microbiol.* **54**, 1415–1421.

Lapage, S. P., Sneath, P. H. A., Lessel, E. F., Jr., Skerman, V. B. D., Seelinger, H. P. R. and Clark, W. A. (eds.), (1992). *International Code of Nomenclature of Bacteria (1990 Revision)"* American Society for Microbiology, Washington, DC.

Levy-Frebault, V. V. and Portales, F. (1992). Proposed minimal standards for the genus *Mycobacterium* and for description of new slowly growing *Mycobacterium* species. *Int. J. Syst. Bacteriol.* **42**, 315–323.

Logan, N. A., Berge, O., Bishop, A. H., Busse, H.-J., De Vos, P., Fritze, D., Heyndrickx, M., Kämpfer, P., Rabinovitch, L., Salkinoja-Salonen, M. S., Seldin, L. and Ventosa, A. (2009). Proposed minimal standards for describing new taxa of aerobic, endospore-forming bacteria. *Int. J. Syst. Evol. Microbiol.* **59**, 2114–2121.

Oren, A., Ventosa, A. and Grant, W. D. (1997). Proposed minimal standards for description of new taxa in the order *Halobacteriales*. *Int. J. Syst. Bacteriol.* **47**, 233–238.

Rosselló-Mora, R. and Amann, R. (2001). The species concept for prokaryotes. *FEMS Microbiol. Rev.* **25**, 39–67.

Schumann, P., Kämpfer, P., Busse, H.-J., and Evtushenko, L. I., for the Subcommittee on the Taxonomy of the Suborder *Micrococcineae* of the International Committee on Systematics of Prokaryotes. (2009). Proposed minimal standards for describing new genera and species of the suborder *Micrococcineae*. Int. J. Syst. Evol. Microbiol. 59, 1823-1849.

Stackebrandt, E., Frederiksen, W., Garrity, G. M., Grimont, P. A. D., Kämpfer, P., Maiden, M. C. J., Nesme, X., Rosselló-Mora, R., Swings, J., Trüper, H. G., Vauterin, L., Ward, A. C. and Whitman, W. B. (2002). Report of the ad hoc committee for the re-evaluation of the species definition in bacteriology. *Int. J. Syst. Evol. Microbiol.* **52**, 1043–1047.

Tindall, B. J., Kämpfer, P., Euzéby, J. and Oren, A. (2006). Valid publication of names of prokaryotes according to the rules of nomenclature: past history and current practice. *Int. J. Syst. Evol. Microbiol.* **56**, 2715–2720.

Tindall, B. J., Rosselló-Móra, R., Busse, H.-J., Ludwig, W. and Kämpfer, P. (2010). Notes on the characterization of prokaryote strains for taxonomic purposes. *Int. J. Syst. Evol. Microbiol.* **60**, 249–266.

Ursing, J. B., Lior, H. and Owen, R. J. (1994). Proposal of minimal standards for describing new species of the family *Campylobacteraceae*. *Int. J. Syst. Bacteriol.* **44**, 842–845.

Vandamme, P., Pot, B., Gillis, M., De Vos, P., Kersters, K. and Swings, J. (1996). Polyphasic taxonomy, a consensus approach to bacterial systematics. *Microbiol. Rev.* **60**, 407–438.

Wayne, L. G., Brenner, D. J., Colwell, R. R., Grimont, P. A. D., Kandler, O., Krichevsky, M. I., Moore, L. H., Moore, W. E. C., Murray, R. G. E., Stackebrandt, E., Starr, M. P. and Trüper, H. G. (1987). Report of the ad hoc committee on reconciliation of approaches to bacterial systematics. *Int. J. Syst. Bacteriol.* **37**, 463–464.

Whitman, W. B., Coleman, D. C. and Wiebe, W. J. (1998). Prokaryotes: the unseen majority. *Proc. Natl. Acad. Sci. U. S. A.* **95**, 6578–6583.

2 How to Describe New Species of Prokaryotes

Fred A. Rainey
Department of Biological Sciences, Louisiana State University, Baton Rouge, Louisiana 70803, USA

♦♦

CONTENTS

♦♦♦♦♦♦ I. DESCRIPTION OF NEW SPECIES OF PROKARYOTES – A POLYPHASIC APPROACH

The aim of this chapter is to outline the steps in the process of characterizing a bacterial or archaeal isolate or group of strains and presenting that information in the form of a taxonomic description. The techniques used to obtain the data required for the determination of the taxonomic status of an isolate and its subsequent description are provided in Chapters 3–19 of this volume. The majority of those wishing to describe a new taxon start out with a single strain or preferably a group of strains which came from a number of sources including isolates from general diversity studies, isolates from selective enrichments studies, or clinical isolates that are causative agents of human or animal disease or opportunistic infections. The preliminary identification of isolates that are to be characterized and described has normally been made from a 16S rRNA gene sequence, from a fatty acid profile or using test strip identification systems.

As pointed out in the introductory chapter of this volume, prokaryotes are currently characterized using a polyphasic approach that brings together a variety of phenotypic, chemotaxonomic and genotypic data that comprise the formal description of a novel taxon. The recent paper of Tindall *et al.* (2010) outlines the data that should be acquired and analysed in characterization studies of prokaryotes. In addition, there are a limited number of publications presenting the minimal standards that have been established by some of the taxonomic subcommittees of the ICSP for specific groups of organisms (see Chapter 1). These minimal standards should be consulted and applied when working with one of the groups for which such standards exist. Many researchers use the papers

METHODS IN MICROBIOLOGY, VOLUME 38
0580-9517 DOI 10.1016/B978-0-12-387730-7.00002-4

describing the closest relatives or other species of the genus in question as reference material and use the data and descriptions therein as templates for their study. This is highly recommended, but a number of issues need to be taken into consideration before embarking on this approach: (i) not all publications or the data they contain are of equivalent high standard, this is especially true for older publications; (ii) the minimal requirements may have changed if the taxonomic group in question has not been added to or emended for many years; (iii) techniques now considered standard may not have been applied in the characterization and description of previously described taxa; and (iv) the amount of information available on reference strains may be very limited and the methods used to obtain the data not well described or defined.

No matter what the source of the strain was or how its presumptive novelty was determined, the first step in any characterization process is to place it into a phylogenetic context and determine its phylogenetic relationship to the type strains of previously described species. Depending on the proximity to other taxa, a strategy for its characterization and description can be designed. The simplest way to establish the phylogenetic position is to determine the 16S rRNA gene sequence of the isolate or group of strains (see Chapter 16).

16S rRNA gene sequence data can provide a good indication of the taxonomic status of a strain. It can also tell the researcher if the strain represents a novel species or is a member of one that has been described previously. This can be determined both from the pairwise sequence similarities to existing taxa and from the taxonomic cluster affiliation of the sequence after phylogenetic analysis and treeing using full 16S rRNA gene sequences. Care should be taken both in the determination of similarity values and in the interpretation of the values obtained. It should be noted that BLAST searches may not use the full sequence in the calculation or may provide similarity values from comparisons to misidentified strains or environmental sequences. One way to overcome this problem is to use the facilities of ExTaxon (www.eztaxon.org) (Chun *et al.*, 2007). This suite of analysis tools and information allows one to obtain pairwise similarity values for the 16S rRNA gene sequence of the novel isolate with the highest quality sequences of all type strains, including those taxa described in arcticles currently 'in press' at *International Journal of Systematic and Evolutionary Microbiology* (IJSEM). It also provides links to other useful information on the taxa to which the novel isolate shows highest similarity.

Interpretation of pairwise 16S rRNA gene sequence similarity values is a critical but yet not a straightforward step and in many cases is dependent on the group of organisms that is being studied. High versus low pairwise similarity values can mean very different things in different or even the same taxonomic groups. In some cases, organisms having 16S rRNA gene sequence pairwise similarity values of >99% are in fact different species based on their phenotype and the results of DNA–DNA hybridization studies. Examples of genera containing species sharing such high sequence similarity values but still representing distinct species include *Bacillus* (e.g. the species *B. amyloliquefaciens*, *B. atrophaeus*, *B. mojavensis*, *B. siamensis*, *B. subtilis*, *B. tequilensis* and *B. vallismortis*, all share >99% 16S rRNA genes sequence similarity) and *Streptomyces* (e.g. the type strain of the species *S. griseus* shares >99.0% 16S rRNA gene sequence similarity with the type strains of 30 other *Streptomyces* species). At the other end of the spectrum, there are genera that contain species sharing very low 16S rRNA gene sequence similarity values. These include *Deinococcus* (e.g. *D. radiodurans*, the type species of the genus, shares <94% [and as low as 87.2%] 16S rRNA gene sequence similarity with 44 of the 47 species of the genus) and *Meiothermus* (e.g. *M. ruber* shares less than 94% 16S rRNA gene sequence similarity with 5 or the 10 species of the genus *Meiothermus*). Consultation of the literature on the taxonomic group in question as well as additional characterization is important before making a decision on the taxonomic status of a new isolate.

When the genus or family affiliation of an isolate has been determined by 16S rRNA gene sequence comparisons, the researcher should consult the literature on the respective taxonomic group before designing a characterization strategy. The website of the 'List of prokaryotic names with standing in nomenclature' (www.bacterio.cict.fr) is an excellent place to start such a literature search. This resource provides information on the number of species of each genus, the citations of the original descriptions or validations of the names of these species, the culture collection holdings and accession numbers of the type strains and their 16S rRNA gene sequence accession numbers.

There are two considerations to be made at this point in the process. The first relates to carrying on with the characterization of one isolate versus another isolate that might belong to a group for which the process might be more suited to the resources of the laboratory or that is not so closely related to an already described taxon. Secondly, if one is working with a single isolate, it might be a good time to attempt to obtain additional isolates that represent strains of the new species that will be characterized and described. The description of a novel species based on data from more than a single strain of a species allows for a more comprehensive species description and may aid in the selection of the most representative strain to be designated as the type strain. This can be done in a number of ways including the application of selective enrichment approaches that might be applicable to the taxonomic group being studied or by selecting and carrying out preliminary identification of a larger number of isolates than first studied. Based on the type of organism being studied, the known characteristics of the closely related taxa and with the aim of presenting as comprehensive a description as possible, one should design a characterization scheme. Before starting a characterization study, ensure that long-term preserves of the isolates are made. As early as possible in the process after enough data have been obtained to allow the designation of a type strain, deposit the type strain and reference strains in two culture collections in two different countries as required for the description of new taxa and the validation of their names (De Vos and Trüper, 2000).

Although data can be taken from published descriptions of closely related taxa and used for comparative purposes, obtaining related type strains and characterizing them at the same time as the new isolates results in a stronger data set and excludes the variation in data sets that might result from interlaboratory variations in materials and conditions of testing. Obtaining these reference type strains from service culture collections can take time due to shipping as well as import/export permitting regulations so they should be ordered early in the process.

The choices of methods applied to the characterization of a new isolate and the types of data collected are in many ways determined by the phylogenetic placement of the strain. Clearly, not all methods and characteristics are applicable to all isolates. As will be seen from a number of other chapters in this volume, not all of the methods described are applicable to all strains, e.g. mycolic acids and teichoic acids are only applicable to certain actinobacteria and other *Firmicutes* strains. In situations where the methodologies required to obtain the necessary data are not available and cannot be realized, even with the detailed protocols provided in the other chapters of this volume, collaborations should be sought with experts in specific techniques or alternatively have these data determined through commercial scientific services. When 16S rRNA gene sequence pairwise similarity values are greater than 98%, the authors should carry out DNA–DNA hybridization studies (see Chapter 15), Multilocus Sequence Analysis (MLSA) (see chapter 17) or whole genome sequencing if genome sequences of reference type strains are available (see Chapter 18).

The importance of using standardized methods for the determination of comparative data cannot be overemphasized. In many species descriptions, there are lists of carbon sources utilized but between studies, these have been determined by different methods using very different conditions and materials. The utilization of carbon sources can be determined by adding the substrate to a minimal medium and detecting the presence or absence of growth or the change in pH of the medium. In other situations, the carbon source is included in a commercial test strip (e.g. API or Biolog), and by adding the organism to the test strip, the utilization of the carbon source can be assessed, normally on the basis of a colour change. Each of these so-called carbon source utilization tests is different and the data from each cannot be compared to the others. Increasingly, it is found that the results from these different methods of determining carbon source utilizations in fact differ greatly for a given strain. Another area where standardization is especially important is in the determination of fatty acid profiles. The media composition, temperature and length of incubation, as well as the phase of growth in which the cells are harvested, all have an impact on the resulting fatty acid profile obtained from a given strain. The reproducibility of characterization tests can best be accessed by repeating tests for type strains of closely related species even when the data are available in the literature and the exact conditions of growth can be replicated.

Figure 1 represents a schema for the characterization of a novel isolate or group of isolates with the aim of describing them as a new taxon.

♦♦♦♦♦♦ II. COMPONENTS OF THE DESCRIPTION OF A NOVEL SPECIES

A. Summary/Abstract

The abstract or summary should contain the main characteristics of the new species including its source, morphology, staining reactions, metabolic type,

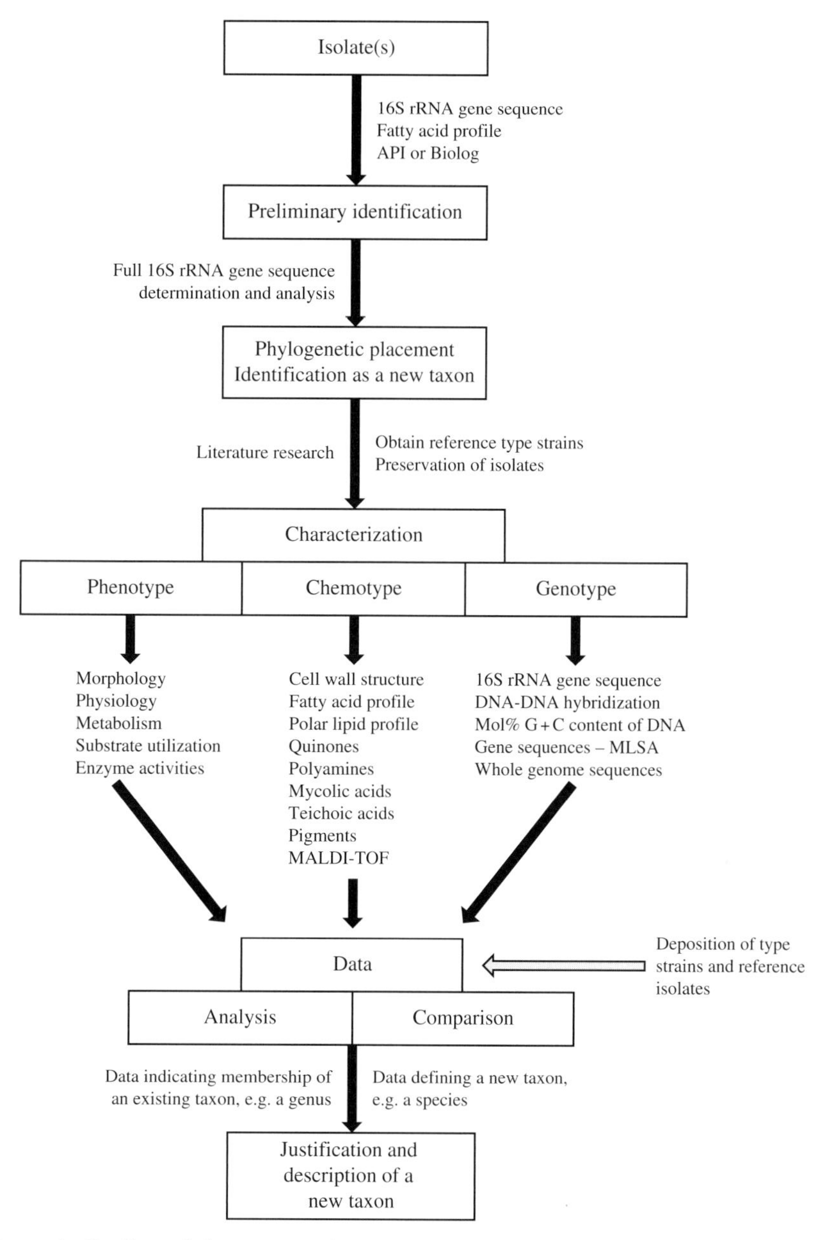

Figure 1. Outline of the steps and processes involved in the characterization and description of a novel prokaryote.

major chemotaxonomic properties, physiological characteristics, G + C content, phylogenetic affiliation listing closest relatives and 16S rRNA gene sequence accession number, the designation of the type strain and accession numbers of the culture collections in which it is deposited.

B. Introduction

The introduction should provide background information of the taxonomic group to which the new isolate or isolates are being assigned. This information should not only inform the reader of the characteristics of the group in question but should also provide a brief review of the taxonomy of the group and its closest relatives. In some journals, this introductory material is being reduced to a bare minimum, which makes for additional literature searches by the reader to familiarize them with the background material. Such introductory material is especially important in cases where organisms are being reclassified or taxonomic restructuring is being proposed.

C. Methods and Materials

The source of the isolates being characterized should be indicated; if they are of environmental origin, information on the location of sample site (preferably with GPS coordinates) should be provided, as well as the nature of the sample and the handling conditions of the samples from collection until used in the laboratory. In the case of human or animal pathogens, the pathology associated with or attributed to the isolate should be described. Detailed methods should be provided or such detailed methods should be referenced for all tests and experiments carried out in the process of characterizing new isolates. This is very important and will allow other researchers to collect a data set that will be comparable to the data being presented as part of the taxonomic description.

D. Results and Discussion

The data collected on the novel isolate can be presented in both text and table formats. Comparisons of the characteristics of the novel isolate with the type strains of previously described taxa can be presented in tables, but data sets should be complete. It is not appropriate to have ND (not determined) entries in such tables that are being used to point out differences between the new isolate and previously described organisms. Photomicrographs of cells of the organism (especially those that show distinctive microscopic characteristics), chromatographs of polar lipid profiles, and structures of unique fatty acids should be included in the manuscript or as supplementary online material.

The discussion should provide a narrative that indicates why the new isolate belongs to the genus it is being assigned to. The authors should state what characteristics the isolate possess that concur with those provided in the original description of the genus to which it is being assigned. The characteristics of the new taxon that do not match with that of the genus description or expand on the genus description should also be pointed out. A justification for the description of a new species should be outlined that includes the differentiating characteristics that actually define the new species. It is important that these differentiating characteristics are clearly stated and discussed with reference to the data available for the closest relatives. The reader should not have to go through the species description and comparative data tables to search for these differentiating characteristics.

E. The Species Description

When the name has been selected and the name and the etymology checked by an expert (see Chapter 19), these should be presented as the first component of the species description.

This should be followed by as much information as possible about the strain or strains that comprise the species being described.

A typical order for the listing of the characteristics would be as follows:

- Morphology, size and distinctive features of the cells
- Staining properties of cells
- Colony morphology, size and colour
- Metabolic type and end products of metabolism
- Physiological characteristics, e.g. temperature, pH and NaCl optima and ranges for growth
- Chemotaxonomic properties of the organism, e.g. cell wall type, fatty acid profile and polar lipid profile
- Substrates utilized and enzyme activities produced indicating if data came from commercial test strips
- Resistance or sensitivities to antimicrobials
- Mole% G + C content of the DNA
- The accession numbers of any DNA sequences that have been deposited in GenBank/EMBL/DDBJ databases
- Habitat or source of the strains of the species
- Designation of the type strain and the accession numbers of the two culture collections at which it is deposited
- List of reference strains of the species and the accession numbers of the culture collections at which they are deposited

It should be pointed out that the IJSEM preferentially uses a 'note format' for the description of new species (Kämpfer *et al.*, 2003). Although the text is not presented as a defined section with headings, the same information should be included.

References

Chun, J., Lee, J.-H., Jung, Y., Kim, M., Kim, S., Kim, B. K. and Lim, Y. W. (2007). EzTaxon: a web-based tool for the identification of prokaryotes based on 16S ribosomal RNA gene sequences. *Int. J. Syst. Evol. Microbiol.* **57**, 2259–2261.

De Vos, P. and Trüper, H. G. (2000). Judicial Commission of the International Committee on Systematic Bacteriology. IXth International (IUMS) Congress of Bacteriology and Applied Microbiology. Minutes of the meetings, 14, 15 and 18 August 1999, Sydney, Australia. *Int. J. Syst. Evol. Microbiol.* **50**, 2239–2244.

Kämpfer, P., Buczolits, S., Albrecht, A., Busse, H.-J. and Stackebrandt, E. (2003). Towards a standardized format for the description of a novel species (of an established genus): *Ochrobactrum gallinifaecis* sp. nov. *Int. J. Syst. Evol. Microbiol.* **53**, 893–896.

Tindall, B. J., Rosselló-Móra, R., Busse, H.-J., Ludwig, W. and Kämpfer, P. (2010). Notes on the characterization of prokaryote strains for taxonomic purposes. *Int. J. Syst. Evol. Microbiol.* **60**, 249–266.

3 Phenotypic and Physiological Characterization Methods

Noel R. Krieg[a] and Penelope J. Padgett[b]

[a] Department of Biological Sciences, Virginia Tech, Blacksburg, Virginia 24061, USA; [b] Environment, Health & Safety, University of North Carolina, North Carolina 27599-1650, USA

E-mail: nrk@vt.edu

CONTENTS

I. INTRODUCTION

Bacterial systematics today uses a polyphasic approach (Gillis *et al.*, 2005), and the results of all methods – rRNA gene sequencing, chemotaxonomic markers, morphological, physiological, pathogenicity and serological – are used to compile a description of a new species. Classification of prokaryotes is now based largely on analysis of the nucleotide sequences of highly conserved genes, especially 16S rRNA genes (Ludwig and Klenk, 2005). However, a comprehensive polyphasic characterization of a new species should still be done, because from even a complete genomic sequence it would be difficult to predict many of the phenotypic features of a new species. In addition to a thorough phenotypic characterization of a new species, it is important to determine which phenotypic features are the ones most useful for identifying new strains of the species. Moreover, the particular methods used for characterizing an organism should always be stated, because the results of phenotypic tests can vary with methodology.

Bear in mind that the phenotypic information for a species has been obtained with cultures grown under artificial conditions. Such conditions may not at all be those that the organism experiences under natural conditions. Therefore, what an organism does in nature may not be what it does in the laboratory.

The results of characterization tests can vary depending on a multitude of physical and environmental factors. The results of characterization tests

METHODS IN MICROBIOLOGY, VOLUME 38
0580-9517 DOI 10.1016/B978-0-12-387730-7.00003-6

obtained by one laboratory often differ from those obtained by another laboratory, although the results within each laboratory may be consistent. Therefore, in the methods section of a species description, the test conditions employed should be specified as precisely as possible. Ideally, it would be desirable to standardize the conditions used for testing physiological characteristics, but this is easier said than done. The use of commercial multi-test systems can increase standardization among various laboratories because of the high degree of quality control exercised over the media and reagents.

Some tests may give results that are not reproducible, weakly positive or subject to varying interpretation. Ideally, a test should give a reproducible result that is clearly positive or negative, without an equivocal reaction. Unfortunately, no such test may exist (although some are far better than others in approaching this ideal).

♦♦♦♦♦♦ II. FUNDAMENTAL PHYSIOLOGICAL TESTS

Some phenotypic characteristics are of such primary importance to a genus or species description that they must appear in every published description. What are the oxidase and catalase reactions expressed by the organism? Is the organism aerobic, anaerobic, microaerophilic or facultatively anaerobic? Over what range of temperatures does the organism grow, and what is the temperature optimum? Are sugars catabolized and, if so, which ones? Are the sugars oxidized or fermented? What carbon and nitrogen sources are used? Is NaCl inhibitory, or is it required at certain concentrations? Is nitrate reduced, and if so, is it used for denitrification?

Many additional, more specific physiological tests are described in Section III of this chapter and may also be useful when describing a new genus or species.

A. Respiratory Enzyme Tests

1. Oxidase test (indophenol oxidase, cytochrome *c* oxidase and cytochrome a_3 oxidase)

This method tests for an enzyme that transfers electrons from a donor molecule to O_2, thereby forming H_2O. Oxidase-positive organisms are usually aerobes or microaerophiles that can use O_2 as their final electron acceptor. The test reagent, *N,N,N′,N′*-tetramethyl-*p*-phenylenediamine (TMPD) acts as an artificial electron acceptor for the oxidase and the reduced form is the coloured compound indophenol blue.

For this test, do not use cultures grown on selective media. Results from old cultures may be unreliable. Do not use cultures grown on media containing fermentable carbohydrates, as acid from fermentation may inhibit oxidase enzyme activity and result in false negatives.

- (Method of Tarrand and Gröschel, 1982) Prepare a 1% (wt/vol) solution of *N,N,N′,N′*-tetramethyl-*p*-phenylenediamine (TMPD; not the HCl salt) in certified-grade dimethylsulfoxide (DMSO). The solution is stable for at least a month under refrigeration.

- Place a piece of Whatman no. 40 ashless filter paper, quantitative grade, in a Petri dish and wet it with 0.5 ml of the TMPD-DMSO solution.
- Use a cotton-tipped swab to pick up one large isolated colony and allow the inoculum on the swab to dry for 5 s. Tamp the swab lightly 10 times on the wet filter paper. Development of a blue-purple colour in 15 s is a positive test.

2. Catalase test

Catalase catalyses the disproportionation reaction $2\ H_2O_2 \rightarrow 2\ H_2O + O_2$, thereby helping to prevent oxidative damage to cells caused by H_2O_2. For this test, do not use cultures grown on blood-containing media, as blood contains catalase; however, cultures grown on a medium containing heated blood, such as chocolate agar, can be used. Some bacteria can make catalase only when provided with heme; these organisms are negative when cultured on media lacking blood but are positive when cultured on chocolate agar. Some bacteria such as lactobacilli and streptococci make a pseudocatalase (a non-heme catalase) when grown on media lacking blood but containing little or no glucose; they are negative for catalase when cultured on media containing 1% glucose. With anaerobic organisms, expose the culture to air for 30 min before performing the test, as some anaerobes have an inducible catalase.

- (Method of Health Protection Agency (2010b)) Add 0.2 ml of a 3% H_2O_2 solution to a screw-cap test tube.
- Using a platinum loop, disposable plastic loop or glass rod, remove some growth from a colony or agar slant and rub the growth on the inner wall of the tube. Cap the tube (to prevent escape of aerosols) and slant it, so that the H_2O_2 solution covers the growth. Effervescence within 10 s indicates a positive reaction.

B. Relation to Oxygen

Aerobes use O_2 as a terminal electron acceptor for an electron transport system, can tolerate a level of O_2 equivalent to or higher than that present in an air atmosphere (21% O_2) and have a strictly respiratory type of metabolism. Some aerobes may also be capable of growing anaerobically with electron acceptors other than oxygen, such as NO_3^-, trimethylamine oxide, AsO_4^{3-} and others.

Anaerobes are incapable of O_2-dependent growth and cannot grow in the presence of 21% O_2 (i.e. air). Some anaerobes may have a strictly fermentative type of metabolism. Others live by anaerobic respiration, in which an electron transport system uses an exogenous terminal electron acceptor other than O_2, such as NO_3^-, trimethylamine oxide, dimethylsulfoxide, AsO_4^{3-} S^0, SO_4^{2-}, Fe^{3+}, Cr^{+6}, fumaric acid or other oxidized substances.

Facultative anaerobes can grow both in the absence of O_2 and in the presence of 21% O_2. Some are capable of growing aerobically by respiring with O_2 and anaerobically either by fermentation or by anaerobic respiration. Others have a strictly fermentative type of metabolism and do not respire with O_2 or any other exogenous terminal electron acceptor.

Microaerophiles respire with O_2 but cannot grow, or grow very poorly, under 21% O_2. They grow best at low O_2 levels; some require levels as low as 1%. Some microaerophiles can also respire anaerobically with electron acceptors other than O_2.

1. Semisolid agar method

- (Method of Smibert and Krieg, 1994) Autoclave a narrow culture tube that has been filled to 60% of its capacity with an appropriate culture medium containing 0.2% agar.
- After the medium has cooled to 45°C, add the inoculum, mix to distribute the organisms uniformly and then allow the agar to solidify. Alternatively, inoculate the medium by stabbing with an inoculating needle after the agar has gelled; this avoids the mixing that otherwise might add dissolved O_2 to the medium.
- Growth occurring only at the surface of the medium suggests that the organism is aerobic. However, a fermentable substrate should be present in the medium because the organism might be a facultative anaerobe that not only respires with O_2 but also grows anaerobically by fermentation.

  Growth occurring only in the bottom region of the tube suggests that the organism is anaerobic. However, some extremely oxygen-intolerant anaerobes may not be able to grow even in the lowest region of the medium, because of the presence of small amounts of O_2 dissolved in the medium during the addition of the inoculum.

  Growth occurring throughout the tube suggests that the organism is a facultative anaerobe. It is important that no potential terminal electron acceptors other than O_2 should be present, as some aerobes can respire anaerobically (e.g. *Pseudomonas aeruginosa* can respire anaerobically with nitrate.)

  Growth occurring only in a disc several millimetres below the surface of the medium suggests that the organism is a microaerophile. Motile microaerophiles usually exhibit negative or positive aerotaxis, which results in their migration to a zone where the rate at which O_2 is diffusing to them matches the rate it is used by the organisms.

2. Manometric method

- (Method of Han *et al.*, 1991) Remove 1 ml of a young culture grown in an appropriate liquid or semisolid medium and mix to yield a homogeneous suspension.
- Spread one loopful of the suspension over the entire surface of a test slant or plate. Incubate at their optimum temperature in sealed vessels equipped with a vent to allow filling with various levels of oxygen.
- After incubation, estimate growth responses turbidimetrically by washing the cells from the agar surface with 5 ml of 0.85% NaCl and measuring the turbidity of the resulting suspension.
- To obtain various O_2 levels, use vacuum desiccator jars or polycarbonate anaerobic culture jars that have been fitted with a vent to allow evacuation and refilling with gases. Connect the vessels to a mercury manometer or very

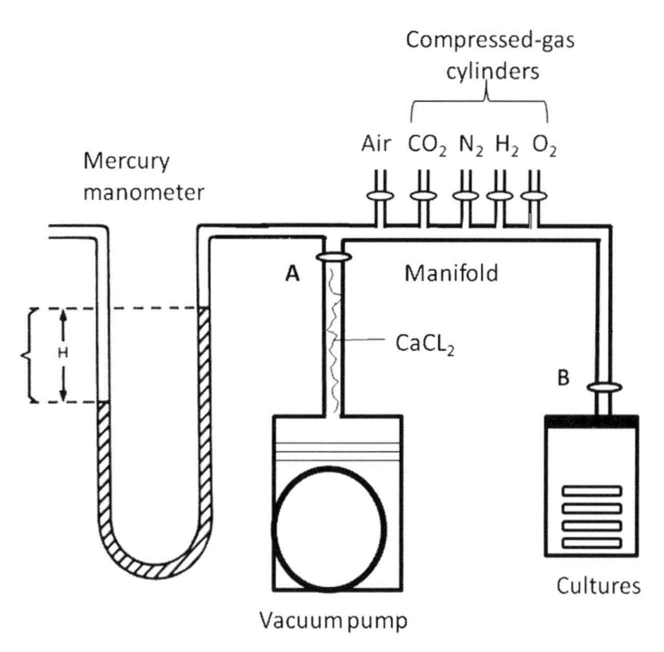

Figure 1. Arrangement of the apparatus to achieve various gas mixtures. Mount a ruler next to the open arm of the manometer, with 0.5 cm marked as 1.0 cm, to avoid having to continually calculate the difference (H) in the height of the two manometer arms. The $CaCl_2$ (anhydrous) prevents water vapour from entering the vacuum pump. In practice, with the vacuum pump turned on, open stopcocks A and B until the desired vacuum is reached. Then close stopcock A and add the appropriate amounts of air or selected gases, closing the valves for the air or gases after addition. After atmospheric pressure has been restored, close stopcock B to seal the culture jar, then remove the jar from the system.

precise vacuum gauge, which is in turn connected to a vacuum pump (Figure 1). Evacuate a portion of the air from the vessel, measure the decrease in pressure and replace the air with N_2. Use the following equation to determine the proper manometer reading before filling with N_2: manometer reading $= Y(21-X)/21$, where X is the desired $\%O_2$ level and Y is the uncompensated barometric pressure at the laboratory. For instance, if 10% O_2 is desired and the barometric pressure is 760 mm Hg, then pump out the air in the vessel until the manometer indicates a reading of $760(21-10)/21 = 391$ mm Hg. Then add N_2 to the vessel until the manometer indicates a return to atmospheric pressure and seal the jar.

- By this method and by use of a manifold to allow addition of various gases from compressed gas cylinders, different levels and various mixtures such as H_2, N_2 and CO_2 can also be obtained (Figure 1). Compressed gas cylinders would contain 100% of these gases, not 21% as with O_2 in air; thus, the '21' in the equation above would be replaced by 100%; i.e. manometer reading $= Y(100-X)/100$.

Note: For growth of strict anaerobes, it may not be sufficient to pump all the air out of the vessel, or even to refill the vessel several times with N_2, because culture media usually contain some dissolved O_2. With organisms inhibited by even very small amounts of O_2, it is best to use pre-reduced media or an anoxic chamber designed especially for anaerobes, as described in detail by Breznak and Costilow (1994).

C. Relation to Carbon Dioxide

Some heterotrophic organisms such as *Campylobacter jejuni* require levels of CO_2 above the 0.03% present in air in order to grow in liquid media or on the surface of solid media; controls lacking CO_2 should show no growth (Vandamme *et al.*, 2005). Semisolid media, however, may support growth in the absence of added CO_2 because the organisms may provide it from their own metabolism so that it accumulates around the cells.

Many other organisms, although not requiring CO_2, grow better when it is provided. Also, some organisms such as *Herbaspirillum autotrophicum* (previously *Aquaspirillum autotrophicum*) are autotrophic and can use CO_2 as their sole carbon source (Aragno and Schlegel, 1978).

1. Manometric method (Han *et al.*, 1991; Smibert and Krieg, 1994)

- (Method based on Han *et al.*, 1991 and Smibert and Krieg, 1994) For aerobes and microaerophiles, use the method described in Section II, B, 2 and in Figure 1. Initially, evacuate some of the air in the vessels, add the desired levels of CO_2 and then add sufficient air to return to atmospheric pressure. One jar in the series should have no CO_2 added.
- For anaerobes, place the inoculated media in each jar; also place fresh or freshly activated commercial palladium catalyst pellets in each jar. Evacuate most of the air from each jar and refill several times with a mixture of 5% H_2 –95% N_2 (which is non-flammable) from a compressed gas cylinder. Evacuate a final time and add the desired amounts of CO_2 manometrically to give the various percentages desired (Figure 1). One jar in the series should have no CO_2 added. Equilibrate to atmospheric pressure with the 5% H_2–95% N_2 mixture.

  Note: High CO_2 levels may cause a decrease in the pH of a medium to inhibitory values unless the medium is well buffered. For maintaining pH values between 6 and 8 at high levels of CO_2, the decrease can be prevented by the addition of a solution of $NaHCO_3$ (sterilized by positive pressure through a membrane filter) to the medium prior to inoculation. The concentration of bicarbonate used must be matched with the level of CO_2 in the atmosphere above the medium. Breznak and Costilow (1994) have summarized the calculations for determining the appropriate concentrations of bicarbonate.

2. Methods for autotrophs

Use a mineral medium devoid of any source of carbon other than CO_2 or HCO_3^-. Controls lacking the latter compounds should show no growth. The energy and electron source may vary considerably, depending on the particular organism.

Some examples of autotrophs and their energy and electron sources are as follows: *Acidithiobacillus thiooxidans* (S^0, $S_2O_3^{2-}$); *Pseudomonas carboxydovorans* (CO); *Ralstonia eutropha* (H_2); *Nitrobacter winogradskyi* (NO_2^-), *Nitrosomonas europaea* (NH_3) and *Gallionella ferruginea* (Fe^{2+}). The following medium was used for growth of the facultative, aerobic, marine autotroph *Salinisphaera hydrothermalis*, which oxidizes $S_2O_3^{2-}$ as its energy and electron source (Crespo-Medina *et al.*, 2009).

- Prepare the mineral medium containing (per litre of distilled water) NaCl, 25.0 g; $(NH_4)_2SO_4$, 1.0 g; $MgSO_4 \cdot 7H_2O$, 1.5 g; $CaCl_2 \cdot 2H_2O$, 0.42 g; and KCl, 0.64 g; Add 2.0 ml of a 0.5% phenol red solution as a pH indicator. Sterilize by autoclaving. Prepare a stock solution of K_2HPO_4, autoclave separately and add to the cooled, autoclaved medium to give a final concentration of 0.05 g/l. Prepare a stock solution of $NaHCO_3$, sterilize by filtration under positive pressure and add to the sterile, cooled medium to a final concentration of 0.046 g/l. Add $Na_2S_2O_3$ from a filter-sterilized stock solution to a final concentration of 20 mM (3.16 g/l). Add vitamin B_{12} from a filter-sterilized stock solution to a final concentration of 1 μM (1.6 mg/l). Add 1.0 ml of filter-sterilized mixed vitamin solution DSMZ 141 and 1.0 ml of filter-sterilized trace-element solution SL-10. A control medium lacking the $NaHCO_3$ should also be prepared .
- The mixed vitamin solution is a component of DSMZ medium 141 (http://www.dsmz.de/microorganisms/media_list.php) and contains the following (per litre of distilled water): biotin, 2.0 mg; folic acid, 2.0 mg; pyridoxine-HCl, 10.0 mg; thiamine · HCl · $2H_2O$, 5.0 mg; riboflavin, 5.0 mg; nicotinic acid, 5.0 mg; Ca-pantothenate, 5.0 mg; vitamin B_{12}, 0.10 mg; *p*-aminobenzoic acid, 5.0 mg; and α-lipoic acid, 5.0 mg. Sterilize by filtration.
- Trace element solution SL-10 contains (per litre of distilled water) the following: conc. HCl, 10.0 ml; $CoCl_2 \cdot 6H_2O$, 190 mg; $CuCl_2 \cdot 2H_2O$, 2 mg; $FeCl_2 \cdot 2H_2O$, 1.5 g; Na_3BO_3, 60 mg; $MnCl \cdot 4H_2O$, 100 mg; $Na_2Mo_7O_{24} \cdot 2H_2O$, 36 mg; $NiCl \cdot 6H_2O$, 24 mg; and ZnCl, 70 mg. Sterilize by filtration.

D. Relation to Hydrogen

Some heterotrophic prokaryotes require H_2 for growth, such as the microaerophile *Campylobacter mucosalis* (Roop *et al.*, 1985). In some instances, sodium formate can replace the H_2 requirement, as with most strains of *C. mucosalis*. The following method was used by Lawson *et al.* (1981) for *C. mucosalis* but can be adapted for use with other organisms. For instance, if using anaerobes, omit the O_2.

1. Manometric method for the heterotroph *Campylobacter mucosalis*

- Place replicate plates of a suitable agar medium (e.g. Brucella agar for *C. mucosalis*) that have been inoculated with a loop and provide gaseous atmospheres in the manner described in Figure 1. Use an atmosphere containing (vol/vol) 77% H_2, 10% N_2, 3% O_2 and 10% CO_2 and an additional atmosphere containing 87% N_2, 3% O_2 and 10% CO_2. (*Note:* H_2 is extremely flammable. Perform all operations in a fume hood.) Incubate the plates in jars under the two atmospheres at an appropriate temperature and examine after 24, 48 and 72 h. Restore the gaseous atmosphere in the jars after each reading. A hydrogen requirement is indicated by growth in the presence of H_2 but not in its absence.

2. Manometric method for *Herbaspirillum autotrophicum*

Some autotrophs are hydrogen autotrophs, able to use H_2 as the energy and electron source for growth with CO_2 or HCO_3^- as the sole carbon source. For example, the following method was used by Aragno and Schlegel (1978) for growing the facultative autotroph *Herbaspirillum autotrophicum*.

- Prepare a mineral medium containing the following (per litre of distilled water): $Na_2HPO_4 \cdot 12H_2O$, 9.0 g; KH_2PO_4, 1.5 g; $MgSO_4 \cdot 7H_2O$, 0.2 g; NH_4Cl, 1.0 g; ferric ammonium citrate, 0.005 g; $CaCl_2 \cdot 2H_2O$, 0.01 g; and trace elements solution SL 6 (solution SL 4 of Pfennig and Lippert, 1966) minus EDTA and iron salt, 3.0 ml; pH 7.1.
- After autoclaving the medium, add a solution of $NaHCO_3$, previously sterilized by filtration under positive pressure, to give a final concentration of 0.05%.
- Incubate the cultures under an atmosphere of 5% O_2–10% CO_2–85% H_2 at 30°C. Controls lacking H_2 (replaced by N_2) should show no growth.

E. Temperature Range and Optima for Growth

- Incubate broth cultures at a range of temperatures, using constant temperature incubators or water baths and measure the growth response of the cultures turbidimetrically. From 5°C above ambient to 100°C, use microbiological air incubators that have fans to circulate the air. For temperatures above 100°C, use screw-cap culture tubes or screw-cap glass bottles and seal the screw caps to prevent the evaporation of the medium. Incubate the cultures in commercial water baths filled with a dimethyl silicone oil.

  For ambient temperatures or below, use a refrigerated growth chamber.
  For temperatures below 0°C, use an ethylene glycol water bath. For marine organisms that require seawater, open ocean seawater has an NaCl concentration of about 35 g/l or 3.5% (wt/vol). In the case of organisms such as *Psychromonas ingrahamii*, which can grow at −12°C, glycerol may need to be added to the culture medium to prevent freezing (Auman *et al.*, 2006).

F. Optimum pH and pH Range for Growth

An essential part of the description of any prokaryote is the range of pH values at which it can grow, as well as the optimal pH for growth. Organisms may differ considerably in the pH range at which they grow.

Table 1. Some biological buffers, their useful range and their pK_a value at 25°C

Buffer	Useful pH range	pK_a at 25°C
Phosphate	1.7–2.9	2.15 (pK_a1)
Maleate	2.7–4.2	3.40
Citrate	2.2–6.5	3.13 (pK_a1)
Malate	2.7–4.2	3.40 (pK_a1)
Citrate	3.0–6.2	4.76 (pK_a2)
Succinate	3.2–5.2	4.21 (pK_a1)
Acetate	3.6–5.6	4.76
Succinate	5.5–6.5	5.64 (pK_a2)
MES [2-(*N*-morpholino)-ethanesulfonic acid][a]	5.5–6.7	6.10
Citrate	5.5–7.2	6.40 (pK_a3)
ACES [*N*-2-acetamido-2-aminoethanesulfonic acid][a]	6.1–7.5	6.78
BES [*N,N-bis*-(2-hydroxyethyl)-2-aminoethanesulfonic acid][a]	6.4–7.8	7.09
MOPS [3-(*N*-morpholino)-propanesulfonic acid][a]	6.7–7.9	7.14
HEPES [*N*-2-hydroxyethylpiperazine-*N′*-2-ethanesulfonic acid][a]	6.8–8.2	7.48
Phosphate	5.8–8.0	7.20 (pK_a2)
Tricine [*N*-(2-hydroxy-1,1-bis(hydroxymethyl)ethyl)glycine][a]	7.4–8.8	8.05
Tris [*N*-tris(hydroxymethyl)methylglycine]	7.5–9.0	8.06
TABS [*N*-tris[hydroxymethyl]-4-amino-butanesulfonic acid][a]	8.2–9.6	8.90
CHES [*N*-cyclohexyl-2-aminoethanesulfonic acid][a]	8.6–10.0	9.50
Glycine	8.8–10.6	9.78 (pK_a2)
CAPS [3-(cyclohexylamino)-1-propanesulfonic acid][a]	9.7–11.1	10.40

[a]Buffers selected by Good *et al.* (1966).

- Measure growth responses from a standardized inoculum using media at various pH values. If broths are used, measure the growth responses turbidimetrically; if solid media are used, wash the growth from the surface and measure the cell density turbidimetrically.

A buffer should be used in most media to maintain a stable pH for growth of the test organism. Buffers are most effective at their pK_a values and should be chosen with this in mind. Some buffers such as citrate, succinate or glycine may be metabolized by the test organism. Others may be toxic; for instance, phosphate concentrations greater than 0.003 M are inhibitory to *Spirillum volutans* (Hylemon *et al.*, 1973). Sometimes a combination of buffers may be helpful; e.g. Wang *et al.* (2009) used a tri-buffer system, containing 50-mM TRICINE [N-(2-hydroxy-1,1-bis(hydroxymethyl)ethyl)glycine], CHES [*N*-cyclohexyl-2-aminoethanesulfonic acid] and CAPS [3-(cyclohexylamino)-1-propanesulfonic acid] for *Alkalimonas amylolytica*.

Some useful biological buffers are listed in Table 1. Phosphate salts are most commonly used because they are effective in the growth range of most bacteria, are usually non-toxic and provide a source of phosphorus for the organism. Certain buffers selected by Good *et al.* (1966) are non-metabolizable, non-toxic, have low reactivity with metal ions and have other desirable features; these buffers are indicated in Table 1.

G. NaCl or Seawater Ranges and Optima for Growth

- Inoculate liquid media containing a range of NaCl concentrations and measure the growth response turbidimetrically. Marine organisms usually require NaCl at levels above 1% and grow with 3% (wt/vol) NaCl or higher. For some marine organisms, NaCl alone may not substitute for filtered seawater, which should be sterilized by filtration and added aseptically to the sterilized medium. Even if seawater is used, it may need to be aged for a few weeks in a glass vessel in the dark to be effective. Seawater contains ca. 3% NaCl and testing marine organisms for growth at levels below this can be done by using various proportions of distilled water to seawater in the medium, or by using artificial seawater in which the level of NaCl can be varied. A simplified formulation of artificial seawater (Pelczar, 1957) contains the following (per litre of distilled water): NaCl, 27.5 g; $MgCl_2$ (anhydrous), 5.0 g; $MgSO_4$ (anhydrous), 2.0 g; $CaCl_2$ (anhydrous), 0.5 g; KCl, 1.0 g; and $FeSO_4$ (anhydrous), 0.001 g. Note that extreme halophiles require NaCl levels of ca. 10% or more and may grow at levels up to 30% (Grant, 2001).
- Freshwater organisms usually are inhibited by 3% NaCl, although they may require low levels of NaCl (e.g. *Spirillum volutans* requires 0.01% NaCl) (Krieg, 2005b).

H. Oxidative and Fermentative Metabolism

1. O/F test

- (Method of Hugh and Leifson, 1953) Use two tubes of a semisolid medium containing the carbohydrate to be tested. The sugar is usually glucose, but some organisms may use other sugars; e.g. *Novispirillum itersonii* uses fructose but not glucose (Hylemon *et al.*, 1974). The medium of Hugh and Leifson (1953) is usually used and contains the following (per litre of distilled water): peptone (pancreatic digest of casein), 2.0 g; NaCl, 5.0 g; K_2HPO_4, 0.3 g; bromothymol blue, 0.03 g; and agar 3.0 g; pH 7.1. Boil to dissolve the agar, dispense 4-ml portions into 13 mm × 100 mm tubes and sterilize by autoclaving, Cool to 45°C, then add sufficient 10% carbohydrate solution (sterilized by filtration) to give a final concentration of 1 g/l and mix.
- Heat the tubes in a boiling water bath for 10 min to expel excess oxygen. After the tubes have cooled, inoculate them vertically with a straight wire. Dispense sterile molten petrolatum into one tube as a seal against oxygen (this is the 'anaerobic' tube).
- After incubation, examine the tubes daily for growth and for a colour change. If growth and an acidic change (yellowing) occur only at or near the surface of the aerobic tube, the organism can oxidize the carbohydrate (an 'O' reaction). If the anaerobic tube shows growth and is acidified, the organism can ferment the carbohydrate (an 'F' reaction). If neither tube becomes acidified but growth occurs, the organism cannot catabolize the carbohydrate. If no growth occurs in either tube, the medium may be lacking some required nutrient; try supplementing it with 0.1% yeast extract.

If bromothymol blue is inhibitory or not sensitive enough to detect small pH changes, substitute with phenol red (0.01 g/l). For marine organisms, use the modified medium of Leifson (1963), which contains the following (per litre): pancreatic digest of casein, 1.0 g; yeast extract, 0.1 g; $(NH_4)_2SO_4$, 0.5 g; Tris buffer, 0.5 g; phenol red, 0.01 g; agar, 3.0 g; and distilled water, 500 ml. Adjust the pH to 7.5, sterilize by autoclaving and cool. Separately autoclave 500 ml of artificial seawater, cool and combine with the first solution. An artificial salt mixture closely resembling the composition of the dissolved salts of ocean water can be obtained commercially (Sigma-Aldrich, Cat. no. S9883).

- Sometimes acid production can be masked by formation of alkaline substances from the peptone; in these instances, the peptone-deficient medium of Board and Holding (1960) may be satisfactory: $(NH_4)H_2PO_4$, 0.5 g; K_2HPO_4, 0.5 g; yeast extract, 0.5 g; bromothymol blue, 0.03 g; mineral salts solution, 20 ml; agar, 5.0 g; and distilled water, 880 ml. Dissolve the phosphates and yeast extract in the water, add the bromothymol blue and a mineral salts solution. After adjusting to pH 7.2, add the agar and boil to dissolve it. Dispense 9-ml portions into tubes, autoclave for 1–2 min at 22 lb/in^2 and cool to 45°C. Add 1.0 ml of a 5% filter-sterilized solution of glucose (or other desired sugar) solution, mix and allow to cool.

  Prepare the mineral salts solution from two solutions, A and B. To prepare Solution A, dissolve 10.0 g of nitrilotriacetic acid in 950 ml of distilled water; then add $MgSO_4$ (anhydrous), 14.45 g; $CaCl_2 \cdot 2H_2O$, 3.335; $(NH_4)_6Mo_7O_{24} \cdot 4H_2O$, 0.00925 g; and $FeSO_4 \cdot 7H_2O$, 0.099 g. Add 50 ml of Solution B and adjust the pH to 6.8 with ca. 7.3 g of KOH. Solution B contains the following (per litre of distilled water): EDTA, 2.5 g; $ZnSO_4 \cdot 7H_2O$, 10.95 g; $FeSO_4 \cdot 7H_2O$, 5.0 g; $MnSO_4 \cdot H_2O$, 1.54 g; $CuSO_4 \cdot 5H_2O$, 0.392 g; $Co(NO_3)_2 \cdot 6H_2O$, 0.248 g; $Na_2B_4O_7 \cdot 10H_2O$, 0.177 g. Add a few drops of sulfuric acid to Solution B to decrease precipitation.

I. Acid Production from Carbohydrates

Some sugars such as arabinose, lactose, maltose, rhamnose, salicin, sucrose, trehalose and xylose are altered or hydrolysed by autoclaving, and even glucose is slowly destroyed in the presence of phosphates (Smith, 1932). In such instances, either sterilize the complete media by filtration or prepare the test media without the sugars and, after the media have been autoclaved and cooled to 45°C, add the sugars from concentrated stock solutions that have been sterilized by filtration. In the case of semisolid media, add the sugar after the basal media (with the agar) have been autoclaved and cooled to 45–50°C.

When testing aerobes, use low concentrations of peptone ($\leq$0.2%) in order to avoid masking the acid production by formation of alkaline substances.

- Demonstrate acid production by incorporating an appropriate pH indicator into the medium. The choice of indicator depends on the pH range desired. Three commonly used indicators are bromothymol blue, pH 6.1 (yellow) to 7.7 (blue); phenol red, pH 6.9 (yellow) to 8.5 (red); and bromocresol purple, pH 5.4 (yellow) to 7.0 (purple). Use these in media at concentrations of

0.01–0.03 g/l. If the indicator is degraded or is toxic to the organisms being tested, prepare the medium without the indicator and add some drops of indicator solution to the cultures after they have grown.

- A more precise method is to insert a narrow combination pH electrode into the cultures after they have grown and measure the pH. For testing a series of cultures grown with various sugars, wash the electrode off with distilled water between each measurement into a beaker and autoclave the beaker and its contents when finished. Disinfect the electrode itself when finished by dipping it into 3% hydrogen peroxide or 70% ethanol.

  When using pre-reduced anaerobically sterilized sugar-containing media, the pH will become lowered due to carbonic acid formation if O_2-free CO_2 has been used to flush the tubes during inoculation. Consequently, the pH of uninoculated media should be measured as a reference point for determining whether acidification has occurred in the inoculated tubes.
- An alternative method is to use the API 50 CH system (bioMérieux), which is a commercial multi-test system to detect acid production from 49 fermentable carbohydrates and one negative control test. While useful for fermentative organisms, acid production by oxidative organisms is often comparatively weak; accordingly, the system might require modification, such as incubating the cultures for longer incubation periods (Kattar *et al.*, 2001).

J. Gas Production from Carbohydrates

- Use a Durham vial (ca. 9 by 30 mm) to capture gas produced from sugar fermentation. Place the vial open end down in the tube of sugar medium; the vial will become completely filled during autoclaving. For thermolabile sugars, prepare the medium without the sugars but with the Durham vial; after autoclaving, cool the medium and add the sugars from concentrated stock solutions that have been sterilized by filtration. Incubate the sterile medium at the incubator temperature (to minimize dissolved gases from air that may otherwise occur in cooled media) for 24–48 h before inoculating to allow the sugar to diffuse completely throughout the medium. A positive test is indicated by accumulation of gas in the Durham vial compared to uninoculated controls.

  For organisms that produce only small amounts of gas, such as heterofermentative lactobacilli, prepare tubes of sugar medium as described above, but before inoculation, add 1 ml of sterile heavy mineral (paraffin) oil to the tube to prevent escape of gas that may be produced (Hayward, 1957).

K. Fermentation Products

Knowing the products of fermentation by an anaerobe or facultative anaerobe is an extremely important part of the phenotypic characterization of that organism. The methods of Holdeman *et al.* (1977), which employ gas-liquid chromatography (GLC) for detection of volatile fatty acids (VFA) and the methyl esters of non-volatile acids (NVA) have been widely used; for details, see Tindall *et al.*

(2007). Although NVA and VFA have frequently been analysed separately on a single gas chromatograph, high-performance liquid chromatography (HPLC) methods offer the advantages of detection of NVA without the need for derivatization, detection of both VA and NVA in a single chromatographic run and increased sensitivity for detection of NVA. Capillary GLC methods do involve simultaneous derivatization of VA and NVA, but both can be separated in a single chromatographic run. Summaries of these methods are given below.

1. Capillary GLC method

- (Method of Richardson *et al.*, 1989) Centrifuge the culture at 3000 × *g* for 15 min and remove 1.0 ml of the supernatant fluid. Add 50 μl of internal standard (2-ethylbutyric acid). Also prepare a standard solution containing the following (millimole per litre of distilled water): formate, 10.0; acetate, 30.0; propionate, 20.0; isobutyrate, 5.0; *n*-butyrate, 20.0; isovalerate, 5.0; *n*-valerate, 5.0; lactate, 10.0 and succinate, 10.0. To the 1.0-ml of sample and also to 1 ml of the standard solution add 0.5 ml of conc. HCl and 2 ml of diethyl ether. Mix for 1 min on a vortex mixer and centrifuge at 3000 × *g* for 10 min, Remove the ether layer and transfer it to a separate capped vial. Add another 1 ml of diethyl ether to the aqueous layer, perform a second extraction and combine the ether extracts. Add 200 μl of *N*-methyl-*N*-t-butyldimethylsilyltrifluoroacetamide (MTBSTFA) and heat at 80°C for 20 min. Incubate the reaction mixture at room temperature for 24 h to ensure complete derivatization of lactic acid.
- For analysis, use a gas chromatograph equipped with a flame ionization detector and split/splitless injector. Use a 30 m × 0.25 mm fused silica capillary column coated with 0.1 μm DB1. Use injector and detector temperatures of 275°C with the column temperature programmed from 63°C for 3 min to 190°C at 10°C/min. Use helium as the carrier gas (head pressure, 135 kPa). Make injections (1 μl) in the split mode (50:1 split). Record peak areas with an integrator.

2. HPLC method I

- (Method of Ehrlich *et al.*, 1981) Centrifuge cultures at 10,000 × *g* for 15 min. Filter the supernatant through a polycarbonate filter (0.45 μm pore diameter). Use the filtered samples immediately for chromatographic analysis. Use a sample of uninoculated medium for comparison. Use standards of known compounds at concentrations of ca. 5–10 mM.
- For analysis, use an HPLC equipped with a solvent delivery system, an injection system, a refractive index (RI) detector, a variable wavelength detector and a printing integrator. Construct the column from 9.5-mm stainless steel tubing (7.8 [inside diameter] by 300 mm) and pack it with 9-μm-diameter HP X-87 cation-exchange resin. Operate the column at 30°C with 0.013 N H_2SO_4 at a flow rate of 0.7 ml/min as the eluent. The order of separation is glucose, pyruvate, lactate/succinate, formate, acetate, propionate, isobutyrate, ethanol, butyrate, isovalerate, butanol and caproate. Note that succinic and lactic acids

are not separated at 30°C but can be separated at 50°C; however, the positions of other peaks may change or overlap.

Deterioration of column performance caused by peak broadening may gradually occur during use of the column. This happens more rapidly with complex media than with defined media consisting only of carbohydrates and growth factors in mineral salt solutions. Column performance can be restored by reversing column flow and pumping at a low flow rate at an elevated temperature for a few hours.

3. HPLC method 2

- (Method of Johansen *et al.*, 1988) Centrifuge the culture and recover the supernatant fluid. Extract the acids from the fluid with ether and then re-extract them from the ether as sodium salts into a small volume of 0.1 N NaOH. The re-extraction into NaOH provides additional purification and concentration of the acids. Make additional preparations from the uninoculated medium as a control, and also from a standard mixture of acids.
- Use an HPLC apparatus consisting of a pump, a controller unit, a diode array detector, a sample injector valve with a fixed 20-μl sample loop, a column such as an Aminex HPX-87H column for organic acids, a microguard column and a holder. Analyse acids at room temperature at a flow rate of 0.5 ml/min, using 11% acetonitrile in 0.008 N sulfuric acid as a column eluent. De-gas the eluent with helium before use. Record chromatograms at 208 nm. Identify acids identified by comparing their retention times with those of standard solutions of acids. The order of appearance of the acids is pyruvic, succinic, fumaric, formic, acetic, propionic, isobutyric, butyric, isovaleric, *p*-hydroxyphenylacetic and phenylacetic acid.

L. Carbon Source Utilization

1. Auxanography (the use of a plate culture in which variable conditions are provided to determine the effect of these conditions on growth)

The following method is adapted from that used by Baldani *et al.* (2005).

- Grow cells in an appropriate liquid medium to early stationary phase. Centrifuge and wash the cells in a sterile, non-nutrient buffer, such as a dilute phosphate buffer.
- Prepare a liquid medium containing all essential nutrients except a carbon source, add 15 g of a purified agar per litre and sterilize by autoclaving. Cool the medium to 45°C.
- Suspend the washed cells in the molten medium to a density of approximately 10^8 cells per millilitre and dispense the mixture into Petri dishes (20 ml per dish).
- After the inoculated medium has solidified, place sterile paper discs whose edges have been dipped into sterile solutions of various carbon sources onto the edge of each plate, using three well-separated discs per dish.

- After incubation for up to 4 days, the development of a halo of growth around a disc indicates utilization of the particular carbon source. If growth should occur over the entire dish, the inoculum was probably too dense.

2. Turbidimetric method

- Use a chemically defined basal medium (DBM) that lacks a carbon source but otherwise is suitable for growth of the organism being tested. The DBM used by Hylemon *et al.* (1973) may be satisfactory (per litre of distilled water): $(NH_4)_2SO_4$, 1.0 g; $MgSO_4 \cdot 7H_2O$, 1.0 g; K_2HPO_4, 0.5 g; $FeCl_3 \cdot 6H_2O$, 0.0047 g; $MnSO_4 \cdot H_2O$, 0.0025 g; $CaCO_3$, 0.001 g; $ZnSO_4 \cdot 7H_2O$, 0.00072 g; $CuSO_4 \cdot 5H_2O$, 0.000125 g; $CoSO_4 \cdot 7H_2O$, 0.00014 g; H_3BO_3, 0.000031 g; and $Na_3MoO_4 \cdot 2H_2O$, 0.000245 g. Adjust the pH to 7.0 with KOH. Add carbon sources to a concentration of 1 g/l and sterilize by autoclaving at 112°C for 10 min. Thermolabile carbon sources should be sterilized separately by filtration and added aseptically to the sterile DBM.
- For marine organisms, instead of distilled water, use a synthetic seawater of the following composition (per litre of distilled water) and sterilize the media by filtration: NH_4NO_3, 0.002 g; H_3BO_3, 0.027 g; $CaCl_2$, 1.14 g; $FePO_4$, 0.001 g; $MgCl_2$, 5.143 g; KBr, 0.10 g; KCl, 0.69 g; $NaHCO_3$, 0.20 g; NaCl, 24.32 g; NaF, 0.003 g; Na_2SiO_3, 0.002 g; Na_2SO_4, 4.06 g; and $SrCl_2$, 0.026 g. The freshwater- and seawater-defined media are clear and do not contain a precipitate.
- Use exponential-phase cells as inoculum and wash the cells to be used at least once in 0.85% saline or, for marine organisms, artificial seawater. Use a standardized inoculum based on turbidity. After growth has occurred, measure the growth response turbidimetrically with a spectrophotometer.

M. Nitrogen Source Utilization

1. Sole nitrogen sources

- Use either the auxanographic method or the turbidimetric method described in Section II, L, 1 and 2. For the auxanographic method, incorporate a utilizable suitable carbon source into the agar medium but omit the nitrogen source. Place sterile paper discs whose edges have been dipped into sterile solutions of a spectrum of amino acids and other organic nitrogen sources, as well as inorganic sources such as ammonium salts and nitrate, onto the edge of each plate, using three or four discs per plate.
- For the turbidimetric method, use the DBM or seawater DBM described in Section II, L, 2 but include a suitable carbon source (1 g/l) and omit the $(NH_4)_2SO_4$.
- To test for compounds that can serve as sources of both carbon and nitrogen, omit carbon and nitrogen sources from the test medium.

2. Nitrogen fixation

The ability to incorporate $^{15}N_2$ into cell material remains the 'gold standard' test for nitrogenase activity but requires a mass spectrometer and a source of the isotope (Daniels *et al.*, 1994). The acetylene reduction method offers a relatively easy and inexpensive alternative, requiring only an isothermal gas chromatograph. The method depends on the ability of nitrogenases (except that of *Streptomyces thermoautotrophicus*) to accept acetylene ($CH{\equiv}CH$) as a substrate and reduce it to ethylene ($CH_2{=}CH_2$).

- Culture the organism to be tested in media devoid of fixed nitrogen sources, but containing a utilizable carbon source and a source of molybdenum (most nitrogenases are of the Mo-Fe type). It may also be helpful to include a small amount of yeast extract (0.005%) in the medium to allow enough initial growth so that nitrogenase can be synthesized.
- Inject pure acetylene into the head space of the culture vessel and determine the formation of ethylene after a period of incubation by means of a gas chromatograph. The synthesis of most nitrogenases is repressed by NH_4^+ (*Rhizobium* species and *S. thermoautotrophicus* are exceptions); so, also test a second culture in which the medium contains an ammonium salt. This control should exhibit little or no ethylene production. Use a control of uninoculated medium as well.

  Note that commercial acetylene usually contains small amounts of ethylene. Use the following procedure to generate pure acetylene. Use a fume hood, as acetylene gas is flammable. Fill a tube or small flask half full with water and stopper it with a one-hole stopper attached to a piece of rubber tubing. Add a lump (ca. 1 g) of calcium carbide to the tube and stopper the tube. Immediately submerge the end of the rubber tubing in a beaker of water and allow the acetylene being generated to flush out the air. After a minute or so, and while the carbide is still bubbling, push the tip of a 26-gauge needle connected to a plastic syringe through the side of the rubber tubing, remove the desired amount of acetylene and inject it into the test system.
- For gas chromatography, use a chromatograph that has a flame ionization detector and a glass column containing Porapak™ T or N packing. Column sizes, oven temperatures and gas flow rates vary among laboratories, but the authors have had good results with a 6-ft by 0.25-in. (approximately 30.5 cm × 0.6 cm) column, a constant oven temperature of 60°C and a carrier gas (helium or nitrogen) flow rate of 85 ml/min. Calibrate the chromatograph with dilutions of commercial ethylene. Use disposable plastic syringes for injecting the ethylene and the culture gas samples. Use a small pliers to bend the tip of the injection needle so that the opening at the end is parallel rather than perpendicular to the needle axis; this will ensure that the tip will slice through the injection port septum instead of punching a hole in it. Ethylene can adhere to plastic, so use a fresh syringe for each injection. After injecting the culture gas samples, the acetylene will take longer than the ethylene to reach the detector; therefore, wait until the acetylene peak from one culture sample appears before injecting the next sample.
- For microaerophiles such as *Azospirillum* (Baldani *et al.*, 2005) and *Prolinoborus* (Krieg, 2005a), the following method can be used. Use a suitable semisolid (0.175% agar) medium devoid of fixed nitrogen and also a similar medium

containing 0.1% $(NH_4)_2SO_4$. Inoculate 0.05 ml of the culture into 3.0 ml of the sterile, molten test medium at 45°C, contained in culture tubes having loosely fitting plastic closures. Mix the inoculum with the medium and allow the medium to solidify. Incubate cultures aerobically for a suitable period, e.g. 3–5 days, at their optimum temperature. Do not agitate the cultures during incubation, as oxygen may reach the growing cells and inactivate the nitrogenase. Replace the plastic closures with rubber vaccine-bottle stoppers and inject pure acetylene through the stoppers to give a final concentration of 10% (vol/vol) in the head space. After 1 h, withdraw 1-ml samples of gas with a plastic syringe from the tubes and inject into the gas chromatograph.

- For aerobes, use a system similar to that for microaerophiles, except that the medium does not need to be semisolid.
- Anaerobes such as *Clostridium pasteurianum* must fix N_2 anaerobically, and facultative anaerobes such as *Klebsiella pneumoniae* and *Bacillus polymyxa* also must do it anaerobically because of the oxygen sensitivity of nitrogenase. Bubble sterile (filtered) O_2-free N_2 through the medium with a sterile pipette to purge any dissolved O_2 and inoculate the medium while purging. Seal the tubes with a sterile vaccine-bottle stopper. After growth has occurred, inject pure acetylene to give a concentration of 10% in the head space.

N. Nitrate Reduction and Denitrification

Many organisms can respire anaerobically by using NO_3^- as a terminal electron acceptor for an electron transport system (nitrate respiration or dissimilatory nitrate reduction). Denitrification is an anaerobic respiratory process whereby NO_3^- is successively reduced to NO_2^-, NO, N_2O and N_2, although some organisms may stop at one of the intermediate steps (but usually not at NO, as this is toxic to cells). Some organisms reduce nitrate to N_2O but cannot accomplish the final step of reduction to N_2. Some others reduce NO_3^- anaerobically to NH_3. If these organisms are fermentative, such as *Clostridium perfringens*, they use the NO_3^- merely as an electron sink rather than for respiration; however, if the reduction generates a proton motive force, as in *Shewanella oneidensis*, then the process is regarded as dissimilatory nitrate reduction to NH_3 or respiratory nitrate ammonification (Cruz-García *et al.*, 2007). Some bacteria such as *Myroides odoratus* reduce NO_2^- to N_2 but do not reduce NO_3^- (Vancanneyt *et al.*, 1996).

1. Nitrate reduction

- (Method of Lányi, 1987) For aerobes and facultative anaerobes, prepare a test medium consisting of the following (per litre of distilled water): beef extract, 3.0 g; peptone, 5 g; KNO_3, 1.0 g; and agar, 1.7 g (to give a semisolid medium); pH 7.3–7.4. (*Note*: in some organisms, the NO_2^- formed from 0.1% KNO_3 may be inhibitory to growth, and it may be necessary to decrease the level of KNO_3 to 0.01%.) Boil to dissolve the agar, dispense 5-ml portions into test tubes containing inverted Durham vials and autoclave. Fastidious bacteria may require the addition of 5% rabbit or ox serum (sterilized by filtration and added to the melted medium cooled to 45°C).

For anaerobes, prepare a test medium containing the following (per litre of distilled water): proteose peptone, 20.0 g; glucose, 1.0 g; $Na_2HPO_4 \cdot 2H_2O$, 2.0 g; KNO_3, 1.0 g; agar, 1.0 g; pH 7.2–7.4. If necessary for growth, add 10-ml hemin solution and 0.2-ml vitamin K solution (see below). Do not add cysteine. Dispense 6–8 ml portions into test tubes containing inverted Durham vials and sterilize by autoclaving. Add a layer of sterile mineral oil immediately after removal from the autoclave. Prepare the hemin solution by dissolving 50 mg of hemin in 1 ml of 1-N NaOH; add distilled water to a final volume of 100 ml and sterilize by autoclaving. Prepare the vitamin K solution by dissolving 20-mg of vitamin K in 20 ml of 95% ethanol.

Reagent A consists of 8.0 g of sulphanilic acid dissolved (by gentle heating) in 1000 ml of 5-N acetic acid (300 ml of glacial acetic acid mixed with 700-ml distilled water). Store at 4°C. Reagent B consists of 5 g of α-naphthylamine or dimethyl-α-naphthylamine dissolved in 1000 ml of 5-N acetic acid. Powdered zinc is also needed for the test.

- Inoculate the test medium and incubate for up to 3 days. (For anaerobes, stab the inoculation needle through the mineral oil.)
- Gas formation in the Durham tube indicates reduction of NO_3^- to N_2 or (less commonly) N_2O. Add 0.1-ml reagent A and 0.1-ml reagent B and look for development of a red colour within 1 min, which indicates the presence of NO_2^-. If no red colour develops, add ca. 15-mg zinc dust. The development of a red colour within 3 min indicates that the NO_3^- has not been reduced. If no red colour develops, and reagents A and B have previously failed to produce a red colour, the NO_3^- has been reduced beyond the nitrite stage, probably to N_2 or (less commonly) N_2O.

2. Nitrite reduction

Some bacteria (e.g. *Myroides odoratus*) do not reduce nitrate, but produce N_2 if NO_2^- is provided as substrate (Vancanneyt *et al.*, 1996). As noted by Lányi (1987), various concentrations of nitrite have been recommended, from 0.001% to 0.1%. For routine use, 0.1% nitrite is the best, because a minute amount of nitrite may be consumed by bacteria otherwise incapable of total denitrification (e.g. *Escherichia coli*). Bacteria that reduce NO_3^- via NO_2^- to N_2 produce gas not only in nitrate but also in nitrite medium; however, gas production may not be visible in a medium with a low NO_2^- concentration.

- Use the same media as those for nitrate reduction but use 0.1% $NaNO_2$ per litre instead of KNO_3. Reagents A and B are also used as described for nitrate reduction, but there is no need to use zinc powder since NO_3^- is not present in the medium.
- After growth occurs, development of a red colour after addition of reagents A and B indicates that the NO_2^- has not been reduced. If a red colour does not develop, the NO_2^- has been reduced to N_2 or N_2O.

3. Detection of N_2O as evidence of denitrification

- (Method of Neyra *et al.*, 1977) This test takes advantage of the inhibition by acetylene of the reduction of N_2O to N_2. Inoculate semisolid media prepared as described for method 1, but without a gas vial. Seal the culture vessel with a rubber vaccine-bottle stopper. (For denitrification under anaerobic conditions, insert a syringe needle, evacuate the head space and refill with helium several times.) Inject pure acetylene to give a final concentration of 10% in the head space after removing 10% of the air (or helium) in the head space. Incubate the culture without agitation.
- To detect N_2O after the culture has grown, remove a sample from the head space and inject it into a gas chromatograph having a Porapak™Q (80–100 mesh) column and thermal conductivity detector. Use helium as the carrier gas and a column temperature of 35°C.

O. Hydrolysis of Polymers

1. Agar hydrolysis

- Although agar is not digested by most terrestrial bacteria, a number of marine bacteria are able to do it. Look for pitting of agar media by colonies, colonies surrounded by craters, or even complete liquefaction after prolonged incubation. For example, Kurahashi and Yokota (2004) isolated six marine strains that began to liquefy agar within 1 week and completely liquefied it after several weeks.

2. Azocoll™ protein hydrolysis

Azocoll™ is an insoluble, finely ground collagen to which a bright-red azo dye has been attached. It is hydrolysed by most types of proteases, has been widely used for the assay of proteolytic enzymes and is commercially available.

a Plate method I

- (Method of Caplan and Fahey, 1982) Prepare Petri plates of nutrient agar containing 0.1% glucose (or other suitable agar medium).
- Suspend finely ground Azocoll in a minimal volume of 95% ethanol and add this suspension to rapidly stirred, autoclaved nutrient agar that has been cooled to 50°C. The Azocoll should be used at a concentration of 5 mg/ml of the agar. Pour 10-ml of the Azocoll-agar mixture on top of the nutrient agar to form an overlay and allow it to solidify. Allow residual ethanol to evaporate from the plates by incubating overnight at 37°C with the lids slightly ajar. The Azocoll should appear as a semiopaque, red film covering the surface of the agar.
- Spread serial dilutions of the organism to be tested over the surface of the plates and incubate. After growth appears, look for zones of clearing around the colonies.

b Quantitative method

- (Method of Chavira *et al.*, 1984) This method was originally used with cell extracts, but it can also be used for exocellular proteinases in culture supernatants. It depends on the release of soluble peptide fragments from Azocoll during proteolysis.
- Prewash a portion of the Azocoll by incubating it in the assay buffer (0.05-M Tris-HCl, 1-mM $CaCl_2$, pH 7.8) at 37°C for 90 min, filter and suspend in fresh buffer immediately before use in an assay.
- Suspend 0.25 g of the prewashed Azocoll in 50 ml of buffer (0.05-M Tris-HCl, 1-mM $CaCl_2$, pH 7.8). With a magnetic stirrer, stir rapidly to give a uniform suspension. Quickly remove 1.0–3.0 ml portions with an open-end plastic syringe (1 ml) and transfer to 13-mm-diameter glass assay tubes. Place the tubes vertically in a rotary water bath shaker at 37°C.
- Warm the tubes for 15 min before initiating the reaction by addition of 20–40 μl of culture supernatant or cell extract. Stop the reaction by rapidly drawing the solution into a 2-ml Pasteur pipette fitted with a small amount of glass wool at the neck to filter out the Azocoll. Transfer the clear filtrate to a clean tube with a second Pasteur pipette.
- As controls, prepare tubes containing Azocoll but add no enzyme. Incubate the controls for the same times as the test samples.
- Read the absorbance of the filtrates at 520 nm, using the buffer as the blank.

3. Cellulose hydrolysis

- Prepare mineral agar culture media in which cellulose is to be provided as a sole carbon source. For instance, Volokita *et al.* (2000) used a medium of the following composition (per litre of distilled water): KNO_3, 0.5 g; K_2HPO_4, 1.0 g; KH_2PO_4, 0.1 g; KCl, 0.5 g; $MgSO_4 \cdot 7H_2O$, 0.5 g; and agar, 15.0 g. Add 0.5 ml of a suitable trace metals solution, such as that given by Vishniac and Santer (1957). Autoclave, cool and dispense into plates.
- Prepare a series of dilutions of the organism to be tested and spread 25–50 μl portions over the surface of the plates.
- Place a sterile disc of lens paper on the seeded surface of the plates and incubate for 3–7 days.
- Look for colonies that form visible holes in the paper. To increase visibility of the holes, stain the paper on the plates or after the paper is removed with 0.2% irgalan black in 2% acetic acid.

4. DNA hydrolysis

- (Method of Smith *et al.*, 1969) Prepare a basal growth medium containing the following (per litre of distilled water): pancreatic digest of casein, 10.0 g; proteose peptone no. 3, 10.0 g; DNA (e.g. calf thymus DNA), 2.0 g; NaCl, 5.0 g; and agar, 15.0 g. Autoclave and cool to 45–50°C.
- Prepare an aqueous 0.5% solution of methyl green. Repeatedly extract this solution with equal volumes of chloroform until the chloroform layer is

colourless. Sterilize the aqueous solution by filtration and add 1.0 ml per 100 ml of basal growth medium. Mix and dispense into Petri dishes.
- Streak or spot the organism to be tested on the medium and incubate at the optimal growth temperature.
- Using a white background, look for colourless zones surrounding the growth on an otherwise green agar.

5. Gelatin and casein hydrolysis

- (Method of Medina and Baresi, 2007) Prepare gelatin agar plates (per litre of deionized water): tryptic soy agar powder, 40.0 g, and gelatin, 16.0 g. For testing casein hydrolysis, use casein agar plates (per litre of deionized water): plate count agar powder, 23.0 g, and dry skim milk, 13.0 g.
- Make a single streak or spot of the microorganism from a stock culture onto a gelatin agar plate and/or casein agar plate and incubate at 30–35°C.
- Prepare a stock solution of 35% (wt/vol) trichloroacetic acid (TCA) in deionized water.
- After incubation for 3 h (or 24 h for the casein hydrolysis test), flood the plate with the TCA solution. Look for occurrence of a clear zone around the growth within at least 4 min. With casein hydrolysis, clear zones may be visible without adding TCA, but the TCA enhances the visibility.

6. Starch hydrolysis

- Any suitable general-purpose agar medium can be used as the base medium (e.g. nutrient agar and tryptic soy agar), provided that it does not contain glucose (Wheater, 1955). Galactose can be substituted for glucose if a carbohydrate is needed for growth. Add 4–10 g of soluble starch per litre, boil the medium to dissolve the agar and sterilize by autoclaving.
- Prepare Gram's iodine solution by grinding 1.0 g of iodine crystals together with 2.0 g of KI in a mortar. Slowly add 300 ml of distilled water while grinding until the iodine is dissolved.
- Make a single streak of the test organisms across a plate of the agar and incubate.
- After growth occurs, flood the plate with the iodine solution. Starch stains blue with iodine, so look for colourless areas around the microbial growth.

P. Pigmentation

1. Fluorescent pigments

a Fluorescent diffusible pigments

Water-soluble pigments produced by bacteria diffuse out of the cells into the surrounding medium. Fluorescent pigments are usually water soluble; the best known examples are the fluorescent pigments produced by some *Pseudomonas* species; these vary in fluorescence from white to blue-green upon excitation with UV light (Palleroni, 1984). Fluorescent pigment production is not limited to pseudomonads; for instance, fluorescent pigments are produced by several freshwater and marine spirilla (Hylemon *et al.*, 1973) and by some *Azotobacter* species (Kennedy *et al.*, 2005).

- Inoculate a single streak of the organism across a plate of a colourless medium such as King medium B, which contains the following (per litre of distilled water): pancreatic digest of casein USP, 10.0 g; peptic digest of animal tissue USP, 10.0 g; K_2HPO_4, 1.5 g; $MgSO_4 \cdot 7H_2O$, 1.5 g; and agar, 14.0 g; pH 7.2.
- After incubation at the appropriate temperature and time, remove the covers from the plates and look for a fluorescent zone surrounding the growth with a UV light (below 260 nm).

b Fluorescent non-diffusible pigments

Coenzyme F_{420} – a fluorescent, non-diffusible, cytoplasmic cofactor – occurs in several distantly related prokaryotes, but only a few such as *Archaeoglobus*, *Ferroglobus* and most methanogens produce sufficient amounts so that the individual cells fluoresce when illuminated at 420 nm (Eirich *et al.*, 1979; Hafenbradl and Stetter, 2001; Huber and Stetter, 2001).

- Examine wet mounts of the cells from a culture with an epifluorescence microscope equipped with a 40 × or 100 × objective and appropriate filters. For instance, for excitation at 420 nm, Doddema and Vogels (1978) used Leitz Ploemopak filter system D (3 mm BG3+KP 425), with barrier filter K 460; under these conditions, the fluorescence was greenish-yellow.

2. Diffusible, non-fluorescent pigments

Diffusible coloured pigments are produced by various bacterial species; some of the best known pigments are the blue pyocyanin produced by *Pseudomonas aeruginosa* and the green (almost insoluble) and orange pigment (freely diffusible) made by *Pseudomonas chloraphis* (Palleroni, 2005). Also, various *Streptomyces* species make diffusible yellow-brown, blue, green, red-orange and violet pigments (Williams *et al.*, 1989).

- Culture the organism on a colourless medium such as King medium A, which contains the following (per litre): peptone, 20.0 g; glycerol, 10 ml; K_2SO_4 (anhydrous), 10.0 g; $MgCl_2$, 1.4 g; and agar, 13.6 g; pH 7.2. After incubation at the appropriate temperature for 24, 48 and 72 h, look for discolouration of the medium. Some pigments are only slightly water soluble and may precipitate out as crystals in the medium; e.g. the green chlororaphin made by *Pseudomonas chloraphis* (Palleroni, 2005).

3. Water-soluble pigments from aromatic amino acids

Some bacteria form diffusible coloured substances when grown with various aromatic amino acids. For instance, several freshwater and marine spirilla produce pigments from one or another of such amino acids (Hylemon *et al.*, 1973).

- Incorporate 0.2% L-tyrosine, L-phenylalanine and L-tryptophan, respectively, into portions of a suitable agar growth medium. After incubating cultures for 1 week at their optimum temperature, look for the presence of a diffusible yellow or brown pigment in the agar surrounding the growth.

4. Water-insoluble pigments

Many prokaryotes exhibit colonies that are coloured. The colour is due to membrane-bound pigments such as carotenoids, flexirubins and bacteriochlorophylls or chlorophylls. Membrane-bound pigments must be extracted from the cell, either by organic or by aqueous means. They can then be separated by using thin-layer chromatography (TLC) or by high-performance liquid chromatography (HPLC). After the pigments have been separated, they can be identified by ultraviolet or visual spectroscopy. Nuclear magnetic resonance (NMR) or mass spectroscopy (MS) can be used to determine the molecular structure and/or elemental composition. See Chapter 12 for further information about pigments.

Phenotypic and Physiological Characterization Methods

Q. Iron Porphyrin Compounds

Iron porphyrin compounds play a central role in the oxidation–reduction systems of biological importance in prokaryotes. They provide a key component of catalases, peroxidases, cytochromes and nitrate reductase.

- (Method of Deibel and Evans, 1960) Dissolve 1.0 g of benzidine dihydrochloride in 20 ml of glacial acetic acid. Add 30 ml of distilled water, heat gently, cool and add 50 ml of 95% ethanol.
- Prepare a fresh 5% working solution of H_2O_2 from a 30% stock solution.
- To an agar plate with ample microbial growth present, add a sufficient amount of the benzidine solution to cover the surface of the plate completely. Immediately add an equal amount of the 5% H_2O_2 solution.
- Look for the rapid development of a blue-green to deep blue colouration of the microbial growth.

♦♦♦♦♦♦ III. OTHER PHYSIOLOGICAL TESTS

Thousands of characterization tests have been described in the microbiological literature. Those that follow are designed for detection of specific enzyme activities, growth responses or metabolic products and they are useful for physiological characterization beyond the more general features of a genus or species.

A. Acetamide Hydrolysis

The reaction for acetamide hydrolysis is $CH_3CONH_2 + H_2O \rightarrow CH_3COOH + NH_3$. The NH_3 renders the medium alkaline, causing the pH indicator phenol red to change a yellow-orange to a red or magenta colour.

- (Method of Greenberg *et al.*, 1985). Prepare acetamide agar (per litre of distilled water): NaCl, 5.0 g; $MgSO_4 \cdot 7H_2O$, 0.5 g; KH_2PO_4, 0.73 g; K_2HPO_4, 1.39 g; acetamide, 10.0 g; phenol red, 0.012 g; and agar, 15.0 g. Adjust the pH to 6.8–7.2. Sterilize at 121°C for 15 min. Cool the medium in a slanted position.
- Inoculate each slant with a drop or loopful of dilute suspension of the organism. Look for a red or magenta colour appearing within 7 days. No change indicates a negative test.

B. Ammonia from Arginine

Bacteria may produce NH_3 from arginine by either of two reactions: L-arginine → L-ornithine + $3NH_3$ + CO_2 or L-arginine → putrescine + urea + CO_2. If the second reaction occurs and the organisms also have a urease, then the urea will be hydrolysed to CO_2 and NH_3. If NH_3 is formed by either route, it can be detected with Nessler's reagent by the formation of a yellow precipitate.

- (Method of Evans and Niven, 1950) Prepare a medium of the following composition (per litre of distilled water): yeast extract, 5.0 g; tryptone, 10.0 g; NaCl, 5.0 g; and L-arginine-HCl, 3.0 g; pH 7.0. Sterilize by autoclaving. Prepare the same medium without the arginine, as a control. Grow the culture for 3 days and test for ammonia production with Nessler's reagent.

  To prepare Nessler's reagent, dissolve 5.0 g of KI in the smallest possible quantity of cold water (5.0 ml). Add a saturated solution of $HgCl_2$ (about 2.2 g in 35.0 ml of water) until an excess is indicated by the formation of a precipitate. Then add 20.0 ml of 5-N NaOH and dilute to 100 ml. Let it settle and draw off the clear liquid. Store in a sealed container in the dark or in low-actinic glassware (the reagent is light sensitive).

C. Aromatic Ring Cleavage

Aromatic compounds can be hydroxylated by monooxygenases to form either catechol or protocatechuate – compounds that have two adjacent –OH groups on the aromatic ring. If subsequent cleavage of the aromatic ring occurs at the carbon–carbon bond between the two –OH groups, it is termed *ortho* cleavage. If the cleavage occurs at the carbon–carbon bond between one of its –OH groups and the adjacent carbon, it is termed *meta* cleavage. The following test can distinguish the type of cleavage used (Palleroni, 1984).

- Prepare a chemically defined solid growth medium containing an aromatic substrate such as 0.1% sodium *p*-hydroxybenzoate as the carbon source. Also prepare yeast extract agar for determining whether the enzymes produced are constitutive or inducible.
- When growth on the solid medium is obtained, remove several colonies and suspend them in 2 ml of 0.02-M Tris buffer (pH 8.0). Add 0.5 ml of toluene, shake and then add 0.2 ml of either 0.1-M catechol or 0.1-M sodium protocatechuate. Look for a bright yellow colour that appears within a few minutes, indicating *meta* cleavage.

- If a yellow colour does not develop, shake the tubes for 1 h at 30°C. Add solid $(NH_4)_2SO_4$ until it the solution is saturated and adjust the pH to ~10 by adding 2 drops of 5 N NH_4OH. Then add 1 drop of freshly prepared 25% (wt/vol) $Na_2[Fe(CN)_5NO] \cdot 2H_2O$. A deep purple colour develops if *ortho* cleavage has occurred.

D. Arylsulfatase Activity

Arylsulphatase catalyses the hydrolysis of the bond between sulfate and the aromatic ring in a compound such as *p*-nitrophenyl sulfate or 5-bromo-4-chloro-3-indolyl sulfate.

1. *p*-Nitrophenyl sulfate method

- (Method of Bürger, 1967) Harvest the cells from a suitable solid or liquid medium and suspend them in 0.85% saline to give a concentration of 4–8 mg dry weight per millilitre.
- Prepare a 2.5-mM solution of the sodium or potassium salt of *p*-nitrophenyl sulfate in 0.5-M sodium acetate–acetic acid buffer (pH 5.8).
- Add 0.3 ml of the substrate solution to 0.3 ml of the cell suspension. Mix and incubate for up to 6 h.
- Add 0.3 ml of 0.04-M glycine-NaOH buffer (pH 10.5). Look for development of a yellow colour, which indicates the presence of free *p*-nitrophenol.

2. 5-Bromo-4-chloro-3-indolyl sulfate method

- Incorporate the chromogenic substrate 5-bromo-4-chloro-3-indolyl sulfate (potassium salt) into a suitable agar medium at a concentration of 0.1 g/l. After incubation, look for colonies that are either green (Wyss, 1989) or blue (Cregut *et al.*, 2009).

E. Bile Solubility

Some bacteria lyse in the presence of bile or bile salts. For example, *Streptococcus pneumoniae* has an autolytic enzyme, which is activated by bile salts and causes the cell wall to dissolve.

1. Colony procedure

- (Method of Health Protection Agency, 2010a) Select a well-isolated colony. Place 1–2 drops of 2% sodium deoxycholate directly on the colony and incubate at 37°C for up to 30 min. Look for disintegration or lysis of the colony.

2. Broth procedure

- (Method of Health Protection Agency, 2010a) Grow the culture in a well-buffered medium at pH 7.8. Centrifuge the cells and prepare a dense suspension in 1 ml of 0.85% saline. Divide into two tubes.

- To one tube add 0.5 ml of 10% sodium deoxycholate. Add 0.5 ml of saline (0.85% NaCl) to the other. Incubate at 37°C for 15 minutes. Look for a clearing of the suspension in the tube with deoxycholate but not in the saline control.

F. Bile Tolerance

Bile or bile salts are inhibitory to some bacteria, whereas others are highly resistant. For example, most species of *Bacteroides* can grow in the presence of 20% bile, but species of *Porphyromonas* cannot. In general, Gram-negative bacteria are more resistant to bile than are Gram-positive bacteria (Levin, 2006). Oxgall (dehydrated bile) is usually used in culture media to test for bile tolerance; a 10% solution of oxgall is equivalent to a fresh bile solution.

1. Solid medium method

- (Method of Cowan, 1974) Prepare bile agar medium (per litre of distilled water): beef extract, 10.0 g; peptone, 10.0 g; and NaCl, 5.0 g. Dissolve by heating. Adjust the pH to 8.0–8.4 with 10 N NaOH and boil for 10 min. If necessary, filter to remove any undissolved components. Adjust the pH to 7.2–7.4 and add 10.0 g of powdered oxgall, readjusting the pH if necessary. Add 15 g of agar and dissolve by boiling. Sterilize by autoclaving, cool to 55°C and aseptically add 50 ml of sterile blood serum. Mix, pour into Petri plates and allow to solidify.
- Inoculate the medium with the organism to be tested. Compare the growth to that on a similar medium which does not contain bile.

2. Broth method for anaerobes

- Use tubes of pre-reduced peptone-yeast-glucose (PYG) broth and PYG broth with 2% oxgall (PYG-bile). These pre-reduced media are prepared according to the instructions given by Holdeman *et al.* (1977); however, they are also commercially available. Inoculate the two media, avoiding the introduction of oxygen, and incubate at the appropriate temperature for growth. Compare the growth response between tubes containing oxgall and tubes of that lack it.

G. Citrate Utilization

Citrate utilization tests the ability of an organism to grow aerobically with sodium citrate as the sole carbon source and with ammonium phosphate as the sole nitrogen source.

- (Method of Simmons, 1926) Prepare the following medium (per litre of distilled water): $(NH_4)H_2PO_4$, 1.0 g; K_2HPO_4, 1.0 g; NaCl, 5.0 g; sodium citrate, 2.0 g; $MgSO_4 \cdot 7H_2O$, 0.2 g; bromothymol blue, 0.08 g; and agar, 15.0 g; pH 6.0. Boil to dissolve the agar. Dispense into tubes and autoclave. Cool to 45–50°C and allow to solidify in a slanted position.

- Select a well-isolated colony from a pure culture and inoculate a slant by touching the tip of an inoculating loop to the colony and then lightly transfer the colony to the surface of the slant. Alternatively, suspend the pure culture in sterile saline and streak this dilute suspension onto the surface of the slant. Incubate for up to 7 days. Make sure the screw-cap on the tube is loose to allow access of air.
- Look for development of a blue colour. A negative test will remain green.

H. Dye Tolerance

The ability to grow in the presence of low concentrations (0.00001–1.0%) of certain dyes may be a useful descriptive feature for a new species. The dyes can be incorporated into liquid or solid culture medium, or dye-soaked paper strips can be placed on inoculated agar plates. Some dyes that have been used are alizarin red, basic fuchsin, brilliant green, crystal violet, Janus green, methylene blue, neutral red, safranin and thionine. For examples of the use of such dyes, see Goodfellow *et al.* (1978); Grimont and de Rosnay (1972); Mallory *et al.* (1977); and Ribeiro and Herr (1990).

I. Esculin Hydrolysis

Hydrolysis of the glycoside esculin yields esculetin and glucose. The esculetin is identified by its ability to react with ferric ions, forming a dark brown to black complex.

1. Agar medium method

- (Method based on Lányi, 1987 and Lennette *et al.*, 1974) Supplement any appropriate nutrient medium with 0.1% esculin and 0.05% ferric ammonium citrate or ferric citrate. The 6,7-dihydroxycoumarin formed by hydrolysis of the esculin forms a black to dark brown complex with the ferric ions. To make the test more specific for Group D streptococci and enterococci, add 4% oxgall (bile) to the medium, as other organisms usually cannot hydrolyse esculin in the presence of bile.

2. Rapid test

- (Method of Qadri *et al.*, 1980) Prepare the following buffered test solution (per litre of distilled water): esculin, 5.0 g; ferric ammonium citrate, 0.5 g; sodium chloride, 8.0 g; K_2HPO_4, 0.4 g; and KH_2PO_4, 0.1 g; pH 5.6. Dispense 0.2-ml portions into 12 by 75-mm tubes. Refrigerate until needed. The solution is stable for up to 3 months, but if a precipitate forms, dissolve it by heating to 60–70°C.
- Suspend 4 colonies of the organism in 0.5 ml of the test solution to give a density similar to that of a no. 3 McFarland nephelometer standard (prepared by adding 0.3 ml of 1.0% $BaCl_2$ to 9.7 ml of 1.0% H_2SO_4). Incubate in a water bath at 35°C for up to 2 h. Look for development of a dark brown to black colour.

J. Glycosidases

The procedures given below are based on the classic ones for glycosidases. However, it should be noted that many commercial multi-test systems (see Section IV) incorporate tests for glycosidases by using microtiter plate wells containing the various chromogenic or fluorogenic substrates. The results can be recorded by automated microtiter plate readers.

1. Chromogenic substrates

The chromogenic substrates used in this test are various sugars that are linked to either *o*-nitrophenol or *p*-nitrophenol by a glycosidic bond. Hydrolysis of the bond frees the nitrophenol, which is detected by its yellow colour. Nitrophenyl derivatives of a large variety of D and L sugars are commercially available, such as *p*-nitrophenyl-α-D-glucopyranoside (for α-D-glucosidase), *p*-nitrophenyl-β-D-glucopyranoside (for β-D-glucosidase), *o*-nitrophenyl-β-D-galactopyranoside (for β-D-galactosidase), *p*-nitrophenyl-α-L-rhamnopyranoside (for α-L-rhamnosidase) etc. Bear in mind that hydrolysis of a chromogenic substrate does not necessarily mean that the glycosidase for the natural substrate is present. For instance, *Staphylococcus aureus* can hydrolyse *o*-nitrophenyl-β-D-galactopyranoside (ONPG) but not lactose (see Kennedy and Scarborough, 1967). Thus results should be reported as 'ONPG hydrolysed', not 'β-galactosidase present'.

- (Method of Edberg and Kontnick, 1986) Prepare Sorensen's phosphate buffer from solutions A and B. Solution A contains 35.6-g $Na_2HPO_4 \cdot 2H_2O$ per litre of distilled water; solution B contains 27.6-g $NaH_2PO_4 \cdot H_2O$ per litre of distilled water. Combine 40.5 ml of solution A with 9.5 ml of solution B to give 50 ml of 0.2-M phosphate buffer and verify that the pH is 7.4. Dilute the stock buffer fourfold to give a 0.05-M working buffer.
- Dissolve the nitrophenyl glycoside in the working buffer to give a concentration of 0.5 mg/ml. Dispense 0.25-ml portions into a 12 mm × 75 mm tube. Inoculate the tube with several colonies of the organism to be tested and incubate in a water bath for 1 h at 35°C. Look for the development of a yellow colour. No colour change should occur in an uninoculated control tube.

2. Fluorogenic substrates

The fluorogenic substrates consist of various sugars that are linked to 4-methylumbelliferone (4-MU) by a glycosidic bond. Hydrolysis of the bond frees the 4-MU, which is detected by its fluorescence under UV light. Examples of commercially available substrates include 4-MU-β-D-glucopyranoside (for β-D-glucosidase), 4-MU-β-D-galactopyranoside (for galactosidase), 4-MU-β-D-glucuronide (for β-D-glucuronidase), 4-MU-2-acetamido-2-deoxy-β-D-glucopyranoside (for *N*-acetyl-β-D-glucosaminidase), 4-MU-α-D-mannopyranoside (for α-D-mannosidase), 4-MU-α-L-rhamnopyranoside (for α-L-rhamnosidase), 4-MU-β-D-fucoside (for β-D-fucosidase) etc.

- (Method of Maddocks and Greenan, 1975) Dissolve 15 μmoles (5 μmoles in the case of 4-MU-β-D-galactopyranoside) of each substrate in 0.2 ml of dimethylsulfoxide (DMSO). Prepare phosphate buffered saline (PBS) to contain (per litre of distilled water): NaCl, 8.0 g; KCl, 0.2 g; Na_2HPO_4, 1.15 g; and KH_2PO_4, 0.2 g; pH 7.3. Add sufficient PBS to the 0.2 ml of DMSO/substrate to give a final volume of 10.0 ml.

- Take a sample of a colony of the organism to be tested with a glass rod and smear the organisms onto a piece of Whatman no. 1 qualitative filter paper. Cover the smear with 2 drops (ca. 0.067 ml) of DMSO/substrate. As a control, make a similar smear nearby and add 2 drops of the DMSO/PBS solution without substrate. Deposit a further 2 drops of DMSO/substrate nearby on the filter paper (but not on a bacterial smear) as an additional control.
- Incubate for 10 min at 37°C. Cover each spot on the filter paper with 2 drops of 0.1-N NaOH as a fluorescence enhancer. Examine the spots in a dark room under UV light (366 nm). Look for light blue fluorescence of the test spot but not with the two controls.

K. Hippurate Hydrolysis

The enzyme hippuricase catalyses the hydrolysis of sodium hippurate to benzoic acid and glycine. The glycine can be detected by its ability to react with ninhydrin to give a purple colour.

- (Method of Yong and Thompson, 1982) Dissolve 0.25 g of sodium hippurate in 25 ml of distilled water and sterilize by filtration.
- With an inoculating loop, remove portions of isolated colonies from a pure culture without touching the loop to the surface of the agar medium and make a moderately dense suspension in 2 drops of the hippurate solution in a narrow tube. Incubate for 1 h at 35°C. Incubate a similar suspension in sterile distilled water as a control.
- To the test and control tubes, add 2 drops of ninhydrin reagent (3.5 g of ninhydrin dissolved in 100 ml of a 1:1 acetone–butanol mixture). Incubate for 15 min. Look for development of a deep purple colour due to the formation of glycine. No colour should be evident in the control; if there is colour, it is probably due to carryover of amino acids in the inoculum, as ninhydrin will react with any α-amino acid.

L. Hydrogen Sulfide Production

Some anaerobic and facultatively anaerobic organisms can produce abundant H_2S by the anaerobic reduction of $S_2O_3{}^{2-}$. The H_2S can be detected by its reaction with iron salts contained in the medium, which form a black precipitate of FeS.

A different type of H_2S production is based on the ability of some organisms to form low levels of H_2S from sulfur-containing amino acids (cysteine, cystine and/or methionine) by means of amino acid desulfurases. The gaseous H_2S so produced is detected by its reaction with lead acetate strips suspended above the surface of the medium.

1. Thiosulfate iron H_2S test

- (Method of Levine *et al.*, 1934) Prepare an appropriate solid or semisolid growth medium containing $S_2O_3^{2-}$. For example, peptone iron agar contains the following (per litre of distilled water): peptone, 15.0 g; proteose peptone, 5.0 g; ferric ammonium citrate, 0.5 g; K_2HPO_4. 1.0 g; $Na_2S_2O_3$, 0.08 g; and agar, 15 g (or 2.0 g for a semisolid medium). Dispense the medium into tubes and autoclave.
- When cool, inoculate by stabbing. Incubate at an appropriate temperature. After growth occurs, look for blackening of the medium. (See also Section III.27 about triple-sugar iron agar reactions.)

2. Paper strip method for amino acid desulfurase activity (Lányi, 1987)

- (Method of Lányi, 1987) Inoculate a suitable semisolid (0.2% agar) growth medium that contains a peptone or other source of sulphur amino acids. Suspend a strip of sterile, lead acetate–impregnated paper about a centimetre above the surface of the culture, fold the upper end over the lip of the tube and hold it in place with the screw cap or cotton plug. During growth of the organisms, H_2S gas reacts with the lead acetate to form black PbS, beginning at the lower part of the strip.

  Lead acetate strips can be prepared by soaking 5-cm strips of filter paper in a 5% aqueous solution of lead acetate, sterilizing them separately in tubes by autoclaving and drying them in an oven.

M. Indole Production

Organisms that possess tryptophanase can carry out the following reaction: L-tryptophan → indole + pyruvic acid + NH_3. The indole can be detected by its ability to react with *p*-dimethylaminobenzaldehyde to form a quinoidal red-violet condensation compound.

1. Kovács' test

- (Method of Kovács, 1928) Inoculate a suitable liquid culture medium supplemented with 0.1–1.0% tryptophan and incubate at the optimal temperature of the organism for 24–72 h.
- Dissolve 5.0 g of *p*-dimethylaminobenzaldehyde in 75 ml of isoamyl alcohol by gentle heating and store at 4°C. Prepare small quantities of Kovács' reagent by mixing 3 volumes of the aldehyde solution with 1 volume of conc. HCl. The reagent can be stored at 4°C for up to 2 months if protected from light.
- Add 0.5 ml of the Kovács' reagent to the broth culture and agitate to extract the indole. Look for development of a deep red colour in the upper layer of liquid. A yellow colour is a negative test.

2. Xylene extraction test

- (Method of Lányi, 1987) This test is more sensitive than the Kovács' test. Grow the test organism in a suitable culture medium supplemented with 0.1–1.0% tryptophan. Avoid using media containing carbohydrates, nitrate or nitrite, as these may interfere with the test. Distribute in 2–3 ml portions and sterilize by autoclaving. When cool, inoculate with the organism to be tested and incubate for up to 3 days.
- Prepare Ehrlich's reagent as follows: 1.0 g of *p*-dimethylaminobenzaldehyde, 95 ml of 95% ethanol and 20 ml of conc. HCl.
- Add 1 ml of xylene to the broth culture, shake vigorously and allow the mixture to stand for about 2 min. Then add 0.5 ml Ehrlich's reagent slowly down the side of the tube so as to form a layer between the medium and xylene. Do not shake the tube after addition of the Ehrlich's reagent. Look for development of a pink or red ring below the xylene layer.

N. Indoxyl Acetate Hydrolysis

Some bacteria contain an esterase that can catalyse the hydrolysis of indoxyl acetate to acetic acid and indoxyl, which is fluorescent. In the presence of air, the indoxyl is oxidized spontaneously to indigo white (which is also fluorescent), and then to indigo blue (which is non-fluorescent but which has a deep blue colour).

- (Method of Mills and Gherna, 1987) Wet paper discs (0.25-in. diameter) with 50 μl of a 10% (wt/vol) solution of indoxyl acetate in acetone and allow to dry. The discs can be stored for up to a year at 4°C in an amber bottle with silica gel desiccant.
- Inoculate a disc with a colony of the test organism and add a drop of sterile distilled water. Look for development of a dark blue colour (indigo) within 5–10 min.

O. 3-Ketolactase from Lactose Oxidation

This test is based on the ability of some bacteria to oxidize lactose to 3-ketolactose, which can be detected by Benedict's reagent.

- (Method of Bernaerts and De Ley, 1963) Prepare lactose agar of the following composition (per litre of distilled water): lactose, 10.0 g; yeast extract, 1.0 g; and agar, 20.0 g. Sterilize by autoclaving, cool and dispense into Petri dishes. Inoculate a loopful from a slant culture of the organism to be tested onto a plate of the lactose agar, so that it forms an area of about 0.5 cm in diameter. Incubate at 28°C for 1–2 days. Flood the plate with a shallow layer of Benedict's reagent and incubate for 1 h at room temperature. Look for development of a yellow ring of Cu_2O around the cell mass.

  To prepare Benedict's reagent, dissolve 173 g of sodium citrate · $2H_2O$ and 100 g of anhydrous Na_2CO_3 in 800 ml of distilled water with warming and stirring. Filter. Separately dissolve 17.3 g of $CuSO_4 \cdot 5H_2O$ in 100 ml of distilled water. Add the $CuSO_4$ solution slowly with stirring to the citrate–carbonate solution. Allow to cool and dilute to 1 l with distilled water.

P. Lactic Acid Optical Rotation

A quantitative method of testing bacterial cultures for production of D-lactic acid and L-lactic acid was described by Cato and Moore (1965). It depends on determining the concentration of L-(+)-lactic acid (obtained by use of NAD-linked L-lactic dehydrogenase) and comparing it to the total concentration of lactic acid. The procedure is laborious, and if the purpose is merely to detect whether D- and/or L-lactic acid is present, a convenient alternative may be to use commercial kits developed for use with foods, dairy products and wines.

Q. Lecithinase

Bacterial lecithinase (phospholipase) causes a clear emulsion of lecithovitellin to become opalescent due to the release of free fat.

- (Method of Lányi, 1987) Prepare a basal medium containing the following (per litre of distilled water): tryptone, 40.0 g; Na_2HPO_4, 5.0 g; NaCl, 2.0 g; $MgSO_4 \cdot 7H_2O$, 0.01 g; glucose, 2.0 g; and agar, 10.0 g; pH 7.6. Dispense 20-ml portions per tube and autoclave.

  To prepare the lecithivitellin solution, scrub and soak antibiotic-free eggs in 95% ethanol for 1 h. Separate the yolks from the whites aseptically and mix the yolks with an equal volume of sterile physiological (0.85%) saline to form a homogeneous emulsion. This can be stored in a refrigerator for up to 3 weeks.

  To prepare the test medium, cool 20 ml of the molten basal medium to 50°C, aseptically add 2 ml of the egg yolk emulsion, mix and pour plates.

- Inoculate the solidified medium as streaks or spots and incubate for 3 days. Look for a zone of turbidity within the agar surrounding the growth.

  Note: Lecithinase activity resulting in the release of fats should not be confused with the action of lipase, which liberates free fatty acids from fats. Whereas the recommended substrate for the lipase test is Tween 80, lipolysis can also be observed on egg yolk agar by oblique lighting as an oily iridescent sheen ('pearly layer') over the growth and in its close vicinity on the surface of the medium. Fatty acids responsible for the pearly layer stain bluish-green if the plate is flooded with saturated copper sulfate solution, the excess solution removed and the plate dried at 37°C for 20 min (Cowan, 1974). Some organisms produce both lecithinase and lipase activity.

R. Lipase

Lipase activity can be shown with egg yolk medium (see above) or by using Tweens, viz., Tween 80 (polyethylene sorbitan monooleate, an oleic acid ester), Tween 60 (a stearic acid ester) and Tween 40 (a palmitic acid ester). Lipolytic organisms split off the fatty acid, and the calcium salts of the fatty acids produce opaque zones around the colonies.

- (Method of Sierra, 1957) Prepare a basal medium containing the following (per litre of distilled water): peptone, 10.0 g; NaCl, 5.0 g; $CaCl_2 \cdot 2H_2O$, 0.1 g; and agar, 9.0 g; pH 7.4. Sterilize by autoclaving. Autoclave the desired Tween separately. Cool the basal medium to 45–50°C, add the Tween to give a final concentration of 1.0%, shake until the Tween is completely dissolved and pour into plates.
- Inoculate the cultures as lines on the surface of the agar. Incubate for up to 7 days, inspecting daily. Look for an opaque halo around the growth.

S. Lysine and Ornithine Decarboxylases

Lysine can be decarboxylated to form CO_2 and cadaverine, and ornithine can be decarboxylated to form putrescine. The free amines produce an alkaline reaction in the medium that is detectable by a pH indicator.

- (Method of Lányi, 1987) Prepare a basal medium composed of the following (per litre of distilled water): peptone, 5.0 g; beef extract, 5.0 g; pyridoxal, 0.005 g; glucose, 0.5 g; bromocresol purple (0.2% solution), 5 ml; cresol red (0.2% solution), 2.5 ml; pH 6.0. (To prepare the bromocresol and cresol red solutions, dissolve 2 g of the indicator in 50 ml of 0.1-N NaOH and add 950 ml distilled water.)

  To prepare the test medium, dissolve 1% L-lysine · HCl or L-ornithine · HCl in the basal medium by gentle heating and readjust to pH 6.0. Distribute the amino acid media and the control basal medium (which lacks the amino acid) in narrow tubes and sterilize by autoclaving. Layer the media with mineral oil (1–1.5 cm) immediately after removal from the autoclave. The colour of the uninoculated medium should be light grey.
- Lightly inoculate the test medium and the control medium through the mineral oil layer and incubate for up to 4 days. Look for development of a violet colour.

T. Malonate Utilization

An organism that simultaneously can utilize disodium malonate as its carbon source and NH_4^+ as its nitrogen source produces an alkaline reaction due to the formation of NaOH. The alkaline reaction is detected with a pH indicator.

- (Method of Lányi, 1987) Prepare a medium containing (per litre of distilled water): $(NH_4)_2SO_4$, 2 g; K_2HPO_4, 0.4 g; NaCl, 2 g; beef extract, 0.5 g or yeast extract, 1.0 g; sodium malonate, 3 g; bromothymol blue, 0.2% solution, 12.5 ml; pH 6.8. (The bromothymol blue solution consists of bromothymol blue, 2.0 g; 0.1-N NaOH, 50 ml; and distilled water, 950 ml.) Distribute 2–3 ml amounts of the medium into narrow tubes and autoclave.
- Inoculate the medium lightly and incubate for 48 h. Look for development of a blue colour, compared with the yellow colour of an uninoculated tube. For malonate utilization, certain bacteria require a small amount of organic nitrogen source (beef or yeast extract). Some authors prefer to add glucose (0.25 g/l) as a further supplement for growth (MacFaddin, 1980).

U. Methyl Red Test

A type of fermentation called the mixed acid fermentation results in the formation of formic acid, acetic acid, lactic acid, succinic acid, ethanol, CO_2 and H_2 in a buffered medium. The combination of acids in the mixed acid fermentation usually lowers the pH of the culture below 4.2. The test is used mainly in the differentiation of enteric bacteria. The organism being tested must be capable of catabolizing glucose.

- (Method of Lányi, 1987) Prepare MR-VP medium containing the following (per litre of distilled water): peptone, 7.0 g; K_2HPO_4, 5.0 g; and glucose, 5.0 g; pH 7.5. Dispense 2–3 ml portions into narrow tubes and sterilize by autoclaving. Inoculate the tubes lightly and incubate for 4 days at the optimum temperature for the organism.
- Add one drop of methyl red reagent (0.25-g methyl red dissolved in 100 ml of ethanol). Look for a red colour (MR positive). A weakly positive test is red-orange and a yellow or orange colour indicates a negative test.

V. Peptidases

The method employs chromogenic substrates, viz., the β-naphthylamide derivatives of amino acids in which the amino acids are linked to β-naphthylamide by a peptide bond. If the bond is broken by a peptidase, the freed β-naphthylamine can be detected by colour-producing reagents. Substrates that are available commercially include the β-naphthylamide derivatives of all of the common L-amino acids.

- (Method of D'Amato, 1978) Dissolve the crystalline β-naphthylamide derivative in a small volume of 50% ethanol or *N,N*-dimethylformamide. Add an appropriate amount of 0.1-M tris(hydroxymethyl)aminomethane (Tris) buffer, pH 8 (but use pH 7.2 for hydroxyproline-β-naphthylamide and pH 7.6 for γ-glutamyl-β-naphthylamide) to give a concentration of 0.001-M β-naphthylamide derivative. Add a 40-μl sample of each β-naphthylamide derivative to the well (280 μl) of a microtiter plate. Prepare control wells with buffer but without the β-naphthylamide substrate. Dry the wells under a vacuum below 50°C and store at 4°C in a desiccator jar.
- Suspend the growth from a suitable agar medium in 0.85% NaCl to achieve a turbidity equivalent to a MacFarland no. 3 nephelometer standard (0.3 ml of 1.0% $BaCl_2$ solution added to 9.7 ml of 1.0% H_2SO_4). Dispense 20–40 μl of this cell suspension into each of the dried microtiter test wells and control wells. Also add 0.85% NaCl (without cells) to dried wells as controls. Incubate the wells for 1–4 h in a humid chamber at 35°C.
- Prepare detection reagent A from 250-g Tris buffer; conc. HCl, 110-ml; sodium dodecyl sulphate (SDS), 100 g; and sufficient distilled water to make 1000 ml. Adjust the final pH to 7.6. Prepare detection reagent B by dissolving 3.0 g of fast blue BB in 1.0 litre of 2-methoxyethanol.
- Add 20–40 μl of reagent A to each well after the incubation period, followed by 20 μl of reagent B. Wait for 15 min for maximum colour development in each well. Any colour change relative to the control wells lacking substrate

and the control wells inoculated with only 0.5% NaCl is considered a positive reaction.

W. Phenylalanine Deaminase

Some organisms can deaminate the amino acid phenylalanine with the formation of phenylpyruvate and NH_3. The phenylpyruvate can be readily detected by its reaction with $FeCl_3$.

- (Method of Lányi, 1987) Use the following medium (per litre of distilled water): yeast extract, 3.0 g; DL-phenylalanine, 2.0 g; $Na_2HPO_4.2H_2O$, 1.0 g; NaCl, 5.0 g; and agar, 12.0 g. Dispense 3-ml portions into narrow tubes, autoclave and cool in a slanted position. Inoculate the slants heavily and incubate them overnight.
- Prepare ferric chloride reagent by dissolving 5.0 g of $FeCl_3$ in 10 ml of 0.1-N HCl. Bring the volume to 100 ml with distilled water.
- Place 3–4 drops of the reagent onto the culture. Development of a green colour on the slant and in the syneresis fluid at the base of the slant within 1 min indicates the presence of phenylpyruvate.

X. Phosphatases

Depending upon their pH optimum, phosphatases are referred to as either acid phosphatases (pH 4–6) or alkaline phosphatases (pH 8.5–9.5). Various substrates for incorporation into culture media have been used for detection of phosphatase activity, including phenolphthalein phosphate, 5-bromo-4-chloro-3-indoxyl and *p*-nitrophenylphosphate.

1. General method for phosphatases

- (Method of Lányi, 1987) Use a non-selective, solid growth medium (such as nutrient agar, pH 6.8) sterilized by autoclaving and cooled to 45–50°C. Add 10 ml of 1.0% phenolphthalein diphosphate solution (sterilized by filtration) per litre of medium. Mix and dispense into Petri plates.
- Inoculate each plate with a single organism to be tested and incubate until colonies develop.
- Invert the culture. To the lid of the dish add 0.1 ml of conc. NH_4OH. Place the culture back on the lid to expose the colonies to the ammonia vapour.
- The development of red colonies within 20–30 s indicates phosphatase activity. The colour is due to the liberation of free phenolphthalein, which turns red under alkaline conditions.

2. Broth method for acid and alkaline phosphatases

- (Method based on Black, 1973 and Bürger *et al.*, 1967) Prepare phosphate buffered saline (PBS) to contain (per 800 ml of distilled water) NaCl, 8.0 g; KCl, 0.2 g; Na_2HPO_4, 1.44 g; and KH_2PO_4, 0.24 g. Dissolve 0.1 g of the sodium salt of phenolphthalein diphosphate in the PBS. Divide the solution into four 200-ml portions and adjust their pH values to 9.4, 7.4, 6.5 and 5.4, respectively. Add distilled water to make the final volume of each portion 250 ml.

- Harvest the cells from an exponential-phase culture by centrifugation. Suspend portions of the cell pellet in 1.0 ml of each of the four PBS preparations. Incubate for 4 h at 37°C.
- Add two drops of 5-N NaOH. The development of a red colour indicates phosphatase activity. Activity at the lower pH values indicates acid phosphatase; activity at the higher pH values indicates alkaline phosphatase.

3. Method for alkaline phosphatase

- (Method of Wolf *et al.*, 1973) Streak the organism on a suitable agar medium and incubate to obtain colonies.
- In a small tube, dissolve 2 mg of 5-bromo-4-chloro-3-indolyl phosphate in 0.25 ml of *N,N*-dimethylformamide. Add this solution to 2.75 ml of 0.2-M propanediol buffer (pH 7.5). Transfer a colony from the agar medium to the tube. Add 0.2 ml of 5-mM $MgCl_2$.
- Incubate the tube at 37°C for 4 h. Look for an initial blue-green precipitate on the bacterial growth settled at the bottom of the tube, followed by development of a blue colour throughout the entire tube.

Y. Poly-β-Hydroxybutyrate (PHB) Formation

PHB is a polymeric ester of β-hydroxybutyrate and is formed by many different kinds of prokaryotes. It serves as an energy and carbon reserve. Optimal conditions for synthesis of PHB depend on the species, but a high ratio of carbon to nitrogen in the medium is often favourable. Low phosphate levels also may help. Some bacteria such as *Ralstonia eutropha* and *Azotobacter beijerinckii* can accumulate up to 80% of their dry cell weight as PHB. Some anaerobes, such as the non-oxygenic phototrophs, synthesize PHB and some Archaea, such as the halobacteria, also synthesize it.

1. Visualization of PHB inclusions in cells

- (Method of Ostle and Holt, 1982) Prepare a 1% aqueous solution of the oxazine dye Nile blue A. Use mild heating to fully dissolve the dye if necessary. Filter the solution before use.
- Heat-fix smears of the cells and stain with the Nile blue A solution at 55°C for 10 min in a staining jar. Wash the slide with water to remove excess stain and then wash with 8% aqueous acetic acid for 1 min. Wash again with water and blot dry with bibulous paper. Remoisten the slide with tap water and cover the preparation with a no. 1 coverslip.
- Examine the slide with an epifluorescence microscope and an oil immersion objective, using an excitation wavelength of approximately 460 nm. Look for intracellular granules that fluoresce bright orange.

2. Chemical analysis

- (Method based on Law and Slepecky, 1961 and Koechlein and Krieg, 1998) Definitive identification of PHB depends on chemical analysis. The following procedure can be used not only for identification but also for quantification of PHB. Harvest the cells from a broth culture by centrifugation, using polypropylene tubes that have been previously washed with ethanol and boiling chloroform to remove plasticizers (chloroform boils at ca. 62°C; boil it by immersion in an 80°C water bath.) Dry the cells in a drying oven at 90°C. Extract the PHB from the dried cells with three portions of boiling chloroform. Transfer the pooled chloroform extracts to a clean glass tube and add two volumes of ice-cold acetone to precipitate the PHB. Centrifuge, discard the acetone and redissolve the PHB in hot chloroform. Repeat the precipitating/redissolving procedure several times and then dry the precipitated PHB thoroughly.
- Add 10.0 ml of conc. H_2SO_4. Cap the tube with a glass marble (to prevent water vapour from entering the tube) and place the tube in a boiling water bath for 10 min. The H_2SO_4 hydrolyses and dehydrates the PHB to form crotonic acid. Cool, mix thoroughly and place an aliquot in a quartz cuvette. Use an ultraviolet spectrophotometer to determine the absorption spectrum from 215 to 255 nm, with concentrated sulfuric acid as a blank. A peak at 235 nm identifies crotonic acid. It may be necessary to make dilutions of the extract in sulfuric acid to obtain a useful absorbance reading (i.e. an A_{235} between 0.05 and 1.0).
- For a method of quantification of PHB, see Daniels *et al.* (1994).

Z. Triple-Sugar Iron Agar Reactions

Triple-sugar iron (TSI) agar is used to characterize an organism's ability to ferment sugars (glucose, lactose and/or sucrose), to produce gas from sugar fermentation and to produce H_2S.

- (Method of Lányi, 1987) Prepare TSI agar (per litre of distilled water): peptone, 20.0 g; beef extract, 3.0 g; yeast extract, 3.0 g; glucose, 1.0 g; lactose, 10.0 g; sucrose, 10.0 g; NaCl, 5.0 g; $Na_2S_2O_3$, 0.3 g; and agar, 13.0 g. Dissolve the ingredients by heating. Add 12 ml of 0.2% phenol red solution (dissolve 0.2 g of phenol red in 8.0 ml of 0.1-N NaOH and 92.0 ml of distilled water). Add 0.2 g of $FeSO_4 \cdot 7H_2O$ dissolved in a small amount of distilled water. Distribute immediately after the addition of the iron salt into cotton-stoppered or loosely screw-capped tubes (access of air to the culture is essential). Sterilize at 115 °C for 20 min; cool to form deep butts and long slants.
- Inoculate by stabbing into the butt and then streaking the slant uniformly without scratching its surface. Incubate for 18–24 h at 37°C.
- If the culture attacks only glucose, acids accumulate in the butt (yellow colour) where anaerobic conditions prevail; the slant usually exhibits an acidic reaction (yellow) at the beginning of multiplication only, then becomes neutral or alkaline (red) because of the oxidation of the glucose and because of alkaline products from peptone utilization.

Owing to the tenfold higher concentration of lactose and sucrose than glucose, organisms that attack either or both of these sugars produce acids in large enough amounts to maintain an acidic reaction (yellow) in the slant as well as in the butt.

Gas produced from the sugars splits the agar or forms bubbles in it.

H_2S production is indicated by a blackening of the butt due to formation of FeS.

AA. Urease

Urease catalyses the reaction $(NH_2)_2CO + H_2O \rightarrow 2NH_3 + CO_2$. The ammonia that is formed causes the medium to become alkaline: $NH_3 + H_2O \rightarrow NH_4^+ + OH^-$. The alkalinity can be detected with a pH indicator.

1. Method 1

- (Method of Christensen, 1946) Prepare the following medium (per litre of distilled water): peptone, 1.0 g; yeast extract, 0.1 g; NaCl, 5.0 g; KH_2PO_4, 2.0 g; phenol red, 0.012 g; glucose, 1.0 g; and agar, 20.0 g; pH. 6.8–6.9. Autoclave and cool to 45°C. Add 100 ml of a 20% (wt/vol) filter-sterilized solution of urea. Mix and dispense 5-ml portions into tubes and slant the tubes so as to leave a butt of about 2.5 cm in depth with a slant of about 4 cm in length. Inoculate the surface of the slant. (The medium can also be prepared as a liquid medium by omitting the agar.)
- Look for development of a red-violet colour compared to an uninoculated control.

2. Method 2

- (Method of Hylemon *et al.*, 1973) This method can be used for other organisms that grow poorly or not at all on Christensen's medium. Use a test medium of the following composition (per litre of distilled water): BES buffer, 1.065 g; urea, 20.0 g; and phenol red, 0.01 g; pH 7.0. Also prepare a control medium lacking urea. Sterilize both media by filtration and dispense 2.0-ml portions into sterile tubes.
- Culture the test organism in a suitable liquid medium. Centrifuge the cells and suspend them in sterile distilled water to a dense concentration. Add 0.5 ml of the suspension to the test medium and the control medium. Incubate the tubes for 24 h at the optimal temperature for the organism.
- Look for the development of a red-violet colour in the test medium but not in the control medium.

AB. Voges-Proskauer (VP) Test

Some fermentative organism catabolize glucose by the butanediol pathway, in which acetoin (acetylmethylcarbinol) occurs as an intermediate in the formation of 2,3-butanediol. In the presence of KOH and O_2, the acetoin is oxidized to diacetyl, which in turn reacts with the guanidine group associated with arginine and other molecules contributed by peptone in the medium to form a pink- to red-coloured product. The α-naphthol intensifies this colour.

- (Method of Lányi, 1987) Prepare MR-VP medium as described under Section III, U. Make reagent A by dissolving 5.0 g of α-naphthol in 100 ml of absolute (100%) ethanol; the reagent must not be darker than straw-colour. Prepare reagent B by dissolving 40.0 g of KOH in 100 ml of distilled water.
- Inoculate the tubes lightly and incubate for two days (routine test) and for 4 days (standard test) at the optimum temperature for the organism being tested.
- Add 0.6 ml of reagent A and agitate to aerate the medium. Add 0.2 ml of reagent B and again agitate the medium. Slant the tube to increase the aeration. Allow to stand for 15–60 min.
- Look for development of a strong cherry red colour at the surface of the medium. A negative reaction shows no colour or a faint pink to copper colour.

AC. X and V Factor Requirements

Some prokaryotes that need iron porphyrins cannot make them from the precursor, δ-aminolevulinic acid (δ-ALA). Such organisms must be supplied with the X factor (protoporphyrin IX), which usually provided in the form of hemin.

Some organisms require the V factor (nicotinamide adenine dinucleotide, NAD), which is destroyed by autoclaving a culture medium. Some organisms require both the X factor and the V factor.

1. Disc method

- (Method of Health Protection Agency, 2011) Use sterile filter paper discs that have been impregnated with the X factor (protoporphyrin IX or hemin), the V factor (nicotinamide adenine dinucleotide, NAD) and both the X + V factor; such discs are commercially available. Prepare a light suspension of the test organism from one or more colonies with a straight inoculating wire and suspend the cells evenly in sterile physiological saline (0.85% NaCl) or sterile distilled water. Spread the suspension over the surface of a blood-free agar medium.
- Place the discs of X, V and X + V factors well separated from one another on the surface of the agar medium.
- After incubation, look for the presence or absence of growth around the discs or strips. The presence of growth around the disc or strip but not elsewhere on the plate indicates a requirement for that particular factor. Growth only around the X + V disc indicates that both factors are required.

2. Rapid test for the ability to synthesize the X factor

- (Method of Kilian, 1974) Prepare 0.1-M sodium phosphate buffer (pH 6.9) containing 2-mM δ-ALA and 0.8-mM $MgSO_4$. Dispense 0.5-ml portions of this test solution into small glass tubes.

Table 2. Some commercial multi-test phenotypic characterization systems for bacterial identification

Manufacturer[a]	Test system	Designed for	Number of tests
Biolog	Gen III	General use; based on reduction of tetrazolium salts when cells are oxidizing a carbon source	94
bioMérieux	An-Ident	Anaerobes	21
	API 20E; API Rapid 20E	Enteric Gram-negative rods	20
	API 20 Strep	Enterococci, streptococci	20
	API 50 CH	General use; based on carbohydrate catabolism	49
	API Campy	Campylobacters	10
	API Coryne	Coryneform rods	20
	API NH	*Haemophilus, Neisseria, Moraxella*	10
	API STAPH	Staphylococci	10
	API ZONE	Non-enteric Gram-negative rods	20
	GNI+	Aerobic and facultative anaerobic Gram-negative rods	28
	Vitek GPI	Gram-positive cocci; coryneform rods	30
	Vitek NHI	*Neisseria, Moraxella*, non-enteric Gram-negative rods	15
BD Diagnostic Systems	Crystal E/NF	Enteric and non-enteric Gram-negative rods	30
	Crystal Gram-Positive ID	Aerobic Gram-positive bacteria	29
	ID TRI Panel	Gram-negative and Gram-positive bacteria	30
	Oxi/Ferm Tub II	Fermentative, oxidase positive and non-fermentative Gram-negative rods	14
	Enterotube II	Enterobacteriaceae and other oxidase negative Gram-negative rods	15
Dade Behring Microscan	Rapid NEG ID3	Enterobacteriaceae and non-enteric Gram-negative rods	36
	Pos ID Type 2	Streptococci, enterococci, staphylococci	27
Remel	RapID ONE	Enteric Gram-negative rods	19
	RapID CB Plus	Coryneform rods	20
	RapID NF Plus	Non-enteric Gram-negative rods	17
	RapID NH	*Haemophilus, Neisseria, Moraxella*	13
	RapID POS ID	Streptococci and enterococci	34
	RapID STR	Enterococci and streptococci	14
Trek Diagnostic Systems	Sensititre AP80	Enterobacteriaceae and non-enteric Gram-negative rods	32
	Sensititre AP90	Enterococci and streptococci	32

[a]Addresses of manufacturers: Biolog, Hayward, CA. (http://www.biolog.com); bioMérieux, Marcy l'Etoile, France (http://www.biomerieux.com/servlet/srt/bio/portail/home); BD Diagnostic Systems, Franklin Lakes, NJ (http://www.bd.com); Dade Behring, Inc., MicroScan Inc., West Sacramento, CA (now owned by Siemens Medical Solutions, Henkestraße 127, Erlangen 91052, Germany) (http://www.medical.siemens.com/webapp/wcs/stores/servlet/SMBridge~q_catalogId~e_-999~a_catTree~e_100001~a_langId~e_-999~a_storeId~e_10001.htm); Remel, Lenexa, Kansas (http://www.remel.com/clinical/microbiology.aspx); Trek Diagnostic Systems, Ltd., East Grinstead, West Sussex, UK (http://www.trekds.com).

- Mix a large loopful of the organism grown on a suitable agar medium (such as chocolate agar) to a tube of the test solution to make a dense cell suspension. Also prepare a similar cell suspension in a tube of the test medium lacking δ-ALA as a control. Incubate the tubes at 37°C for 4 h.
- Examine the tubes under a UV light at a wavelength of 360 nm in a dark room. In the tube containing δ-ALA, look for a red fluorescence in either the bacterial sediment or the supernatant fluid. Such fluorescence indicates porphyrin synthesis and thus the absence of a requirement for X factor. Lack of fluorescence indicates that the bacterium requires the X factor for growth. The control tube should show no fluorescence.

Phenotypic and Physiological Characterization Methods

♦♦♦♦♦♦ IV. COMMERCIAL MULTI-TEST SYSTEMS

Many convenient and rapid multi-test systems for identifying bacteria are commercially available. These systems offer the advantages of miniaturization and are usually used in conjunction with a computerized system for identification of the organisms. As mentioned earlier, the use of these systems can increase standardization among various laboratories because of the high degree of quality control exercised over the media and reagents.

It is important to realize that most such systems are designed for identification of particular taxa and not for determining the physiological features of other taxa or new taxa. Indeed, a particular system may not even be applicable to other taxa. For instance, a commercial system for identifying species of Enterobacteriaceae would be useless if applied to *Campylobacter* species. With these precautions in mind, multi-test systems can provide useful information about the physiological characteristics of other organisms. For describing new taxa, the characterization systems that are used, as well as the inoculum age and size and the incubation temperature, must always be stated because reactions may not always agree with the results from classical characterization tests or with the results with other multi-test systems.

A summary of some of the currently available systems is given in Table 2, but new systems are being developed continually. Each manufacturer provides charts, tables, coding systems and characterization profiles for use with the particular multi-test system being offered.

References

Aragno, M. and Schlegel, H. G. (1978). *Aquaspirillum autotrophicum*, a new species of hydrogen-oxidizing, facultatively autotrophic bacteria. *Int. J. Syst. Bacteriol.* **28**, 112–116.

Auman, A. J., Breezee, J. L., Gosink, J. J., Kämpfer, P. and Staley, J. T. (2006). *Psychromonas ingrahamii* sp. nov., a novel gas vacuolate, psychrophilic bacterium isolated from Arctic polar sea ice. *Int. J. Syst. Evol. Microbiol.* **56**, 1001–1007.

Baldani, J. I., Krieg, N. R., Baldani, V. L. D., Hartmann, A. and Döbereiner, J. (2005). The genus *Azospirillum*. In *Bergey's Manual of Systematic Bacteriology* (G. M. Garrrity, D. J. Brenner, N. R. Krieg, and J. T. Staley, Eds.) 2nd edn., vol. 2, Part C, pp. 7–26. Springer-Verlag, New York.

Bernaerts, M. J. and De Ley, J. (1963). A biochemical test for crown gall bacteria. *Nature (London)* **197**, 406–407.

Black, F. T. (1973). Phosphatase activity in T-mycoplasmas. *Int. J. Syst. Bacteriol.* **23**, 65–66.

Board, R. G. and Holding, A. J. (1960). The utilization of glucose by aerobic gram-negative bacteria. *J. Appl. Bacteriol.* 23. xi.

Breznak, J. A. and Costilow, R. N. (1994). Physicochemical factors in growth. In *Methods for General and Molecular Microbiology* (P. Gerhardt, R. G. E. Murray, W. A. Wood, and N. R. Krieg, eds), pp. 137–154. American Society for Microbiology, Washington, D.C.

Bürger, H. (1967). Biochemische Leistungen nichtproliferierender Mikroorganismen. *Zentralbl. Bakteriol. Parasitenkd. Infektionskr. Hyg. Abt. 1 Orig.* **202**, 395–401.

Bürger, H., Doss, M., Mannheim, W. and Schüler, A. (1967). Studien zur biochemischen Differenzierung von Mycoplasmen. *Z. Med. Mikrobiol. Immunol.* **153**, 138–148.

Caplan, J. A. and Fahey, J. W. (1982). Plate-clearing technique to screen mixed microbial populations for protein degraders. *Soil Biol. Biochem.* **14**, 373–375.

Cato, E. P. and Moore, W. E. C. (1965). A routine determination of the optically active isomers of lactic acid for bacterial classification. *Can. J. Microbiol.* **11**, 319–324.

Chavira, R., Jr., Burnett, T. J. and Hageman, J. H. (1984). Assaying proteinases with Azocoll. *Analyt. Biochem.* **136**, 446–450.

Christensen, W. B. (1946). Urea decomposition as a means of differentiating *Proteus* and paracolon cultures from each other and from *Salmonella* and *Shigella* types. *J. Bacteriol.* **52**, 461–466.

Cowan, S. T. (1974). *Cowan and Steele's Manual for the Identification of Medical Bacteria.* 2nd ed. Cambridge University Press, London.

Cregut, M., Piutti, S., Vong., P.-C., Slezack-Deschaumes, S., Crovisier, I. and Benizri, E. (2009). Density, structure, and diversity of the cultivable arylsulfatase-producing bacterial community in the rhizosphere of field-grown rape and barley. *Soil Biol. Biochem.* **41**, 704–710.

Crespo-Medina, M., Chatziefthimiou, A., Cruz-Matos, R., Pérez-Rodríguez, I., Barkay, T., Lutz, A., Starovoytov, V. and Vetriani, C. (2009). *Salinisphaera hydrothermalis* sp. nov., a mesophilic, halotolerant, facultatively autotrophic, thiosulfate-oxidizing gammaproteobacterium from deep-sea hydrothermal vents, and emended description of the genus *Salinisphaera*. *Int. J. Syst. Evol. Microbiol.* **59**, 1497–1503.

Cruz-García, C., Murray, A. E., Klappenbach, K. A., Stewart, V. and Tiedje, J. M. (2007). Respiratory nitrate ammonification by *Shewanella oneidensis* MR-1. *J. Bacteriol.* **189**, 656–662.

D'Amato, R. F., Eriquez, L. A., Tomfohrde, K. M. and Singerman, E. (1978). Rapid identification of *Neisseria gonorrhoeae* and *Neisseria meningitidis* by using enzymatic profiles. *J. Clin. Microbiol.* **7**, 77–81.

Daniels, L., Hanson, R. S. and Phillips, J. A. (1994). Chemical analysis. In *Methods for General and Molecular Bacteriology* (P. Gerhardt, R. G. E. Murray, W. A. Wood, and N. R. Krieg, eds), pp. 512–554. American Society for Microbiology, Washington, D.C.

Deibel, R. H. and Evans, J. B. (1960). Modified benzidine test for the detection of cytochrome-containing respiratory systems in microorganisms. *J. Bacteriol.* **79**, 356–360.

Doddema, H. J. and Vogels, G. D. (1978). Improved identification of methanogenic bacteria by fluorescence microscopy. *Appl. Environ. Microbiol.* **36**, 752–754.

Edberg, S. C. and Kontnick, C. M. (1986). Comparison of 3-glucuronidase-based substrate systems for identification of *Escherichia coli*. *J. Clin. Microbiol.* **24**, 368–371.

Ehrlich, G. G., Goerlitz, D. F., Bourell, J. H., Eisen, G. V. and Godsy, E. M. (1981). Liquid chromatographic procedure for fermentation product analysis in the identification of anaerobic bacteria. *Appl. Environ. Microbiol.* **42**, 878–885.

Eirich, L. F., Vogels, G. D. and Wolfe, R. S. (1979). Distribution of coenzyme F_{420} and properties of its hydrolytic fragments. *J. Bacteriol.* **149**, 20–27.

Evans, J. B. and Niven, C. F., Jr. (1950). Comparative study of known food-poisoning staphylococci and related varieties. *J. Bacteriol.* **59**, 545–550.

Gillis, M., Vandamme, P., De Vos, P., Swings, J. and Kersters, K. (2005). Polyphasic taxonomy. In *Bergey's Manual of Systematic Bacteriology* – 2nd ed. (D. J. Brenner, N. R. Krieg, and J. T. Staley, eds), Vol. 2A, pp. 43–48. Springer, New York.

Good, N. E., Winget, G. D., Winter, W., Connolly, T. N., Izawa, S. and Singh, R. M. M. (1966). Hydrogen ion buffers for biological research. *Biochemistry* **5**, 467–477.

Goodfellow, M., Orlean, P. A. B., Collins, M. D., Alshamaony, L. and Minniken, D. E. (1978). Chemical and numerical taxonomy of strains received as *Gordona aurantiaca*. *J. Gen. Microbiol.* **109**, 57–68.

Grant, W. D. (2001). Genus I. *Halobacterium*. In *Bergey's Manual of Systematic Bacteriology* 2nd edn. (G. M. Garrity, D. R. Boone, and R. W. Castenholz, eds), Vol. 1, pp. 301–305. Springer-Verlag, New York.

Greenberg, A. E., Trussel, R. R. and Clesceri, L. C. (eds), (1985). *Standard Methods for the Examination of Water and Wastewater* 16th edn. American Public Health Association, Washington, D.C.

Grimont, P. A. D. and de Rosnay, H. L. C. D. (1972). Numerical study of 60 strains of *Serratia*. *J. Gen. Microbiol.* **72**, 259–268.

Hafenbradl, D. and Stetter, K. O. (2001). Genus I. *Ferroglobus*. In *Bergey's Manual of Systematic Bacteriology* 2nd edn. (G. M. Garrity, D. R. Boone, and R. W. Castenholz, eds), Vol. 1, pp. 352–353. Springer-Verlag, New York.

Han, Y.-H., Smibert, R. M. and Krieg, N. R. (1991). *Wolinella recta, Wolinella curva, Bacteroides ureolyticus*, and *Bacteroides gracilis* are microaerophiles, not anaerobes. *Int. J. Syst. Bacteriol.* **41**, 218–222.

Hayward, A. C. (1957). Detection of gas production from glucose by heterofermentative lactic acid bacteria. *J. Gen. Microbiol.* **16**, 9–15.

Health Protection Agency (2010a). Bile solubility test, Issue no: 2.1. Standards Unit, Department for Evaluations, Standards and Training, Centre for Infections, London. (http://www.hpa-standardmethods.org.uk/documents/bsopTP/pdf/bsoptp5.pdf).

Health Protection Agency (2010b). Catalase test, Issue no: 2.1. Standards Unit, Department for Evaluations, Standards and Training, Centre for Infections, London. (http://www.hpa-standardmethods.org.uk/documents/bsopTP/pdf/bsoptp8.pdf).

Health Protection Agency. (2011). X and V factor test, Issue No.2.3, Standards Unit, Department for Evaluations, Standards and Training, Centre for Infections, London. (http://www.hpa-standardmethods.org.uk/documents/bsopTP/pdf/bsoptp38.pdf).

Holdeman, L. V., Cato, E. P. and Moore, W. E. C. (1977). *Anaerobe Laboratory Manual*. 4th edn.Virginia Polytechnic Institute and State University, Blacksburg.

Huber, H. and Stetter, K. O. (2001). Genus I. *Archaeoglobus*. In *Bergey's Manual of Systematic Bacteriology* 2nd edn. (G. M. Garrity, D. R. Boone, and R. W. Castenholz, eds), Vol. 1, pp. 349–352. Springer-Verlag, New York.

Hugh, R. and Leifson, E. (1953). The taxonomic significance of fermentative versus oxidative metabolism of carbohydrates by various gram-negative bacteria. *J. Bacteriol* **66**, 22–26.

Hylemon, P. B., Krieg, N. R. and Phibbs, P. V., Jr. (1974). Transport and catabolism of D-fructose by *Spirillum itersonii*. *J. Bacteriol.* **117**, 144–150.

Hylemon, P. B., Wells, J. S., Jr., Krieg, N. R. and Jannasch, H. W. (1973). The genus *Spirillum*: a taxonomic study. *Int. J. Syst. Bacteriol.* **23**, 340–380.
Johansen, H., Ingar Olsen, I. and Kerekes, K. (1988). Differentiation between *Bacteroides gingivalis, Bacteroides endodontalis* and *Bacteroides asaccharolyticus* by means of HPLC analysis of nonderivatized free metabolic acids. *Oral Microbiol. Immunol.* **3**, 42–45.
Kattar, M. M., Cookson, B. T., Carlson, L. C., Stiglich, S. K., Schwartz, M. A., Nguyen, T. T., Daza, R., Wallis, C. K., Yarfitz, S. L. and Coyle, M. B. (2001). *Tsukamurella strandjordae* sp. nov., a proposed new species causing sepsis. *J. Clin. Microbiol.* **39**, 1467–1476.
Kennedy, C., Rudnick, P., MacDonald, M. L. and Melton, T. (2005). The genus *Azotobacter*. In *Bergey's Manual of Systematic Bacteriology* 2nd edn. (G. M. Garrity, D. J. Brenner, N. R. Krieg, and J. T. Staley, eds), Vol. 2, pp. 384–402. Springer-Verlag, New York. Part B.
Kennedy, E. P. and Scarborough, G. A. (1967). Mechanism of hydrolysis of *o*-nitrophenyl-β-galactoside in *Staphylococcus aureus* and its significance for theories of sugar transport. *Proc. Nat. Acad. Sci. USA* **58**, 225–228.
Kilian, M. (1974). A rapid method for the differentiation of *Haemophilus* strains. The porphyrin test. *Acta Pathol. Microbiol. Scand. Sect. B* **82**, 835–842.
Koechlein, D. J. and Krieg, N. R. (1998). Viable but nonculturable forms if *Prolinoborus fasciculus* (*Aquaspirillum fasciculus*). *Can. J. Microbiol.* **44**, 910–912.
Kovács, N. (1928). Eine vereinfachte Methode zum Nachweis der Indolbildung durch Bakterien. *Z. ImmunForsch. Exp. Ther.* **55**, 311–315.
Krieg, N. R. (2005a). The genus *Prolinoborus*. In *Bergey's Manual of Systematic Bacteriology* (G. M. Garrrity, D. J. Brenner, N. R. Krieg, and J. T. Staley, Eds.) 2nd edn., vol. 2, Part C, pp. 841–843. Springer-Verlag, New York.
Krieg, N. R. (2005b). The genus *Spirillum*. In *Bergey's Manual of Systematic Bacteriology* (G. M. Garrrity, D. J. Brenner, N. R. Krieg, and J. T. Staley, Eds.) 2nd edn., vol. 2, Part C, pp. 870–880. Springer-Verlag, New York.
Kurahashi, M. and Yokota, A. (2004). *Agarivorans albus* gen. nov., sp. nov., a gammaproteobacterium isolated from marine animals. *Int. J. Syst. Evol. Microbiol.* **54**, 693–697.
Lányi, B. (1987). Classical and rapid identification methods for medically important bacteria. In *Methods in Microbiology* (R. R. Colwell and R. Grigorova, eds), Vol. 19, pp. 1–67. Academic Press, New York.
Law, J. H. and Slepecky, R. A. (1961). Assay of poly-β-hydroxybutyric acid. *J. Bacteriol.* **82**, 33–36.
Lawson, G. H. K., Leaver, J. L., Pettigrew, G. W. and Rowland, A. C. (1981). Some features of *Campylobacter sputorum* subsp. *mucosalis* subsp. nov., nom. rev. and their taxonomic significance. *Int. J. Syst. Bacteriol.* **31**, 385–391.
Leifson, E. (1963). Determination of carbohydrate catabolism of marine bacteria. *J. Bacteriol.* **85**, 1183–1184.
Lennette, E. H., Spaulding, E. H., and Truant, J. P. (1974). *Manual of Clinical Microbiology*. 2nd. edn. American Society for Microbiology, Washington, D.C.
Levin, R. E. (2006). Food microbiology. In *Food BioTechnology* (K. Shetty, G. Paliyath, A. Pometto, and R. E. Levin, eds), 2nd edn. pp. 3–18. CRC Press, Boca Raton, Florida.
Levine, M., Epstein, S. S. and Vaughn, R. H. (1934). Differential reactions in the colon group of bacteria. *Am. J. Public Health Nations Health* **24**, 505–510.
Ludwig, W. and Klenk, H.-P. (2005). Overview: a phylogenetic backbone and taxonomic framework for procaryotic systematics. In *Bergey's Manual of Systematic Bacteriology* (G. M. Garrrity, D. J. Brenner, N. R. Krieg, and J. T. Staley, Eds.) 2nd edn., vol. 2, Part A, pp. 49–65. Springer-Verlag, New York.

MacFaddin, J. F. (1980). *Biochemical Tests for the Identification of Medical Bacteria.* 2nd edn. Williams & Wilkins, Baltimore.

Maddocks, J. L. and Greenan, M. J. (1975). A rapid method for identifying bacterial enzymes. *J. Clin. Pathol.* **28**, 686–687.

Mallory, L. M., Austin, B. and Colwell, R. R. (1977). Numerical taxonomy and ecology of oligotrophic bacteria isolated from the estuarine environment. *Can. J. Microbiol.* **23**, 733–750.

Medina, P. and Baresi, L. (2007). Rapid identification of gelatin and casein hydrolysis using TCA. *J. Microbiol. Meth.* **69**, 391–393.

Mills, C. K. and Gherna, R. L. (1987). Hydrolysis of indoxyl acetate by *Campylobacter* species. *J. Clin. Microbiol.* **25**, 1560–1561.

Neyra, C. A., Döbereiner, J., LaLande, R. and Knowles, R. (1977). Denitrification by N_2 fixing *Spirillum lipoferum*. *Can. J. Microbiol.* **23**, 300–305.

Ostle, A. G. and Holt, J. G. (1982). Nile blue A as a fluorescent stain for poly-β-hydroxybutyrate. Appl. Environ. Microbiol. **44**, 238–241.

Palleroni, N. J. (1984). The genus *Pseudomonas*. In *Bergey's Manual of Systematic Bacteriology* (N. R. Krieg, and J. G. Holt, eds), Vol. 1, pp. 141–199. Williams & Wilkins, Baltimore.

Palleroni, N. J. (2005) The genus *Pseudomonas*. In *Bergey's Manual of Systematic Bacteriology* (G. M. Garrrity, D. J. Brenner, N. R. Krieg, and J. T. Staley, Eds.) 2nd edn., vol. 2, Part B, pp. 323–379. Springer-Verlag, New York.

Pelczar, M. J., Jr. (ed). (1957) *Manual of Microbiological Methods*, McGraw-Hill, New York.

Pfennig, N. and Lippert, K. D. (1966). Über das Vitamin-B_{12}-Bedürfnis phototropher Schwefelbakterien. *Arch. Mikrobiol* **55**, 245–256.

Qadri, S. M. H., Desilva, M. I. and Zubairi, S. (1980). Rapid test for determination of esculin hydrolysis. *J. Clin. Microbiol.* **12**, 472–474.

Ribeiro, L. M. and Herr, S. (1990). The use of filter paper discs impregnated with thionin acetate, basic fuchsin and thionin blue in the identification of *Brucella* species. *Onderstepoort J. Vet. Res.* **57**, 197–199.

Richardson, A. J., Calder, A. G., Stewart, C. S. and Smith, A. (1989). Simultaneous determination of volatile and non-volatile acidic fermentation products of anaerobes by capillary gas chromatography. *Lett. Appl. Microbiol.* **9**, 5–8.

Roop, R. M., II, Smibert, R. M., Johnson, J. L. and Krieg, N. R. (1985). *Campylobacter mucosalis* (Lawson, Leaver, Pettigrew and Rowland 1981) comb. nov.: emended description. *Int. J. Syst. Bacteriol.* **35**, 189–192.

Sierra, G. (1957). A simple method for the detection of lipolytic activity of microorganisms and some observations on the influence of the contact between cells and fatty substrates. *Antonie van Leeuwenhoek* **23**, 15–22.

Simmons, J. S. (1926). A culture medium for differentiating organisms of the typhoid-colon-aerogenes groups and for isolation of certain fungi. *J. Infect Dis.* **39**, 209–214.

Smibert, R. M. and Krieg, N. R. (1994). Phenotypic characterization. In *Methods for General and Molecular Bacteriology* (P. Gerhardt, R. G. E. Murray, W. A. Wood, and N. R. Krieg, eds), pp. 607–654. American Society for Microbiology, Washington, D.C.

Smith, M. L. (1932). CLXXIII. The effect of heat on sugar solutions used for culture media. *Biochem. J.* **26**, 1467–1472.

Smith, P. B., Hancock, G. A. and Rhoden, D. L. (1969). Improved medium for detecting deoxyribonuclease-producing bacteria. *Appl. Microbiol.* **18**, 991–993.

Tarrand, J. J. and Gröschel, D. H. M. (1982). Rapid, modified oxidase test for oxidase-variable bacterial isolates. *J. Clin. Microbiol.* **16**, 772–774.

Tindall, B. J., Sikorski, J., Smibert, R. M. and Krieg, N. R. (2007). Phenotypic characterization and the principles of comparative systematics. In *Methods for General and*

Molecular Microbiology (T. J. Beveridge, J. A. Breznak, G. A. Marzluf, T. M. Schmidt, and L. R. Snyder, eds), 3rd edn. pp. 330–393. ASM Press, Washington, D.C.

Vancanneyt, M., Segers, P., Torck, U., Hoste, B., Bernardet, J.-F., Vandamme, P. and Kersters, K. (1996). Reclassification of *Flavobacterium odoratum* (Stutzer 1929) strains to a new genus, *Myroides*, as *Myroides odoratus* comb. nov. and *Myroides odoratimimus* sp. nov. *Int. J. Syst. Bacteriol.* **46**, 926–932.

Vandamme, P., Dewhirst, F. E., Paster, B. J. and On, S. L. W. (2005). Genus I. *Campylobacter*. In *Bergey's Manual of Systematic Bacteriology* (G. M. Garrrity, D. J. Brenner, N. R. Krieg, and J. T. Staley, Eds.) 2nd edn., vol. 2, Part C, pp. 1147–1160. Springer-Verlag, New York.

Vishniac, W. and Santer, M. (1957). The thiobacilli. *Bacteriol. Rev.* **21**, 195–213.

Volokita, M., Abeliovich, A. and Soares, M. I. M. (2000). Detection of microorganisms with overall cellulolytic activity. *Curr. Microbiol.* **40** 136–136

Wang, Q., Han, H., Xue, Y., Qian, Z., Meng, B., Peng, F., Wang, Z., Tong, W., Zhou, C., Wang, Q., Guo, Y., Li, G., Liu, S. and Ma, Y. (2009). Exploring membrane and cytoplasm proteomic responses of *Alkalimonas amylolytica* N10 to different external pHs with combination strategy of *de novo* peptide sequencing. *Proteomics* **9**, 1254–1273.

Wheater, D. M. (1955). The characteristics of *Lactobacillus acidophilus* and *Lactobacillus bulgaricus*. *J. Gen. Microbiol.* **12**, 123–132.

Williams, S. T., Goodfellow, M. and Alderson, G. (1989). The genus *Streptomyces*. In *Bergey's Manual of Systematic Bacteriology* (J. G. Holt, S. T. Williams, and M. E. Sharpe, eds), Vol. 4, pp. 2452–2492. Williams & Wilkins, Baltimore.

Wolf, P. L., von der Muehll, E. and Praisler, K. (1973). A test for bacterial alkaline phosphatase: use in rapid identification of *Serratia* organisms. *Clin. Chem.* **19**, 1248–1249.

Wyss, C. (1989). *Campylobacter-Wolinella* group organisms are the only oral bacteria that form arylsulfatase-active colonies on a synthetic indicator medium. *Infect. Immun.* **57**, 1380–1383.

Yong, D. C. T. and Thompson, J. S. (1982). Rapid microbiochemical method for identification of *Gardnerella (Haemophilus) vaginalis*. *J. Clin. Microbiol.* **16**, 30–33.

4 Microscopy

Manfred Rohde
Department of Medical Microbiology, Helmholtz Centre for Infection Research (HZI), Braunschweig, Germany
E-mail: manfred.rohde@helmholtz-hzi.de

CONTENTS

I. INTRODUCTION

The word microscope is derived from the Greek words micros (small) and skopeo (look at). From the dawn of science, biologists examined the morphological structures of animals, plants and tissues by the naked eye. With the advent of the first light microscopes, most probably invented from the Galilean telescope during the seventeenth century, scientists were able to look at much smaller objects like protozoa and bacteria. The name of Antonie van Leeuwenhoek (1632–1723) stands for this hallmark in microscopy with respect to the visualization of microorganisms. Later on, the invention of so-called compound light microscopes set a pioneering landmark. These microscopes consisted of an objective lens and an eyepiece, which magnified the image produced by the objective lens together with a focusing aid, a mirror or a source for light and a specimen table for positioning and holding the specimen. These microscopes later on allowed scientists to magnify up to 1000×. Those early light microscopy studies unravelled the astonishing diversity of bacterial cell shapes and the binary fission of bacteria. By 1876, Ernst Abbe had optimized the light microscope and analysed the effects of diffraction on image formation. However, due to the physics of light, these microscopes were limited to a resolution of around 0.2 μm. This physical theoretical limitation of compound light microscopes was reached in the early 1930s. In addition, there has always been a scientific desire to look at much smaller details of the interior morphology inside a eukaryotic cell, like mitochondria or the

METHODS IN MICROBIOLOGY, VOLUME 38
0580-9517 DOI 10.1016/B978-0-12-387730-7.00004-8

nucleus, or to be able to visualize the flagella in moving bacteria. The transmission electron microscope (TEM) was the first electron microscope to be developed by Knoll and Ruska (1932a, 1932b) in Lichtenfels, Berlin, in 1931/1932. The construction scheme of the TEM followed that of the compound light microscopes with the exception that a focused beam of electrons was used instead of light to examine and visualize the interior of the specimen. The resolution was pushed to a limit of around 0.1 μm at that time. Nowadays, modern TEMs have a resolution power reaching up to 0.05 nm. The other common type of electron microscope is the scanning electron microscope (SEM), which was invented in the late 1930s and 1940s. Knoll (1935) and von Ardenne (1938, 1939) were involved in the first description and partly building of an SEM. In 1942, the Americans, Zworykin, Hillier and Snijder built the first SEM with a resolution power of around 50 nm. The first commercially available instrument was brought to the market in 1965 by the company Cambridge, in England. Modern SEMs have a resolution that is around 1 nm, thus allowing the visualization of appendages of bacteria like flagella or secretion systems of Gram-negative bacteria (Rohde *et al.*, 2003b; Chakravortty *et al.*, 2005).

In addition to the introduction of transmission electron microscopy (TEM), the development of preparative techniques for biological material that stabilize the fragile microstructures to withstand the high vacuum in the microscope column was required. The progress in technical and preparative techniques made TEM very popular in the 1950s and 1960s with biologists and microbiologists (Luft, 1961; Palade, 1952; Palade and Siekevitz, 1956). These methods include the widely used negative-staining procedure with heavy metals (Valentine *et al.*, 1968), the ultrathin sectioning procedure to gain insight into the internal structures of microorganisms (Richarson *et al.*, 1960), freeze-fracture (Bullivant and Ames, 1966) and freeze-etching (Moor and Mühlethaler, 1963; Gross *et al.*, 1978) methods and rapid freezing with subsequent freeze-substitution of the specimen (Graham and Beveridge, 1990; Kellenberger, 1991). For the localization of proteins, immune-EM methods carried out as pre-embedding labelling or post-embedding labelling (Roth *et al.*, 1978), as well as applying cryo-ultrathin sections, were used (Griffiths *et al.*, 1984; Tokuyashu, 1986). These methods have contributed to our understanding of bacterial morphological features like cell membrane composition of Gram-negative bacteria, Gram-positive bacteria, Archaea and mycobacteria. They have aided in the detection of bacterial structures like sheathed or non-sheathed flagella, pili, fimbriae or stalks. During recent years, other cryo-methods like high-pressure freezing (Hohenberg *et al.*, 1994) with subsequent freeze-substitution or cyro-sectioning have been introduced. These methods allow the observation in a fully hydrated state of the specimen without chemical fixation or contrast-enhancing metal stains. A very recently introduced technique is cryo-electron tomography (CET) allowing visualization of intact bacteria in a nearly native state with high resolution to obtain a three-dimensional image of the entire bacterium or interior structures (Baumeister *et al.*, 1999). Nevertheless, CET needs a "state of the art" high-voltage (300 kV) TEM and sophisticated image acquisition and processing equipment. TEM has been widely used throughout the microbiology community compared to SEM during the last several decades. However, with the advent of high-resolution field emission scanning electron microscopes (FESEMs) in the early 1990s and due to increased research on adherence to and invasion of pathogenic bacteria into host cells, FESEM has become a more widely used and powerful tool in imaging bacterial ultrastructure and interaction of bacteria within the ecological environment (Rohde *et al.*, 2003a, 2003b, 2011; Chakravortty *et al.*, 2005; Kwok *et al.*, 2007). In this chapter, an overview of methods is presented that are commonly used to image bacteria according to morphological criteria for taxonomic determination based on light and electron microscopy methods. Procedures will be described representing the most commonly used light microscopy methods for the characterization of bacterial morphology. For higher resolution description of microorganisms, electron microscopy techniques like negative-staining, embedding and ultrathin sectioning for TEM and preparation schemes for high-resolution scanning electron microscopy are presented.

♦♦♦♦♦♦ II. LIGHT MICROSCOPY

The light microscope is an essential tool in every microbiology laboratory orientated toward the description and identification of microorganisms. It goes back to a long-standing use in microbiology, starting with the first scientific investigations by van Leeuwenhoek using a single crude lens for visualization of microorganisms. Light microscopy was also heavily involved in the discovery of the two fundamental categories in the bacterial world, the Gram-positive and Gram-negative bacteria by Christian Gram in the mid-1880s. In 1932, Zernicke invented the phase-contrast microscope resulting in much clearer images, especially of microorganisms. Nowadays, this technique is widely used when imaging bacteria without any additional staining. Bacteria have been classified by light microscopic examination on the basis of shape, size and staining characteristics. Although unstained preparations can be used when applying phase-contrast imaging, stained preparations generally provided a better and clear identification. The shape and size of bacteria sometimes may be sufficient for presumptive identification, but bacteria also possess certain characteristic structures such as capsules, flagella, spores or polyphosphate granules. Their presence or absence is used as a tool for identification, and by simple staining procedures the presence of these structures can be demonstrated. For example, the presence of capsules may be demonstrated by negative staining using India ink, nigrosin or dark field microscopy. Some bacteria possess flagella and their number, size and position are characteristic traits for the identification of some taxa. The presence of flagella cannot be detected easily by light microscopy due to the small diameter of flagella, although motility can be seen under a light microscope. To see the flagella, it is necessary to increase their apparent diameter by first coating these structures with a mordant such as tannic acid and then applying one of the special stains such as the Leifson's stain. In this section, a number of imaging techniques and staining methods are described, which should allow for the publication of a more complete description of newly isolated microorganisms.

Books on light microscopy are readily available and give a good introduction to the basic principles, physics and techniques involved. Satisfying light microscopy is best learned through direct practical experience and the art of preparing a worthwhile specimen. Always try to prepare more specimens than you would usually prepare since another slide with the same specimen might give an even better image or have stained better than the previous one.

A. Illumination Techniques in Light Microscopy

1. Transmitted light microscopy

Transmitted light microscopy is the general term used for any type of microscopy where the light is transmitted from a source on the opposite side of the specimen to the objective lens. The light is passed through a condenser to focus it on the specimen for maximum illumination. After the light passes through the specimen, it travels through the objective lens for magnification of the sample image and then to the oculars, where the enlarged image is viewed. To get reasonable images, the specimen must be properly illuminated. The optimum set-up for specimen illumination and image generation is known as Köhler illumination. The microscopy techniques requiring a transmitted light path include bright field, dark field, phase contrast, polarization and differential interference contrast optics.

2. Bright field (Köhler illumination) microscopy

The Köhler system of illumination is an absolute must for high-resolution imaging of microorganisms. It represents the normal light microscopy when no other optical contrast technique is employed. Köhler illumination is achieved by focusing an image of the light source at the level of the condenser diaphragm, when the condenser is in the correct position relative to the specimen. Under these illumination conditions, an objective lens in focus on the specimen will be fully illuminated regardless of the size of the light source. The condenser is used to focus parallel rays of light on the specimen, thereby giving the advantages of an evenly illuminated field resulting in a bright image without or with minimized glare.

3. Dark field microscopy

The specimen is illuminated obliquely, with no direct light entering the objective. This is achieved by illuminating the specimen with a hollow cone of light such that only the light diffracted by the objects in the field of view is transmitted through the microscope to the eye piece or camera. Other transmitted light is at too low an angle to be detected by the objective lens, even with a high numerical aperture. Features in the specimen plane, which scatter light, can clearly be seen against a dark background. Dark field illumination is provided by either a dark field element in a phase-contrast condenser or a purpose-built dark field condenser. The latter is required for high-resolution objectives to prevent the oblique rays entering the wide aperture of the objective.

4. Phase-contrast microscopy

Phase-contrast microscopy is a technique used for gaining contrast in a translucent specimen without staining the specimen. One major advantage is that phase-contrast microscopy can be used with high-resolution objectives, but it requires a specialized condenser and more expensive objectives. The optical methodology was introduced by Zernicke in 1932. This technique exploits the fact that light slows slightly when passing through biological specimens. The specimen is illuminated by a hollow cone of light coming through a phase annulus in the condenser. Phase-contrast objectives that have a corresponding phase plate must be used. Light rays passing through the specimen are slightly retarded, and further retardation takes place in the phase plate. When these rays combine with rays that have not taken this other path, degrees of constructive and destructive interference occur, which produce the characteristic light and dark features in the image. Specimens for phase-contrast microscopy should be as small as possible, and they must be mounted in a fluid or gel for an even grey background in the image.

5. Fluorescence microscopy

Some biological substances, such as chlorophyll and some oils and waxes, have primary fluorescence (auto-fluorescence). However, most biological molecules do not fluoresce on their own. One exception is the fluorescence of coenzymes like factor 350 and factor 420 in methanogenic bacteria (Doddema and Vogels, 1978). Therefore, for visualization of those non-fluorescent molecules they have to be linked with fluorescent molecules so-called fluorochromes. The fluorochromes emit light of a given wavelength when excited by incident light of a different (shorter) wavelength. Specimens labelled with a fluorochrome such as fluorescein or green fluorescent protein (GFP) are illuminated with the blue wavelength of light and emit the energy as a longer wavelength (in this example as a green fluorescence). Nearly all modern fluorescence microscopes are designed as epi-fluorescence microscopes in which a dichroic mirror is built in. The dichroic mirror cube comprises three components: a dichroic beam splitter (partial mirror), an excitation filter and a barrier filter. Specific filters are used to isolate the excitation and emission wavelengths for each fluorochrome. The dichroic mirror reflects shorter wavelengths of light, allows longer wavelengths to pass and is required because the objective acts as both the condenser lens (excitation light) and the objective lens (emission light). Therefore, the beam splitter isolates the emitted light from the excitation wavelength. This epi-illumination type of light path is required to create a dark background so that emitted fluorescence can be easily detected. As a bright light source for producing the correct wavelengths for excitation, normally a 50-W or 100-W mercury arc lamp is used.

6. Confocal laser scanning microscopy

One of the main problems in fluorescence microscopy is the out-of-focus blur degrading the image by obscuring fine structures in the specimen, particularly in thicker specimens, like in a biofilm. In conventional microscopy, not only is the plane of focus in a certain part in the specimen illuminated, but much of other parts of the specimen above and below this focus point are also illuminated resulting in an out-of-focus blur from these areas. This out-of-focus light leads to a reduction in image contrast and brightness as well as to a decrease in resolution. In the confocal microscope, all out-of-focus structures are minimized by a diaphragm, the confocal aperture. Thus, the confocal aperture, also called detection pinhole, works at the image acquisition and formation. The detection pinhole does not permit rays of light from out-of-focus points to pass through it, resulting in no blurry images. The wavelength of light, the numerical aperture of the objective and the diameter of the diaphragm (a wider detection pinhole reduces the confocal effect) affect the depth of the focal plane. When using confocal microscopy to view fluorescence, where up to 90% of the emission light can be filtered out by the pin hole, specific wavelength lasers are used as light sources, since these are extremely bright and monochromatic. For imaging the point of light, the laser beam is scanned across the specimen by scanning mirrors. The emitted light passing through the detector pinhole is transformed into electrical signals by a photomultiplier and displayed on a computer monitor. One major advantage of confocal microscopy is the fact that single sections through a specimen can be made, the number depending on the thickness of the specimen, the so-called Z-series. These individual images can be merged into a single image, resulting in an image of the entire specimen that is always in focus.

B. Preparation Methods for Light Microscopy

1. Living cell suspensions

The examination of living bacterial cells can be performed easily by spreading a drop of natural specimen or a culture on a microscope slide.

- Dilute or, if possible, concentrate a bacterial suspension to faint turbidity with sterile buffer like phosphate buffered saline (PBS, 50 mM sodium-phosphate, 50 mM potassium phosphate, 90 mM NaCl, pH 7.0).
- Place a drop of ~5 µl on a microscope slide.
- Apply a cover slip; use a cover slip with No. 1 thickness (~0.15 mm thick) to allow the usage of oil-immersion objectives for higher magnifications.
- If no immediate examination can be done, cover the edges of the cover slip with clear nail polish, air-dry and then examine at a later time.

For solid or semi-solid material, e.g. from colonies of agar plates

- Place a drop of ~10 µl of sterile buffer on a microscope slide.
- Transfer a minimal quantity of bacteria from a colony with the aid of a sterile needle.

- Resuspend the bacteria gently in the drop of buffer.
- Take 5 μl of the suspension and dilute further with sterile buffer.
- Cover the undiluted and diluted suspensions with a cover slip.

For all examinations, be aware of the fact that culture conditions and age of the culture may influence the shape, size and length of bacteria. Furthermore, most flagella are only expressed at a certain growth rate. The formation of intracellular granules is also growth rate dependent.

2. Immobilization of motile bacteria

Examination of very motile bacteria sometimes requires immobilization. One possibility is to add 15–25% gelatin to the bacterial suspension to reduce motility due to higher viscosity of the suspension. An additional method is as follows:

- Melt 1.5–2% high-quality agar in distilled boiling water.
- Transfer the melted suspension onto a microscope slide covering an area about the size of the cover slip.
- Allow the agar to solidify for 10 min.
- Dry the slide in a dust-free oven at 40–50°C.
- Place a drop of 25–50 μl of the bacterial suspension onto a cover slip.
- Turn the cover slip around and place on the dried agar on the microscope slide.
- Dried agar will start to swell and press the bacteria against the cover slip, thereby minimizing their motility.

3. Fixation of suspensions or smears

For staining of bacteria, it is recommended to fix bacteria to prevent alterations in shape or loss of appendages like flagella or pili. If one deals with pathogenic bacteria, it also is worthwhile to apply fixed bacteria to minimize the infection risk. Bacteria can be fixed either physically by heat or chemically with aldehydes prior to staining procedures.

- Spread a smear of bacteria or 5 μl of suspension with faint turbidity onto a microscope slide.
- Let the microscope slide air-dry completely.
- Pass the underside of the slide (bacteria facing upwards) several times over the flame of a Bunsen burner, avoid overheating. As a rule of thumb, the microscope slide should still be touchable with the fingers after heat fixation.
- By doing so bacteria are induced to adhere firmly to the glass of the microscope slide.
- After cooling down, the slide can be used for staining.

Important: When morphological features are being examined, be aware of the fact that heating and drying can cause detrimental effects on shape and size of bacteria and most probably lead to distortions of the bacterial cell. For that reason, chemical fixation is preferred.

- Prepare a 5% formaldehyde solution in sterile buffer.
- Alternatively prepare a fixation solution consisting of 5% formaldehyde and 2% glutaraldehyde.
- Resuspend bacteria in the fixation solution until a faint turbidity is reached.
- Allow to react for 10–15 min at room temperature.
- Place on a microscope slide that is covered with poly-L-lysine (see Section IV.A.2.).
- Air-dry completely and perform staining procedures.
- Alternatively, do not allow to dry at all during the following staining steps after placing bacteria on the poly-L-lysine-coated microscope slides.

4. Negative staining of capsules and layers

Negative-staining with nigrosin or India ink is a quick and easy method to gain information about the presence or absence of capsules or any other layers around bacteria. An organism with a capsule will show a halo around the cell.

- Place a drop (~25 μl) of a 7% nigrosin solution or Indian ink on a microscope slide; depending on the specimen, sometimes Indian ink has to be diluted with distilled water.
- Resuspend a small quantity of biomass in the staining solution.
- Apply a cover slip.
- Soak up redundant staining solution with filter paper until a thin layer of staining solution is left under the cover slip and examine.
- Alternatively, place a drop (~15 μl) of the bacterial suspension on a microscope slide.
- Cover with a cover slip.
- Place 7% nigrosin or the water-diluted Indian ink at one edge of the cover slip.
- Soak the staining solution through the suspension by applying a piece of a filter paper on the opposite side of the cover slip.
- This will result in an uneven distribution of the dye solution, and the best area of dye thickness can then be examined.

5. Gram-staining

Numerous modifications of the Gram-staining have been published since the first stains by C. Gram in 1884. Gram-staining is the most important stain used to subdivide bacteria into the two groups: the Gram-positive bacteria and the Gram-negative bacteria. In addition, there are also bacteria that are referred to as Gram-variable. The principle of the Gram-stain involves staining of bacteria with crystal-violet and then with an iodine solution as a mordant. Bacteria are then washed with ethanol. Gram-positive bacteria retain the colour of the stain, whereas Gram-negative bacteria do not. Here the protocol according to Bartholomew and Mittwer (1952) and Bartholomew (1962) will be described.

- Prepare a solution of 2.0 g crystal-violet in 20 ml ethanol (solution A).
- Prepare a solution of 0.8 g ammonium oxalate in 80 ml distilled water (solution B).
- Mix solution A and solution B to obtain the crystal-violet staining solution.
- Store the solution for at least 24 h and filter it through a paper filter shortly before use.
- Prepare the mordant solution consisting of 1.0 g iodine and 2.0 g potassium iodide in 300 ml distilled water. It is useful to grind the iodine and potassium iodide before making the solution.
- Store the mordant in amber bottles in the dark.
- Prepare the decolourizing solution of 95% ethanol.
- Prepare the counterstaining solution consisting of 10 ml safranin O (2.5% in 95% ethanol) in 100 ml distilled water.
- Put the microscope slide with heat-fixed bacteria in the crystal-violet staining solution for 1 min.
- Wash the slide gently with an indirect stream of tap water for 5 s.
- Put the slide in the mordant solution for 1 min.
- Wash the slide gently with an indirect stream of tap water for 5 s.
- Blot the slide dry on filter paper.
- Put the slide in 95% ethanol with agitation for 30 s.
- Blot the slide dry with filter paper.
- Put the slide in the counterstaining solution of safranin O for 15 s.
- Wash the slide gently with an indirect stream of tap water until no colour is visible in the water stream anymore.
- Blot the slide dry with filter paper and examine.

Important: Always run a control of known Gram-positive and Gram-negative bacteria to judge the performance of the staining procedure and the reliability of the results.

6. Flagella staining

Although the width of flagella is small, they can be detected by light microscopy and staining methods have been introduced to enhance the visibility of flagella (Figure 1). All of these staining procedures have in common the fact that they do not represent consistently reliable methods. Molecular Probes® (Invitrogen, Darmstadt, Germany) offer NanoOrange® as a fluorescent dye, which can be used as an effective method for visualizing flagellation of bacteria. Here I will firstly describe the staining method according to Leifson (1951) or Clark (1976), which represents one of the most established flagella staining procedures.

- Prepare a solution of 1.5 g sodium chloride in 100 ml distilled water (solution A).
- Prepare a solution of 3.0 g tannic acid in 300 ml distilled water (solution B).
- Prepare a solution of 0.9 g pararosaniline and 0.3 g pararosaniline hydrochloride in 100 ml 95% ethanol (solution C).
- Mix equal volumes of solution A and solution B.

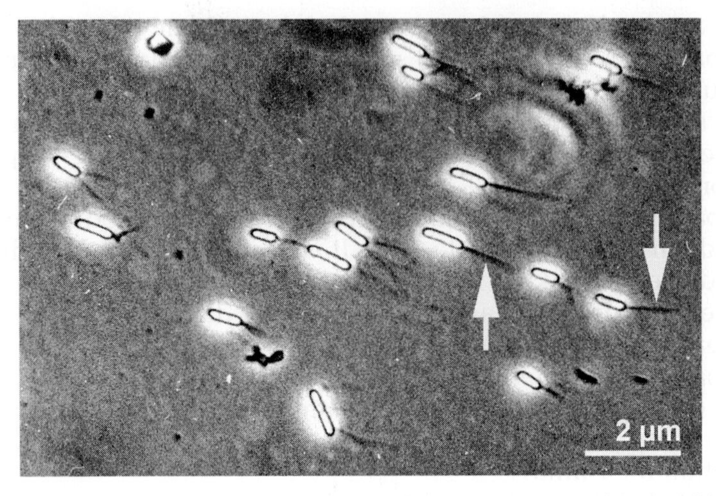

Figure 1. Light microscopic image of flagella (arrows) staining of *Escherichia coli.*

- Take two volumes of this mixture (A+B) and add to it one volume of solution C.
- The mixed staining solution can be used for a few weeks, but immediate usage is recommended.
- Prepare an air-dried microscope slide with areas of attached bacteria.
- Use a wax pencil to draw a line around the bacteria.
- Fill the inside the area circled by the wax lines with the dye solution.
- Leave the dye solution for 5–10 min, the optimal time has to be determined experimentally.
- Observe the dye solution until a golden film develops on the surface and precipitates form inside the dye solution.
- Take the slide and rinse gently with an indirect stream of tap water.
- Air-dry, apply a cover slip and examine. Flagella should stain red.

Another staining method for flagella according to Heimbrook *et al.* (1989) can be used in suspension and has been applied successfully over years in a practical course for students.

- Prepare a solution consisting of 2% tannic acid in 10 ml 5% aqueous phenol and 10 ml of saturated potassium aluminium sulfate (solution A).
- Set up a solution of saturated crystal-violet in 95% ethanol (solution B).
- Take 10 volumes of solution A and 1 volume of solution B, mix and pass through a 0.2 μm filter.
- Fill a 1 ml syringe with the staining solution.
- Place a drop (5 or 10 μl) of the bacterial suspension on a microscope slide.
- Apply a cover slip.
- Allow to stand for 5–10 min.
- On the edge of the cover slip, place a small drop of staining solution; the dye will diffuse under the cover slip. Alternatively, soak through by placing a filter paper at the opposite side of the cover slip.

- Wait for 5 min and examine starting at the edge of the cover slip where the dye has been placed.

7. Acid-fast staining

Bacteria of the genera *Mycobacterium* and *Nocardia* are medically relevant and some cause infectious diseases. Many of these bacteria have unusual cell walls that are waxy and nearly impermeable due to the presence of mycolic acids and large amounts of fatty acids, waxes and complex lipids. These organisms are difficult to stain with water-based stains, such as those used in Gram-staining. Because the cell wall is so resistant to permeation by most compounds, these bacteria require a special staining technique involving heat in order to drive the stain into their complex cell wall.

- Dissolve 0.3 g of basic fuchsin in 10 ml of 95% ethanol.
- Add 5 ml of heated, melted phenol crystals to 95-ml distilled water.
- Mix both solutions and let stand for several days and filter through a paper filter (carbol fuchsin stain).
- Add 3 ml of concentrated HCl to 97 ml of 95% ethanol (decolourizing solution).
- Dissolve 0.3 g methylene blue chloride in 100 ml distilled water (counterstain solution).
- Prepare an air-dried and heat-fixed bacteria suspension on a microscope slide.
- Cut a piece of filter paper to fit the size of the slide and soak the filter paper in the carbol fuchsin stain.
- Place the stained filter paper onto the slide.
- Heat the stain until it starts to steam by putting the underside of the microscope slide on a hot plate or above a steaming water bath.
- While steaming for 5 min keep the sample moist with stain by regularly reapplying stain.
- Remove the filter paper.
- Wash the slide gently with an indirect stream of tap water until no colour appears in the washing water.
- Hold the slide with forceps and wash it with the decolourizing solution.
- Wash immediately with tap water as above.
- Repeat the previous two steps if the bacterial the bacterial suspension looks faintly pink.
- Cover the suspension with the counterstain (methylene blue) for 20–30 s.
- Wash with tap water.
- Apply a cover slip and examine.
- Acid-fast bacteria appear red and non-acid-fast bacteria in blue.

8. Endospore staining

Normal water-based techniques, such as the Gram-stain, will not stain these tough, resistant structures. To stain endospores, malachite green must be forced into the spore using heat, in much the same way that carbol fuchsin is forced through the waxy mycolic acid layer of mycobacterial cells in the acid-fast stain.

- Prepare a 0.5% malachite green solution in water.
- Prepare the counterstaining solution consisting of 10-ml safranin O (2.5% in 95% ethanol) in 100 ml distilled water.
- Prepare a heat-fixed suspension of bacteria on a microscope slide.
- Cut a piece of filter paper to fit the size of the slide and soak the filter paper in the malachite stain solution.
- Place the stained filter paper on the slide.
- Heat the stain until it starts to steam by putting the underside of the microscope slide on a hot plate or above a steaming water bath.
- While steaming for 5 min keep the sample moist with stain and by regularly reapplying stain.
- Remove the filter paper.
- Wash the slide with an indirect stream of tap water until no colour appears in the washing water.
- Counterstain the slide with safranin O for 30 s.
- Wash the slide with an indirect stream of tap water.
- Air-dry the slide and examine.
- Endospores appear bright green, whereas vegetative cells appear brownish-red.

9. Cytoplasmic inclusions staining

When grown under certain growth conditions, some bacteria produce intracellular deposits termed inclusion bodies. These inclusion bodies are evident by phase-contrast microscopy because of their refractility. For identification, specialized staining procedures have to be performed. Here staining for poly-β-hydroxyalkanoate (also known as poly-β-hydroxybutyrate [PHB]), polyphosphate, glycogen and lipids will be described.

a Poly-β-hydroxyalkanoate

- Prepare a solution of 0.3% Sudan black B in ethylene glycol (solution A).
- Prepare a solution of 0.5% safranin O in distilled water (solution B).
- Prepare a heat-fixed film of bacteria.
- Immerse the slide in solution A for 10–15 min.
- Wash the microscope slide with an indirect stream of tap water until no colour appears in the washing water.
- Air-dry the specimen.
- Afterwards, immerse and withdraw the microscope slide several times in xylene.
- Always blot the slide dry on filter paper between each step.
- Wash the slide with an indirect stream of tap water until no colour appears in the washing water.
- Blot dry.
- Poly-β-hydroxyalkanoate granules appear as blue-dark granules in a pink cytoplasm.

b Polyphosphate granules

- Prepare a saturated solution of methylene blue chloride by adding 1.6 g of methylene blue chloride to 95% ethanol (solution A).
- Prepare a solution of 0.01% potassium hydroxide (solution B).
- Mix 30 ml of the supernatant of solution A with 100 ml of solution B. The resulting staining solution is termed Loeffler's dye.
- Alternatively, prepare a solution of 1% toluidine blue in distilled water.
- Prepare a heat-fixed film of bacteria.
- Immerse the slide in Loeffler's dye or toluidine blue dye for 10–15 s.
- Wash the slide with an indirect stream of tap water until no colour appears in the washing water.
- Blot dry on filter paper.
- With Loeffler's dye, polyphosphate granules stain deep blue to violet, whereas toluidine blue stains the granules red and the cytoplasm appears blue.

c Glycogen

- Prepare a solution of 0.3% iodine and 0.6% potassium iodide in distilled water (Lugol's reagent); grind iodine before setting up the solution.
- Prepare a slide with a 5 μl drop of the bacteria solution.
- Apply a cover slip.
- Place a drop of Lugol's reagent on one side of the cover slip.
- Soak the dye through the cover slip with filter paper by placing the filter paper on the opposite side of the cover slip.
- Observe in bright field mode (not phase contrast).
- Starch granules appear black/blue, granulose blue/violet or brown/violet, whereas glycogen appears reddish-brown.
- If this staining is not successful, periodate-Schiff staining (PAS) should be performed.

PAS

- Prepare a solution of 20 ml of 4% aqueous periodic acid.
- Add 10 ml of 0.2 M aqueous sodium acetate and 70 ml of 95% ethanol.
- Protect the solution from light, solution A (periodate solution).
- Prepare a solution of 10 g of potassium iodide and 10 g sodium thiosulfate pentahydrate in 200 ml distilled water.
- Add first 300 ml of 100% ethanol and then 5 ml of 2 M HCl to the potassium iodide/sodium thiosulfate solution.
- Stir for 5 min, allow the sulfur precipitate to settle and then decant the supernatant, solution B (reducing solution).
- Dissolve 2 g basic fuchsin in 400 ml of boiling distilled water.
- Cool to 50°C and filter through a paper filter.
- Add 10 ml of 2 M HCl and 4 g potassium metabisulfite, stir slightly.
- Fix stopper tightly and allow to stand for 12 h in a refrigerator.
- Add about 10 ml of 2 M HCl until the reagent does not show a pink tint after drying on a glass slide, store in the dark, solution C (Schiff reagent).

- Prepare a solution of 2 g potassium metabisulfite in 500 ml distilled water and add 5 ml of concentrated HCl, solution D (washing solution).
- Prepare a solution of 0.002% aqueous malachite green, solution E (counterstain solution).
- Prepare a smear of heat-fixed specimen on a glass slide.
- Flood the slide for 5 min in solution A (periodate solution).
- Wash the slide with 70% ethanol.
- Flood the slide for 5 min with solution B (reducing solution).
- Wash the slide with 70% ethanol.
- Stain for 15–45 min with solution C (Schiff reagent), time of staining has to be determined experimentally.
- Wash the slide several times with solution D.
- Wash the slide with tap water.
- Counterstain with solution E for 2–5 s.
- Wash the slide with tap water, blot dry.
- Examined under oil immersion, glycogen or other polysaccharides appear red, whereas cytoplasm appears green.

d Lipids

- A reliable method works with BODIPY® 493/503 Molecular Probes® (Invitrogen, Darmstadt, Germany).
- Dissolve BODIPY® 493/503 in 100% ethanol to give a stock solution of 1 mg/ml.
- Store in the dark for immediate use or freeze at −20°C for longer storage.
- Staining can be performed with unfixed or 1% formaldehyde-fixed bacteria.
- Rinse fixed bacteria several times in PBS.
- Resuspend the last pellet in a 1:250, 1:500 and 1:1000 dilution of BODIPY 493/503 in PBS. The best dilution has to be tested out empirically.
- Shake for 15 min.
- Wash three times with PBS.
- Place a 5-μl drop on a microscope slide, apply a cover slip and seal with nail polish.
- Examine using a fluorescence microscope with blue excitation light (as for fluorescein isothiocyanate, FITC).
- Take a phase-contrast image together with the fluorescence image and merge the two images.
- Lipid droplets light up green.

10. Immunofluorescence labelling of a bacterial cell surface-bound antigen

The following labelling protocol describes the immune labelling of a bacterial cell surface-bound antigen. A prerequisite is that antibodies against the antigen have been raised.

- Fix bacteria with 1% formaldehyde on ice for 1 h.
- Wash in PBS containing 10-mM glycine for quenching free aldehyde groups.
- Alternatively, the labelling protocol can be carried out with unfixed bacteria.

- Centrifuge 2 ml of the unfixed/fixed bacteria solution in an Eppendorf tube.
- Resuspend bacteria in 500 μl PBS.
- Add purified antibodies to a final concentration of 50 μg in 500 μl PBS.
- Incubate for 1 h at 30°C.
- Wash three times with PBS.
- Resuspend the last pellet in a 1:150 or 1:200 dilution of anti-antibodies (depending in which animal the antibodies were raised; if raised in rabbits a goat anti-rabbit-antibody is used) coupled to a fluorescence dye like Alexa Fluor® 488 (green) or Alexa Fluor® 568 (red) (Invitrogen, Darmstadt, Germany).
- Incubate for 30 min at 30°C.
- Wash three times with PBS.
- Resuspend the last pellet in 100 μl PBS.
- Place a 5 μl drop on a microscope slide, apply a cover slip and seal with clear nail polish.
- If Alexa Fluor® dyes are used, no antifade chemical is needed. If FITC- or rhodamine-coupled anti-antibodies are applied, an antifade chemical has to be used to prevent bleaching under the fluorescence microscope.
- Examine using a fluorescence microscope at an excitation wavelength appropriate for the dye used.

♦♦♦♦♦♦ III. TRANSMISSION ELECTRON MICROSCOPY (TEM)

The greatest challenge to examine biological material in a TEM is the non-physiological condition to which specimen must be exposed. First, since specimens are examined in a high vacuum, they have to be dry. Secondly, the image contrast in a TEM results from electron scattering, and contrast correlates directly with the amount of scattered electrons. The higher the atomic number, the more electrons are scattered and the greater the contrast. Unfortunately, most biological elements like carbon, hydrogen, oxygen or nitrogen do not scatter electrons to gain a good contrast. To improve contrast, heavy metals have to be added during the preparation steps. Thirdly, electrons have a limiting power to penetrate the specimen. This requires a preparation of a very thin specimen or slicing the specimen into thin sections of 50–100 nm. In summary, the preparation of biological specimen while retaining the native structural morphology was, is and will be a challenge for TEM imaging. Here two of the main methods are described in detail (negative-staining and embedding of specimens for preparing ultrathin sections) to obtain images for use in taxonomic classification.

A. Negative-Staining Methods

Negative staining is the most rapid, reproducible and easy-to-perform method for analysing bacteria, phages or enzyme molecules in the TEM. Since negative staining involves deposition of heavy atom, using stains like uranyl acetate on the specimen with subsequent air-drying, structural artefacts such as flattening are commonly observed. It is assumed that the heavy-metal salt solutions occupy hydrated regions of the specimen and penetrate into cavities within the sample. The TEM image represents a projection of the sample, which is electron translucent, whereas the surrounding areas of the sample are grey to dark. Nowadays, cryo-electron microscopy is currently regarded as the best method to view small native specimens after flash-freezing the samples in a thin film of vitreous ice and imaging without staining. Nevertheless, negative staining is the method of choice to characterize bacteria and to analyse surface appendages like flagella, pili or fimbriae (Figures 2 and 3).

1. Preparation of carbon film on mica

- Cleave a piece of mica (3 cm × 3 cm) with a scalpel to gain a clean and even surface.
- Place the mica with the newly cleaved side facing up in a Petri dish and fix with tape and place in the carbon evaporation apparatus.
- Carbon films are produced by resistance evaporation of carbon rods, which are sharpened to points and afterwards blunted to a diameter of 1 mm. The blunted ends, touching each others, are then mounted in the evaporation apparatus held together by a pressure spring. By passing an electric current through the carbon rods, carbon is evaporated and the carbon film forms on mica. Instead of using carbon rods, carbon thread can be evaporated, which usually results in a more regular carbon film. The procedure used is based on the available apparatus in the laboratory.
- Carbon-coated mica is stored in a Petri dish and sealed with Parafilm™ to avoid humidification.

2. Negative staining with carbon film

Specimen and supporting films require a mechanical support in the form of a metal grid for imaging in the TEM. The grid itself exhibits an electron opaque component, the bars and an electron translucent part, the open area. The number of grid bars and percentage of open area can be varied depending of the grid types used; 100–400 mesh (lines/inch.) grids are commonly used. Most EM grids are made of copper because they are the cheapest and their non-ferromagnetic character results in minimal distortion of the magnetic field of the objective lens.

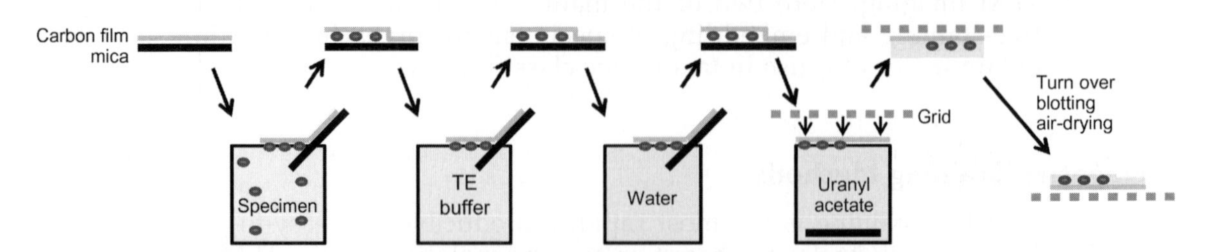

Figure 2. Scheme for negative-staining applying carbon film on mica. (see colour insert)

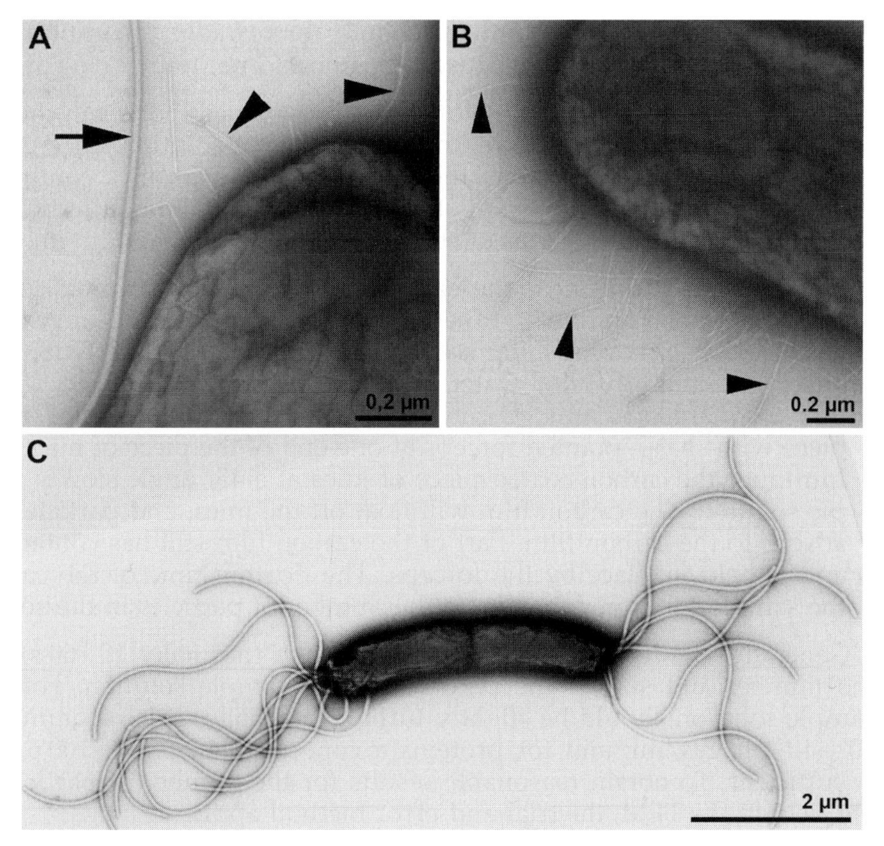

Figure 3. Negative-staining of *Escherichia coli* and *Helicobacter pylori* with 2% uranyl acetate. (A) Part of a flagellum (arrow) and pili (arrow heads) is depicted. (B) Pili (arrow heads) are visible covering the surface of *E. coli*. (C) *H. pylori* displays an amphitrichous flagellation (at both poles of the bacterium).

In addition, the copper mesh also conducts heat away from the support film and helps prevent thermal expansion, and hence movement of the specimen under the electron beam can be minimized. For immune labelling studies, nickel grids are used since copper grids develop copper rust with time when incubated with high-salt buffers, like PBS.

The following three negative-staining solutions containing heavy-metal solutions are commonly used:

- Aqueous 0.5–4% uranyl acetate solution, pH around 4.5, if pH needs to be adjusted to the more basic side, add 0.1 M KOH carefully, above pH 5.0 non-soluble $UO_2(OH)_2$ is formed, by adding 1 mM EDTA, the pH can be raised to 7.5, but the grain size of the stain will increase (van Bruggen *et al.*, 1960, 1962).
- Aqueous 0.5–3% phosphotungstic acid solution or corresponding Na and K salts, pH is adjustable with 1 M KOH or NaOH (Brenner and Horne, 1959).

- Aqueous 1–10% ammonium molybdate, for pH adjustment titrate with 0.1 M KOH, NaOH or ammonia (Muscatello and Horne, 1968).

Important: Store all staining solutions in brown glass bottles in the dark.

Uranyl acetate is the preferred negative-staining solution since it is "wetting" the specimen much better than the other stains and forms a continuous film of stain on the carbon film. Phosphotungstic acid and ammonium molybdate tend to form preferentially patches of stain instead of a staining film on carbon film.

- Pipette 50-μl droplets of the sample, washing and staining solutions on a clean piece of Parafilm™. Take care that a pronounced convex meniscus is always visible. For washing solutions, TE-buffer (20 mM TRIS, 1 mM EDTA and pH 6.9) and distilled water can be used.
- Cut a square of 2 mm × 2 mm from the carbon-coated mica and hold the piece with sharp, pointed forceps at one end of the piece of mica.
- Introduce the carbon-coated piece of mica at a 45° angle slowly into the sample solution. The carbon film will float off the mica, and particles will start to adsorb to the carbon film. Part of the carbon film still has contact to the mica and is held in place by the forceps. The floating time of the carbon film on the sample solution depends on the number of particles in the solution.

Generally, the length of the adsorption time (normally 10–60 s) depends on the number and size of the particles in the sample solution. For bacteria the sample solution should be slightly turbid; for phages it is recommended to use 10^7–10^9 phages/ml; and for proteins a concentration of 50–100 μg protein/ml is sufficient. To obtain reasonable results for the number of particles per viewing field in the TEM, the trial-and-error method applies.

- Remove the piece of mica slowly from the sample solution, and the carbon film will fall back into its original position. Remove excess sample solution by blotting on filter paper, but do not blot dry.
- Using the same procedure as described above, transfer the carbon film onto TE-buffer and distilled water, always remove excess fluid by blotting on filter paper.
- Completely float the mica off onto the staining solution by moving the piece of mica under the surface of the staining solution drop, open the forceps, and the piece of mica will fall to the bottom and the carbon film floats on the staining solution.
- Pick up the carbon film with the adsorbed specimen using a 300 mesh copper grid. The grid should be cleaned before with acetone. Since grids have the tendency to be hydrophobic, it is recommended to put the grid with the rough side on a drop of distilled water, leave for a few seconds and blot on a filter paper and immediately put the grid onto the carbon film. Slightly press on the grid with the forceps to get better contact of the film to the grid and to remove the occasionally build air bubbles between the carbon film and the grid.
- Pick up the grid perpendicular from the staining solution drop, turn it around and blot the staining solution from the grid with filter paper. Put the filter paper at the edge of the grid and let the solution creep up the filter paper. If the staining solution does not flow anymore into the filter paper,

remove the filter paper immediately. Using this procedure, one usually gets a gradient of shallow stain and deeper stain on the carbon film. Blotting of staining solution is the most crucial step in the entire negative staining procedure and needs a bit of training to reach the desired staining thickness.
- Air-dry the grid with the attached carbon film with the help of a warm light globe by placing the grid under the globe at a distance of 1–2 cm for 10–15 s.
- Examine by TEM.

3. Preparation of carbon-coated plastic films on grids

Formvar films are widely used (Drummond, 1950; Bradley, 1954) and prepared from 0.5–1% polyvinylformaldehyde in chloroform. Butvar films are prepared from 0.5–2% polyvinylbutyral in chloroform. Both of these plastic films are relatively stable under the electron beam and, when freshly prepared, exhibit a hydrophilic character, which makes them suitable for attaching particles out of aqueous solutions. A third plastic film can be made from a 1–2% collodion solution in isoamyl acetate and is called a collodion film. All three types of films can be coated with carbon to gain stability.

(a) Formvar/Butvar film

- Clean a glass microscope slide with detergent and air-dry. Place the slide in a cylindrical glass funnel with a tap and a drain tube that has been filled with 1% Formvar or Butvar in water-free chloroform. The slide should not be completely submersed; 1 cm of the upper edge should be still in air to facilitate handling of the slide.
- Cover the funnel with a lid and allow to stand for 30–60 s.
- Open the tap and let the Formvar/Butvar solution drain out of the funnel; the faster it flows the thicker the remaining film is.
- Wait another 30 s, open the lid and remove the slide with forceps and air-dry in a sealed beaker for 5 min.
- With a razor-blade scratch the edges of the slide and cut through the film on the upper part of the slide.
- Float off the Formvar/Butvar film on a clean distilled water surface in a large Petri dish. This is best done by slowly immersing the slide with its short edge at a 45° angle into the water. The Formvar/Butvar film will float off.
- Place the desired grids onto the Formvar/Butvar film with the shiny side up. Start at the corners of the film and then place the grids onto the entire film surface.
- Take a suitable-sized piece of Parafilm™ backing paper (not Parafilm™ itself), which is larger in size than the Formvar/Butvar film and gently lower it down onto the grid-covered Formvar/Butvar film, gently press on the backing paper to remove air bubbles and wait until the backing paper is completely wet. Alternatively, filter paper can be used to pick up the film.
- With forceps take the entire sandwich of backing paper, grids and Formvar/Butvar film from the water surface in one fast movement.
- Turn the sandwich around and place it on filter paper and air-dry for a day.

(b) Collodion film

- Clean a Büchner funnel with a tap or similar glassware with distilled water and fill with water.
- Place a fine mesh wire gauze (distance between the bars less than 2 mm) or filter paper (5 cm) on the bottom.
- Place grids on the wire gauze or filter paper with the shiny side down.
- Fill the tip of a glass pipette with collodion solution in water-free isoamyl acetate.
- Place one drop of collodion solution on the water surface.
- The drop will spread, and when the surface is illuminated under an 45° angle interference colours of the evaporating collodion solution can be visualized on the nascent film.
- When interference has stopped, small wrinkles can often be detected on the collodion surface indicating that all the solvent has evaporated.
- The collodion film is slowly lowered on the grids by draining off the water through the tap.
- Take the gauze or filter paper, place it on filter paper for drying and air-dry the collodion coated grids for a day at room temperature.
- Formvar/Butvar and collodion films can be used uncoated or coated with carbon as described in Section III.A.2.

4. Negative staining with carbon-coated Formvar/Butvar films

- Pipette 50 μl droplets of the sample, washing and staining solutions on a clean piece of Parafilm™. Take care that a pronounced convex meniscus is always visible. Use TE-buffer (20 mM TRIS, 1 mM EDTA and pH 6.9) and distilled water as washing solutions.
- Place the carbon-coated Formvar/Butvar grid on the sample solution (the carbon side facing into the solution). For making the carbon-coated plastic films more hydrophilic, they can be glow discharged for 30 s before usage.
- Leave the grid for 30–60 s on the sample solution depending on the number of particles.
- Take the grid with forceps and blot the excess liquid on filter paper, do not dry completely.
- Wash in TE-buffer by placing the grid on a drop of TE-buffer, blot onto filter paper.
- Wash in distilled water, blot again on filter paper.
- Stain the sample by putting the grid on the staining solution of 1–4% aqueous uranyl acetate.
- After 30–60 s, take the grid from the staining solution and remove excess liquid in the same way as described in Section III.A.2. for establishing a gradient of shallow and deep stain on the grid.
- Air-dry.
- Examine by TEM.

B. Metal-Shadowing

Besides negative-staining metal-shadowing with platinum, iridium, tungsten or tantalum provides an effective method for visualization of bacterial appendages like flagella, pili or fimbriae. Contrast is achieved by deposition of particles of these metals at an angle of between 30° and 45°. One side of the appendage is coated with the metal, whereas the other side has less metal or even no metal. Therefore, this side appears brighter in the TEM compared to the metal-covered side (Heinmets, 1949). Resolution of metal-shadowing depends on the metal used, the method of evaporation and the evaporation temperature. Resistance heating results in larger grain sizes than evaporation with an electron beam gun.

- Fix the specimen with 3% formaldehyde and 1% glutaraldehyde on ice, for 1 h, centrifuge the specimen at the lowest possible *g*-value to form a fluffy pellet.
- Gently resuspend the pellet with TE-buffer, centrifuge again.
- Resuspend in a small amount of TE-buffer.
- Apply specimen solution onto a carbon-coated Formvar/Butvar grid.
- Allow to stand for 30–60 s and carefully blot dry onto filter paper.
- Wash grid with three drops of distilled water.
- Air-dry or freeze-dry the specimen.
- Depending of the evaporation unit, place the grid on a carrier.
- Install in the evaporation unit and metal-shadow.

Platinum-iridium wire is normally used because the metal film is relatively stable to humidity; in case of a tungsten–tantalum film, grids have to be observed immediately after metal-shadowing or stored in a desiccator.

C. Embedding and Ultrathin Sectioning

Nowadays, a wide variety of methods can be applied to prepare biological material for examination of the intracellular morphology in the TEM. The common aim is to obtain morphological information by reproducible methods, which can be repeated in other laboratories as well. Therefore, the results obtained by one laboratory are strictly comparable with those obtained by another. Technical methodology has reached a point where reproducibility of specimen preparation is now possible in different laboratories. Reproducibility strengthens the belief that micrographs are faithful reflections of the native state of the specimen. The risk of introduction of technique-induced artefacts is lower as preparation techniques become more refined. It often helps to study a given sample with a variety of preparation techniques, as each can provide independent and self-consistent information. It is a good practice to combine a conventional embedding method with a cryo-based method, like high-pressure freezing and freeze-substitution, since the latter should allow examination of the specimen in an even closer native state.

1. Conventional embedding of specimen for ultrathin sectioning

The most popular technique for examining the intracellular morphology of biological materials is to embed the specimen in a liquid plastic resin, and after polymerization of the resin, ultrathin sections are cut in the range of 50–100 nm. These sections can be examined in a TEM. Biological specimen is chemically fixed with aldehydes, such as formaldehyde or glutaraldehyde, contrasted with solutions of heavy-metal salts like osmium tetroxide and uranyl acetate, dehydrated in ethanol or acetone and embedded in plastic resins like epoxy resins or acrylate resins. Subsequently, ultrathin sections are cut with glass or diamond knives using an ultramicrotome, transferred to specimen support grids, counterstained with heavy-metal salt solutions like uranyl acetate or a combination of uranyl acetate and lead citrate before examination in the TEM (for a comprehensive overview refer to the books of Hoppert and Holzenburg (1998); Bozzola and Russel (1999); Glauert and Lewis (2000); and Hoppert (2003)).

2. Fixation

The overall purpose of fixation is to maintain the native state, especially the morphology of the specimen, so it can resist the effects of subsequent steps in the preparative procedures. The common aim is to preserve every detail of cellular ultrastructure right down to the molecular level exactly as it appears in the native state. Nevertheless, it is well documented that chemical fixation and dehydration will introduce some preparation artefacts. The first step is to perform fixation of bacteria in their normal living condition, i.e., it is recommended to fix bacteria in the growth medium. The primary purpose of the fixative is to solidify the morphology of the sample as fast as possible. Proteins floating in the cytoplasm and bound to cellular membranes should be kept at their location, and the spatial relationships of the various proteins and compartments of bacteria should be preserved. Fixation should also render all the other chemical constituents of the sample, such as nucleic acids, nucleoproteins, carbohydrates and lipids, insoluble. As mentioned above, the aim of fixation is to preserve the native state of the specimen as well as possible by chemical fixation. A commonly used fixation procedure is a mixture of 1–5% formaldehyde (prepared from paraformaldehyde) and 1–3% glutaraldehyde, followed by an osmification step using osmiumtetroxide. It is recommended to fix the specimen in the growth culture media, thus avoiding centrifugation and resuspending of bacteria in a buffer, which might be of a detrimental effect on bacterial morphology.

- Prepare a 25% stock solution of formaldehyde by adding distilled water to 25 g of paraformaldehyde to reach a 100 ml volume. Heat the solution to 60°C with stirring and add drop by drop 10 M NaOH until the paraformaldehyde is dissolved. Leave the solution for 1 week at room temperature and take the clear supernatant, pass it through a paper filter and centrifuge the solution at 13,000 rpm.

- Fixation of the specimen can be routinely performed in the culture medium with 5% formaldehyde and 2% glutaraldehyde as the final concentration. Sometimes it is advantageous to perform the formaldehyde fixation at first for 5 min, followed by the glutardialdehyde fixation.

Important: Use EM-grade glutaraldehyde.

- Keep the specimen for at least 1 h at 4–8°C and centrifuge and resuspend the pellet in an appropriate buffer. We use very often cacodylate buffer (0.1 M cacodylate, 0.01 M $CaCl_2$, 0.01 M $MgCl_2$, 0.09 M sucrose and pH 6.9), alternatively a HEPES buffer (pH 7.0) can be used.
- Repeat the centrifugation and washing step twice.
- Resuspend the last pellet in 1% or 2% osmium tetroxide either in distilled water or in an appropriate buffer like cacodylate buffer.
- Perform the osmification at room temperature for 1–2 h, afterward wash with buffer.

Important: Always perform osmification in a hood since osmiumtetroxide is very volatile and toxic, always wear eye protection.

- Resuspend the pellet in 2% aqueous agar, which is kept at 45°C, mix the pellet with equal amounts of agar, soak the mixture in a glass Pasteur pipette and pour the content after a partial solidification of the agar mixture on a glass slide, put the glass slide on ice to aid further solidification of the agar, cut small pieces of 2–3 mm and transfer in an Eppendorf tube. These steps allow handling of the specimen without performing a centrifugation after every preparation step so that the dehydration fluid and resin are replaced by soaking and refilling the Eppendorf tube.

3. Dehydration

The aim of this preparation step is to replace all the free water in the sample with a fluid that is miscible both with water and with the embedding resin monomer. Usually ethanol, methanol and acetone are applied for dehydration. Acetone, e.g., is the preferred agent for dehydration if the epoxy resin according to Spurr (1969) is used. The duration of each dehydration step is kept as short as possible to prevent extraction of components, especially lipids, resulting in subsequent shrinkage of the sample. This is carried out on ice to prevent more pronounced extraction processes. Dehydration and embedding in the epoxy resin formula according to Spurr will be described.

- Add the graded series of acetone (10%, 30%, 50%) to an Eppendorf tube with the specimen and allow to incubate for 10–15 min for each step on ice.
- Fill the Eppendorf tube with 2% uranyl acetate in 70% acetone and leave overnight at 4–7°C.
- Continue the dehydration with 90% and 100% acetone steps on ice.
- Remove the specimen from the ice and bring to room temperature, repeat the 100% acetone step twice.

4. Embedding with resin

Properties of an ideal embedding resin include it being soluble in acetone, ethanol and methanol as well as in the resin monomers before polymerization. The resin does not chemically modify itself, the specimen or physically disrupt or distort the specimen. It should polymerize and harden uniformly, thereby producing a sample block hard enough, yet plastic enough, to cut into ultrathin sections. Last but not least, it should be stable under electron irradiation. None of the available embedding resins fulfil all these characteristics. Nowadays, a wide variety of resins can be used, which are grouped into the following main categories: (a) water-soluble resins like glycolmethacrylate and Durcupan or water-tolerating resins like the hydrophilic Lowicryl resins KM4 and KM11, (b) methacrylate resins like GMA, HPMA or JB-4, (c) acrylic resins like LR White, LR Gold or Unicryl and (d) epoxy resins like Epon 812, Araldite or Spurr's resin (Rosenberg *et al.*, 1960; Luft, 1961; Kushida, 1964; Armbruster *et al.*, 1982; Carlemalm *et al.*, 1982). The epoxy resin formula according to Spurr is often used due to its low viscosity and therefore good penetration capabilities into the sample (Spurr, 1969). In addition, Spurr, as well as Epon 812, is convenient because the hardness of the final block is controlled by the proportions of the two hardeners used in the monomer mixtures. If other resins are chosen, refer to the manufacturer's recommendations.

- Infiltrate with the resin following the scheme:
  1 part/volume acetone:1 part/volume Spurr resin for 12 h
  1 part/volume acetone:2 parts/volumes Spurr resin for 12 h
  pure Spurr resin for 12 h with one to two changes
  pure Spurr resin containing the catalyst for 24 h with one to two changes.
- Transfer the specimen to the bottom of a gelatine drug capsule (0.5 ml) and fill the capsule with pure resin to three quarters.
- Put the gelatine capsules containing the specimen in a vacuum chamber and apply a slight vacuum (0.6–0.8 bar) for 3–5 min. This step will release air in the resin mixture, which might have been incorporated during the mixing process. If this step is omitted, polymerized resin can be brittle and too hard to perform proper sectioning.
- Before sealing the gelatine capsules with the lid, place a small notice in the upper part of the capsule written on paper with a pencil to describe the sample. This notice will be polymerized together with the sample.
- Polymerize the specimen for 12 h at 70°C. If the ultrathin sectioning is not satisfactory, polymerize for another 12 h.

5. Embedding for visualization of intracellular membranes

Since many bacteria, e.g., phototrophic bacteria or spore forming bacteria form membranes intracellularly, a special fixation protocol is required. Aldehyde fixation is followed by osmification, and on block-staining with uranyl acetate embedding sometimes does not preserve the membranes properly. To gain sufficient contrast of intracellular membranes, the thiocarbohydrazide ferrocyanide-reduced osmium method can be applied (Willingham and Rutherford, 1984).

- Prepare a 1% thiocarbohydrazide solution in 58°C distilled water.
- Bring to room temperature.
- Filter using normal filter paper.
- Prepare a 1.5% ferrocyanide solution in 1% osmiumtetroxide in distilled water shortly before usage of the solution.
- Fix the specimen with 5% formaldehyde and 2% glutaraldehyde for 30 min.
- Wash in an appropriate buffer like cacodylate buffer.
- Resuspend the pellet in the 1.5% ferrocyanide/1% osmiumtetroxide solution and incubate for 30 min at room temperature.
- Centrifuge and resuspend the pellet in 1% thiocarbohydrazide for 5 min.
- Centrifuge and resuspend the pellet in 1.5% ferrocyanide/1% osmiumtetroxide for 5 min.
- Centrifuge and wash once with an appropriate buffer, e.g. cacodylate.
- Follow the standard embedding protocol for embedding in Spurr resin (see Section III.C.3. onwards).

6. Embedding for immunocytochemistry applying the progressive lowering of temperature (PLT) method and Lowicryl resins

Lowicryl resins are the most popular resins for immunocytochemistry because of their suitability for low-temperature embedding (Roth *et al.*, 1978, 1981). This has the advantage that cellular ultrastructure is very well preserved and the antigenicity of proteins is retained. In addition, low-temperature embedding reduces the extraction of lipids during the embedding procedure. Since Lowicryls are methacrylate resins that are volatile and toxic, adequate fume hoods, appropriate gloves (e.g. 4H gloves), and protective glasses should always be used. Other disadvantages of this method are the technical efforts that are required to maintain the specimens at low temperatures while infiltrating, embedding and polymerizing them. Lowicryl resin is available in four types: two polar hydrophilic resins, K4M (used at −35°C and above) and K11M (used at −60°C and above) and two non-polar hydrophobic resins, HM20 (used at −70°C and above) and HM23 (used at −80°C and above).

- Fixation of the specimen is performed with either 1% formaldehyde on ice for 1 h or a fixation solution of 0.2% glutaraldehyde and 0.5% formaldehyde on ice for 1 h.
- Wash the specimen with an appropriate buffer containing 10 mM glycine for quenching unbound aldehyde.
- Immobilize the specimen in aqueous agar (see Section III.B.2.).
- Dehydrate the specimen with 30% ethanol/methanol on ice for 15 min.
- Dehydrate the specimen with 50% ethanol/methanol at −20°C for 30 min, always cool down the dehydration step solution before use.
- Dehydrate the specimen with 70%, 90%, 100% ethanol/methanol at −35°C for K4M (or at −60°C to −80°C for the other Lowicryl resins) for 30 min for each step.
- Repeat the 100% ethanol/methanol step.

- Infiltrate with 1 part ethanol/methanol/volume:1 part/volume Lowicryl K4M at −35°C for 3 h.
- Infiltrate with 1 part ethanol/methanol/volume:2 parts/volumes Lowicryl K4M at −35°C for 12 h.
- Infiltrate with pure resin at −35°C for 12 h with one change of resin.
- Infiltrate with pure resin at −35°C for 24 h with one change.
- Infiltrate with pure resin at −35°C for 3 h.
- Transfer the specimen in Lowicryl K4M-filled gelatin drug capsules at −35°C, fill capsules completely, put on the lid of the gelatin capsule.
- Keep the specimen at −35°C and polymerize with 360-nm UV-light for 12 h or longer.
- Polymerize specimen for another 2–3 days with 360-nm UV-light.

If the technical tools for low-temperature embedding are not commercially available, reasonable home-made models can be designed. Since it is very important to avoid inhaling the vapours, a foam box that fits in a fume hood filled with dry ice can act as a cool chamber. A metal block with holes drilled in it to accommodate 2-ml cryovials is a useful holder. For adjusting temperatures from −20°C to −80°C, various amounts of ethanol/methanol can be added to adjust the temperature. A thermometer sitting within a cyrovial is a useful way to monitor the temperature. For UV polymerization, a foam box lined with aluminium foil attached to a hand-held UV-lamp and placed in a freezer at the desired temperature is a good alternative and quite effective.

7. Embedding for immunocytochemistry using LRWhite resin

LRWhite is an acrylic resin and is especially useful for localization of proteins in bacteria. It is easy to work with as it is purchased as a complete mixture with low viscosity. Only the accelerator has to be added at the last infiltration steps. LRWhite resin is less toxic compared to methacrylate resins. The disadvantages of LRWhite are that it is usable from 0°C to room temperature and requires heat (45°C) for polymerization; therefore heat-sensitive epitopes may not retain sufficient antigenicity.

1. Fixation of the specimen is performed either with 1% formaldehyde on ice for 1 h or with a fixation solution of 0.2% glutaraldehyde and 0.5% formaldehyde on ice for 1 h.
2. Wash the specimen with an appropriate buffer containing 10 mM glycine for quenching unbound aldehyde.
3. Dehydrate the specimen with a graded series of ethanol (10%, 30%, 50%, 70%, 90%, 100%) for 15 min each step on ice.
4. Repeat the 100% ethanol step twice.
5. Infiltrate with 1 part/volume LRWhite resin without accelerator:1 part/volume 100% ethanol on ice for 2 h.
6. Infiltrate with 2 parts/volume LRWhite resin without accelerator:1 part/volume ethanol on ice for 2 h.
7. Infiltrate with pure LRWhite resin without accelerator on ice for 12 h.

8. Infiltrate with LRWhite resin containing the accelerator for 12 h with two changes.
9. Transfer the specimen into gelatin drug capsules, fill the capsule completely, put the lid of the gelatin capsule on.
10. Polymerize at 40°C or 45°C for 1 day.

8. Embedding after high-pressure freezing (HPF) and freeze-substitution (FS)

Chemical fixation applying aldehyde, osmium tetroxide and uranyl acetate has the drawback that not all cellular components are simultaneously stabilized. During the embedding process, certain cellular components can be lost due to dehydration and infiltration of the resin at ambient temperature. In contrast to chemical fixation, cryofixation enables a simultaneous fixation of all cellular components. In addition, a much faster fixation rate can be achieved and the cell structure is preserved in its native and physiologically active state. Successful cryofixation, followed by FS, introduces dehydrating agents and fixatives at low temperature and allows proper crosslinking of cellular components. Taken together, HPF and FS reveal a superior morphological preservation of the specimen and allow for an accurate immune localization of antigens at high

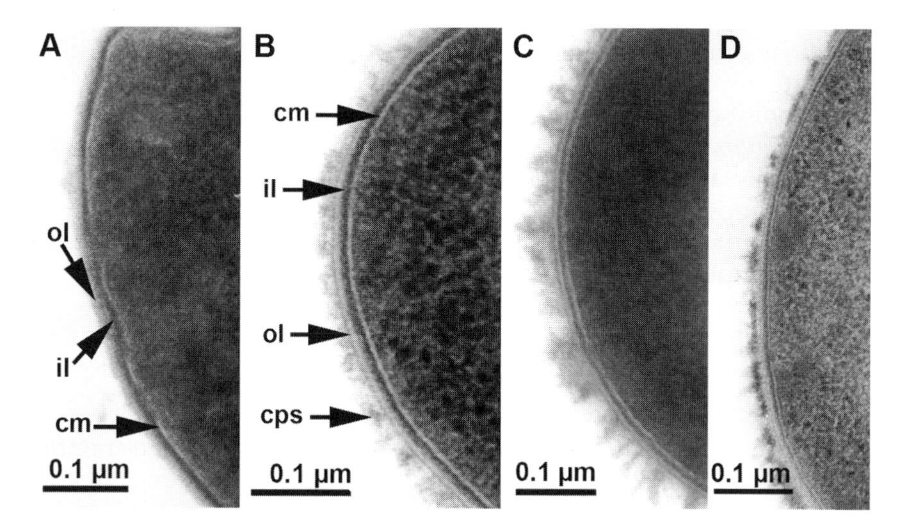

Figure 4. Ultrathin sections of Group A, B and C streptococci after chemical fixation and HPF and FS followed by embedding in an epoxy resin according to Spurr.
(A) Chemical fixation and standard embedding reveal partial visibility of the cytoplasmic membrane (cm). The Gram-positive cell wall exhibits an electron dense inner lamina (il) and a more translucent outer lamina (ol). Note the absence of any capsular material or "hairy-like" structures of the M-proteins. (B–D) HPF Group A (B), Group B (C) and Group C streptococci (D) after FS exhibit a more pronounced visibility of the cytoplasmic membrane. In contrast to chemical-fixed Group A streptococci (A, identical strain as in B), HPF-frozen bacteria exhibit capsular material (cps) covering the outer lamina (ol). In addition, the outer lamina looks morphologically very distinct when compared with the chemical-fixed bacteria.

resolution (Figure 4). This method is recommended if delicate microorganisms, like Archaea, are embedded and ultrathin sectioned (Rachel *et al.*, 2010).

- Grow the organism and centrifuge to form a fluffy pellet.
- Alternatively, the specimen can be grown within cellulose capillaries.
- Depending on the manufacturer of the high-pressure freezing unit, soak the pellet in capillaries and freeze in copper tubes for the Leica EM-Pact/Pact2 or use brass hats as for the Bal-Tec HPF 010. If brass hats are used, perform filling and freezing quickly to avoid air-drying at the top of the sample.
- Freeze the specimen and store in liquid nitrogen or transfer immediately into the freeze-substitution unit.

Freeze-substitution is the process of dissolution of ice in the frozen specimen by an organic solvent at a temperature below −70°C to prevent secondary ice crystal formation. Normally a second fixative is added to the substitution medium. Once substitution is complete, the specimen can be warmed up to room temperature and embedded in resins. Alternatively, the specimen can be embedded at low temperature (−60°C to −35°C) in Lowicryl resins. Since numerous protocols exist for freeze-substitution, one standard protocol is given resulting in good ultrastructural preservation.

- Prepare a substitution media consisting of pure acetone supplemented with 2% osmium tetroxide plus up to 4% water (varies for the specimen examined, addition of water enhances visibility of membranes) and 0.25% uranyl acetate.
- Pre-cool the substitution media to −90°C.
- Cover the specimen with the substitution medium.
- Substitute at −90°C for 12 h or longer.
- Raise temperature to −60°C, substitute for 12 h or longer.
- Raise temperature to −30°C, substitute for 12 h or longer.
- Raise temperature to 0°C.
- Rinse twice with pure acetone.
- Bring to room temperature.
- Embed the substituted specimen in Spurr resin (see Section III.C.4.) or any other suitable resin.

For immunocytochemical localization studies, the above-mentioned substitution medium can be applied without osmium tetroxide. Substituted samples can be embedded at room temperature in LRWhite resin (see Section III.C.7.) or at lower temperatures depending on the Lowicryl resin used (see Section III.C.6.).

9. Ultramicrotomy

Ultrathin sections of specimen should be 60–90 nm thick to allow proper imaging under the TEM. Therefore, specimens embedded in gelatin capsules have to be trimmed to obtain a small flat-topped pyramid with an area of 0.2–1 mm^2

and ultrathin sectioned using a diamond knife or a glass knife. Since several different types of ultramicrotomes are on the market, the following scheme describes only the main steps in obtaining ultrathin sections.

- Trim the specimen with a new razor-blade or a rotating milling cutter with a diamond cutter.
- Area of the flat-topped pyramid should be between 0.2–1 mm^2.
- Mount the trimmed specimen in the specimen holder of the ultramicrotome, check that the specimen is held firmly by the holder.
- Insert the specimen holder in the ultramicrotome.
- Insert the knife with a clearance angle of 4–6°.
- Position the specimen parallel to the knife edge.
- Approach the specimen to the knife edge.
- Perform one to three thick sections of 0.5–1 μm.
- Fill the trough with distilled water.
- Start cutting with a section thickness of 150 nm.
- Observe interference colour of the section.
- Adjust section thickness to achieve a light yellow/gold interference colour of the ultrathin sections, section thickness is around 60–90 nm.
- Observe the section thickness and floating of the sections from the knife edge.
- Move sections onto the water surface with the aid of a mounted eyelash; arrange two to three sections together.
- Pick up sections from the water surface with a Formvar/Butvar grid by slowly approaching the sections from above, press the grid onto sections, allow sections to attach for a few seconds and then withdraw the grid. Alternatively, sections can be picked up by inserting the grid with forceps under the water surface in the trough and slowly moving the grid upwards towards the sections at a 45° angle. Then move the grid slowly out of the water, and sections will adhere to it.

Remove excess liquid with filter paper and air-dry.

- Post-staining of ultrathin sections with 4% uranyl acetate for 3–10 min, omit direct sunlight or illumination.
- Rinse in three drops of distilled water.
- Stain in lead citrate in a CO_2-free atmosphere (place NaOH pellets into a Petri dish for CO_2 absorption) for 1–5 min (Venable and Coggeshall, 1965).

Alternatively, the Ultrostain II (Leica, Wetzlar, Germany) can be applied. This lead citrate solution can be used immediately, it causes only very minor precipitation and is stable for years when kept in the dark at ambient temperature. For easy use, fill a 5 ml syringe with staining solution from the stock solution. Seal with a yellow tip and cover the end of the tip with Parafilm™ to prevent CO_2 influx into the syringe.

- Rinse inside the Petri dish with drops of boiled distilled water.
- Rinse with boiled water in a beaker.
- Air-dry and examine by TEM.

♦♦♦♦♦♦ IV. FIELD EMISSION SCANNING ELECTRON MICROSCOPY (FESEM)

A. Preparation Steps for FESEM

1. Fixation

The aim of fixation is to preserve the native state of the specimen as well as possible by chemical fixation. A commonly used fixative is a mixture of 1–5% formaldehyde (prepared from paraformaldehyde) and 1–3% glutaraldehyde. It is recommended to fix the specimen in the growth culture media, thus avoiding centrifugation and resuspending of bacteria in a buffer, which might have a detrimental effect on the morphology of the specimen.

- Prepare a 25% stock solution of formaldehyde by adding distilled water to 25 g of paraformaldheyde to reach a 100 ml volume. Heat the solution to 60°C with stirring and add drop by drop 10 M NaOH until the paraformaldehyde is dissolved. Leave the solution for 1 week at room temperature and take the clear supernatant, pass it through a paper filter and centrifuge the solution at 13,000 rpm.
- Fixation of the specimen is performed in the culture medium with 5% formaldehyde and 2% glutaraldehyde as end concentration. It is important to use EM-grade glutaraldehyde.
- Incubate the specimen for at least 1 h at 4–8°C, centrifuge and resuspend the pellet in TE-buffer, repeat the centrifugation and washing step with TE-buffer. The washing step with TE-buffer reduces the formation of salt crystals on the specimen.

2. Support for bacteria in FESEM

Due to the small size of bacteria, they require a support to be imaged. Different types of filters (paper filter, Nucleopore filter) can be used as a support. The main problem with using filters is the irregular background, which might mask some morphological features of the specimen when imaged in the FESEM. The Nucleopore filters are better in this respect since they provide a "clean" background, but still the numerous pores are visible in the images as well as the specimen. For these reasons, cover slips (12 mm in diameter) are used, which are coated with poly-L-lysine (Figures 5 and 6).

- Add a drop (50 μl) of 0.01-M poly-L-lysine on a cover slip and leave for 10 min.
- Wash the cover slip with distilled water and air-dry.
- Pipette 50 μl of the fixed and TE-buffer-washed specimen solution onto the cover slip. If a lower cell density of bacteria should be examined, dilute the fixed specimen solution with TE-buffer.
- Allow the fixed specimen solution to stand for 5–10 min (depending on the size of the bacteria) so that the specimen can settle and is "glued" onto the poly-L-lysine.

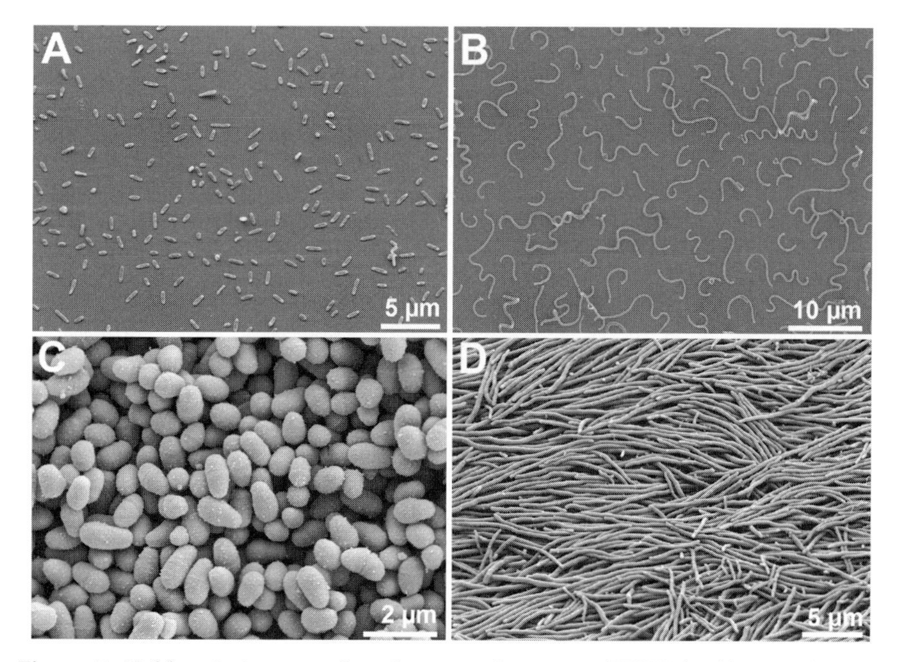

Figure 5. Field emission scanning electron microscopy (FESEM) of bacterial isolates prepared on poly-L-lysine-coated glass cover slips at different dilutions.
Owenweeksia hongkongensis (A) and *Runella slithyformis* (B) were resuspended and then further diluted to obtain low numbers of bacteria on the cover slip. (C and D) *Dietzia maris* and *Catenulispora acidiphila* were applied directly after resuspension of the pellet, resulting in a "lawn" of bacteria on the cover slip.

- Transfer the cover slip into a fixation solution of 3% glutaraldehyde in TE-buffer, allow to stand for 10–15 min at room temperature and subsequently wash with TE-buffer.

Important: Be especially careful not to air-dry the specimen at any step during these preparation steps.

3. Dehydration

For dehydration, a graded series of acetone or ethanol is applied. It is recommended to dry acetone either with $CaCl_2$ or using a molecular sieve before use. Ethanol should be used from a freshly opened bottle.

- Transfer the cover slip in a small glass Petri dish or alternatively into a plastic cell culture plate on ice (depending on the number of cover slips , a 4-well or 24-well plate can be used).
- Add the graded series of acetone/ethanol (10%, 30%, 50%, 70%, 90%) and allow to stand for 10–15 min, when changing the solutions do not pipette the solutions directly on the cover slip.

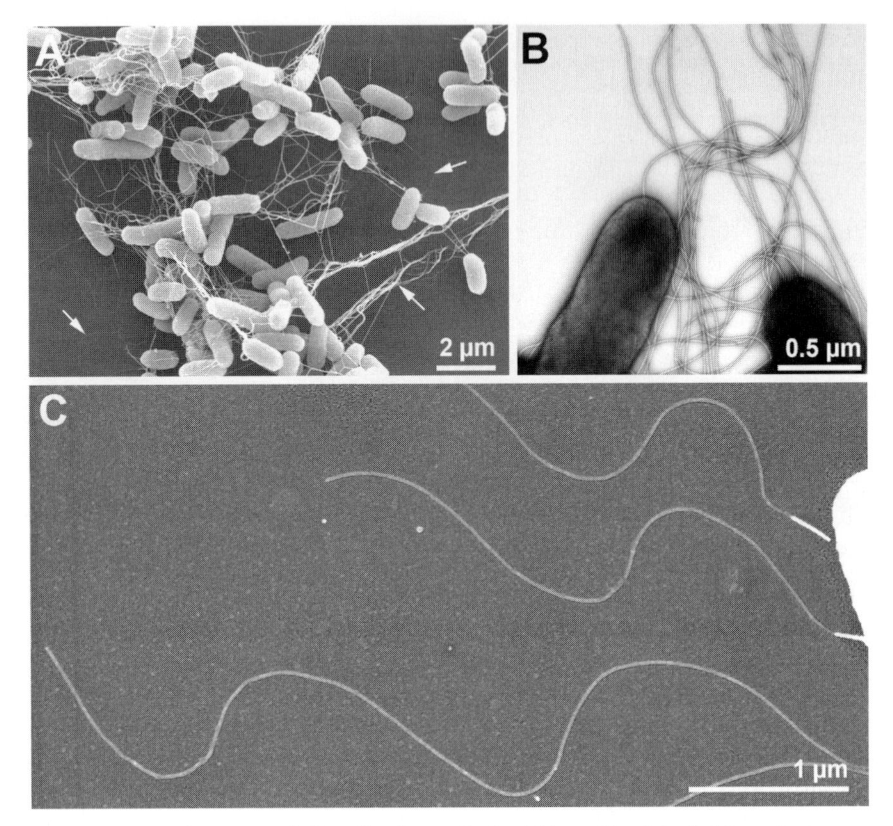

Figure 6. Visualization of flagella from *Escherichia coli* with field emission scanning electron microscopy (FESEM) and negative-staining with 2% uranyl acetate.
(A) Imaging was performed with the Everhart–Thornley SE-detector for topographical contrast and the Inlens SE-detector for resolution. Both signals were mixed in a 50:50 ratio. White arrows point to flagella. (B) Negative-staining of the identical sample reveals the long appendages attached to the bacteria are flagella. (C) FESEM of flagella, which are directly adsorbed to the cover slip.

Important: If dehydration is performed with acetone in a plastic culture plate, transfer the cover slips from the 90% acetone step into glass Petri dishes because 100% acetone will dissolve the plastic Petri dish.

- Transfer the glass Petri dish or culture plate to room temperature.
- Add 100% acetone/ethanol, leave for 15 min, repeat the step.

Important: Be careful not to air-dry the specimen at this preparation step since acetone will evaporate rapidly from the specimen in a warm environment.

4. Critical-point drying

The critical-point drying (CPD) step has to be performed to obtain a very well prepared specimen for high-resolution imaging. For this step in the preparation, a CPD apparatus is used. For performing CPD, a number of different apparatuses are available commercially. In principle, the specimens, which are prepared in dehydration medium consisting of 100% acetone/ethanol, are transferred in a pressure chamber filled with 100% acetone/ethanol at 8°C. The chamber is set under pressure, and liquid CO_2 is used as a transitional medium. In several exchanges, acetone is replaced with liquid CO_2. Transition of the liquid CO_2 to gaseous CO_2 is done at the critical pressure for CO_2 (73.8 bar) and the critical temperature at 31°C. Above the critical point, the densities of the drying medium in its liquid and gaseous phases are identical. Therefore, a phase boundary no longer exists. If the temperature is kept above the critical point, the gas phase can be slowly vented of the apparatus by a needle valve and the specimen are dried without causing any surface tensions.

Alternatively, if no CPD apparatus is available, a chemical drying procedure with hexamethyldisilazane (HMDS) can be applied. This methods works well, especially for Gram-positive bacteria and some Gram-negative bacteria, but for delicate organisms like Archaea or unknown species it is not recommended.

- Fixation is the same as for CPD, but dehydration of the specimen has to be performed in ethanol.
- Once the specimen is in 100% ethanol, it must be transferred to 100% HMDS in a graded series of ethanol–HMDS mixtures, just cover the specimen with the liquid, perform the infiltration with HMDS at room temperature.
- 1 part/volume:1 part/volume, 100% ethanol:100% HMDS for 10 min.
- 1 part/volume:2 parts/volumes, 100% ethanol:100% HMDS for 10 min.
- 1 part/volume:3 parts/volumes, 100% ethanol:100% HMDS for 10 min.
- 1 part/volume:4 parts/volumes, 100% ethanol:100% HMDS for 10 min, keep in mind that incomplete transition from ethanol into HMDS is a bad source of problems.
- 100% HMDS with two changes for 5 min each step; be very cautious that the specimen is not yet air-drying.
- Exchange once again with 100% HMDS.
- Let the specimen completely air-dry, depending on the sample size, for 10–30 min.

5. Mounting the specimen

Due to the hydrophilic character of most biological specimens, they should be mounted and sputter coated immediately. For mounting, circles of adhesive tape fitting the size of the sample stub are commonly used and work very well for cover slips.

6. Sputter coating of the specimen

In general, biological specimens are poor electrical conductors. Therefore, conductivity must be achieved by coating the specimen with conductive material to perform imaging in a FESEM or other SEMs. In most cases, specimens are sputter-coated with a thin layer of gold (around 5–10 nm, depending on the specimen) since gold is a very good secondary electron emitter. As other materials for sputter coating, gold–palladium or platinum targets can be used. In addition, carbon coating is also applicable but lacks a good emission of secondary electrons. If immune FESEM is performed, specimens have to be coated with carbon to receive the good secondary electron signals of bound gold-nanoparticles.

B. Immune FESEM

For some time, the identification of cellular components or proteins with immunogold labelling by SEM was restricted due to limitations in the resolving power of the SEMs for imaging gold-nanoparticles in the range of 5–15 nm (Hoyer *et al.*, 1979; de Harven *et al.*, 1984). The improved resolving power of FESEM instruments and the use of backscattered electron imaging helped this situation considerably, and the method is now considered to be extremely sensitive and specific for accurate localization of proteins or structures like the capsule on bacterial cells. Since FESEM images the surface of the specimen, only surface structures or proteins exposed on the bacterial surface can be localized with antibodies and gold-nanoparticles. It is preferred to apply polyclonal IgG antibodies instead of monoclonal antibodies because the polyclonal antibodies exhibit a better binding capacity to the antigen. In addition, IgA or IgM is sometimes also suitable. It is recommended to purify the antibodies using affinity chromatography like a protein A or a protein G Sepharose column.

1. Fixation of the specimen

- Fixation of the specimen is performed with 1% formaldehyde to retain sufficient antigenicity of, e.g. a specific protein.
- The specimen is washed with PBS buffer containing 0.01 M glycine for quenching free aldehyde groups and therefore minimizes some background staining.

Important: As a rule of thumb, use a 2 ml Eppendorf tube filled with a turbid solution of the fixed specimen; the resulting pellet after centrifugation should measure around 15–20 μl in volume.

2. Incubation with antibody and protein gold-nanoparticles

- Dissolve the last pellet of the washing step in 200 μl PBS and add the specific antibody. The concentration of the IgG antibodies should be in the range 50–100 μg IgG/ml; antibodies are always applied in access.

Important: It is strongly recommended to purify antibodies using a Sepharose protein A or protein G column before usage.

- Incubate for 1 h at 30°C with occasional shaking.
- Wash three times with PBS buffer.

- Dissolve the last pellet in PBS containing 1 mg/ml polyethylene glycol (PEG), PEG provides better stability for the protein gold complexes.
- Add 5–10 μl of the stock solution of the protein gold complexes, the protein A complexes are always used in access and the solution should exhibit a slightly pink colour after adding the gold complexes.

Protein A-, protein G- and protein A/G-coated gold-nanoparticles or second antibody-coated gold-nanoparticles are commonly used. These gold markers are commercially available in a variety of gold-nanoparticle diameters ranging from 5 to 30 nm, which reflects a suitable diameter for FESEM analysis. It is recommended to begin with 15-nm-diameter gold-nanoparticles for the first labelling studies.

- Incubate for 30 min at 30°C with occasional shaking.
- Wash three times with PBS buffer.

3. Second fixation step

- Dissolve the pellet of the last washing step in PBS buffer with 2% glutaraldehyde.
- Wash with PBS buffer and TE-buffer.

4. Mounting of the specimen

- Specimens can be placed onto poly-L-lysine-coated cover slips and critical point dried (see Sections IV.A.2. and IV.A.4.)
- Alternatively, specimen can be mounted onto plastic film-coated grids.
- Place 2 μl of the sample onto a Butvar- or Formvar-coated 300 mesh grid.
- Allow to stand for 5 min.
- Blot the grid at a 45° angle onto filter paper and transfer immediately onto a drop of TE-buffer to wash it.
- Blot the grid again onto filter paper and transfer onto a drop of distilled water.
- Wash again in distilled water by blotting and transferring the grid onto another drop of water.
- Blot the grid onto filter paper and air-dry.
- Grids are mounted onto circles of adhesive tape, which fit the size of the sample stub, press the edge of the grid firmly with scissors on the adhesive tape for conductivity.

5. Coating of the specimen

- Coat the specimen with carbon (see Section III.A.1.).
- Or observe uncoated specimen.

For imaging gold-nanoparticles under the FESEM, specimens have to be coated with carbon or uncoated specimen can also be examined. The latter gives rise to surface charge accumulation. This phenomenon can be partially overcome by illuminating parts of the sample at a low magnification for 5–10 min. In most cases, an equilibrium of impacting electrons from the primary beam and emitted electrons will be build up to allow imaging.

6. Imaging of the specimen

- Start imaging with an acceleration voltage of 5 kV using the Everhart–Thornley SE-detector (secondary electrons detector) and the In-lens SE-detector in a 75:25 ratio.
- Once gold-nanoparticles are visible, change the acceleration voltage to a higher or lower voltage to get the best images of the gold-nanoparticles and the bacteria.
- If the SEM is equipped with a BSE-detector (backscattered electrons detector), this detector can be used for detecting the gold signal (compositional contrast) and the other detectors for imaging the bacteria (topographical contrast).
- If the SEM is equipped with an EsB detector (energy selective backscatter detector), this detector gives an excellent contrast of the gold-nanoparticles.

Important: Small gold-nanoparticles can only be resolved in a FESEM due to the high-resolution power. In a standard SEM, gold-nanoparticles can only be imaged if a backscattered electron detector (BSE-detector) is used for imaging and larger gold-nanoparticles (20 nm or 30 nm in size) should be applied for the labelling studies.

◆◆◆◆◆◆ V. SUPPLIERS OF LIGHT AND ELECTRON MICROSCOPIC EQUIPMENT AND CHEMICALS

The following is a list of main suppliers of chemicals, supplies, accessories and equipment for light and electron microscopy:

(a) Electron Microscopy Sciences (EMS) serves light and electron microscopy.
P.O. Box 550, 1560 Industry Road, Hatfield, PA 19440, USA
Tel.: ++01 21541284000, http://www.emsdiasum.com, catalogue also in Chinese, Japanese, Russian, French, German and Spanish

(b) Plano GmbH serves light and electron microscopy.
Ernst Befort Str, 12, 35578 Wetzlar, Germany
Tel.: ++49 6441-97650, http://www.plano-em.de, catalogue only in German

(c) Agar Scientific serves electron microscopy.
Unit 7, M11 Business Link, Parsonage Lane, Stansted, Essex CM24 8GF, England
Tel.: ++44 1279813519, http://www.agarscientific.com

(d) Ted Pella Inc. serves electron microscopy.
P.O.Box 492477, Redding, CA, USA
Tel.: ++015302432200, http://www.tedpella.com

(e) Science Services GmbH serves electron microscopy.
Görzer Str. 70, 81549 München, Germany
Tel.: ++49 8915980280, http://www.scienceservices.de, catalogue only in German

♦♦♦♦♦♦ VI. CONCLUDING REMARKS

Many people might describe new microorganisms by light microscopy because they have access to light microscopy equipment and not to any electron microscopic equipment. Such techniques still have a merit nowadays. The light microscopy procedures described in the chapter should provide sufficient information for a morphological description of an isolated microorganism. On the other hand, the use of electron microscopy has declined considerably over the past decades despite the incredible ability of a wide array of EM methods to provide ultrastructural information, usually not revealed by other approaches. In this aspect, it seems that the loss of interest in the use of electron microscopy began as new technology emerged for light microscopic techniques, especially live imaging microscopy and high-resolution confocal laser scanning microscopy. A short glance in cell biology- or microbiology-orientated journals undoubtedly reveals that EM image plates have been more and more replaced by colourful confocal image plates. Nevertheless, electron microscopy remains the only technique with sufficient high resolution and the only method to gain morphological information about the intracellular organization at a glance of a newly isolated microorganism when considering ultrathin sections. The ideal approach would be to combine light microscopy with TEM and SEM/FESEM for a comprehensive morphological description of newly isolated microorganisms. New developments in respect to electron microscopic preparation methods like high-pressure freezing and cryo electron tomography (CET) and the newly invented TEMs with the possibility of performing elemental analysis with electron spectroscopic imaging (ESI) and electron energy loss spectroscopy (EELS) will allow for an even more sophisticated description of microorganisms. In addition, new FESEMs are equipped with much more sensitive detectors for SE or BSE or with new detectors like the energy selective backscattered detector (EsB) for identification of immune gold labelling. These new technologies will open up new fields of insights into the world of microorganisms. In summary, the described methodologies within this chapter should provide a useful guideline to reach such goals independently if light microscopy or electron microscopy – or both –are applied to gain insight in morphological characteristics of a newly isolated microorganism.

References

Armbruster, B. L., Carlemalm, E., Chiovetti, R., Garavito, L. R. M., Hobot, J. A., Kellenberger, E. and Villinger, W. (1982). Specimen preparation for electron microscopy using low temperature embedding resin. *J. Microsc.* **126**, 77–85.

Bartholomew, J. W. (1962). Variables influencing results, and the precise definition of steps in Gram staining as a means of standardizing the results obtained. *Stain Technol.* **37**, 139–155.
Bartholomew, J. W. and Mittwer, T. (1952). The Gram stain. *Bacteriol. Rev.* **16**, 1–29.
Baumeister, W., Grimm, R. and Walz, J. (1999). Electron tomography of molecules and cells. *Trends Cell Biol.* **9**, 81–85.
Bozzola, J. J. and Russel, L. D. (1999). *Electron Microscopy Principles and Techniques*. 2nd ed. Jones and Bartlett, Boston, MA.
Bradley, D. E. (1954). Evaporated carbon films for use in electron microscopy. *Br. J. Appl. Phys.* **5**, 65–66.
Brenner, S. and Horne, R. W. (1959). A negative staining method for high resolution electron microscopy of viruses. *Biochim. Biophys. Acta* **34**, 103–110.
Bullivant, S. and Ames, A. (1966). A simple freeze-fracture replication method for electron microscopy. *J. Cell Biol.* **29**, 435–447.
Carlemalm, E., Garavito, R. M. and Villiger, W. (1982). Resin development for electron microscopy and an analysis of embedding at low temperature. *J. Microsc.* **126**, 123–143.
Chakravortty, D., Rohde, M., Jäger, L., Deiwick, J. and Hensel, M. (2005). Formation of a novel surface structure encoded by *Salmonella* pathogenicity island 2. *EMBO J.* **24**, 2043–2052.
Clark, W. A. (1976). A simplified Leifson flagella stain. *J. Clin. Microbiol.* **3**, 632–634.
de Harven, E., Leung, R. and Christensen, H. (1984). A novel approach for scanning electron microscopy of colloidal gold-labeled cell surfaces. *J. Cell Biol.* **99**, 53–57.
Doddema, H. J. and Vogels, G. D. (1978). Improved identification of methanogenic bacteria by fluorescence microscopy. *Appl. Environ. Microbiol.* **36**, 752–754.
Drummond, D. G. (1950). The practice of electron microscopy. *J. R. Micosc. Soc.* **70**, 1–141.
Glauert, A. M. and Lewis, P. R. (2000). Biological specimen preparation for transmission electron microscopy. In *Practical Methods in Electron Microscopy* (A. M. Glauert, ed), Vol. 17, Portland Press, London.
Graham, L. L. and Beveridge, T. J. (1990). Evaluation of freeze-substitution and conventional embedding protocols for routine electron microscopic processing of eubacteria. *J. Bacteriol.* **171**, 2141–2149.
Griffiths, G., McDowall, A. F., Back, R. and Dubochet, J. (1984). On the preparation of cryosections for immunechemistry. *J. Ultrastruct. Res.* **89**, 65–78.
Gross, H., Bas, E. and Moor, H. (1978). Freeze-etching in ultrahigh vacuum at −196°C. *J. Cell Biol.* **76**, 712–728.
Heimbrook, M., Wang, W. L. and Campbell, G. (1989). Staining bacterial flagella easily. *J. Clin. Microbiol.* **27**, 2612–2615.
Heinmets, F. (1949). Modification of silica replica techniques for study of biological membranes and application of rotary condensation in electron microscopy. *J. Appl. Phys.* **20**, 384–389.
Hohenberg, H., Mannweiler, K. and Müller, M. (1994). High-pressure freezing of cell suspensions in cellulose capillary tubes. *J. Microsc.* **175**, 34–43.
Hoppert, M. (2003). *Microscopic Techniques in Biotechnology*. Wiley-VCH, Weinheim.
Hoppert, M. and Holzenburg, A. (1998). *Electron Microscopy in Microbiology*. BIOS Scientific, Oxford.
Hoyer, L. C., Lee, J. C. and Bucana, C. (1979). Scanning immunoelectron microscopy for the identification and mapping of two or more antigens on cell surfaces. *Scanning Electron Microsc.* **3**, 629–636.
Kellenberger, E. (1991). The potential of cryofixation and freeze-substitution: observations and theoretical considerations. *J. Microsc.* **161**, 183–203.

Knoll, M. (1935). Aufladepotential und Sekundäremission elektronenbestrahlter Körper. *Zeitschrift für technische Physik* **16**, 467–475.
Knoll, M. and Ruska, E. (1932a). Beitrag zur geometrischen Elektronenoptik. *Ann. Physik* **12**, 607–661.
Knoll, M. and Ruska, E. (1932b). Das Elektronenmikroskop. *Zeitschrift für Physik* **78**, 318–339.
Kushida, H. (1964). Improved methods for embedding with Durcupan. *J. Electron Microsc.* **13**, 139–144.
Kwok, T., Zabler, D., Urman, S., Rohde, M., Hartig, R., Wessler, S., Misselwitz, R., Berger, J., Sewald, N., König, W. and Backert, S. (2007). *Helicobacter* exploits integrin for type IV secretion and kinase activation. *Nature* **449**, 862–866.
Leifson, E. (1951). Staining, shape, and arrangement of bacterial flagella. *J. Bacteriol.* **62**, 377–389.
Luft, J. H. (1961). Improvements in epoxy embedding materials. *J. Biophys. Biochem. Cytol.* **9**, 409–414.
Moor, H. and Mühlethaler, K. (1963). Fine structure of frozen etched yeast cells. *J. Cell Biol.* **17**, 609–628.
Muscatello, U. and Horne, R. W. (1968). Effect of the tonicity of some negative-staining solutions on elementary structure of membrane-bounded systems. *J. Ultrastruct. Res.* **25**, 73–83.
Palade, G. E. (1952). A study of fixation for electron microscopy. *J. Exp. Med.* **95**, 285–298.
Palade, G. E. and Siekevitz, P. (1956). Pancreatic microsomes: An integrated morphological and biochemical study. *J. Biophys. Biochem. Cytol.* **2**, 171–200.
Rachel, R., Meyer, C., Klingl, A., Gürster, S., Heimerl, T., Wasserburger, N., Burghardt, T., Küper, U., Bellack, A., Schopf, S., Wirth, R., Huber, H. and Wanner, G. (2010). Analysis of the ultrastructure of Archaea by electron microscopy. In *Electron Microscopy of Model Systems* (T. Müller-Reichert, ed), pp. 47–69. Elsevier, Boston, MA.
Richarson, K. C., Jarret, L. and Finke, R. H. (1960). Embedding in epoxy resins for ultrathin sectioning in electron microscopy. *Stain Technol.* **35**, 313–323.
Rohde, M., Müller, E., Chhatwal, G. S. and Talay, S. R. (2003a). Host cell caveolae act as an entry-port for group A streptococci. *Cell. Microbiol.* **5**, 323–342.
Rohde, M., Püls, J., Buhrdorf, R., Fischer, W. and Haas, R. (2003b). A novel sheated surface organelle of the *Helicobacter pylori* cag type IV secretion system. *Mol. Microbiol.* **49**, 219–234.
Rohde, M., Graham, R. M., Borchers, P., Preuss, C., Branitzki-Heinemann, K., Schleicher, I., Zähner, D., Fulde, M., Talay, S., Dinkla, K. and Chhatwal, G. S. (2011). Differences in the aromatic domain of homologous streptococcal fibronectin-binding proteins trigger distinct cell invasion mechanisms and survival rates. *Cell Microbiol.* **13**, 450–468.
Rosenberg, M., Bartl, P. and Lesko, J. (1960). Water-soluble methacrylate as an embedding medium for the preparation of ultrathin sections. *J. Ultrastruct. Res.* **4**, 298–303.
Roth, J., Bendayan, M. and Orci, L. (1978). Ultrastructural localization of intracellular antigens by the use of Protein A-gold complex. *J. Histochem. Cytochem.* **26**, 1074–1081.
Roth, J., Bendayan, M., Carlemalm, E., Villiger, W. and Garavito, M. (1981). Enhancement of structural preservation and immunocytochemical staining in low temperature embedded pancreatic tissue. *J. Histochem. Cytochem.* **29**, 663–669.
Spurr, A. R. (1969). A low-viscosity epoxy resin embedding medium for electron microscopy. *J. Ultrastruct. Res.* **26**, 31–43.
Tokuyashu, K. T. (1986). Application of cryoultramicrotomy to immunecytochemistry. *J. Microsc.* **143**, 139–149.

Valentine, R. C., Shapiro, B. M. and Stadtman, E. R. (1968). Regulation of glutamine synthetase. XII. Electron microscopy of the enzyme from *Escherichia coli*. *Biochemistry* **7**, 2143–2152.

van Bruggen, E. F. J., Wiebinga, E. H. and Gruber, M. (1960). Negative-staining electron microsocpy of proteins at pH values below their isoelectric points. Its application to hemocyanin. *Biochim. Biophys. Acta* **42**, 171–172.

van Bruggen, E. F. H., Wiebinga, E. H. and Gruber, M. (1962). Structure and properties of hemocyanin. I. Electron micrographs of hemocyanin and apohemocyanin from *Helix pomatia* at different pH-values. *J. Mol. Biol.* **4**, 1–7.

Venable, J. H. and Coggeshall, R. (1965). A simplified lead citrate stain for use in electron microscopy. *J. Cell Biol.* **25**, 407–408.

von Ardenne, M. (1938). Das Elektronen-Rastermikroskop. Praktische Ausführung. *Zeitschrift für technische Physik* **19**, 407–416.

von Ardenne, M. (1939). Das Elektronen-Rastermikroskop. Theoretische Grundlagen. *Zeitschrift für Physik* **108**, 553–572.

Willingham, M. C. and Rutherford, A. V. (1984). The use of osmium-thiocarbyhydrazide-osmium (OTO) and ferrocyanide-reduced osmium methods to enhance membrane contrast and preservation in cultured cells. *J. Histochem. Cytochem.* **32**, 455–460.

5 Peptidoglycan Structure

Peter Schumann
Leibniz-Institut DSMZ-Deutsche Sammlung von Mikroorganismen und Zellkulturen GmbH, Braunschweig, Germany
E-mail: psc@dsmz.de

◆◆

CONTENTS

◆◆◆◆◆◆ I. INTRODUCTION

Peptidoglycan (less commonly referred to as 'murein') forms a mesh-like layer outside the cytoplasmic membrane, is responsible for rigidity and shape of bacterial cells and protects them from osmotic disruption. It is a polymer that occurs in cell walls of both Gram-positive and Gram-negative bacteria but has not been found in Archaea. In Gram-positive bacteria it represents *ca.* 30–70% of the cell wall besides polysaccharides, teichoic or teichuronic acids. It is only a minor component (<10%) in Gram-negative cell walls that mainly consist of lipopolysaccharides and lipoproteins. Since its first analysis, the peptidoglycan structure has received wide interest from those investigating the action of antibiotics and mechanisms of resistance, phage susceptibility and serological behaviour, immune responses, and last but not the least for the classification and identification of bacteria. The analysis of the peptidoglycan is hampered by the fact that it is a three-dimensionally cross-linked, and hence an insoluble polymer that must be either hydrolysed to amino acids, peptides, and amino sugars or digested to muropeptides (Young, 1998) in order to conclude structural information from its constituents. This analytical approach requires careful purification from associated polymers as well as cytoplasmic and membrane material because many contaminating polymers may give rise to the same degradation products as those resulting from the peptidoglycan itself.

While Gram-negative bacteria are uniform in their peptidoglycan structure (Schleifer and Kandler, 1972), there is a bewildering diversity in the structure of peptidoglycans of Gram-positive bacteria (see Table 1). For this reason, the peptidoglycan structure is an important taxonomic criterion for their

METHODS IN MICROBIOLOGY, VOLUME 38
0580-9517 DOI 10.1016/B978-0-12-387730-7.00005-X

Table 1. Nomenclature of peptidoglycan structures of cross-linkage types A and B

Type A of cross-linkage		
A11	A1α	L-Lys-direct
A11.gly	A1α	L-Lys-direct; α-carboxyl group of D-Glu substituted by Gly
A11.pep	A2	L-Lys peptide subunit
A11.1	A3α	L-Lys-Gly
A11.2	A3α	L-Lys-Gly_{5-6}
A11.3	A3α	L-Lys-Gly_{2-4}-L-Ser_{1-2}-Gly
A11.4	A3α	L-Lys-L-Ala
A11.5	A3α	L-Lys-L-Ala_2
A11.6	A3α	L-Lys-L-Ala_3
A11.7	A3α	L-Lys-L-Ala_4
A11.8	A3α	L-Lys-L-Ala-Gly_{4-5}
A11.9	A3α	L-Lys-L-Ala-L-Ala(L-Ser)
A11.10	A3α	L-Lys-L-Ala_2-L-Ala(L-Ser)
A11.11	A3α	L-Lys(L-Orn)-L-Ala_2-L-Ser
A11.12	A3α	L-Lys-L-Ala_2-Gly-L-Ala
A11.13	A3α	L-Lys-L-Ala-L-Ser
A11.14	A3α	L-Lys-L-Ser
A11.15	A3α	L-Lys-L-Ala_2-L-Ser
A11.16	A3α	L-Lys-L-Ala(L-Ser)-L-Ser
A11.17	A3α	L-Lys-L-Ala-L-Thr-L-Ala
A11.18	A3α	L-Lys-L-Ser-L-Ala
A11.19	A3α	L-Lys-L-Ser-L-Ser(L-Ala)
A11.20	A3α	L-Lys-L-Ser-L-Ala_2
A11.21	A3α	L-Lys(L-Orn)-L-Ala(L-Ser)-L-Ala_2
A11.22	A3α	L-Lys-L-Ser-L-Ala_{2-3}; α-carboxyl group of D-Glu substituted by glycine amide
A11.23	A3α	L-Lys-L-Ser-L-Thr-L-Ala
A11.24	A3α	L-Lys-L-Thr-Gly
A11.25	A3α	L-Lys-L-Thr-L-Ala
A11.26	A3α	L-Lys-L-Thr-L-Ala; α-carboxyl group of D-Glu substituted by alanine amide
A11.27	A3α	L-Lys-L-Thr-L-Ala_2
A11.28	A3α	L-Lys-L-Thr-L-Ala_3
A11.29	A3α	L-Lys-L-Thr-L-Ser(L-Ala)
A11.30	A3α	L-Lys-L-Thr-L-Ser-L-Ala_2
A11.31	A4α	L-Lys-D-Asp
A11.32	A4α	L-Lys(L-Orn)-D-Asp
A11.33	A4α	L-Lys-D-Glu
A11.34	A4α	L-Lys-L-Ala-D-Asp
A11.35	A4α	L-Lys-L-Ala-L-Glu
A11.36	A4α	L-Lys-L-Ser-D-Asp
A11.37	A4α	L-Lys-D-Ser-D-Asp
A11.38	A4α	L-Lys-D-Ser-D-Glu
A11.39	A4α	L-Lys-L-Ser_2-D-Glu
A11.40	A4α	L-Lys-Gly-D-Glu
A11.41	A3α	L-Lys-Gly-L-Ala_3
A11.42	A4α	L-Lys-Gly-D-Asp

(*Continued*)

Table 1. (Continued)

A11.43	A4α	L-Lys-D-Glu_2
A11.44	A3α	L-Lys-Gly(L-Ser)
A11.45	A4α	L-Lys-L-Thr-D-Asp
A11.46	A3α	L-Lys-L-Ala-Gly-L-Ala_2
A11.47	A3α	L-Lys-L-Ala-Gly; α-carboxyl group of D-Glu substituted by alanine amide
A11.48	A4α	L-Lys-L-Ser-D-Glu
A11.49	A3α	L-Lys-L-Ser-Gly
A11.50	A3α	L-Lys-L-Ala-Gly
A11.51	A5α	L-Lys-L-Lys-D-Glu
A11.52	A5α	L-Lys-L-Lys-D-Asp
A11.53	A5α	L-Lys-L-Ala-L-Lys-D-Glu
A11.54	A4α	L-Lys-L-Glu
A11.55	A5α	L-Lys(L-Orn)-L-Lys-D-Glu
A11.56	A4α	L-Lys-Gly-L-Glu
A11.57	A4α	L-Lys-L-Thr-D-Glu
A11.58	A4α	L-Lys-L-Ser-L-Glu
A11.59	A4α	L-Lys-L-Ala-D-Glu
A11.60	A4α	L-Lys-D-Asp; α-carboxyl group of D-Glu substituted by Gly
A11.61	A3α	L-Lys-L-Ala(L-Ser)
A11.62	A3α	L-Lys-L-Ala_2-Gly_{2-3}-L-Ala(Gly)
A11.63	A3α	L-Lys-L-Ala-Gly_3, L-Ser_{0-1}
A12.1	A3α′[a]	L-Lys-Gly
A12.2	A4α′	L-Lys-D-Glu
A12.3	A4α′	L-Lys-Gly-D-Asp
A12.4	A3α′	L-Lys-L-Ser
A21.1	A3β	L-Orn-Gly_{2-3}
A21.2	A3β	L-Orn(L-Lys)-L-Ala_{2-3}
A21.3	A3β	L-Orn-L-Ser-L-Ala-L-Thr-L-Ala
A21.4	A4β	L-Orn-D-Asp
A21.5	A4β	L-Orn-D-Glu
A21.6	A4β	L-Orn(L-Lys)-D-Glu
A21.7	A4β	L-Orn-D-Ser-D-Asp
A21.8	A4β	L-Orn(L-Lys)-D-Ser-D-Asp
A21.9	A4β	L-Orn-L-Ser-D-Glu
A21.10	A3β	L-Orn-β-Ala
A21.11	A5β	L-Orn-L-Lys-D-Glu
A21.12	A4β	L-Orn-D-Ser-D-Glu
A21.13	A4β	L-Orn-Gly_2-D-Glu
A21.14	A4β	L-Orn-L-Ala-Gly-D-Asp
A22.1	A4β′	L-Orn-D-Asp
A23.1	A4β	L-Orn-L-Glu; position 1 of the peptid subunit is L-Ser
A31	A1γ	*meso*-Dpm-direct[b]
A31.1	A4γ	*meso*-Dpm-D-Glu_2
A31.2	A4γ	*meso*-Dpm-D-Glu_2; α-carboxyl group of D-Glu substituted by Gly
A31.3	A4γ	*meso*-Dpm-D-Asp-D-Glu; α-carboxyl group of D-Glu substituted by Gly
A32.1	A1γ′	*meso*-Dpm-direct
A41.1	A3γ	LL-Dpm-Gly

(Continued)

Table 1. (Continued)

A41.2	A3γ	LL-Dpm-Gly_3; α-carboxyl group of D-Glu substituted by Gly
A42.1	A3γ′	LL-Dpm-Gly
A51	A1δ	Lan-direct
Type B of cross-linkage		
B1	B1α	{Gly} [L-Lys] D-Glu(Hyg)-Gly-L-Lys[c]
B2	B1β	{Gly} [L-Hse] D-Glu(Hyg)-Gly_2-L-Lys
B3	B1γ	{Gly} [L-Glu] D-Glu(Hyg)-Gly_2-L-Lys
B4	B2α	{L-Ser} [L-Orn] D-Glu-D-Lys(D-Orn)
B5	B2β	{Gly} [L-Hse] D-Glu-D-Orn
B6	B2β	{Gly} [L-Hse] D-Glu(Hyg)-Gly-D-Orn
B7	B2γ	{Gly} [L-Dab] D-Glu-D-Dab
B8	B1β	{Gly} [L-Hse] D-Glu-Gly-L-Dab
B9	B2α	{Gly} [L-Orn] D-Glu-D-Orn
B10	B2β	{Gly} [L-Hse] D-Glu-D-Dab
B11	B2δ	{Gly} [L-Ala] D-Glu-D-Dab-L-Thr
B12	B2α	{Gly} [L-Orn] D-Glu-Gly-D-Orn
B13	B1δ	{L-Ser} [L-Ala] D-Glu-L-Asp-L-Lys
B14	B2β	{Gly} [L-Hse] D-Glu-D-Lys
B15	B1δ	{L-Ser} [L-Ala] D-Glu-Gly-L-Lys-L-Lys

[a]Extension by an apostrophe (e.g. A3α′) denotes that the L-Ala residue found normally at position 1 of the peptide subunit in type A is replaced by Gly.
[b]*Abbreviations*: Dab, 2,4-diaminobutyric acid; Dpm, 2,6-diaminopimelic acid; Hse, homoserine; Hyg, *threo*-3-hydroxy-glutamic acid; Lan, lanthionine.
[c]{ }, position 1; [ ], position 3.

differentiation. Pioneering work on the elucidation of the peptidoglycan structure dates back to the 1960s (e.g. Ghuysen, 1968; Schleifer and Kandler, 1967). The achievements of these studies and the relevance of the peptidoglycan diversity for bacterial taxonomy were summarized in the review of Schleifer and Kandler (1972). Methods for isolation of peptidoglycan and analyses of its structure for bacterial taxonomy are compiled in additional reviews (Hancock, 1994; Komagata and Suzuki, 1987; Rosenthal and Dziarski, 1994; Schleifer, 1985; Schleifer and Seidl, 1985). Information on the peptidoglycan structure is considered a recommended criterion for the description of new taxa of Gram-positive bacteria (Tindall *et al.*, 2010), in particular of members of the suborder *Micrococcineae* (Schumann *et al.*, 2009), staphylococci (Freney *et al.*, 1999) and aerobic endospore-forming bacteria (Logan *et al.*, 2009). As currently only few laboratories worldwide are able to fulfil these requirements in all details, the aim of this chapter is to provide a set of detailed and feasible protocols for using contemporary techniques for the analysis of the peptidoglycan structure.

♦♦♦♦♦♦ II. PRIMARY STRUCTURE OF THE PEPTIDOGLYCAN

Peptidoglycan is a heteropolymer that consists of glycan strands that are cross-linked by peptides. The glycan backbone is composed of alternating units of *N*-acetylglucosamine and *N*-acetylmuramic acid linked by β-1,4-glycosidic bonds. The only taxonomically relevant variation within the glycan strands is

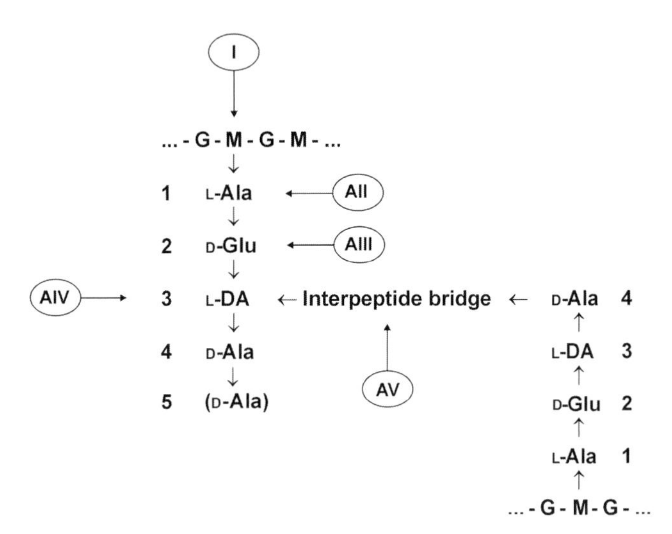

Figure 1. Scheme of the structure of the peptidoglycan A-type (cross-linked between positions 3 and 4). Labels I and AII to AV indicate the sites of structural variations.
I: partial substitution of *N*-acetyl by *N*-glycolyl groups at muramic acid residues
AII: substitution of L-Ala by Gly
AIII: α-carboxyl group of D-Glu either free or bound to Gly, glycinamide, D-alaninamide, cadaverine or putrescine
AIV: L-DA, L-diamino acid; *meso*-2,6-diaminopimelic acid, LL-2,6-diaminopimelic acid, L-Lys, L-Orn, lanthionine, 2,6-diamino-3-hydroxypimelic acid, hydroxylysine or *threo*-β-hydroxyornithine
AV: 1–7 amino acid residues, e.g. Gly, L-Ala, L-Ser, L-Thr, β-Ala, Glu, Asp, L-Lys

the partial substitution of *N*-acetyl by *N*-glycolyl groups of muramic acid residues (Figures 1 and 2), which was originally observed by Adam *et al.* (1969). After setting up colourimetric tests for the screening of bacterial cells for glycolate, the presence of *N*-glycolyl groups in the peptidoglycan became an important criterion for differentiation of genera (Uchida and Aida, 1977, 1979, 1984; Uchida and Seino, 1997; Uchida *et al.*, 1999). The lactyl groups of the muramic acid residues are the anchoring points for the peptide subunits that contain four alternating L- and D-amino acids (Figures 1 and 2). Occasionally, a second C-terminal D-alanine residue is found at position 5 of the peptide subunit. In the majority of bacteria positions 2 and 3 of the peptide subunit are occupied by amino acids with three functional groups (D-glutamic acid or *threo*-3-hydroxy-glutamic acid [Hyg] at position 2 [Figure 2], an L-diamino acid at position 3 [Figure 1]), which may serve as 'T-piece connectors' for cross-linkage with the adjacent peptide subunit. Peptidoglycans were divided into two main types (A and B) based on the way of their cross-linkage (Schleifer and Kandler, 1972). In the case of type A (Figure 1), the cross-linkage is accomplished by linking the ω-amino group of the diamino acid at position 3 to the carboxylic group of D-alanine at position 4 of the adjacent peptide subunit either directly (type A1; Schleifer and Kandler, 1972) or by means of an interpeptide bridge consisting of 1–7 amino acid residues (Figure 1).

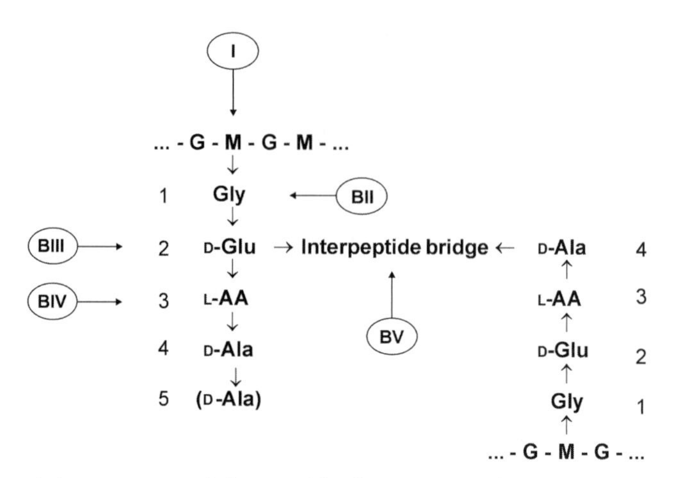

Figure 2. Scheme of the structure of the peptidoglycan B-type (cross-linked between positions 2 and 4). Labels I and BII to BV indicate the sites of structural variations.
I: partial substitution of *N*-acetyl by *N*-glycolyl groups at muramic acid residues
BII: substitution of Gly by L-Ser
BIII: substitution of D-Glu by *threo*-3-hydroxyglutamic acid
BIV: L-AA, L-amino acid; L-homoserine, L-Ala, L-Glu, L-Orn, L-Lys, L-2,4-diaminobutyric acid
BV: type B1 containing an L-diamino acid (L-Lys or L-2,4-diaminobutyric acid); type B2 containing a D-diamino acid (D-Orn, D-Lys or D-2,4-diaminobutyric acid); residues of Gly, L-Thr or L-Asp may occur in addition

In the case of type B the α-carboxylic group of D-glutamic acid or Hyg (Schleifer *et al.*, 1967) at position 2 is involved in the cross-linkage (Figure 2). This type of cross-linkage always requires the presence of a diamino acid in the interpeptide bridge (BV; Figure 2) in order to provide a free amino group for linkage to the carboxylic group of the adjacent peptide subunit.

Several elements of the peptidoglycan may undergo characteristic structural modifications that give rise to more than 100 structural types (see Table 1; Figures 1 and 2; Schleifer and Kandler, 1972; Schleifer and Seidl, 1985). Because the nomenclature system proposed by Schleifer and Kandler (1972) did not allow a biunique affiliation of peptidoglycan variants and of novel structures unravelled more recently, Norbert Weiss (curator at DSMZ until his retirement in 2003) established the 'Key to Murein Types' (www.peptidoglycan-types.info; Table 1) for the DSMZ—Catalogue of Strains.

♦♦♦♦♦♦ III. ANALYTICAL APPROACHES FOR THE ELUCIDATION OF THE PEPTIDOGLYCAN STRUCTURE

A. Information from Whole-Cell Hydrolysates

In most cases, the investigation of the variable sites of the peptidoglycan structure requires the time-consuming isolation and purification of peptidoglycan.

However, by rather simple methods important information for the classification and identification can already be obtained from whole bacterial cells: information on the presence of isomers of 2,6-diaminopimelic acid (Dpm) and of 2,6-diamino-3-hydroxypimelic acid (OH-Dpm), on the occurrence of glycolyl residues in the glycan backbone and on diagnostic whole-cell sugars.

1. Detection of Dpm isomers and of OH-Dpm

Two chemical features of Dpm isomers (*meso*- and LL-Dpm) and OH-Dpm allow their detection already in whole-cell hydrolysates: (1) Dpm and OH-Dpm occur exclusively in the peptidoglycan, and (2) they can be easily separated from all proteinogenic amino acids and other hydrolysis products of bacterial cells by thin-layer chromatography (TLC). Protocol 1 is recommended to screen bacterial cells for the presence of Dpm isomers or OH-Dpm but is suited for the analysis of amino acids in peptidoglycan preparations (Section III.C.1), too.

Protocol 1: Analysis of Dpm isomers and OH-Dpm in whole-cell hydrolysates

- Add either *ca.* 2–3 mg lyophilized cells, two loops of wet biomass or 1 mg peptidoglycan to 200 μl 4.0 N HCl in a 2 ml glass ampoule (Gerresheimer Querétaro, Mexico, P/N 12050U-2/untreated).
- Heat-seal the ampoule and keep it for 16 h at 100°C in a drying oven.
- Open the ampoule after cooling and filter its content through a tiny amount of charcoal (tip of a micro-spatula) for removing coloured by-products. For this purpose a pipette tip (Eppendorf, P/N 0030000.919) is closed by a small bead made of filter paper before adding the charcoal. Press the hydrolysate gently through the charcoal bed by using a pipette bulb.
- Dry down the hydrolysate in a gentle air-stream at 35°C. Re-dissolve the residue in 200 μl distilled water and dry again. Repeat this procedure to remove all traces of acid.
- Dissolve the residue in 100 μl of distilled water. Apply *ca.* 2 μl of the solution to a spot on the baseline of a 20 cm long stripe of a cellulose TLC plate (Merck, P/N 1.05577) using a glass capillary. Two diamino acid standards, the first consisting of Dpm (Fluka, P/N 33240; containing both *meso*- and LL-Dpm), DL-lysine dihydrochloride (Aldrich, P/N 62910) and DL-ornithine monohydrochloride (Sigma, P/N O-2250), the second consisting of DL-2,4-diaminobutyric acid dihydrochloride (Sigma, P/N D-3758) (all 5 mg/ml distilled water) are applied in the same way. The standard of OH-Dpm can be obtained by hydrolysis of cells of, e.g. *Rugosimonospora acidiphila* DSM 45227^T (=NBRC 104874^T).
- Fill a TLC tank lined with filter paper with the developing solvent of Rhuland *et al.* (1955) consisting of 32 ml methanol (Sigma–Aldrich, P/N 32213), 4 ml pyridine (Sigma, P/N P3776), 7 ml distilled water and 1 ml 12.0 N HCl (Sigma–Aldrich, P/N 320331) for equilibration at ambient temperature the day before use.
- Develop the chromatogram until the solvent front reached almost the top of the plate. The TLC plate is dried in a fume hood at room temperature for 1 h before spraying with ninhydrin reagent (1.25 mg ninhydrin [Fluka, P/N

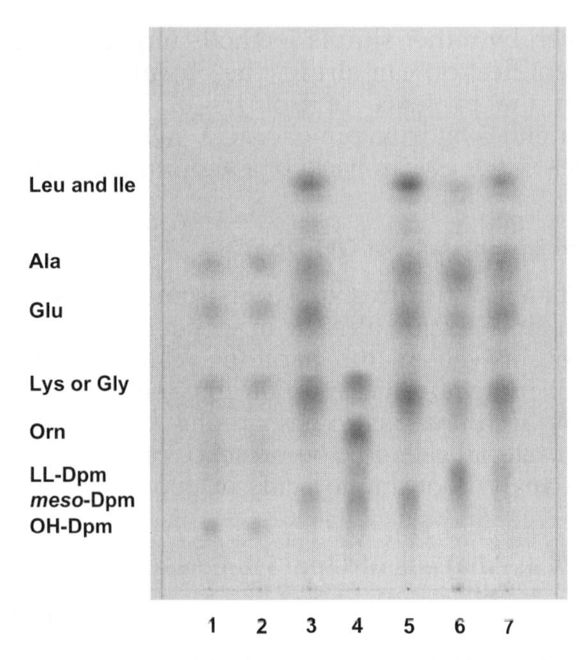

Figure 3. Thin-layer chromatogram of hydrolysates (4.0 N HCl, 100°C, 16 h) of peptidoglycan preparations (lanes 1 and 2) and whole cells (lanes 3, 5–7) and of a standard (lane 4) consisting of *meso*-Dpm, LL-Dpm, Orn and Lys (each 5 mg/ml). Chromatographic conditions see Protocol 1.
Lane 1: *Rugosimonospora acidiphila* DSM 45227^{T} (containing OH-Dpm)
Lane 2: *Rugosimonospora africana* DSM 45228^{T} (containing OH-Dpm)
Lane 3: *Brevibacterium frigoritolerans* DSM 8801^{T} (containing *meso*-Dpm)
Lane 5: *Bacillus subtilis* subsp. *subtilis* DSM 10^{T} (containing *meso*-Dpm)
Lane 6: *Kribbella antibiotica* DSM 15501^{T} (containing LL-Dpm)
Lane 7: *Kribbella flavida* DSM 17836^{T} (containing LL-Dpm)
Abbreviations: OH-Dpm, 2,6-diamino-3-hydroxypimelic acid; Dpm, 2,6-diaminopimelic acid. (see colour insert)

33437], 233 ml water-saturated 1-butanol [Sigma–Aldrich, P/N 360465], 17 ml acetic acid [Sigma–Aldrich, P/N 242853]). The spots appear after heating at 100°C for 5 min in an oven.

Isomers of Dpm and OH-Dpm migrate much slower than other amino acids. They can be identified by low Rf values (Figure 3) and by their characteristic green-greyish colour after reaction with ninhydrin. After storing the developed TLC plates in the dark for at least one day, the colour of the spots of Dpm turn to yellow and are rather durable while the spots of other amino acids appear in the colour shade of blue to pinkish and fade quickly. LL-Dpm runs slightly faster than *meso*-Dpm on the cellulose plates. OH-Dpm is migrating slower than both Dpm isomers and occurs usually in combination with *meso*-Dpm (see Figure 3; Rf values: OH-Dpm, *ca.* 0.13; *meso*-Dpm, *ca.* 0.17; LL-Dpm, *ca.* 0.22).

The presence of Dpm isomers or OH-Dpm is a very important criterion and supports the affiliation of unknown isolates to genera when combined with data of a polyphasic characterization (Tindall *et al.*, 2010). The occurrence of LL-Dpm allows already the tentative affiliation of isolates to certain genera of the families *Streptomyceataceae*, *Nocardioidaceae*, *Intrasporangiaceae* and *Propionibacteriaceae* but gives no unambiguous identification of the peptidoglycan structure, as there are three variations of the peptidoglycan type A3γ: A41.1, A41.2 and A42.1 (Table 1). The occurrence of *meso*-Dpm has been reported up till now only for the directly cross-linked peptidoglycan types A1γ = A31 and A1γ' = A32.1 (Kawamoto *et al.*, 1981) as well as for the three variations A31.1, A31.2 and A31.3 of the peptidoglycan type A4γ based on *meso*-Dpm combined with an interpeptide bridge (Table 1). The variations of type A4γ have been found so far exclusively in members of the genera *Brachybacterium*, *Devriesea* and *Dermabacter* (Schumann *et al.*, 2009). If a close relationship to members of the family *Dermabacteraceae* can be excluded by 16S rRNA gene sequence analysis, the organism with *meso*-Dpm under study shows either the peptidoglycan type A31 or A32.1.

2. Detection of glycolic acid

Glycolic acid occurs in bacterial cells exclusively linked to muramic acid of the glycan backbone and its molar amount is equivalent to those of other constituents of the peptidoglycan like, e.g. muramic acid, glucosamine and D-glutamic acid. Only *N*-glycolyl but no *O*-glycolyl residues have been found so far (Uchida and Seino, 1997). For these reasons it is possible to examine whole-cell hydrolysates of bacteria for the taxonomically important 'acyl type'. It is worth mentioning that results are easier to evaluate when peptidoglycan preparations are available for performing the test as the red-purple colour indicating the presence of glycolic acid is less disturbed by brownish contaminations occasionally resulting from whole-cell hydolysates. The following protocol was adapted from the diethylether-extraction method described by Uchida *et al.* (1999).

Protocol 2: Analysis of glycolic acid in whole-cell hydrolysates

- Hydrolyse either 2 mg lyophilized cells or 1 mg peptidoglycan with 100 μl 6.0 N HCl in a sealed 2 ml glass ampoule at 100°C for 2 h.
- Open the ampoule after cooling and transfer the hydrolysate to a test tube (Fiolax 10 × 100; Schott P/N 261100606). Extract with 1 ml of water-saturated diethylether (Sigma–Aldrich, P/N 346136) twice. The first etheric extract (upper phase) usually contains contaminations and is discarded. The second extract is collected in a separate test tube.
- Add 2 μl of 0.1 N NaOH to the extract, and evaporate the ether in a gentle stream of nitrogen (not air, to exclude formation of explosive peroxides!).
- Dissolve the residue in 500 μl of a freshly prepared 2,7-dihydroxynaphthalene (Sigma, P/N D-8628) solution (0.02%, w/v, in conc. H_2SO_4 [Sigma–Aldrich, P/N 320501]).
- Heat the samples in a boiling water bath for 10 min. The red-purple colour indicates a positive glycolyl test.

Actinobacteria with the 'glycolyl type' contain up to *ca.* 150 nmol glycolyl residues per 1 mg cells, those of the 'acetyl type' less than *ca.* 10 nmol. The glycolyl test is considered positive for a colour intensity that corresponds to 35 nmol or more glycolyl residues/mg cells (Uchida and Seino, 1997).

The test results can be evaluated by visual comparison with the colour obtained for reference organisms (glycolyl type: e.g. *Rhodococcus*, *Micromonospora*, *Microbacterium*; acetyl type: e.g. *Streptomyces*, *Cellulomonas*, *Brevibacterium*) or quantitatively by measurement of the absorbance at 530 nm on the basis of a calibration curve obtained by using a sodium glycolate standard solution. All standards must be subjected to the same sample preparation as used for the bacterial cells.

3. Analysis of whole-cell sugars

While Dpm isomers, OH-Dpm, and glycolic acid always originate from peptidoglycan, sugars occurring in whole-cell hydrolysates may also come from a number of sources including cell wall associated or capsular polysaccharides, glycolipids, nucleic acids or carbohydrate storage products. The composition of whole-cell sugars may vary depending on cultivation conditions, and therefore they are less stable chemotaxonomic markers than those derived exclusively from the structurally conserved peptidoglycan. Despite this fact, the whole-cell sugar composition is frequently reported in taxonomic descriptions and serves as one of the criteria of the definition of 'cell-wall chemotypes' for actinobacteria (Lechevalier and Lechevalier, 1970). The sugar composition of purified cell walls is possibly a more conservative characteristic but there is only a limited amount of published data for comparison.

Xylose, galactose, arabinose and madurose (3-*O*-methyl-D-galactose, Lechevalier and Gerber, 1970) are the 'diagnostic sugars' for actinobacteria (Lechevalier and Lechevalier, 1970). In addition, rhamnose, glucose, ribose, mannose, fucose (6-deoxy-galactose), tyvelose (3,6-dideoxy-D-mannose) and 2-*O*-methyl-D-mannose are whole-cell sugars reported in descriptions of new taxa. Whole cells should be hydrolysed using sulfuric acid instead of hydrochloric acid in order to exclude the formation of deoxychlorosugars. Sulfuric acid can be removed prior to chromatography by precipitation with barium hydroxide (Lechevalier and Gerber, 1970) or conveniently by extraction with *N,N*-dioctylmethylamine in chloroform (Whiton *et al.*, 1985). The analysis of whole-cell sugars is performed using paper chromatography (PC; Lechevalier and Gerber, 1970), TLC on cellulose plates (Staneck and Roberts, 1974), gas chromatography (GC; Saddler *et al.*, 1991) or HPLC (Yokota and Hasegawa, 1988). Protocol 3 for analysis of whole-cell sugars by TLC is adapted from the method of Staneck and Roberts (1974). Protocol 4 describes the GC analysis of alditol acetates obtained by borohydride reduction and subsequent acetylation of sugars as described by Saddler *et al.* (1991). The TLC method is suited for a rapid screening of isolates for their whole-cell sugars. The GC method is more sensitive, allowing the quantification of sugars and their identification by GC/MS.

Protocol 3: Analysis of whole-cell sugars by TLC

- Hydrolyse 15 mg lyophilized cells (or 3–4 loops wet biomass) with 1 ml of 1.0 N H_2SO_4 at 100°C for 2 h in a screw-cap vial with teflon-lined septum.
- Add 4 ml distilled water after cooling and mix.
- Add 5 ml 20% (v/v) *N*-methyldioctylamine (Aldrich, P/N 42430) in chloroform and mix using a Vortex mixer.
- Centrifuge the suspension at 1700 × *g* (*g*, acceleration of gravity) for 15 min. Remove and discard the chloroform phase (bottom). Check the neutral reaction of the hydrolysate (upper phase) by pH 1–12 test paper (duotest; Macherey-Nagel, P/N 90301) and repeat the extraction in the case that an acidic reaction indicates residual H_2SO_4.
- Dry down the hydrolysate in a gentle air-stream at 35°C. Process the residue as follows or use it for sample preparation for GC analysis (Protocol 4).
- Dissolve the residue in 100 μl distilled water. Apply *ca.* 1 μl of the solution 2, 4 or 6 times to each of three spots at the baseline of a 20 cm × 20 cm cellulose TLC plate (Merck, P/N 1.05577).
- Apply *ca.* 3 μl of two standards adjacent to the sample spots. Standard I contains rhamnose, arabinose and galactose. Standard II contains mannose, ribose, glucose and xylose (all analytical grade; 1 mg/ml distilled water).
- Fill a TLC tank lined with filter paper with the developing solvent consisting of 50 ml 1-butanol, 30 ml pyridine and 20 ml distilled water for equilibration at ambient temperature the day before use.
- Develop the chromatogram until the solvent front reached almost the top of the plate. Dry the TLC plate in a hood for 1 h before spraying with aniline phthalate reagent (0.8 g phthalic acid [Sigma–Aldrich, P/N 402915], 2 ml distilled water, 24 ml 1-butanol, 24 ml diethylether, and 0.5 ml aniline [Sigma–Aldrich, P/N 242284]). The spots become visible after heating at 100°C for 10 min.

The sugars migrate in the following order (increasing Rf values): galactose, glucose, arabinose/mannose, xylose, ribose and rhamnose. Due to co-migrating arabinose and mannose the use of two standard mixtures (I and II, see Protocol 3) is recommended. Detection with aniline phthalate reagent allows the differentiation of pentoses (reddish brown spots) and hexoses (yellowish brown spots). Madurose and fucose show the same Rf value as xylose but can be distinguished by the yellowish brown colour of their spots. The differentiation of madurose and fucose requires additional analyses, e.g. by HPLC (Promnuan *et al.*, 2011) or GC/MS of derivatives obtained by Protocol 4.

Protocol 4: Analysis of whole-cell sugars by GC

- Use the neutralized sugar hydrolysate (see Protocol 3) and dissolve it in 0.2 ml 28–30% ammonium hydroxide solution (Sigma–Aldrich, P/N 320145) in a screw-cap tube with teflon-lined septum.
- Add 0.1 ml of a freshly prepared solution of 10% (m/v) sodium borohydride (Sigma–Aldrich, P/N S9125) in 3 M ammonium hydroxide solution, seal tightly, mix and allow to react at 37°C for 1 h.

- Cool to ambient temperature, open the screw-cap tube precautiously (increased pressure, foaming!) and add acetic acid dropwise until there is no more hydrogen formation.
- Transfer 0.4 ml of the mixture to a fresh screw-cap tube and cool in an ice bath.
- Add 0.3 ml 1-methylimidazol (Aldrich, P/N M50834).
- Add 2 ml acetic anhydride (Sigma–Aldrich, P/N 539996) precautiously and dropwise (exothermic reaction!) and seal with a screw-cap and teflon-lined septum tightly.
- Mix and allow reaction to proceed at ambient temperature for 15 min.
- Cool in an ice bath and add 5 ml of distilled water.
- Extract with 1 ml dichloromethane (Aldrich, P/N 414751) by using an overhead-mixer for 3 min. Allow phases to separate. Remove the dichloromethane phase (bottom) and dry it over anhydrous sodium sulfate. Discard the upper phase.
- Subject the solution of alditol acetates in dichloromethane to capillary gas chromatography under the following conditions: column PERMABOND-FFAP 25 m $\times$ 0.25 mm ID (Macherey-Nagel, P/N 723116.25); temperature program: hold for 10 min at 220°C before raising to 230°C by a rate of 1 K/min; carrier gas, helium, 20 cm/s; injector temperature, 270°C; temperature of the flame ionization detector, 300°C; sample volume, 1 μl.

The derivatives of sugars elute in the following order (increasing retention times): Rhamnose, ribose, arabinose, xylose, mannose, madurose, galactose, and glucose. The separation of the peaks of galactose and glucose may be incomplete, depending on the separating capacity of the column.

B. Preparation of Peptidoglycan

When isomers of Dpm and OH-Dpm cannot be detected in whole-cell hydrolysates, the content of peptidoglycan in the bacterial cell wall is either too low for the detection of these diaminoacids or the peptidoglycan of the strain under study contains another diamino acid instead of Dpm or OH-Dpm. In both cases, it is recommended to isolate the peptidoglycan.

The isolation of the peptidoglycan requires steps to remove contaminating components that originate from the cytoplasma and the cytoplasmatic membrane and to detach covalently linked polymers. It depends on the required information on the peptidoglycan structure whether all associated polymers (e.g. nucleic acids, proteins, teichoic acids, lipoteichoic acids, teichuronic acids, polysaccharides) must be completely removed by laborious procedures. As taxonomic decisions are mainly based on the qualitative and quantitative content of amino acids as well as on their arrangement, lower amounts of remaining nucleic acids or polysaccharides may be tolerated but it is essential to remove all amino acid containing contaminations, in particular proteins and teichoic acids. The product obtained by Protocols 5, 7 and 8 is referred to as 'peptidoglycan' in this chapter but be aware of the possibility that it might contain remaining associated polymers. Excessive amounts of alanine or glutamic acid in protein-free peptidoglycan preparations indicate the presence of higher amounts of teichoic acids, which can be removed by cleavage of their phosphodiester bonds with hydrogen fluoride (see Protocol 8).

Erroneous reports on peptidoglycan structures are in most cases due to incomplete removal of proteins. The presence of non-peptidoglycan amino acids, e.g. leucine, isoleucine or phenylalanine in hydrolysates of peptidoglycan preparations indicates contaminating proteins (compare chromatograms of whole-cell hydrolysates and hydolysates of peptidoglycan preparations in Figure 3).

I. Gram-negative bacteria

The preparation of peptidoglycan of Gram-negative bacteria is rather uncomplicated, because it is covalently attached only to a lipoprotein or even occurs as free polymer *in situ* (Rosenthal and Dziarski, 1994). Peptidoglycan is usually collected as an insoluble material remaining after Gram-negative cells are dissolved in hot detergents, which can be purified by proteinase treatment and washing. However, the remarkably lower content of peptidoglycan in comparison to Gram-positive bacteria must be considered for the amount of starting material. According to Rosenthal and Dziarski (1994), the production of 50 mg peptidoglycan of gonococci requires an amount of cells obtained from nine 1.5 l cultures. Special care is needed to protect the low amounts of peptidoglycan of Gram-negative cells from hydrolytic autolysins by a rapid harvest using a refrigerated centrifuge (4°C) and ice-cooling before treatment with the detergent. Protocol 5 is adapted from the methods of Schleifer (1985) and Young (1998).

Protocol 5: Preparation of peptidoglycan of Gram-negative bacteria

- Suspend the cells to 0.2 g (wet-weight)/ml in ice-cold water.
- Add the cell suspension dropwise to an equal volume of boiling 8% (w/v) sodium dodecyl sulfate (SDS; Sigma, P/N L4390) solution under vigorous stirring. The temperature of the mixture must not fall below 90°C.
- Boil and stir for 30 min and let the suspension cool to ambient temperature overnight.
- Pellet the rough peptidoglycan by centrifugation at 78,000 × *g* at 20°C for 20 min. The pellet should be translucent. White opaque material must be retreated with boiling SDS solution.
- Wash with distilled water and centrifugation at 20°C until there is no more foaming (at least three times).
- Resuspend the pellet in 2.5 ml 0.01 M Tris–HCl, 0.01 M NaCl (pH 7.5) and add imidazol (Sigma–Aldrich, P/N 12399) in 0.01 M Tris–HCl to a final concentration of 0.32 M to inhibit hydrolases.
- Add Pronase E (Merck, P/N 1.07433) from a stock solution (10 mg/ml) to give a final concentration of 200 μg/ml. The pronase stock solution must be pretreated for 2 h at 60°C to inactivate lysozyme contaminations.
- Incubate 2 h at 60°C to release covalently bound lipoproteins.

- Add the sample to an equal volume of boiling 8% SDS solution and boil for 15 min.
- Cool to room temperature and wash with distilled water and centrifuge at 20°C until there is no more foaming (at least three times).
- Lyophilize the pellet overnight.

2. Gram-positive acid-fast bacteria

Acid-fast bacteria (e.g. members of the families *Corynebacteriaceae, Mycobacteriaceae* and *Nocardiaceae*) typically exhibit the peptidoglycan type A1γ = A31 (Schleifer and Kandler, 1972) based on *meso*-Dpm. They represent a special group of Gram-positives that raises additional problems for the preparation of peptidoglycan because of its covalent linkage to a diverse series of lipophilic constituents (e.g. arabinogalactan mycolate), carbohydrates and polypeptides (Hirschfield *et al.*, 1990; Petit *et al.*, 1975). Lipid layers inhibit the disintegration of cells as well as the access of proteolytic enzymes and should be removed either by stirring with chloroform/methanol (2:1, v/v) overnight at room temperature (Minnikin *et al.*, 1977) or by using Protocol 6 that is based on the procedure of Card (1973).

Protocol 6: Removal of lipids prior to preparation of purified cell walls of Gram-positive bacteria

- Suspend *ca.* 1 g wet-weight of cells in 10 ml of aqueous 0.3% (w/v) NaCl solution and add the suspension to 100 ml methanol.
- Reflux over a boiling water bath for 5 min.
- Cool and add 50 ml chloroform and 30 ml 0.3% NaCl solution in order to obtain a final ratio of chloroform:methanol:NaCl solution of 1:2:0.8 (v/v).
- Stir the suspension at ambient temperature for at least 2 h before transferring it to a separatory funnel.
- Add chloroform and 0.3% NaCl solution to give final concentrations of chloroform–methanol–NaCl solution of 2:2:1.8 (v/v). Shake and allow the phases to separate.
- Remove the chloroform phase (bottom). Collect the cells suspended in the aqueous phase by centrifugation at 10,000 × *g*. Use the cell pellet for Protocol 7.

3. Gram-positive non-acid-fast bacteria

The methods for isolation and purification of the highly diverse and hence taxonomically distinctive peptidoglycans of Gram-positive non-acid-fast bacteria are described in more detail. The high resistance to mechanical forces hampers the preparation of peptidoglycan of Gram-positive cells. Efficient mechanical disruption is a crucial prerequisite for the success of the subsequent steps, i.e. enzymatic digestion of proteins and extraction processes. Peptidoglycan preparations obtained by extraction procedures without mechanical disruption (e.g. by treatment with hot trichloroacetic acid solution; Park and Hancock, 1960; Schleifer and Kandler, 1972) are usually not pure enough for elucidation of the structure. Therefore, the isolation and purification of peptidoglycan of Gram-positive bacteria require a set of consecutive procedures that are discussed in the following sections.

a Growing the cells

The fact that variations in the cultivation conditions usually have no significant influence on the peptidoglycan structure, underlines the suitability of the peptidoglycan structure as taxonomic marker. The cultivation conditions and the composition of media must support high yields of biomass from the harvest of bacteria at the end of the exponential phase. Excessive amounts of certain amino acids in the culture media should be avoided because culture-dependent modifications of the interpeptide bridge of staphylococci and of substituents at the α-carboxyl group of D-glutamic acid may be caused by media enriched in glycine, L-serine, or L-alanine content (Schleifer and Kandler, 1972).

b Inactivation of autolytic enzymes

Autolytic enzymes may alter the structure and reduce the yield of peptidoglycan (Weidel *et al.*, 1963). These enzymes can be inhibited by boiling the cells for 10 min (Schleifer, 1985) or autoclaving them at 120°C for 20 min. The latter procedure might be helpful to inactivate endo-β-*N*-acetylglucosaminidase resistant to 100°C and avoids risk of infection when analysing pathogenic organisms.

c Mechanical disruption of cells

Though several methods have been suggested for the disruption of cells (e.g. French pressing, ultra-sonication, freezing–thawing cycles) the most efficient procedure seems to be the shaking of wet biomass with glass beads of 0.1–0.3 mm diameter at 10–15°C using commercial devices (e.g. Vibrogen Cell Mill, Edmund Bühler or MSK Cell Homogenizer, B. Braun International). The conditions (diameter of beads, duration of shaking) must be optimized for the size and shape of cells. Small coccoid cells are most difficult to disintegrate, requiring the smallest beads and the longest shaking period (up to *ca.* 30 min). The consistency of the cell–beads–water mixture has a remarkable effect on the efficiency of the cell disruption, which can be monitored either by phase-contrast microscopy or by the loss of positive Gram-staining. Optimally, the wet cell mass is mixed with the beads to give an evenly stiff paste before starting the disintegration process. The fraction of undisrupted cells can be separated from the cell debris by slow (*ca.* 350 × *g*) centrifugation. The crude cell walls in the supernatant are sedimented by high-speed centrifugation (>20,000 × *g*).

d Removing peptidoglycan-associated polymers

The effective degradation of proteins is very important and requires optimal conditions for digestion with proteases like trypsin or pepsin. Polymers covalently linked to the peptidoglycan can be removed with boiling SDS or trichloroacetic acid solutions, or with hot formamide or aqueous phenol (Schleifer,

1985). It must be taken into consideration that reagents used for extraction may degrade the peptidoglycan partially or modify its structure (Brown *et al.*, 1976; Perkins, 1965; Rosenthal and Dziarski, 1994). The author's laboratory is using a combination of trypsin digestion with repeated intense washing steps and extraction with a boiling 2% (w/v) SDS solution. The latter step turns out to be very efficient, but requires care as some cell-wall preparations dissolve completely in boiling SDS solution within 1 min. The sensitivity of the cell walls towards the SDS treatment must be pre-tested. It is recommendable to treat even sensitive cell walls by the SDS solution at least at ambient temperature (see Protocol 7). The treatment with hydrogen fluoride (if necessary; see above) is most efficient when peptidoglycan is used before lyophilization (see Protocol 7). However, the necessity of a HF treatment usually becomes obvious only after the amino acid analysis of lyophilized peptidoglycan (see Protocol 10).

Protocol 7: Preparation of purified cell walls of Gram-positive bacteria

- Approximately 3 g wet-weight of cells (centrifugal pellet) is needed for the preparation of purified cell walls. The cells must be either boiled in water for 10 min or autoclaved at 120°C for 20 min. Wash the cells twice by suspending in distilled water and centrifugation.
- Transfer the cell pellet to the 15 ml V4A beaker of a Vibrogen Cell Mill and mix it with distilled water and glass beads using a thin steel rod to an evenly stiff paste. To prevent foaming, one drop of 1-octanol (Sigma–Aldrich, P/N 293245) can be added. For the disruption of most bacterial cells it is suggested to mix equal masses of glass beads (Sartorius Stedim Biotech) of the diameters 0.10–0.11 mm (P/N BBI-8541400) and 0.17–0.18 mm (P/N BBI-8541507). (The glass beads can be regenerated by rinsing with a detergent solution, boiling and washing with distilled water and drying in an oven at 100°C.)
- Shake the cell-beads paste vigorously while cooling with tap water in the Cell Mill. The progress of cell disruption is monitored by phase-contrast microscopy. Grey cell debris instead of dark intact cells indicates almost complete disruption of the cells, which takes usually 20–30 min.
- Separate the cell debris from the glass beads by washing with distilled water on a glass frit (porosity 1; Duran, P/N 258522102) and collect the suspension in a filter flask.
- Centrifuge the suspension at low speed ($350 \times g$) at 10°C, transfer the supernatant (containing the cell debris) to another centrifuge tube and discard the pellet (intact cells).
- Sediment the cell debris by centrifugation at $23{,}000 \times g$ at 18°C.
- Resuspend the pellet in 25 ml distilled water by using an Ultra-Turrax and centrifuge at $23{,}000 \times g$. Repeat this washing step. The Ultra-Turrax will be used for resuspension, washing will be done with a volume of 25 ml and centrifugation will be carried out at $23{,}000 \times g$ in all following steps.
- Wash the pellet with 0.1 M phosphate buffer pH 7.9. Resuspend the pellet after centrifugation in 10 ml phosphate buffer containing 2 mg trypsin (Serva, P/N 37291).
- Shake the suspension after addition of one drop toluene (to inhibit growth of microorganisms) at 37°C overnight.

- Centrifuge and wash the pellet three times with distilled water.
- Resuspend the pellet in 10 ml distilled water. Use 1 ml of this suspension for a pre-test: Add 0.1 ml 20% (w/v) SDS solution and incubate for 3 min in a boiling water bath. When the suspension becomes clear, either omit the next two steps (if a too low yield is expected) or incubate at room temperature for up to 20 min instead of boiling.
- If the suspension endures the pre-test, add 0.9 ml 20% (w/v) SDS solution to 9 ml suspension and incubate in a boiling water bath for maximal 3 min.
- Wash with distilled water and centrifugation at 20°C until there is no more foaming (usually three to four times).
- Remove the supernatant after centrifugation and subject the pellet to either lyophilization overnight or HF treatment (Protocol 8).

Protocol 8: HF treatment of cell walls

- Approximately 15 mg of lyophilized (or the corresponding amount wet-weight) purified cell walls are suspended in 1 ml of 48% (w/v) hydrofluoric acid (Sigma–Aldrich, P/N 30107; pre-cooled at 8°C) in a 2 ml Eppendorf tube by using a Vortex mixer. Observe safety instructions, HF is very toxic and highly corrosive.
- Keep the suspension in an ice bath for 8 h.
- Sediment cell walls by centrifugation at 13,500 × *g* for 2 min and remove the HF supernatant.
- Resuspend the pellet in 1.5 ml distilled water by using a Vortex mixer and centrifuge at 13,500 × *g* for 2 min.
- Repeat the washing steps until universal pH 1–12 test paper indicates the absence of acid.
- Lyophilize the peptidoglycan overnight.

C. Analyses of Peptidoglycan Preparations

1. Qualitative amino acid composition

The qualitative amino acid composition of the peptidoglycan, in particular the diagnostic diamino acid, is of high taxonomic relevance and allows in combination with 16S rRNA gene sequence data the tentative affiliation of a strain to a genus. The identification of the diamino acids L-lysine, L-ornithine, L-2,4-diaminobutyric acid (L-Dab), lanthionine (Kato *et al.*, 1981), hydroxylysine (Smith and Henderson, 1964) and *threo*-2-hydroxyornithine (Schleifer *et al.*, 1983) requires the analysis of protein-free peptidoglycan. When replacing whole cells by 1 mg peptidoglycan, Protocol 1 can be used for the analysis of peptidoglycan amino acids by TLC (see Figure 3, peptidoglycan hydrolysates; note the absence of leucine/isoleucine and the reduced number of amino acid spots when compared to whole-cell hydrolysates). The conditions for 'total hydrolysis' in Protocol 1 are appropriate to cleave most peptide bonds. However, peptides consisting of aspartic acid or threonine linked to the distal amino group of lysine or ornithine are rather resistant (Schleifer, 1985; see peptide 1 in strain DS-3[T], total hydrolysis [TH] in Figure 4) and their

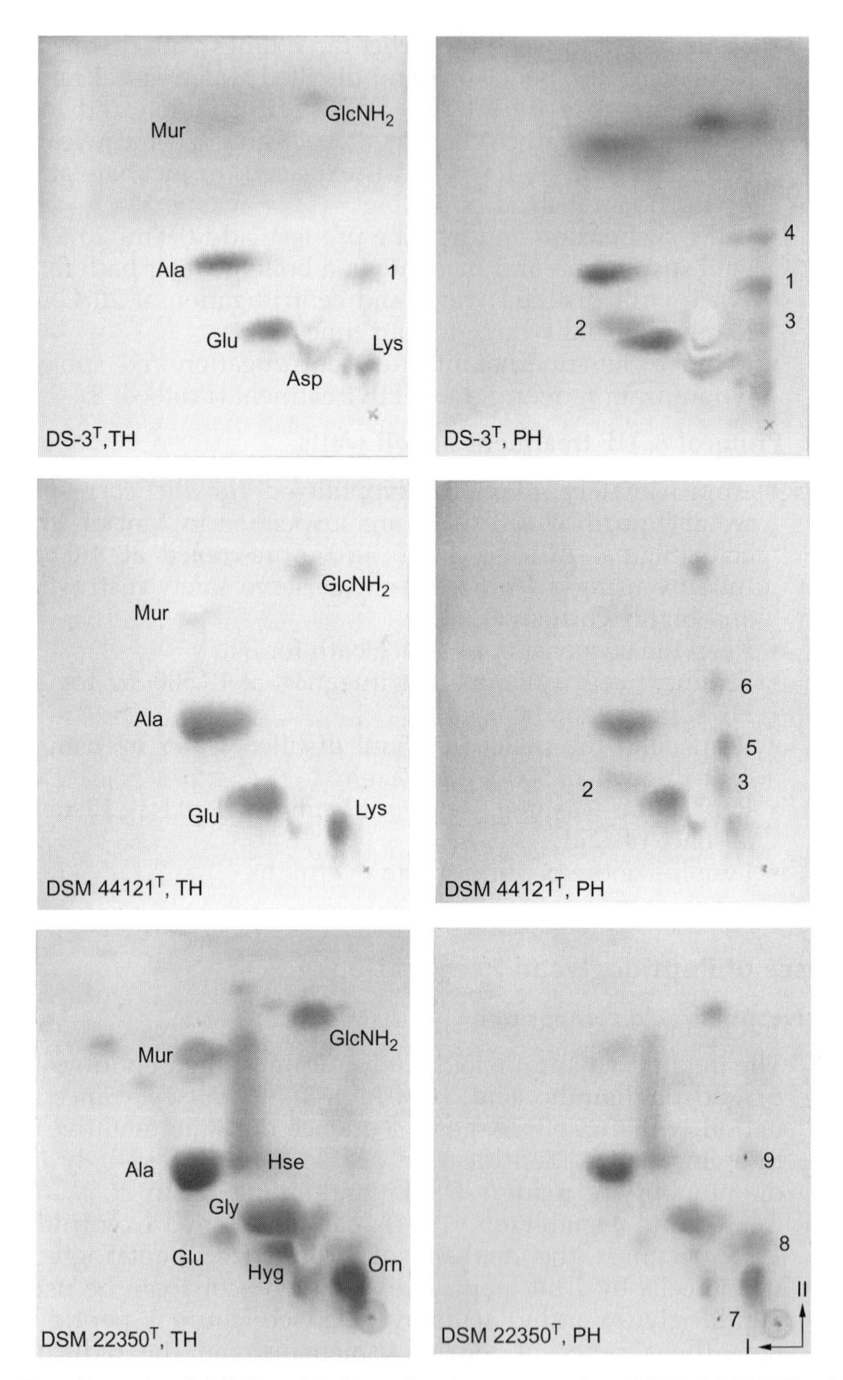

Figure 4. Two-dimensional cellulose thin-layer chromatograms of total (TH; 4.0 N HCl, 100°C, 16 h) and partial (PH; 4.0 N HCl, 100°C, 0.75 h) hydrolysates of the peptidoglycan of *Isoptericola dokdonensis* DS-3^T (A11.31; A4α L-Lys-D-Asp; Yoon *et al*., 2006), *Promicromonospora sukumoe* DSM 44121^T (A11.59; A4α L-Lys-L-Ala-D-Glu; Stackebrandt and Schumann, 2004) and *Marisediminicola antarctica* DSM

overnight hydrolysis requires at least 6.0 N HCl and a temperature of 120°C. On cellulose TLC plates developed by using the solvent system of Rhuland *et al.* (1955), peptidoglycan amino acids migrate in the following order (increasing Rf values): OH-Dpm, *meso*-Dpm, LL-Dpm, Dab, Asp/Orn, Lys/Gly, Ser, Glu, Thr, Ala (compare to data for PC in Table 2 of Schleifer, 1985). The strong point of the TLC method adapted from the method of Rhuland *et al.* (1955) is the good resolution of the diamino acids and Dpm isomers. The lack of separation of the pairs Lys/Gly and Asp/Orn as well as the inability to detect the amino acids homoserine (Hse) and Hyg are obvious disadvantages of the one-dimensional TLC system. Characteristic colours of the spots after spraying with ninhydrin and their changes after storage of the chromatograms in the dark support the identification of some amino acids: Orn appears bluish and later turns to pinkish colour, while Asp first appears pinkish, turns to bluish and fades easily. Dab shows a characteristic grey-bluish colour. To obtain more information about the qualitative amino acid composition, it is necessary to supplement one-(1D-) by two-dimensional (2D-) TLC (Protocol 9, Figure 4) and/or by gas chromatography/mass spectrometry (GC/MS) after derivatization according to Protocol 10 (Figure 5).

Protocol 9: Qualitative analysis of amino acids by 2D-TLC

- Prepare a TLC cellulose plate of 10 cm × 10 cm with the origin at the lower left corner, 1 cm distant from both edges.
- Apply *ca.* 1–2 μl of cell-wall hydrolysate using a glass capillary to the origin and allow it to dry.
- Develop the chromatogram in a TLC tank lined with filter paper and filled at least a day before with the solvent system I consisting of isopropanol (Sigma–Aldrich, P/N 190764), acetic acid and distilled water (75:10:14, v/v/v; Schleifer and Kandler, 1972). The solvent front should reach the upper edge.

Figure 4. (*Continued*) 22350^T (B6; B2β, position 1: Gly, position 3: L-Hse, interpeptide bridge: D-Glu (Hyg)-Gly-D-Orn; Li et al. (2010); see also Schleifer, 1970). Chromatographic conditions see Protocol 9.
Abbreviations: Mur, muramic acid; $GlcNH_2$, glucosamine; Hse, homoserine; Hyg, *threo*-3-hydroxy-glutamic acid; I and II, indicating the direction of development with solvent systems I and II.
Peptides:
1. D-Asp-L-Lys
2. L-Ala–D-Glu
3. L-Lys–D-Ala
4. D-Asp–L-Lys–D-Ala
5. L-Ala-L-Lys
6. L-Ala–L-Lys–D-Ala
7. Gly–Hyg, Gly–Hyg–Gly or Gly–Hyg–Hse
8. D-Orn–D-Ala
9. Gly–D-Orn–D-Ala

Peptide 1 is stable under conditions of total hydrolysis. Peptides 7 and 9 appear yellow shortly after spraying with ninhydrin reagent and heating before they turn to bluish. (see colour insert)

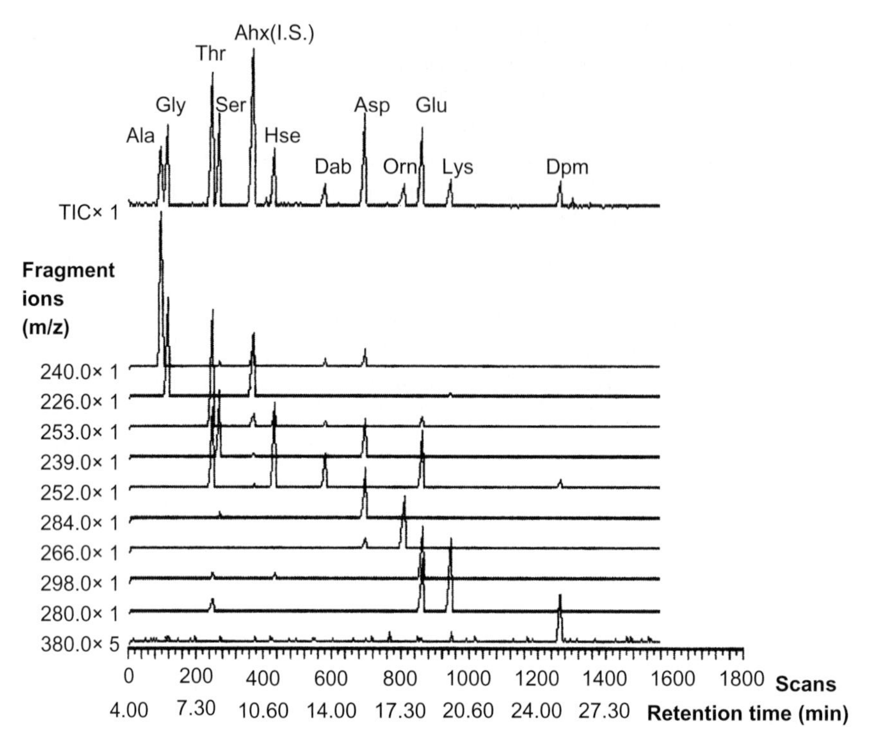

Figure 5. Electron impact mass chromatogram of a standard mixture (2.5 μmol/ml) for identification of *N*-heptafluorobutyryl amino acid isobutylesters by their retention times and characteristic fragment ions (gas chromatographic conditions see Protocol 10). *Abbreviations*: Ahx, 2-aminohexanoic acid (used as internal standard [I.S.]); Hse, homoserine; Dab, 2,4-diaminobutyric acid; Dpm, 2,6-diaminopimelic acid; TIC, total ion chromatogram; 1 and 5, intensity factors.

- Remove the TLC plate and allow it to dry in a fume hood for 1 h before repeating the development in the same direction using solvent system I.
- Remove the TLC plate and allow it to dry in a fume hood overnight.
- Turn the TLC plate 90° counter-clockwise and develop it twice as described above in a TLC tank, equilibrated with solvent system II consisting of 2-picoline (Aldrich, P/N 10,983-5), 25% ammonium hydroxide solution (Fluka, P/N 09860) and distilled water (70:2:28, v/v/v; Schleifer and Kandler, 1972).
- Remove the TLC plate and allow it to dry in a fume hood overnight before spraying with ninhydrin reagent and visualization of spots as described in Protocol 1.

For identification of spots see examples in Figure 4, analyse mixtures consisting of authentic peptidoglycan-relevant amino acid standards and compare to migration data for PC reported by Schleifer and Kandler (1972) and Schleifer (1985). The inability to identify diamino acids is a disadvantage of 2D-TLC following Protocol 9. However, 2D-TLC allows differentiation of glycine and lysine, of ornithine and aspartic acid and detection of the amino acids Hse and Hyg (Figure 4). Characteristic colours and specific features of the spots support their identification (e.g. aspartic acid, steel-blue [see DS-3^{T}, TH and PH in

Figure 4]; both Hse [see DSM 22350^T, TH and PH in Figure 4] and Dab give rise to diffused smears instead of well-defined spots).

2. Quantitative analysis of amino acids

Not only the differentiation of several peptidoglycan types (see Table 1, e.g. the pairs A11.6/A11.7; A11.13/A11.15; A11.27/A11.28) but also the confirmation of types concluded from 2D-TLC require the quantification of amino acids, which can be done by an amino acid analyzer (e.g. Schleifer and Kandler, 1967), GC (e.g. O'Donnell *et al.*, 1982) or by HPLC (Lee, 2007). Protocol 10 for quantification of *N*-heptafluorobutyryl amino acid isobutylesters by GC and GC/MS is based on the method of MacKenzie (1984, 1987). The volatile amino acid derivatives are obtained by esterification with isobutanol and subsequent *N*-acylation with heptafluorobutyric anhydride.

Peptidoglycan Structure

Protocol 10: Quantitative analysis of amino acids by GC

- Filter the cell-wall hydrolysate through a pipette tip closed by a small bead from filter paper as described in Protocol 1 but without charcoal and collect it in a 1 ml V-vial (Wheaton, P/N W986254) with screw-cap and Tuf-Bond disc (Thermo Scientific, P/N TS-12712).
- Dry in a vacuum desiccator containing KOH pellets.
- Remove residual water azeotropically by adding 100 μl dichloromethane and evaporate using a gentle stream of nitrogen. Process the dry hydrolysate by the following steps or use it for Protocol 11.
- Prepare the acylation reagent by adding 270 μl acetyl chloride (Fluka, P/N 00990) cautiously and dropwise to 1 ml pre-cooled isobutanol (Fluka, P/N 58450) in an ice bath (exothermic reaction!).
- Add 100 μl acylation reagent to the dried hydrolysate, seal the V-vial, mix using a Vortex mixer and heat in a heating block at 120°C for 10 min. Mix again and allow the reaction to proceed at 120°C for an additional 20 min.
- Open the vial after cooling and evaporate the reagent in a vacuum desiccator.
- Add 50 μl heptafluorobutyric anhydride (Fluka, P/N 77253), seal, mix and heat the V-vial at 150°C for 10 min. Evaporate the reagent in a vacuum desiccator containing KOH pellets.
- Subject the solution of *N*-heptafluorobutyryl amino acid isobutylesters in 200 μl ethyl acetate (Sigma–Aldrich, P/N 650528) to capillary gas chromatography or GC/MS under the following conditions: column OV-1: 25 m × 0.25 mm ID (Macherey-Nagel, P/N 723116.25); temperature program: raise the temperature from 90 to 260°C by a rate of 5 K/min; carrier gas, hydrogen, 30 cm/s; injector temperature, 270°C; temperature of the flame ionization detector, 300°C; sample volume, 1 μl.

The peaks are identified by comparing the retention times with those of authentic standard substances and/or on the basis of characteristic fragment ions recorded by electron impact mass spectrometry (Figure 5). The standard mixture of 10 peptidoglycan-relevant amino acids (see Figure 5) contains 2.5 μmol of each component per 1 ml of 0.1 N HCl. Aliquots of 500 μl are dried in 2 ml ampoules and can be stored at -20°C for several years. The standard mixture must be hydrolysed and derivatized in the same way like the peptidoglycan sample. If the standard is supplemented with peptidoglycan-relevant amino acids and their gas chromatographic separation is optimized, the EZ:fast Amino Acid Analysis Kits (Phenomenex, P/N KGO-7167, KGO-7168) may be a fast and convenient alternative to Protocol 10.

3. Enantiomeric analysis of amino acids

The differentiation of peptidoglycan structures that differ in the configuration of certain amino acids (e.g. the pairs A11.33/A11.54, A.11.38/A11.58 and A11.40/A11.56) or the decision whether alanine originates from positions 3 or 4 (e.g. B11) require enantiomeric analyses of amino acids by enzymatic methods (Schleifer and Kandler, 1967), optical rotary dispersion of 2,4-dinitrophenyl (DNP) derivatives (Kandler *et al.*, 1968), HPLC using chiral reagents (Sasaki *et al.*, 1998) or GC on chiral columns (Groth *et al.*, 1997). The presence of stereoisomers of amino acids can also be concluded from the occurrence of certain diastereomeric peptides (e.g. L-Ala-D-Glu and D-Ala-D-Glu) that can be separated by 2D-TLC (see III.C.4.b and Table IV of Schleifer, 1985). Protocol 11 for chiral GC is based on the method of Frank *et al.* (1980). Because of the temperature-sensitive chiral stationary phase it is recommended to replace the *N*-heptafluorobutyryl amino acid isobutylesters of Protocol 10 with the homologous propyl derivatives, which have lower boiling points and can be synthesized in an analogous two-step reaction.

Protocol 11: Analysis of enantiomeric amino acids by GC

- Prepare the acylation reagent by cautiously adding 327 μl acetyl chloride dropwise to 1 ml of pre-cooled isopropanol in an ice bath (exothermic reaction!).
- Add 200 μl of acylation reagent to the dried hydrolysate (see Protocol 10), seal the V-vial, mix by using a Vortex mixer and heat in a heating block at 110°C for 10 min. Mix again and allow reaction to proceed at 110°C for a further 50 min.
- Open the vial after cooling and evaporate the reagent in a vacuum desiccator.
- Dissolve the residue in 100 μl dichloromethane. Add 50 μl pentafluoropropionic anhydride (Aldrich, P/N 25,238-7) and heat the sealed V-vial at 150°C for 10 min.
- Evaporate the reagent in an vacuum desiccator.
- Subject the solution of *N*-pentafluoropropionyl amino acid isopropyl esters in 200 μl ethyl acetate to capillary gas chromatography under the following conditions: column PERMABOND-L-Chirasil-Val 25 m × 0.25 mm ID (Macherey-Nagel, P/N 723730.25); temperature program: raise the temperature from 80 to 190°C by a rate of 2 K/min; carrier gas, hydrogen, 30 cm/s; injector temperature, 250°C; temperature of the flame ionization detector, 260°C; sample volume, 1 μl.

L-Amino acids interact stronger with the L-valine bound to the stationary phase of the chiral column and elute somewhat later than the D-isomers (see

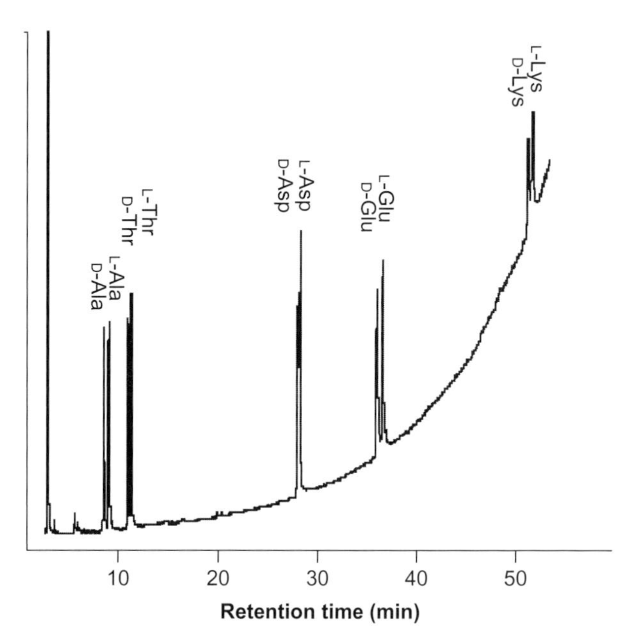

Figure 6. Gas chromatographic separation of a standard mixture of racemic *N*-pentafluoropropionyl amino acid isopropyl esters (2.5 μmol/ml) on a chiral capillary column (gas chromatographic conditions see Protocol 11).

Figure 6). Peaks are identified by comparison with those of racemic amino acids of a standard mixture (2.5 μmol per 1 ml 0.1 N HCl), which is treated in the same way as described for peptidoglycan. The complete separation of the optical isomers of the aspartic acid derivative is a challenge for the quality of the chiral column. Diamino acid derivatives elute only at temperatures close to the thermal limit of the capillary column (see Lys; Figure 6) and it is recommended to use the HPLC method of Sasaki *et al.* (1998) for their enantiomeric analysis.

4. Arrangement of amino acids within the peptidoglycan

The qualitative and quantitative content as well as information on the configuration of amino acids are still not sufficient to circumscribe peptidoglycan structures comprehensively. There are peptidoglycan structures that differ by the arrangement of amino acids despite identical ratios of optical isomers of amino acids, e.g. the pairs A11.pep/A11.1 or A11.13/A11.18. When taking into consideration the possibilities of protein contaminations, decomposition of amino acids or incomplete hydrolysis of the peptidoglycan, the quantitative data cannot be precise enough to conclude unambiguously on the stoichiometric ratio of amino acids within the peptidoglycan. It is therefore necessary to explore the arrangement of amino acids, which can be done by two simple methods: (1) labelling of the N-terminal amino acids in the peptidoglycan and (2) partial hydrolysis of the peptidoglycan to deduce the arrangement of amino acids from the resulting short peptides.

a *Dinitrophenylation of* **N-terminal** *amino acids*

As the cross-linkage of the peptide units is never complete, *ca.* 40% of the interpeptide bridges possess a free N-terminal ('last') amino acid that can be labelled chemically. Amino groups of diamino acids that are not involved in peptide bonds will also be labelled and thus provide information on their arrangement within the peptidoglycan. For detection of N-terminal amino acids, the intact peptidoglycan is treated with 1-fluoro-2,4-dinitrobenzene, then subjected to hydrolysis and the dinitrophenylated amino acids are analysed in the hydrolysate. Protocol 12 is based on the method described by Schleifer (1985).

Protocol 12: Detection of N-terminal amino acids in the peptidoglycan

- Add 8 ml reagent I and 4 ml reagent II to 10–15 mg purified cell walls in a brown GL 25 glass flask (Schott, P/N 2180614) with screw cap and septum.
  Reagent I: 120 μl 1-fluoro-2,4-dinitrobenzene (Sigma, P/N D-1529) in 8 ml ethanol.
  Reagent II: 1.25% (w/v) aqueous $NaHCO_3$ solution.
- Agitate using a magnetic stirrer at 37°C in the dark for 16 h.
- Spin down in a centrifuge (500 × *g*, 20 min) and wash the pellet successively with diethylether, 96% (v/v) ethanol and 50% (v/v) ethanol. Repeat the last step until the supernatant becomes colourless.
- Wash the yellow pellet with distilled water, centrifuge and lyophilize in a tube wrapped in aluminium foil for protection against light.
- Hydrolyse 3 mg dinitrophenylated cell walls in a 5 ml brown-glass ampoule (Gerresheimer Querétaro, Mexico, P/N 12040G-2/amber/untreated) with 0.5 ml 4.0 N HCl at 100°C for 16 h. Protect the hydrolysate and all extracts from light.
- Filter the hydrolysate as described in Protocol 1 under omission of charcoal.
- Extract three times with 0.5 ml diethylether. Combine the extracts and evaporate the ether in a gentle stream of nitrogen (not air!). Keep the bottom phase (hydrolysate) for 1-butanol/ethyl acetate extraction.
- Dissolve the residue in 150 μl ethyl acetate and purify the DNP-amino acids by passing through a solid-phase extraction silica cartridge (Varian, P/N 12102010) equilibrated with ethyl acetate. Collect the yellow fraction and dry it in a stream of nitrogen.
- Dilute the hydrolysate with 0.5 ml distilled water and extract once with 0.5 ml of 50% (v/v) 1-butanol in ethyl acetate. Dry the extract in a gentle stream of nitrogen.
- Dissolve the ether-extracted DNP-amino acids and the butanol-extracted mono-DNP-diamino acids each in one drop of methanol and apply the solutions together with those of authentic DNP-amino acids (Sigma–Aldrich) to spots on the baseline of a 10 cm × 10 cm silica gel 60 F_{254} TLC plate (Merck, P/N 1.05554).

- Develop the TLC plate in a tank equilibrated for 2 h with a solvent system consisting of 65 ml chloroform, 25 ml methanol, 3 ml acetic acid and 6.5 ml distilled water. DNP-amino acids are visible as yellow spots.

The yellow spots are identified by comparison of the migration behaviour with those of commercially available DNP-amino acids. DNP-serine and DNP-glutamic acid are not separated by the system used in Protocol 12 but can be differentiated by TLC on a cellulose plate developed in 1.5 M phosphate buffer of pH 6.0.

In contrast to the ω-mono-DNP derivatives the α-mono-DNP derivatives of diamino acids are photolabile (Russell, 1963) when exposed to sunlight. This feature is important for the differentiation of B1 type peptidoglycans from those of the B2 type where the D-diamino acid is linked to the carboxyl group of D-alanine by its α-amino group (Schleifer and Kandler, 1972).

b Partial hydrolysis and 2D-TLC of short peptides

When peptidoglycan preparations are hydrolysed under milder conditions than in Protocol 1, not all peptide bonds are cleaved and di-, tri- and oligo-peptides can be detected besides the amino acids. These short peptides reveal the structural vicinity of certain amino acids and provide valuable information on the arrangement of amino acids within the intact peptidoglycan. Hydrolysis of 2 mg of peptidoglycan using 4.0 N HCl in closed glass ampoules in an oil bath at 100°C for 45 min provides a good yield of di- and tripeptides. The 'partial hydrolysate' is filtered as described in Protocol 1 but without charcoal. The residue obtained after drying in a stream of air at 37°C is re-dissolved in one drop of distilled water and 2D-TLC is performed as described in Protocol 9. It is recommended to perform 2D-TLC of total and partial hydrolysates of a peptidoglycan, each on a TLC plate, simultaneously in the same tank.

The spots can be identified by their migration behaviour and characteristic colours after spraying with ninhydrin reagent. Figure 4 shows 2D-TLC plates of partial hydrolysates of peptidoglycans representing the types A11.31, A11.59 and B6 as examples. Despite the differences in the migration behaviour of peptides between 2D-PC and 2D-TLC, it is recommended to plot the R_{Ala} values of Table IV of Schleifer (1985) on graph paper for guidance in identification and also to study the PC images in early publications of Schleifer and co-workers. The colour development on TLC plates appears to be less characteristic than described for PC. However, it is important to observe the appearance of spots from the beginning of heating of the ninhydrin-sprayed plates because, e.g. the characteristic yellow colour of peptides containing N-terminal glycine is visible only for the first few minutes before turning to bluish. Peptides that cannot be identified this way must be isolated by preparative descending PC as described by Schleifer (1985). Subsequently, the purified peptides are analysed with respect to their N-termini, qualitative and quantitative composition as well as stereoisomerism of the amino acids under appropriate modifications of the protocols described in this chapter. It also appears that the use of contemporary mass spectrometric approaches like electrospray or MALDI-TOF mass spectrometry for the analysis of peptides in peptidoglycan hydrolysates have promising potential.

♦♦♦♦♦♦ IV. CONCLUDING REMARKS

The intense investigations of the peptidoglycan structure in the 1960s to 1980s contributed remarkably to the knowledge of the diversity of Gram-positive bacteria and hence improved their taxonomy. The plausibility of early sequence analyses of conservative genes was evaluated against the background of chemotaxonomic data including those on the peptidoglycan structure. The increasing and legitimate trust in sequence data in the following era led to the unfortunate decline in the application of chemotaxonomic techniques, first and foremost in the laborious studies of the peptidoglycan structure. Considering the high value of these structures for the taxonomy of Gram-positive bacteria, the reader should feel encouraged to apply the set of down-to-earth protocols of this chapter instead of hoping that one day in future the complex peptidoglycan structure might be concluded from genome sequences.

♦♦♦♦♦♦ ACKNOWLEDGMENTS

The author is indebted to Norbert Weiss for his practical introduction to the elucidation of peptidoglycan structures. He wishes to express his cordial thanks to Erika M. Tóth and Erko Stackebrandt for critically reading the manuscript and their invaluable suggestions for its improvement. The skilful assistance of Anika Wasner in peptidoglycan analyses and her review of the protocols is gratefully acknowledged.

References

Adam, A., Petit, J. F., Wietzerbin-Falszpan, J., Sinay, P., Thomas, D. W. and Lederer, E. (1969). L'acide *N*-glycolylmuramique, constituent des parois de *Mycobacterium smegmatis*: identification par spectrométrie de masse. *Fed. Eur. Biochem. Soc. Lett.* **4**, 87–92.

Brown, W. C., Doyle, R. J. and Streips, U. N. (1976). Comparison of various procedures for removing proteins and nucleic acids from cell walls of *Bacillus subtilis*. *Prep. Biochem.* **6**, 479–488.

Card, G. L. (1973). Metabolism of phosphatidylglycerol, phosphatidylethanolamine, and cardiolipin of *Bacillus stearothermophilus*. *J. Bacteriol.* **114**, 1125–1137.

Frank, H., Rettenmeier, A., Weicker, H., Nicholson, G. J. and Bayer, E. (1980). A new gas chromatographic method for determination of amino acid levels in human serum. *Clin. Chim. Acta* **105**, 201–211.

Freney, J., Kloos, W. E., Hajek, V., Webster, J. A., Bes, M., Brun, Y. and Vernozy-Rozand, C. (1999). Recommended minimal standards for description of new staphylococcal species. *Int. J. Syst. Bacteriol.* **49**, 489–502.

Ghuysen, J. M. (1968). Use of bacteriolytic enzymes in determination of wall structure and their role in cell metabolism. *Bacteriol. Rev.* **32**, 425–464.

Groth, I., Schumann, P., Rainey, F. A., Martin, K., Schütze, B. and Augsten, K. (1997). *Bogoriella caseilytica* gen. nov., sp. nov., a new alkaliphilic actinomycete from a soda lake in Africa. *Int. J. Syst. Bacteriol.* **47**, 788–794.

Hancock, I. C. (1994). Analysis of cell wall constituents of Gram-positive bacteria. In: *Chemical Methods in Prokaryotic Systematics* (M. Goodfellow and A. G. O'Donnell., eds.), pp. 63–84. John Wiley and Sons, Chichester.
Hirschfield, G. R., McNeil, M. and Brennan., P. J. (1990). Peptidoglycan-associated polypeptides of *Mycobacterium tuberculosis*. *J. Bacteriol.* **172**, 1005–1013.
Kandler, O., Koch, D. and Schleifer, K. H. (1968). Die Aminosäuresequenz eines glycinhaltigen Mureins einiger Stämme von *Lactobacillus bifidus*. *Arch. Mikrobiol.* **61**, 181–186.
Kato, K., Umemoto, T., Fukuhara, H., Sagawa, H. and Kotani, S. (1981). Variation of dibasic amino acid in the cell wall peptidoglycan of bacteria of genus *Fusobacterium*. *FEMS Microbiol. Lett.* **10**, 81–85.
Kawamoto, I., Oka, T. and Nara, T. (1981). Cell wall composition of *Micromonospora olivoasterospora*, *Micromonospora sagamiensis*, and related organisms. *J. Bacteriol.* **146**, 527–534.
Komagata, K. and Suzuki, K.-I. (1987). Lipid and cell-wall analysis in bacterial systematics. In: *Methods in Microbiology* (R. R. Colwell and R. Grigorova, eds.), Vol. 19, pp. 161–207. Academic Press, London.
Lechevalier, M. P. and Gerber, N. N. (1970). The identity of madurose with 3-O-methylgalactose. *Carbohydr. Res.* **13**, 451–454.
Lechevalier, M. P. and Lechevalier, H. (1970). Chemical composition as a criterion in the classification of aerobic actinomycetes. *Int. J. Syst. Bacteriol.* **20**, 435–443.
Lee, S. D. (2007). *Labedella gwakjiensis* gen. nov., sp. nov., a novel actinomycete of the family *Microbacteriaceae*. *Int. J. Syst. Evol. Microbiol.* **57**, 2498–2502.
Li, H.-R., Yu, Y., Luo, W. and Zeng, Y.-X. (2010). *Marisediminicola antarctica* gen. nov., sp. nov., an actinobacterium isolated from the Antarctic. *Int. J. Syst. Evol. Microbiol.* **60**, 2535–2539.
Logan, N. A., Berge, O., Bishop, A. H., Busse, H.-J., De Vos, P., Fritze, D., Heyndrickx, M., Kämpfer, P., Rabinovitch, L., Salkinoja-Salonen, M. S., Seldin, L. and Ventosa, A. (2009). Proposed minimal standards for describing new taxa of aerobic, endospore-forming bacteria. *Int. J. Syst. Evol. Microbiol.* **59**, 2114–2121.
MacKenzie, S. L. (1984). Amino acids and peptides. In: *Gas chromatography/mass spectrometry applications in microbiology* (G. Odham, L. Larsson, and P. Mardh, eds.), pp. 157–204. Plenum Publishing Corporation, New York.
MacKenzie, S. L. (1987). Gas chromatographic analysis of amino acids as the *N*-heptafluorobutyryl isobutyl esters. *J. Assoc. Off. Anal. Chem.* **70**, 151–160.
Minnikin, D. E., Patel, P. V., Alshamaony, L. and Goodfellow, M. (1977). Polar lipid composition in the classification of *Nocardia* and related bacteria. *Int. J. Syst. Bacteriol.* **27**, 104–117.
O'Donnell, A. G., Minnikin, D. E., Goodfellow, M. and Parlett, J. H. (1982). The analysis of actinomycete wall amino acids by gas chromatography. *FEMS Microbiol. Lett.* **15**, 75E–78E.
Park, J. T. and Hancock, R. (1960). A fractionation procedure for studies of the synthesis of cell-wall mucopeptide and of other polymers in cells of *Staphylococcus aureus*. *J. Gen. Microbiol.* **22**, 249–258.
Perkins, H. R. (1965). The action of hot formamide on bacterial cell walls. *Biochem. J.* **95**, 876–882.
Petit, J. F., Wietzerbin, J., Das, B. C. and Lederer, E. (1975). Chemical structure of the cell wall of *Mycobacterium tuberculosis* var. *bovis*, strain BCG. *Zeitschrift für Immunitätsforschung, experimentelle und klinische Immunologie* **149**, 118–125.
Promnuan, Y., Kudo, T., Ohkuma, M., and Chantawannakul, P. (2011). *Actinomadura apis* sp. nov., isolated from a honey bee (*Apis mellifera*) hive in Thailand and the

reclassification of *Actinomadura cremea* subsp. *rifamycini* Gauze et al. 1987 as *Actinomadura rifamycini* (Gauze et al. 1987) sp. nov., comb. nov. *Int. J. Syst. Evol. Microbiol.* **61**, 2271–2277.

Rhuland, L. E., Work, E., Denman, R. F. and Hoare, D. S. (1955). The behavior of the isomers of α,ε-diaminopimelic acid on paper chromatograms. *J. Am. Chem. Soc.* **77**, 4844–4846.

Rosenthal, R. S., and Dziarski, R. (1994). Isolation of peptidoglycan and soluble peptidoglycan fragments. *In Methods in Enzymology* (P. M. Bavoil, and V. L. Clark, Eds.), Vol. 235, pp. 253–285. Academic Press, London.

Russell, D. W. (1963). Studies on the photochemical behaviour of 2,4-dinitrophenyl derivatives of some amino acids and peptides. *Biochem. J.* **87**, 1–4.

Saddler, G. S., Tavecchia, P., Lociuro, S., Zanol, M., Colombo, L. and Selva, E. (1991). Analysis of madurose and other actinomycete whole-cell sugars by gas chromatography. *J. Microbiol. Meth.* **14**, 185–191.

Sasaki, J., Chijimatsu, M. and Suzuki, K.-I. (1998). Taxonomic significance of 2,4-diaminobutyric acid isomers in the cell wall peptidoglycan of actinomycetes and reclassification of *Clavibacter toxicus* as *Rathayibacter toxicus* comb. nov. *Int. J. Syst. Bacteriol.* **48**, 403–410.

Schleifer, K. H. (1970). Die Mureintypen in der Gattung *Microbacterium. Arch. Mikrobiol.* **71**, 271–282.

Schleifer, K. H. (1985). Analysis of the chemical composition and primary structure of murein. *In Methods in Microbiology* (T. Bergan, Ed.), Vol. 18, pp. 123–156. Academic Press, London.

Schleifer, K. H. and Kandler, O. (1967). On the chemical composition of the cell wall of streptococci. I. The amino acid sequence of the murein of *Str. thermophilus* and *Str. faecalis. Arch. Mikrobiol.* **57**, 335–364.

Schleifer, K. H. and Kandler, O. (1972). Peptidoglycan types of bacterial cell walls and their taxonomic implications. *Bacteriol. Rev.* **36**, 407–477.

Schleifer, K. H. and Seidl, H. P. (1985). Chemical composition and structure of murein. In: *Chemical Methods in Bacterial Systematics* (M. Goodfellow and D. E. Minnikin, eds.), pp. 201–219. Academic Press, London.

Schleifer, K. H., Plapp, R. and Kandler, O. (1967). Identification of *threo*-3-hydroxyglutamic acid in the cell wall of *Microbacterium lacticum. Biochem. Biophys. Res. Commun.* **28**, 566–570.

Schleifer, K. H., Hayn, I., Seidl, H. P. and Firl, J. (1983). *Threo*-beta-hydroxyornithine: a natural constituent of the peptidoglycan of *Corynebacterium* species Co 112. *Arch. Microbiol.* **134**, 243–246.

Schumann, P., Kämpfer, P., Busse, H.-J. and Evtushenko, L. I., and for the Subcommittee on the Taxonomy of the Suborder *Micrococcineae* of the International Committee on Systematics of Prokaryotes. (1983). Proposed minimal standards for describing new genera and species of the suborder *Micrococcineae. Int. J. Syst. Evol. Microbiol.* **59**, 1823–1849.

Smith, W. G. and Henderson, L. M. (1964). Relationships of lysine and hydroxylysine in *Streptococcus faecalis* and *Leuconostoc mesenteroides. J. Biol. Chem.* **239**, 1867–1871.

Stackebrandt, E. and Schumann, P. (2004). Reclassification of *Promicromonospora pachnodae* Cazemier et al. 2004 as *Xylanimicrobium pachnodae* gen. nov., comb. nov. *Int. J. Syst. Evol. Microbiol.* **54**, 1383–1386.

Staneck, J. L. and Roberts, G. D. (1974). Simplified approach to identification of aerobic actinomycetes by thin-layer chromatography. *Appl. Microbiol.* **28**, 226–231.

Tindall, B. J., Rosselló-Móra, R., Busse, H.-J., Ludwig, W. and Kämpfer, P. (2010). Notes on the characterization of prokaryote strains for taxonomic purposes. *Int. J. Syst. Evol. Microbiol.* **60**, 249–266.
Uchida, K. and Aida, K. (1977). Acyl type of bacterial cell wall: its simple identification by colorimetric method. *J. Gen. Appl. Microbiol.* **23**, 249–260.
Uchida, K. and Aida, K. (1979). Taxonomic significance of cell-wall acyl type in *Corynebacterium-Mycobacterium-Nocardia* group by a glycolate test. *J. Gen. Appl. Microbiol.* **25**, 169–183.
Uchida, K. and Aida, K. (1984). An improved method for the glycolate test for simple identification of the acyl type of bacterial cell walls. *J. Gen. Appl. Microbiol.* **30**, 131–134.
Uchida, K. and Seino, A. (1997). Intra- and intergeneric relationships of various Actinomycete strains based on the acyl types of the muramyl residue in cell wall peptidoglycans examined in a glycolate test. *Int. J. Syst. Bacteriol.* **47**, 182–190.
Uchida, K., Kudo, T., Suzuki, K. and Nakase, T. (1999). A new rapid method of glycolate test by diethyl ether extraction, which is applicable to a small amount of bacterial cells of less than one milligram. *J. Gen. Appl. Micobiol.* **45**, 49–56.
Weidel, W., Frank, H. and Leutgeb, W. (1963). Autolytic enzymes as a source of error in the preparation and study of Gram-negative cell walls. *J. Gen. Microbiol.* **30**, 127–130.
Whiton, R. S., Lau, P., Morgan, S. L., Gilbart, J. and Fox, A. (1985). Modifications in the alditol acetate method for analysis of muramic acid and other neutral and amino sugars by capillary gas chromatography-mass spectrometry with selected ion monitoring. *J. Chromatogr. A* **347**, 109–120.
Yokota, A. and Hasegawa, T. (1988). The analysis of madurose, an actinomycete whole-cell sugar, by HPLC after enzymic treatment. *J. Gen. Appl. Microbiol.* **34**, 445–449.
Yoon, J.-H., Schumann, P., Kang, S.-J., Jung, S.-Y. and Oh, T.-K. (2006). *Isoptericola dokdonensis* sp. nov., isolated from soil. *Int. J. Syst. Evol. Microbiol.* **56**, 2893–2897.
Young, K. D. (1998). Techniques for analysis of peptidoglycans. In: *Methods in Microbiology* (P. H. Williams, J. Ketley, and G. Salmond, eds.), Vol. 27, pp. 277–286. Academic Press, London.

6 Cell Wall Teichoic Acids in the Taxonomy and Characterization of Gram-positive Bacteria†

Natal'ya V. Potekhina[1], Galina M. Streshinskaya[1], Elena M. Tul'skaya[1] and Alexander S. Shashkov[2]

[1] Department of Biology, M. V. Lomonosov Moscow State University, Moscow, Russia; [2] N. D. Zelinsky Institute of Organic Chemistry, Russian Academy of Sciences, Moscow, Russia

†Dedicated to the memory of Professor L.V. Backinowsky (1938–2011) – friend and colleague, whose inspiration and insight problems contributed to the writing of this chapter.

CONTENTS

I. INTRODUCTION

Teichoic acids are anionic polymers comprising polyol phosphates as indispensable components of the main chain. They are the most abundant and the best studied class of Gram-positive cell wall components. About 10 genes encoding enzymes involved in teichoic acid biosynthesis have been characterized (Neuhaus and Baddiley, 2003; Swoboda *et al.*, 2010). Our knowledge about the structure of teichoic acids, their distribution, localization in the cell wall, the pathway for their biosynthesis as well as their functioning and biological properties were summarized in a number of reviews (Archibald, 1974; Baddiley, 1972; Naumova *et al.*, 2001; Naumova and Shashkov, 1997; Neuhaus and Baddiley, 2003; Potekhina, 2006; Swoboda *et al.*, 2010; Weidenmaier and Peschel, 2008). Teichoic acids have been found in microorganisms occupying diverse ecological niches such as soils, including those polluted with wastes of chemical industry and from salt-producing

METHODS IN MICROBIOLOGY, VOLUME 38
0580-9517 DOI 10.1016/B978-0-12-387730-7.00006-1

works, in permafrost and Antarctic ices, in seawater and sewage, as well as in clinical specimens and in foods (Deak, 2011; Gavrish *et al.*, 2004; Kroppenstedt and Evtushenko, 2006).

Cell wall teichoic acids (CWTAs) are characterized by their immense structural diversity (see Section II). The anionic nature of these compounds, due to the presence of phosphoric acid residues, determines their functioning in the cell: ion exchange and control of the activity of autolytic enzymes. Teichoic acids are localized on the cell surface and participate in interactions of bacteria with other microorganisms, plants, animals, proteins and antibodies (Swoboda *et al.*, 2010; Weidenmaier and Peschel, 2008).

The involvement of CWTAs in vital cell functions and the great variety of their structures, as well as the wide occurrence in Gram-positive bacteria, predetermined their significance as potential chemotaxonomic markers and suggested their applicability in the systematics of microorganisms.

A major contribution to the investigation into the range and abundance of teichoic acids among the representatives of the order Actinomycetales and the taxonomic value of CWTAs was made by Prof. I. B. Naumova (1931–2003).

The main types, methods of isolation and structural investigation of CWTAs are presented in this chapter. The use of these polymers in the taxonomy of members of the order Actinomycetales and the genus *Bacillus* is discussed.

♦♦♦♦♦♦ II. CELL WALL TEICHOIC ACIDS: OCCURRENCE AND STRUCTURAL DIVERSITY

Among the CWTAs studied, the following five structural types can be distinguished, depending on the composition of the main chains (Naumova *et al.*, 2001; Potekhina *et al.*, 2011): the polymers with a chain consisting of only polyol, glycosylpolyol or acylglycosylpolyol residues linked by phosphodiester bonds (types I–III, respectively), and the polymers of mixed structures, i.e. those containing glycosyl 1-phosphate in addition to polyol phosphate residues (type IV), or alternating polyol phosphate and glycosylpolyol phosphate units (type V).

Several factors contribute to the structural diversity of teichoic acids, such as the nature of the polyol and sugar (amino sugar) residues that might serve as both integral parts of the main chain and branches linked to the polyol residues by glycosidic bonds, different localization of the phosphodiester bonds and various *O*-acyl residues. Chains devoid of glycosyl and acyl substituents are also known.

The main types of teichoic acids structures, diversity and their occurrence among different groups of microorganisms are summarized in Table 1.

A. Poly(polyol phosphates)—Teichoic Acids of Type I

Polymers of type I contain glycerol, ribitol, erythritol, mannitol, arabinitol, and 6-*O*-methylgalactitol as polyol residues in the main chain (Naumova and Shashkov, 1997). Nine subtypes can be distinguished within this type based on variations of the polyols and the localization of the phosphodiester bonds.

Table 1. Types and subtypes of cell wall teichoic acids

Polyol	Subtype of main chain[a]	Glycosyle substituents[a]	Other substituents	Occurrence of teichoic acids in bacterial genera
		Type I, poly(polyol phosphates)		
Glycerol	-1)-Gro-(3-*P*-	α-, β-Glc; α-Gal; α-, β-GlcNAc; α-GalNAc; α-Rha; α-Rha3OMe; disaccharide: β-Gal3OMe-(1→3)-α-GalNAc-(1→; tetrasaccharide: β-Glc-(1→3)-α-GalNAc-(1→3)-β-GalNAc-(1→6)-α-GlcNAc-(1→	Gro-P, Rib-ol-P, D-Ala, L-Lys, L-Glu, AcOH, H_2SO_4	*Arthrobacter* (1), *Bacillus* (2), *Brevibacterium* (3), *Clavibacter* (1), *Glycomyces* (1), *herbidospora* (4), *Lactobacillus* (5), *Microbispora* (6), *Nocardiopsis* (7), *Nonomuraea* (1), *Planotetraspora* (4), *Staphylococcus* (5, 8), *Spirilliplanes* (9), *Streptomyces* (1, 10, 11, 12, 13), *Thermobifida* (6)
Glycerol	-2)-Gro-(3-*P*-	α-, β-Glc; α-Gal; disaccharide: α-Gal-(1→3)-β-GalNAc-(1→	none	*Bacillus* (2), *Glycomyces* (1), *Planotetraspora* (4), *Streptomyces* (1, 10, 14)
Ribitol	-1)-Rib-ol-(5-*P*-	α-, β-Glc; α-Gal; α-, β-GlcNAc; α-Rha; α-Rha3OMe; tetrasaccharide: α-Rha-(1→3)-α-Rha-(1→3)-β-Glc*p*NAc-(1→2)-α-Rha-(1→	Gro-P, Rib-ol-P, Lac-P, D-Ala, L-Lys, AcOH, Suc, (*R*)-Pyr	*Agromyces* (1), *Bacillus* (2), *Lactobacillus* (5, 8, 15), *Listeria* (7), *Nocardioides* (16), *Nocardiopsis* (7), *Staphylococcus* (5, 8), *Streptomyces* (1, 14)
Ribitol	-3)-Rib-ol-(5-*P*-	none	none	*Kroppenstedtia* (17), *Nocardiopsis* (7)
Mannitol	-1)-Man-ol-(6-*P*-	β-Glc, α-Rha	(*S*)-Pyr, H_3PO_4	*Brevibacterium* (3, 18, 19)
Mannitol	-4)-Man-ol-(6-*P*-	none	none	*Bifidobacterium* (7)
Erythritol	-1)-Ery-ol-(4-*P*-	α-, β-GlcNAc, β-GlcNAcA	AcOH	*Brachybacterium* (20), *Glycomyces* (1)
Arabinitol	Unknown	trisaccharide: α-6-deoxy-Tal-(1→3)-β–GlcNAc-(1→2)-α-Rha-(1→	none	*Agromyces* (1)
6-*O*-methyl-galactitol	Unknown	probably—GalNAc	none	*Erysipelothrix* (21)

(Continued)

Table 1. (Continued)

Polyol	Subtype of main chain[a]	Glycosyle substituents[a]	Other substituents	Occurrence of teichoic acids in bacterial genera
		Type II, poly(glycosylpolyol phosphates)		
Glycerol	-6)-Sugar-(1→1)-Gro-(3-*P*- R^1 R^2	α-, β-Glc; α-, β-Gal; disaccharide: -6)-α-Gal- -(1→3)-β-GalNAc-(1→ (basic chain); α-Man (branching, R^1); β-Glc (branching, R^2)	Gro-P	*Actinoplanes* (1, 22), *Bacillus* (2), *Lactobacillus* (5), *Streptomyces* (14, 22)
Glycerol	-6)-Sugar-(1→2)-Gro-(3-*P*- R	β-Glc, α-, β-Gal, β-Gal3OMe, α-GlcNAc (basic chain); α-, β-Glc (branching, R)	none	*Actinomadura* (1), *Actinocorallia* (4), *Bacillus* (2), *Spirilliplanes* (9), *Streptomyces* (13)
Glycerol	-4)-Sugar-(1→1)-Gro-(3-*P*-	β-Glc	none	*Streptomyces* (1)
Glycerol	-3)-Sugar-(1→1)-Gro-(3-*P*- R	α-, β-Gal, β-Glc (basic chain); α-GlcNAc, α-, β-Glc and disaccharide: α-Rha-(1®b-3)-GlcNAc-(1®b-, (branching, R)	(*S*)-Pyr; AcOH	*Actinomadura* (23), *Bacillus* (2), *Brevibacterium* (24), *Nocardioides* (25), *Streptomyces* (26)
Ribitol	-3)-Sugar-(1→1)-Rib-ol-(5-*P*- R	none	*Agromyces* (1)	
Ribitol	-3)-Sugar-(1→4)-Rib-ol-(1-*P*-	β-GlcNAc	none	*Bacillus* (2)
Ribitol	-4)-Sugar-(1→2)-Rib-ol-(5-*P*- R	β-GlcNAc (basic chain); β-Glc, α-Gal, α-GlcNAc (branching, R^1)	none	*Listeria* (27)
Ribitol	-4)-Sugar-(1→4)-Rib-ol-(5-*P*-	β-GlcNAc	none	*Listeria* (28)
Ribitol	-6)-Sugar-(1→1)-Rib-ol-(5-*P*-	oligosaccharides of the complex composition	EtN-P, Cho-P	*Streptococcus* (1, 29)
Glucitol	Unknown	oligosaccharide of the complex composition	none	*Streptococcus* (1)

Type III, poly(acylglycosylpolyol phosphates)				
Glycerol	-2)-GroA-(1→4)-Sugar-(1→1) – Gro–(3-*P*-	β-GroA4NQui	none	*Bacillus* (2)
Type IV, poly(polyol phosphate-glycosyl phosphates)				
Glycerol	-1)-Gro-(3-*P*-4)-Sugar-(1-*P*-	α-GlcNAc	D-Ala	*Staphylococcus* (7, 8)
Type V, poly(polyol phosphate-glycosylpolyol phosphates)				
Glycerol	-1)-Gro-(3-*P*-**3)**-Sugar-(1→2)-Gro-(3-*P*-	β-GalNAc	(*S*)-Pyr	*Nocardiopsis* (7, 30)
Glycerol	-1)-Gro-(3-*P*-**4)**-Sugar-(1→2)-Gro-(3-*P*-	β-GalNAc	Suc	*Nocardiopsis* (7, 30)

Abbreviations: Suc, succinic acid; Pyr, pyruvic acid; (*R*) and (*S*) refer to the absolute configuration of the pyruvic acid ketal; EtN-P, ethanolamine phosphate; Cho-P, choline phosphate; Lac-P, lactyl phosphate; Gro-P, glycerol phosphate; Rib-ol-P, ribitol phosphate; D-Ala, D-alanine; L-Lys, L-lysine; L-Glu, L-glutamic acid; Tal, talose; Rha, ramnose; AcOH, acetic acid; GroA, glyceric acid; 4NQui, 4-amino-4,6-dideoxyglucose (4-amino-4-deoxyquinovose); Rha3OMe, 3-*O*-methylramnose; Gal3OMe, 3-*O*-methylgalactose.

Numbers in parentheses correspond to the following references: (1) Naumova and Shashkov, 1997; (2) Potekhina *et al* 2011; (3) Potekhina *et al.*, 2004; (4) Streshinskaya *et al.*, 2002; (5) Baddiley, 1972; (6) Potekhina *et al.*, 2003b; (7) Naumova *et al.*, 2001; (8) Archibald, 1974; (9) Shashkov *et al.*, 2001; (10) Tul'skaya *et al.*, 1997; (11) Kozlova *et al.*, 1999; (12) Shashkov *et al.*, 1998; (13) Streshinskaya *et al.*, 2005; (14) Naumova, 1988; (15) Tomita *et al.*, 2009; (16) Shashkov *et al.*, 2000; (17) Evtushenko *et al.*, 2007; (18) Potekhina *et al.*, 2003a; (19) Potekhina *et al.*, 2003c; (20) Schubert *et al.*, 1993; (21) Schubert and Fiedler, 2001; (22) Kozlova *et al*, 2000; (23) Shashkov *et al.*, 1999a; (24) Shashkov *et al.*, 2004; (25) Shashkov *et al.*, 1999b; (26) Streshinskaya *et al.*, 2007; (27) Fiedler, 1988; (28) Uchikawa *et al.*, 1986; (29) Bergström *et al.*, 2003; (30) Tul'skaya *et al.*, 2007b.

[a]Since in certain studies, the ring size of the monosaccharide constituents (*p*-pyranose or *f*-furanose) and their absolute configuration (D or L) have not been determined, these characteristics are not mentioned. The absolute configuration of glycerol (*sn*-, e.g. L-glycerol) and mannitol (D) is omitted.

B. Poly(glycosylpolyol phosphates)—Teichoic Acids of Type II

Teichoic acids with the poly(glycosylpolyol phosphate) structure contain a glycosyl component in the main chain together with a polyol. The phosphodiester bond links the OH groups at C3 of glycerol or C5 of ribitol residues and C3, C4 or C6 of the sugar moiety. The glycosyl residues are usually attached to the OH group at C1, C2 of glycerol or C1, C2, C4 of ribitol of the main chain. Ten subtypes of poly(glycosylpolyol phosphate) polymers can be defined based on the combination of certain polyol residues (glycerol, ribitol, glucitol) and the position of phosphodiester bond between the polyol and sugar residues and glycosidic bond at the polyol. Polymers that contain a disubstituted glycerol residue, -6)-Sugar-(1 → 1/2)-[Sugar-(1 → 1/2)]-Gro-(3-*P*-, are also known (Potekhina *et al.*, 2011).

Polymers of types I and II are the most widely occurring bacterial CWTAs.

C. Poly(acylglycosylpolyol phosphates)—Teichoic Acids of Type III

This novel structural type of teichoic acids (thus far identified only in *B. subtilis* VKM B-762) has an unusual linkage pattern and glyceric acid (GroA) as the acyl component of the repeating unit. Phosphodiester bonds link OH groups at C3 of the glycerol residue and C2 of glyceric acid, which *N*-acylates the amino sugar (4-amino-4-deoxyquinovose) residue of the main chain: -2)-GroA4NQui-(1 → 1)-*sn*-Gro-(3-*P*-. Unlike other types of teichoic acids, the constituents of its polymeric chain are linked through three types of inter-residue bonds, i.e. amide, glycosidic, and phosphodiester (Shashkov *et al.*, 2011).

D. Poly(polyol phosphate-glycosyl phosphates)—Teichoic Acids of Type IV

Teichoic acids of this type, containing alternating glycerol phosphate and sugar 1-phosphate residues in the integral portion of the chain, have been identified only in some *Staphylococcus* strains (*S. lactis* 13 (Archibald, 1974), *S. hyicus* subsp. *hyicus* NCTC 10350 and *S. sciuri* DSM 20352 (Endl *et al.*, 1983)).

E. Poly(polyol phosphate-glycosylpolyol phosphates)—Teichoic Acids of Type V

Teichoic acids of this type have been found exclusively in *Nocardiopsis* strains. The main chain of the type V polymer consists of alternating units of glycerol phosphate and glycosylglycerol phosphate, where the glycosyl residue is *N*-acetyl-β-galactosamine. To date, two subtypes of the type V teichoic acids are known: 3,3-linked polymers [-1)-Gro-(3-*P*-3)-Sugar-(1 → 2)-Gro-(3-*P*-] with the phosphodiester bond between C3 of glycerol and C3 of the sugar residues, and 3,4-linked polymers [-1)-Gro-(3-*P*-4)-Sugar-(1 → 2)-Gro-(3-*P*-] with the phosphodiester bond between C3 of glycerol and C4 of the sugar residues.

♦♦♦♦♦♦ III. CELL WALL TEICHOIC ACIDS IN THE TAXONOMY OF GRAM-POSITIVE BACTERIA

A. Order Actinomycetales

1. Presence/absence of cell wall teichoic acids as a chemotaxonomic characteristic of members of the order Actinomycetales

Over the years, the possibility of employing CWTAs in the taxonomy of Gram-positive bacteria has been disputed. Initially, CWTAs were used to classify *Lactobacillus* and *Staphylococcus* species while the absence of teichoic acids from *Micrococcus* species is an important chemotaxonomic characteristic that distinguishes them from *Staphylococcus* species (Potekhina *et al.*, 2011). In a similar manner, one can classify strains of the genera *Saccharothrix* and *Nocardiopsis*. The transfer of certain species (*N. coeruleofusca*, *N. longispora*, and *N. syringae*) to the genus *Saccharothrix* (Grund and Kroppenstedt, 1990) correlates positively with the absence of CWTAs in their cell walls (Naumova and Shashkov, 1997). Currently, the characteristic presence/absence of CWTAs is a compulsory marker used in the description of certain genera of Gram-positive cocci (Holt *et al.*, 1997); CWTAs are absent in the members of the genera *Nesterenkonia*, *Kytococcus*, *Dermacoccus*, *Kocuria* (Stackebrandt *et al.*, 1995), and *Macrococcus*, and they are present in species of the genus *Staphylococcus* (Potekhina *et al.*, 2011).

Studies on the distribution of CWTAs among 500 strains, belonging to 345 species, 25 families, and 9 suborders of the order *Actinomycetales*, have shown that in the majority of cases the characteristic presence/absence of CWTAs allows rigorous discrimination of suborders and families of Actinobacteria. CWTAs have been found in representatives of the suborders Glycomycineae and Streptosporangineae; they are absent in the members of the suborders Corynebacterineae, Pseudonocardineae and Frankineae. This characteristic is variable for the organisms of the suborders Streptomycineae, Micromonosporineae, Propionibacterineae and Micrococcineae, and differentiates lower-rank taxa, e.g. separates the families *Nocardioidaceae* and *Propionibacteriaceae* of the suborder Propionibacterineae; it differentiates genera and groups of genera in the families *Streptomycetaceae* and *Micromonosporaceae*. The characteristic (presence/absence) differentiates species or groups of species inside the genera *Agromyces*, *Arthrobacter*, and *Brachybacterium* (families *Microbacteriaceae*, *Micrococcaceae*, and *Dermabacteriaceae*, respectively).

CWTAs are typical in actinomycetes with well-developed aerial mycelia and complex reproductive structures; they are less common in organisms with irregular rods/cocci morphology. These polymers were found in the overwhelming majority of actinomycetes that contain L,L-diaminopimelic acid as the peptidoglycan constituent.

Their proportion in the asporogenic organisms of the family *Nocardioidaceae* is usually smaller than in spore-forming ones, for instance, *Streptomycetaceae*.

No CWTAs were detected in actinomycetes with coccoid cells as well as in anaerobes of the family *Propionibacteriaceae*, or in organisms with arabinogalactan and mycolic acids in the cell walls (suborders Corynebacterineae,

Pseudonocardineae). CWTAs are also absent in organisms with peptidoglycan of group B (family *Microbacteriaceae*), with the exception of sporadic mycelial strains of the genus *Agromyces*. Besides, they are not found in the absolute majority of organisms (including sporangial) with lysine and ornithine as peptidoglycan constituents, for instance, members of the genera *Arthrobacter* (*'Arthrobacter globiformis* group') and *Microbacterium* (Takeuchi and Yokota, 1989). The exception forms the representatives of the genus *Rarobacter* (family *Rarobacteraceae*) and certain phylogenetically isolated species of the genus *Arthrobacter* (*'A. nicotianae* group', family *Micrococcaceae*) that contain CWTAs.

Thus, the characteristic presence/absence of CWTAs correlates with the assignment of actinomycetes to taxa of different levels as well as with the morphology of the microorganism and the qualitative composition of the other cell wall polymers.

2. Cell wall teichoic acids as a species-specific marker for members of the order Actinomycetales

The importance of the structural features of CWTAs in the taxonomy of prokaryotes becomes evident when considering similarity/dissimilarity of polymers within one phylogenetic group of organisms, such as genera and phenetic clusters. The data published for some genera (e.g. *Actinomadura, Agromyces, Brevibacterium, Glycomyces, Nocardioides, Nocardiopsis, Nonomuraea, Streptomyces*) have demonstrated that the structures of the CWTAs can be species-specific and indicate the phylogenetic clusters.

(a) Cell wall teichoic acids of Nocardiopsis species and subspecies

Cell wall teichoic acids have been studied in 32 strains belonging to 12 *Nocardiopsis* species and subspecies with validly published names (Euzéby, 2011), and verified by data on DNA–DNA relatedness (Kroppenstedt and Evtushenko, 2006). The indicative features of these polymer structures are summarized in Table 2. The structures and combinations of CWTAs in the *Nocardiopsis* species correlate well with the phylogenetic grouping of strains determined by 16S rRNA sequence analysis (Naumova *et al.*, 2001; http://www.bergeys.org/outlines/bergeys_vol_5_roadmap_outline.pdf). *N. dassonvillei* subsp. *dassonvillei* (18 strains including VKM Ac-797^T and VKM Ac-836^T), *N. dassonvillei* subsp. *albirubida* VKM Ac-1882^T, *N. synnemataformans* VKM Ac-2518^T and *N. halotolerans* VKM Ac-2519^T with CWTAs of type V fall into a common phylogenetic group with *N. dassonvillei*. The remaining species studied contain diverse CWTAs in different combinations and belong to other phylogenetic groups (Naumova *et al.*, 2001; Tul'skaya *et al.*, 2007b).

Taking into account the significant differences in the ^{13}C-NMR spectra of CWTAs (Figure 1), the latter may be considered as fingerprints of Nocardiopsis species and subspecies and successfully used for the identification of new strains of the genus. The current intrageneric classification is therefore linked to studies of CWTAs structures as chemotaxonomic markers for the species and subspecies of the genus *Nocardiopsis*.

Table 2. Cell wall teichoic acids of *Nocardiopsis* species and subspecies (VKM strains)

Teichoic acid	*N. dassonvillei subsp. dassonvillei* Ac-797^T, Ac-836^T and 16 other strains	*N. synnemataformans* Ac-2518^T	*N. halotolerans* Ac-2519^T	*N. dassonvillei subsp. albirubida* Ac-1882^T	*N. alba* Ac-1883^T, Ac-1879 Ac-1884	*N. listeri* 1881^T	*N. metallicus* Ac-2522^T	*N. prasina* Ac-1880^T	*N. lucentensis* Ac-1962^T	*N. compostus* Ac-2521^T, Ac-2520	*N. trehalosi* Ac-942^T	*N. tropica* Ac-1457^T
	Group of *N. dassonvillei*				Other groups							
(a) Poly(glycerol phosphate-*N*-acetyl-β-D-galactosaminyl-glycerol phosphate)/bond -(3-*P*-3)-	♦	♦♦										
(b) Poly(glycerol phosphate-*N*-acetyl-β-D-galactosaminyl-glycerol phosphate)/bond -(3-*P*-4)-		♦	♦									
(c) Poly(glycerol phosphate-*N*-acetyl-β-D-galactosaminyl -glycerol phosphate)/bond -(3-*P*-3)-, with (*R*)-pyruvate ketal groups			♦♦									
(d) Poly(glycerol phosphate-*N*-acetyl-β-D-galactosaminyl-glycerol phosphate)/bond -(3-*P*-4)-, with *O*-succinyl residues				♦								
(e) Unsubstituted 3,5-poly (ribitol phosphate)					♦			♦				
(f) 1,3-Poly(glycerol phosphate) with *N*-acetyl-α-D-glucosamine					♦	♦♦	♦	♦	♦	♦		
(g) 1,5-Poly(ribitol phosphate) with pyruvate ketal group[a]					♦	♦	♦♦					

(*Continued*)

Table 2. (Continued)

Teichoic acid	*N. dassonvillei subsp. dassonvillei* Ac-797^T, Ac-836^T and 16 other strains	*N. synnemataformans* Ac-2518^T	*N. halotolerans* Ac-2519^T	*N. dassonvillei subsp. albirubida* Ac-1882^T	*N. alba* Ac-1883^T, Ac-1879 Ac-1884	*N. listeri* 1881^T	*N. metallicus* Ac-2522^T	*N. prasina* Ac-1880^T	*N. lucentensis* Ac-1962^T	*N. compostus* Ac-2521^T, Ac-2520	*N. trehalosi* Ac-942^T	*N. tropica* Ac-1457^T
	Group of *N. dassonvillei*				Other groups							
(h) 1,3-Poly(glycerol phosphate) with β-D-glucopyranose												♦
(i) 1,5-Poly(ribitol phosphate) with side chains of glycerol phosphate oligomers												♦

♦, presence of teichoic acid in the cell wall; ♦♦, the predominant teichoic acid; the CWTAs structures are denoted by letters and their ^{13}C NMR spectra are shown in Figure 2.

Abbreviations: ^{T}, type strain; VKM, All-Russian Collection of Microorganisms.

[a]Absolute configuration of the pyruvic ketal group (*R* or *S*) not detected.

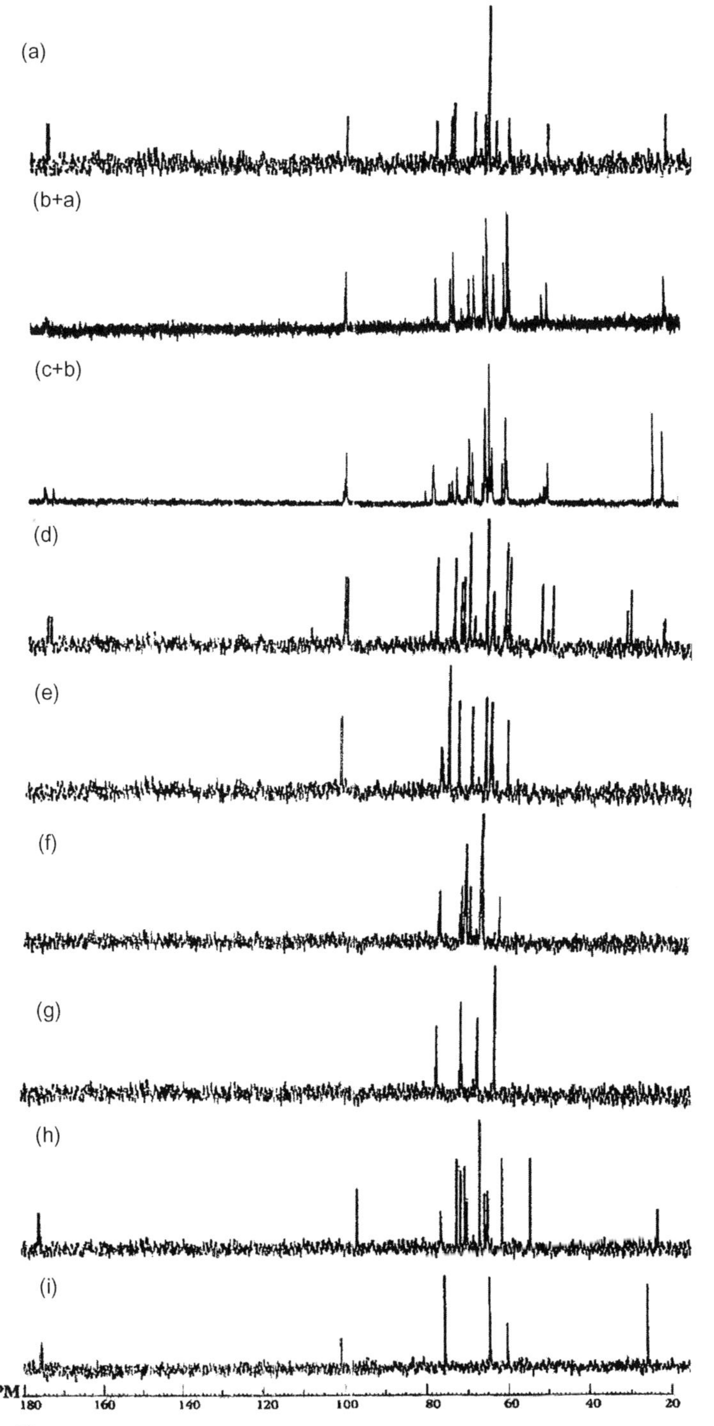

Figure 1. ^{13}C NMR spectra of cell wall teichoic acids of *Nocardiopsis* species and subspecies (symbols are as given in Table 2).

(b) Cell wall teichoic acids of Brevibacterium species

A comparative analysis of CWTAs of five *Brevibacterium* species with validly published names (Euzéby, 2011)—*B. linens* VKM Ac-2112^T, *B. iodinum* VKM Ac-2106^T, *B. permense* VKM Ac-2280^T, *B. antiquum* (VKM Ac-2118^T and Ac-2281), and *B. aurantiacum* VKM Ac-2111^T—showed each type strain to contain a distinctive set of anionic polymers (Table 3). The differences in the structures of the teichoic acids of *B. permense* VKM Ac-2280^T, *B. antiquum* VKM Ac-2118^T and *B. aurantiacum* VKM Ac-2111^T agree well with the segregation of these strains from *B. linens* into a new species, based on data of DNA–DNA hybridization as well as physiological and biochemical characteristics (Gavrish *et al.*, 2004). A characteristic feature of brevibacteria is the presence of CWTAs with

Table 3. Cell wall teichoic acids of *Brevibacterium* species (VKM strains)

Teichoic acid/substitution	*B. permense* Ac-2280^T	*B. antiquum* Ac-2118^T, Ac-2281	*B. iodinum* Ac-2106^T	*B. aurantiacum* Ac-2111^T	*B. linens* Ac-2112^T
1,6-Poly(mannitol phosphate) with α-L-rhamnopyranose and (*S*)-pyruvate ketal groups	♦				
1,6-Poly(mannitol phosphate) with phosphate groups		♦			
1,6-Poly(mannitol phosphate) with β-D-glucopyranose and (*S*)-pyruvate ketal groups			♦		
1,3-Poly(glycerol phosphate) with *N*-acetyl-α-D-galactosamine and (*R*)-pyruvate ketal groups			♦		
1,3-Poly(glycerol phosphate) with *N*-acetyl-α-D-galactosamine and L-glutamic acid				♦	
1,3-Poly(glycerol phosphate) with β-D-glucopyranose					♦
Poly(galactosylglycerol phosphate)[a]				♦	
Poly(galactosylglycerol phosphate)[b]					♦
1,3-Poly(glycerol phosphate)	♦	♦	♦	♦	♦

Abbreviations: VKM, ^{T}, *sn*-, D-, *S*-, *R*-, *p*, are given in Tables 1 and 2.
[a] 4)-α-D-Gal*p*-(1→2)-*sn*-Gro-(3-*P*-.
[b] -6)-α-D-Gal*p*-(1→2)-*sn*-Gro-(3-*P*-.

mannitol as a polyol (Potekhina *et al.*, 2003a, 2003b, 2004), and detection of the latter in cell wall hydrolysates can facilitate the identification of new strains of the genus *Brevibacterium*.

(c) Cell wall teichoic acids of *Agromyces* species and subspecies

Type strains of four species of the genus *Agromyces* with validly published names (Euzéby, 2011) contain different CWTAs with glycerol, ribitol and arabinitol as a polyol (Naumova and Shashkov, 1997) (Table 4). Thus, the CWTAs and their constituents can be used as a chemotaxonomic characteristic in the strain attribution of the genus *Agromyces*.

(d) Cell wall teichoic acids of *Nocardioides* species

Currently, the genus *Nocardioides* has 49 species with validly published names (Euzéby, 2011). The presence of CWTAs was shown for type strains *N. simplex*, *N. jensenii* (Takeuchi and Yokota, 1989), *N. albus* and *N. luteus* (Shashkov *et al.*, 1999b, 2000). The structure of the CWTA of *N. simplex* VKM Ac-1118^{T} was determined as 1,3-poly(glycerol phosphate) bearing *O*-acetic residues by chemical methods (Naumova, 1988). The full structure of CWTAs of the two mycelium-forming species (*N. albus* and *N. luteus*) was established, and the taxonomic value of CWTAs structures has been demonstrated for these species. The strain *N. luteus* VKM Ac-1246^{T} and seven strains with white colonies (VKM Ac-562, Ac-563, Ac-564, Ac-565, Ac-566, Ac-567, Ac-568) previously assigned to the species *N. albus* exhibited between 63 and 74.2% DNA relatedness (Tul'skaya *et al.*, 2003), and their cell walls contained identical 1,5-poly(ribitol phosphates) completely substituted at O4 with α-galactopyranosyl residues carrying a 4,6-(*R*)-pyruvate ketal group (Shashkov *et al.*, 2000). The strain *N. albus* VKM Ac-805^{T} possesses a poly(galactosylglycerol phosphate) in which the β-galactopyranosyl residues are substituted at O4 with β-glucopyranose bearing a

Table 4. Cell wall teichoic acids of *Agromyces* species and subspecies (strains of VKM)

Teichoic acid/substitution	*A. fucosus* subsp. *fucosus* Ac-1345^{T}	*A. hippuratus* Ac-1352^{T}, Ac-1353	*A. cerinus* subsp. *nitratus* Ac-1351^{T}	*A. cerinus* subsp. *cerinus* Ac-1340^{T}
1,3-Poly(glycerol phosphate) with *N*-acetyl-β-D-glucosamine	♦			
1,5-Poly(ribitol phosphate) with a tetrasaccharide[a]		♦		
Poly(ribofuranosylribitol phosphate)[b]			♦	
Poly(arabinitol phosphate) with a disaccharide[c] or a trisaccharide[d]				♦

Abbreviations: VKM, ^{T}, *sn*-, D-, *p*, and *f* are given in Tables 1 and 2.
[a] α-L-Rha*p*-(1→3)-α-L-Rha*p*-(1→3)-β-D-Glc*p*NAc-(1→2)-α-L-Rha*p*-(1→.
[b] -3)-[α-D-Gal*p*-(1→2)]-β-D-Rib*f*-(1→1)-Rib-ol-(5-*P*-.
[c] β-D-Glc*p*NAc-(1→2)-α-L-Rha*p*-(1→.
[d] α-6-deoxy-L-Tal*p*-(1→3)-β-Glc*p*NAc-(1→2)-α-Rha*p*-(1→.

4,6-(*S*)-pyruvate ketal group (Shashkov *et al.*, 1999b). On the basis of the data obtained, it was proposed to transfer the above-mentioned seven strains considered to be *N. albus* to the species *N. luteus* and to consider CWTAs and their structural components as important diagnostic characteristics of these species (Tul'skaya *et al.*, 2003). Thus, CWTAs are species-specific for three species of the genus *Nocardioides*.

(e) Cell wall teichoic acids of Glycomyces species

Currently, the genus *Glycomyces* has 10 species with validly published names (Euzéby, 2011), and representatives of three of them have been studied: *G. tenuis* VKM Ac-1250^T, *G. rutgersensis* VKM Ac-1248^T and *G. harbinensis* (VKM Ac-1247^T and NRRL 16897). *G. tenuis* VKM Ac-1250^T contained a unique CWTA with erythritol as a polyol: 1,4-poly(erythritol phosphate) with *N*-acetyl-α-glucosaminyl and acetyl substituents. The CWTA of *G. rutgersensis* VKM Ac-1248^T was represented by 1,3-poly(glycerol phosphate) bearing α-glucopyranosyl side-groups. Two polymers, 1,3- and 2,3-poly(glycerol phosphates) with α-glucopyranosyl residues, were identified in *G. harbinensis* VKM Ac-1247^T and NRRL 16897 (Naumova and Shashkov, 1997; Potekhina *et al.*, 1998).

Thus, the representatives of the type strains of three species of the genus *Glycomyces* contained different sets of CWTAs with dissimilar structures, while identical polymers were identified in the cell walls of two strains of *G. harbinensis*.

(f) Cell wall teichoic acids of Actinomadura and Nonomuraea species

Screening for the teichoic acids of 29 species (37 strains) of the genus *Actinomadura* was carried out (Naumova, 1988) and the complete structures were established for CWTAs of *A. cremea* INA 292^T, *A. madura* VKM Ac-809^T and *A. viridis* (VKM Ac-1315^T, VKM Ac-631), *A. roseoviolacea* subsp. *carminata* INA 4281^T, *A. polychroma* INA 2755^T and *A. recticatena* VKM Ac-940^T. After reclassification of the genus *Actinomadura* (Zhang *et al.*, 1998), first three species studied have been left as the members of this genus, while last three ones were transferred to the genus *Nonomuraea* (Euzéby, 2011). The polymers of the *Actinomadura* species belonged to type II and each of the strains of the three species studied possessed its own CWTA structure. Thus, 3,6-linked poly[galactopyranosyl-(β1→1)-glycerol phosphate] with lateral glycerol phosphate units attached at O3 of galactopyranose was detected in the cell wall of *A. cremea* INA 292^T. In *A. madurae* VKM Ac-809^T, 3,6-linked poly[glycosyl-(β1→2)-glycerol phosphate] polymers with galactopyranose and 3-*O*-methyl-galactopyranose as a glycosyl residue in the basic chain were identified (Naumova and Shashkov, 1997). The CWTAs of two strains of *A. viridis* (VKM Ac-1315^T and VKM Ac-631) are represented by 3,3-linked poly[galactopyranosyl-(β1→1)-glycerol phosphate] with 3-*O*-methyl-β-galactopyranose and β-glucopyranose as substituents at O4 and/or O6 of galactopyranose, respectively, along with unsubstituted 1,3-poly(glycerol phosphate) (Shashkov *et al.*, 1999a).

That the same type of CWTAs is characteristic of polymers of *A. japonica* JCM 3396, *A. formosensis* VKM Ac-1954^T, *A. kijaniata* VKM Ac-874^T, *A. rubrobrunea*

VKM Ac-775^T, and *A. livida* VKM Ac-908^T was inferred from their NMR spectroscopic data, though their structures have not been elucidated so far. Likewise, acid hydrolysis data allow assignment of this type of CWTAs to the polymers of *A. macra* ATCC 31286^T, *A. fulvescens* INA 3321^T, *A. cremea* subsp. *rifamycini* INA 1349^T, and *A. coeruleoviolacea* INA 3564^T (Naumova, 1988).

Thus, CWTAs with the poly(glycosylglycerol phosphate) structure are abundant in the representatives of the genus *Actinomadura* (Naumova, 1988), and the structures of the described CWTAs of the type strains studied of this genus are evidence of their species specificity.

Species-specificity of CWTAs was also shown for representatives of the genus *Nonomuraea*. Three species studied were found to contain 1,3-poly (glycerol phosphates), though each of them possesses its own features. Thus, *N. roseoviolacea* subsp. *carminata* INA 4281^T has a CWTA with *N*-acetyl-α-D-galactosamine and a disaccharide, β-D-Gal*p*3OMe-(1→3)-α-D-Gal*p*NAc. The disaccharide substituent is also identified in the CWTA of *N. polychroma* INA 2755^T, while the CWTA of *N. recticatena* VKM Ac-940^T contained only *N*-acetyl-α-D-galactosamine as a substituent (Naumova and Shashkov, 1997). Previous data (Naumova, 1988) together with the present findings allow us to propose that CWTAs of type I are abundant in the members of the genus *Nonomuraea*, which distinguish them from the members of the genus *Actinomadura*.

Cell Wall Teichoic Acids in the Taxonomy and Characterization

(g) Cell wall teichoic acids of *Streptomyces* species

The genus *Streptomyces* includes a large number of species with validly published names, many of them being synonyms (Anderson and Wellington, 2001). Williams *et al.* (1983) used a numerical taxonomic approach based on phenetic characters which resulted in a reduction in the number of described *Streptomyces* species. On the basis of phenetic characters, the genus *Streptomyces* was subdivided into species-groups or taxonomic clusters and the taxonomic status of representatives of the major clusters groups defined by Williams *et al.* (1983) was supported by sequence analysis of 16S rRNA genes (γ region) (Anderson and Wellington, 2001). Each cluster could be regarded as single species despite the high diversity observed within some clusters, and these, therefore, were regarded as species-groups (Anderson and Wellington, 2001). However, subsequent studies of DNA–DNA homology of the organisms pertaining to separate phenetic clusters have shown their heterogeneity (Labeda, 1992, 1996) and served as the basis for reclassification of several species.

Highlighting the value of a polyphasic approach, the set and structure of cell wall polymers may be one of the phenotypic traits for delimitation of species within the genus *Streptomyces* on one hand and serve the characteristics for members of species-groups on the other hand.

The results of the study of CWTAs among members of the phenetic *Streptomyces* clusters (*S. cyaneus*, *S. fulvissimus*, *S. violaceusniger*) and the *Streptoverticillium* species group are presented in the sections below.

The *Streptomyces cyaneus* cluster (Williams *et al.*, 1983)

Cell walls of the cluster members (13 species—*S. afghaniensis* Ac-703, *S. janthinus* Ac-208, *S. purpurascens* Ac-755^T, *S. roseoviolaceus* Ac-1901, *S. violatus* Ac-532, *S. azureus* Ac-719, *S. bellus* Ac-573, *S. caelestis* Ac-1822, *S. coeruleorubidus* Ac-576^T, *S. curacoi* Ac-621, *S. violarus* Ac-528, *S. cyaneus* Ac-1307, and *S. hawaiiensis* Ac-1761 from the All-Russian Collection of Microorganisms, VKM) contain two teichoic acids of the same structures—β-glucosylated 1,5-poly(ribitol phosphate) and unsubstituted 1,3-poly(glycerol phosphate). The first five strains are characterized by a high degree (67–88%) of DNA homology (Labeda and Lyons, 1991) and have completely substituted 1,5-poly(ribitol phosphates). The other eight strains have a lower level of DNA homology in the 54–76% range (Labeda and Lyons, 1991) and ribitol teichoic acids with different degrees of β-glucosylation.

Thus, the CWTA structures suggest homogeneity of the *S. cyaneus* cluster and confirm the genetic relatedness of the constituent species (Streshinskaya *et al.*, 2003).

The *Streptomyces fulvissimus* cluster (Williams *et al.*, 1983)

The investigated cluster members *S. fulvissimus* VKM Ac-994^T, *S. longispororuber* VKM Ac-1735^T, *S. aureoveticillatus* VKM Ac-48^T and *S. spectabilis* INA 00606 exhibited a low level of DNA relatedness (Labeda, 1998) and contain identical 1,3-poly(glycerol phosphates) substituted with L-glutamic acid, L-lysine and *N*-acetylated α-D-glucosamine (Shashkov *et al.*, 2006). Despite genetic differences in the cluster members and certain quantitative differences in the structures of the polymers, one may consider the details of the CWTA structures and, specifically, the presence of L-glutamic acid residues as a chemotaxonomic characteristic for the representatives of this cluster.

The *Streptomyces violaceusniger* cluster (Williams *et al.*, 1983)

The members of this cluster for which CWTAs have been examined include *S. melanosporofaciens* VKM Ac-1864^T, *S. hygroscopicus* subsp. *hygroscopicus* VKM Ac-831^T, *S. violaceusniger* VKM Ac-583^T, *S. endus* VKM Ac-1331^T, *S. endus* VKM Ac-129 and *S. castelarensis* VKM Ac-832^T. They are phenotypically and phylogenetically closely related species that share a 16S rRNA gene similarity of 99.3–99.7% (Kumar and Goodfellow, 2008) and contain CWTAs with identical structures and the Kdn-containing polymer (Figure 2). 1,3-Poly(glycerol phosphate) carrying *O*-acetyl and *O*-L-lysyl groups and substituted with α-D-glucosamine, only part of which are *N*-acetylated, is the quantitatively major polymer; unsubstituted 1,3- and 2,3-poly(glycerol phosphates) are minor polymers (Tul'skaya *et al.*, 2007a). The above-mentioned species differ in the number of α-D-glucosaminyl substitutes and in the degree of their *N*-acetylation in the predominant teichoic acid. These differences are seen by the comparison of ^{13}C NMR spectra of CWTAs (Figure 2).

Thus, the structure and set of CWTAs suggest the homogeneity of the *S. violaceusniger* cluster and can be used as a characteristic typical of its representatives.

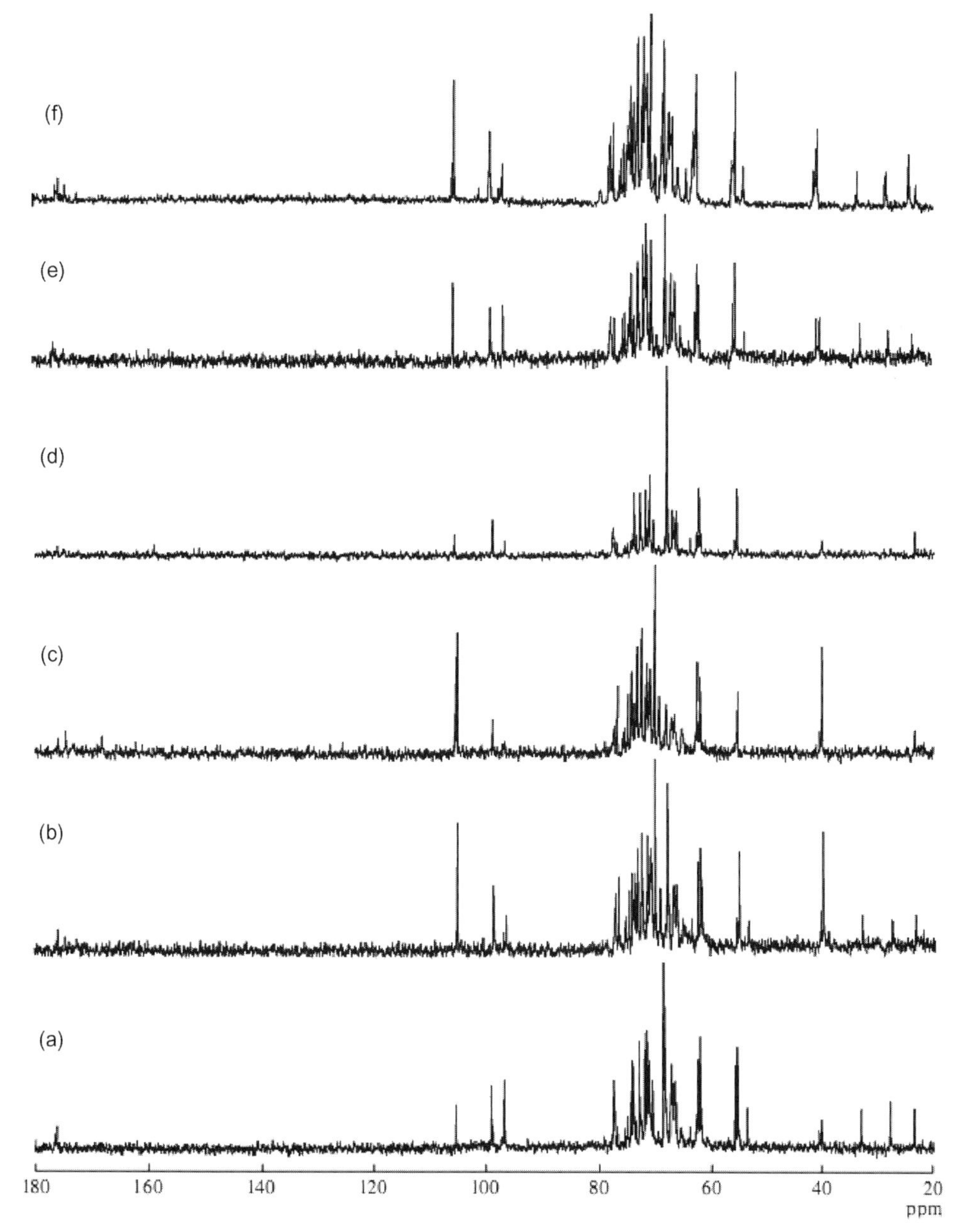

Figure 2. ^{13}C NMR spectra of cell wall glycopolymers found in the species of the cluster *S. violaceusniger*. (a) *S. endus* VKM Ac-1331^T, (b) *S. endus* VKM Ac-129, (c) *S. hygroscopicus* subsp. *hygroscopicus* VKM Ac-831^T, (d) *S. castelarensis* VKM Ac-832^T, (e) *S. violaceusniger* VKM Ac-583^T, (f) *S. melanosporofaciens* VKM Ac-1864^T (Tul'skaya *et al.*, 2007a).

The *Streptoverticillium* species group

The *Streptoverticillium* strains belong to a historically isolated group of the genus *Streptomyces* characterized by their whorl production. Twenty-five strains form four genospecies in accordance with the DNA–DNA relatedness (70–75%, Labeda, 1996) and three groups (Table 5) in compliance with the structures of the cell wall polymers (Kozlova *et al.*, 1999) and general similarity of their phenotypic characteristics (unpublished data). A characteristic feature of *Streptoverticillium* species cell wall is the presence of neutral phosphate-containing polymers where the amino sugar with the free amino group ($-NH_3^+$) neutralizes the negative charge of the chain phosphate group. These data indicate

Table 5. Cell wall polymers of *Streptomyces* species belonging to the *Streptoverticillium* group (strains of VKM)

Strains	1,3-Poly(glycerol phosphate) with		Poly(sugar 1-phosphate)[c]
	GlcNAc[a]	$GlcNH_3^+$ [b]	
Group 1 (genospecies 1 and 2)			
S. flavopersicus Ac-947	♦	♦	
S. kentuckensis Ac-877	♦	♦	
S. netropsis Ac-820	♦	♦	
S. ladakanum Ac-879	♦	♦	
S. mobaraensis Ac-928	♦	♦	
S. distallicus Ac-948	♦	♦	
Group 2 (genospecies 3)			
S. baldacci Ac-821[T]		♦	
S. biverticillatus Ac-891[T]		♦	
S. roseoverticillatus Ac-880		♦	
S. aureoversilis Ac-884		♦	
S. griseoverticillatus Ac-883		♦	
S. salmonis Ac-944		♦	
S. blastmyceticus Ac-766		♦	
S. eurocidicus Ac-903		♦	
S. hiroshimensis Ac-902		♦	
S. fervens BKM Ac-882		♦	
S. olivoverticillatus Ac-890		♦	
S. mashuensis Ac-949		♦	
S. ardum Ac-930		♦	
Group 3 (genospecies 4)			
S. hachijoensis Ac-191[T]			♦
S. abikoensis Ac-946			♦
S. cinnamoneus Ac-876			♦
S. cinnamoneus subsp. *azacoluta* Ac-606[T]			♦
S. griseocarneus Ac-881			♦
S. lavenduligriseus Ac-934			♦

Abbreviations: VKM, [T] and *p* are given in Tables 1 and 2.
[a]GlcNAc, 2-acetamido-2-deoxy-α-D-glucopyranose.
[b]$GlcNH_3^+$, 2-amino-2-deoxy-α-D-glucopyranose.
[c]-6)-α-D-Glc*p*NH_2-(1→6)-α-D-Glc*p*NAc-(1-*P*-.

specificity of the phosphate-containing polymers in the strains of *Streptoverticillium* group, despite the heterogeneity of this group.

B. The Genus *Bacillus*

1. Presence/absence of cell wall teichoic acids as a chemotaxonomic characteristic for species of the genus *Bacillus*

Strains of *B. anthracis*, *B. cereus*, and *B. thuringiensis* belonging to the *B. cereus* group are devoid of teichoic acids, while strains studied of the *B. subtilis* group contain them. Thus, strains of two groups, *B. cereus* and *B. subtilis*, that form phylogenetically different clusters also differ in characteristics such as the presence/absence of CWTAs (Potekhina *et al.*, 2011).

2. Cell wall teichoic acids as a species-specific marker for the representatives of the *Bacillus subtilis* group

The determination of the taxonomy of some species clusters of the genus *Bacillus* has been impeded by the high level of similarity of sequences of the 16S rRNA gene as well as similar phenotypic and biochemical characteristics (Rooney *et al.*, 2009). Recent studies have shown that the presence of certain CWTAs may be used to discriminate closely related strains within the *B. subtilis* group (unpublished data) in addition to its use for members of the order Actinomycetales. Investigations into the structures of CWTAs have been performed for a number of type strains of the *B. subtilis* group. Representatives of each species/subspecies differed in the structures and the set of phosphate-containing cell wall polymers (Table 6). Thus, the later can serve as markers for the subdivision of the strains of bacilli at the species/subspecies level.

♦♦♦♦♦♦ IV. METHODS FOR THE ISOLATION AND STRUCTURAL INVESTIGATION OF CELL WALL TEICHOIC ACIDS

Approaches for the use of CWTAs in the taxonomy of the members of the order Actinomycetales and the genus *Bacillus* may vary, depending on the aims of the study. They may include different chemical and physicochemical methods and combinations of those.

Conventional methods, e.g. cell wall hydrolysis with 2 M HCl, can be sufficient to answer the question of the presence/absence of CWTAs. Specific hydrolysis products such as polyol phosphates, which can be detected by paper electrophoresis, and polyols identifiable by paper chromatography are specific markers of CWTAs. These simple methods of chemical analysis allow one to quickly determine these characteristics for the organisms studied.

Elucidation of the complete structure is required to use CWTAs as a species-specific characteristic. This includes localization of the phosphodiester, *O*-ester,

Table 6. Cell wall teichoic acids of *Bacillus* species and subspecies (VKM strains)

Teichoic acid/substitutions	*B. vallismortis* B-2653^{T}	*B. subtilis* subsp.		*B. sonorensis* B-2652^{T}	*B. amyloliquefaciens* B-2582^{T}	*B. pumilus* B-508^{T}	*B. mojavensis* B-2588^{T}	*B. licheniformis* B-511^{T}
		subtilis B-501^{T} (=168)[a]	*inaquosorum* B-2572^{T}					
1,3-Poly(glycerol phosphate)				♦	♦	♦	♦	♦
1,3-Poly(glycerol phosphate) with D-Ala				♦	♦	♦	♦	♦
1,3-Poly(glycerol phosphate) with β-D-glucopyranose			♦	♦				
1,3-Poly(glycerol phosphate) with α-D-glucopyranose		♦			♦		♦	
1,3-Poly(glycerol phosphate) with *N*-acetyl-β-D-glucosamine	♦							
1,3-Poly(glycerol phosphate) with *N*-acetyl-α-D-glucosamine						♦		
Poly(glucosylglycerol phosphate)[b]							♦	
1,5-Poly(ribitol phosphate) with β-D-glucopyranose and D-Ala			♦					
1,5-Poly(ribitol phosphate) with *N*-acetyl-β-D-glucosamine and D-Ala	♦							
Poly(galactosylglycerol phosphate)[c]								♦

Abbreviations: VKM, ^{T}, *sn*-, D- and *p* are given in Tables 1 and 2.
[a]The set of cell wall polymers from the strain *B. subtilis* 168 (*B. subtilis* subsp. *subtilis* NRRL B-744^{T}) is identical with that in the strain VKM B-501^{T} (Potekhina *et al.*, 2011).
[b]-4)-α-D-[α-D-Glc*p*NAc-(1→3)]-Glc*p*-(1→2)-*sn*-Gro-(3-*P*-.
[c]-6)-α-D-Gal*p*-(1→2)-*sn*-Gro-(3-*P*-.

and glycosidic bonds and the determination of the configuration of the latter, as well as the nature and absolute configuration of the constituent polyols, monosaccharides, amino acids, and organic acids.

Formerly, CWTAs have been isolated from whole cells and purified from membrane components by gel chromatography. Structure elucidation of CWTAs required recourse to laborious and time-consuming chemical and enzymatic degradation methods. With the help of ultrasonic disintegrators, considerable amounts of cell walls became accessible, which allowed direct isolation of teichoic acids from them. The content of contaminants in the obtained preparations became much smaller so that additional purification became, as a rule, unnecessary.

Progress in studies of CWTAs in the 1970–1980s was associated with the introduction of non-destructive methods of analysis, such as ^{13}C, ^{1}H, and ^{31}P NMR spectroscopy, into practice. The chemical methods became secondary, though they cannot be abandoned entirely. Thus, gel chromatography allows separation of both a mixture of polymers and their hydrolysis products. The use of 48% hydrofluoric acid for dephosphorylation and access to glycosides followed by acetylation of the latter make it possible to localize the position of the polyol glycosylation. Determination of the absolute configuration of one monosaccharide (by chemical methods) in a disaccharide fragment of a chain allows determination of the absolute configuration of the second monosaccharide based on the value of the glycosylation effect (in NMR experiments). Structural studies of teichoic acids in polymer mixtures by NMR spectroscopy can be accomplished without isolation of individual components, and the use of diffusion-ordered NMR spectroscopy (DOSY) makes it possible to infer the presence of structurally different teichoic acids in the preparation. The existing database of NMR spectra, which is continuously expanded, favours structural elucidation of teichoic acids, and specific features of polymer ^{13}C NMR spectra offer yet another way to identify the organisms of a genus based on their CWTA composition.

A. Cultivation Conditions

Cells for teichoic acid analysis may be grown on any suitable liquid media supplemented with phosphates (e.g. KH_2PO_4) indispensable for the polymer biosynthesis.

1. Cultivation conditions of actinobacteria

Produce cell mass by growing the culture aerobically at 28°C, with shaking, in a peptone-yeast medium of the following composition: 5.0 g peptone, 3.0 g yeast extract, 0.2 g K_2HPO_4, 5.0 g glucose, 1000 ml tap water, pH 7. The biomass is collected at the exponential phase of growth (when the quantity of CWTAs is maximum); the culture purity should be checked by microscopy and by plating on peptone-yeast medium agar. The cells are harvested by centrifugation, washed with cold 0.95% NaCl, stored at −18°C, and used for cell wall isolation.

2. Cultivation conditions of *Bacillus* species

To obtain biomass, cells are grown aerobically in shake flasks for 36 h at 28°C in medium containing aminopeptide (a pre-digested casein), 60 ml; tryptone, 5 g; yeast extract, 1 g; soy extract, 30 ml; K_2HPO_4, 0.2 g; distilled water to 1000 ml, pH 7.2 before sterilization. After sterilization, 20 ml of 50% molasses is added to the medium.

B. Isolation of Cell Walls

Biomass (about 50 g wet weight in 70 ml of distilled water) is disrupted by several two-minutes cycles of sonication at 22–30 kHz at 4°C. Suspension is centrifuged at 10,000 × *g* for 15–20 min at 40°C, the resulting pellet is resuspended in 2% sodium dodecyl sulfate heated for 5 min at 100°C and centrifuged as above. The lower dense portion of the pellet (unbroken cells) is discarded and the upper loose layer (native cell walls) is collected with a spatula, washed several times with distilled water, and freeze-dried. The cell walls are used for extraction of teichoic acids.

C. Extraction of Teichoic Acids

There are a few methods of CWTA isolation described in the literature (Archibald, 1972). The most suitable is the fractional extraction with 10% trichloroacetic acid (TCA).

- Extract teichoic acids from the freeze-dried cell wall (about 500 mg) with aqueous 10% (w/v) TCA (~10 ml) for 24 h at 4°C with constant stirring.
- Centrifuge the mixture at 10,000 × *g* at 40°C for 15 min and collect the supernatant.
- Repeatedly (one or two times) treat the remaining cell walls as above.
- Combine the supernatants, dialyse against distilled water, and freeze-dry to yield preparation I.
- Additionally, treat the remaining cell walls following extraction with cold 10% TCA with 5% TCA for 5 min at 100°C with constant stirring and centrifuge as above.
- Dialyse the supernatant obtained against distilled water and freeze-dry to yield preparation II.

Precipitation by portionwise addition of ethanol is one of the methods for fractionation of teichoic acids and is carried out as follows:

- Following extraction with cold 10% TCA, mix the combined supernatant with two volumes of 96% ethanol, and left for 24 h at 4°C.
- Collect the resulting precipitate by centrifugation at 10,000 × *g* at 40°C for 15 min.
- Dissolve the precipitate in cold distilled water (~0.5 ml/20 mg), centrifuge at 10,000 × *g* at 40°C for 15 min, and discard the insoluble material. Repeat the procedure 2–3 times and combine the supernatants.

- Dialyse the combined supernatant against distilled water at 4°C to a pH of ~6.
- Lyophilize to yield preparation III.
- Add two more volumes of 96% ethanol to the supernatant obtained after first precipitation by ethanol, and process as above to obtain preparation IV.
- Use freeze-dried preparations I–IV for later investigation by chemical and NMR spectroscopic methods.

D. Chemical Methods of Teichoic Acids Analysis

1. Acid hydrolysis

Acid hydrolysis of cell walls and preparations I–IV (2–10 mg) is carried out in sealed tubes under different conditions.

To detect teichoic acid degradation products (polyols, monosaccharides, phosphoric esters, amino sugars), hydrolysis is performed in 2 M HCl for 3 h at 100°C. After removal of HCl by evaporation, the samples are examined by paper chromatography and electrophoresis.

Dephosphorylation of teichoic acid (~ 50 mg) with 48% aqueous hydrofluoric acid (~0.5 ml) is carried out at 4°C for 16 h, followed by evaporation of the volatiles with a stream of air. The dephosphorylated products are separated on a column (90 × 1.5 cm) of TSK HW-40S gel (Toyopearl, Japan), eluted with 1% acetic acid using a differential refractometer (Knauer, Germany) to monitor elution. The obtained glycosides are characterized by NMR spectroscopic methods. The glycosides can be acetylated with an acetic anhydride pyridine mixture (1:1, by volume, 22°C, 18 h), followed by evaporation in vacuum. The preparation obtained is dissolved in deuteriochloroform (0.4 ml) and investigated by NMR spectroscopic methods.

2. Descending paper chromatography

Descending paper chromatography and electrophoresis are performed on Filtrak (Germany) FN-3 paper, by comparison with authentic samples. The following solvent systems are used: (i) pyridine:benzene:butan-1-ol:water (3:1:5:3, by volume) to separate polyols, glycosides, and sugars; (ii) butan-1-ol:acetic acid:water (4:1:5, by volume) to separate amino acids and amino sugars; (iii) pentanol:5 M formic acid (1:1, by volume) to separate organic acids.

3. Paper electrophoresis

Paper electrophoresis is performed in pyridine-acetate buffer (glacial acetic acid 3.5 ml, pyridine 11.8 ml per 1000 ml distilled water, pH 5.6, using a potential gradient of 20 V/cm for 3–4 h) to separate the cell wall teichoic acids (preparations I–IV) and products of their acid degradation; glycerol phosphate, a neutral- and aminosugars are used for control.

4. Detection of compounds

Degradation products are detected using the following spray reagents:

(i) Molybdate reagent for phosphorus-containing compounds (teichoic acids and phosphate esters of polyols):
- To prepare the reagent, 1 g $(NH_4)_2MoO_4$ is dissolved in 3 ml of 2 M HCl, then 3 ml 60% $HClO_4$ and 8 ml distilled water are sequentially added, and the solution is adjusted to 100 ml with acetone.
- The dry electropherogram is sprayed with the molybdate reagent, dry in a fume hood and heated at 100°C for 3–5 min, and then treated with steam. Organic phosphorus-containing compounds appear as blue spots.

(ii) Ninhydrin reagent for amino sugars and amino acids:
- To prepare the reagent, 1 g of ninhydrin is dissolved in 8 ml of distilled water, 2 ml glacial acetic acid is added, and the volume is adjusted to 200 ml with acetone.
- The dry chromatogram or electropherogram is sprayed with the ninhydrin reagent, dry in a fume hood and heated at 100°C for 3–5 min. Amino group-containing compounds are visualized as pink (purple) spots.

(iii) $AgNO_3$ reagent in aqueous ammonia for detection of polyols, sugars, and glycosides:
- To prepare the reagent, aqueous ammonia is added to 5 g $AgNO_3$ until the solution is clear; then the volume of solution is adjusted to 100 ml with distilled water.
- The dry chromatogram or electropherogram is sprayed with the $AgNO_3$ reagent, air-dried and heated at 100°C for 5 min. Polyols are manifested in the form of black spots, glycosides and monosaccharides as brown ones.

(iv) Aniline hydrogen phthalate for reducing sugars:
- To prepare the reagent, 1.66 g of phthalic acid is dissolved in 100 ml of water-saturated butan-1-ol (mix equal volumes of butane-1-ol and distilled water in a separating funnel and discard the lower aqueous layer after phase separation), and add 0.75 ml of aniline.
- Spray the dry chromatogram with the aniline hydrogen phthalate reagent, dry in a fume hood, and heat at 100°C for 5 min. Hexoses are manifested as brown spots and pentoses as pink spots.

(v) Aniline xylose for organic acids:
- To prepare the reagent, 5 ml of aniline and 5 g of xylose are dissolved in 100 ml of 50% aqueous ethanol.
- Spray the dry chromatogram with the reagent, air-dry in a fume hood and heat at 125–130°C for 5 min. Organic acids appear as brown spots.

5. Gel chromatography

Gel chromatography of the preparations is performed on a column (2.6 × 70 cm) of Sephadex G-50S in 0.05 M pyridine-acetate buffer, pH 4.5. The elution is monitored with a differential refractometer (Knauer, Germany). The polymer fractions are freeze-dried and investigated by NMR spectroscopic methods.

6. Determination of the absolute configuration of the polymer components

(a) The absolute configuration of glutamic acid

The absolute configuration of glutamic acid (Glu) is determined as follows:

- Hydrolyse a preparation of teichoic acid (3 mg dry weight) with 2 M trifluoroacetic acid at 120°C for 2 h.
- Dry with a stream of air.
- Add 15 μl of anhydrous trifluoroacetic acid and 0.2 ml of (+)-octan-2-ol to the dry residue and heat at 120°C for 16 h.
- Evaporate the sample under a stream of air.
- Acetylate the sample with acetic anhydride:pyridine (1:1, by volume) at 100°C for 1 h.
- Concentrate the acetylated sample to dryness under a stream of air.
- Prepare standards from L-Glu with (+)-octan-2-ol and (±)-2-octan-2-ol.
- Separate the compounds by gas–liquid chromatography, using a Hewlett-Packard apparatus (model 5880) and an Ultra-1 capillary column (0.2 mm × 25 m) with methyl-silicone (OV-1) phase, and a temperature gradient from 120 to 290°C, using nitrogen as the carrier gas and authentic standards prepared from L-Glu.

(b) The absolute configuration of lysine

The absolute configuration of lysine is determined as follows:

- Hydrolyse a preparation of teichoic acid (3–5 mg dry weight) with 2 M trifluoroacetic acid at 120°C for 2 h.
- Dry by evaporation.
- *N*-acetylate the sample with acetic anhydride in aqueous saturated $NaHCO_3$ (30 min at 0°C, then 30 min at 20°C).
- Deionize with IRA-120 resin (H^+-form), then heat with (+)-octan-2-ol (0.1 ml) and trifluoroacetic acid (15 μl) for 12 h at 120°C.
- Evaporate the liquid, acetylate the residue (to prepare the sample for subsequent analysis) as described above (Section IV.D.6a) and analyse by gas–liquid chromatography using a Hewlett-Packard apparatus (model 5880) and an Ultra-1 capillary column (0.2 mm × 25 m) with SE-54, using authentic samples prepared from L-lysine with (+)-octan-2-ol and (±)-octan-2-ol as standards.

(c) The absolute configuration of amino sugars

The absolute configuration of amino sugars is determined as follows:

- Hydrolyse a preparation of teichoic acid (2–3 mg dry weight) with 0.2 ml of 2 M trifluoroacetic acid for 2 h at 120°C.
- Evaporate the acid under a stream of air.
- Add dropwise 5% aqueous ammonia to neutralize the pH, and concentrate to dryness.
- Acetylate the sample with acetic anhydride (0.1 ml) at room temperature for 4 h.

- Remove excess of acetic anhydride under a stream of air.
- Treat the sample with 0.2 ml of optically active (*S*)-butan-2-ol in the presence of 15 μl of acetyl chloride overnight at 80°C.
- Remove the butanol under a stream of air.
- Acetylate the dry residue with 0.4 ml of pyridine:acetic anhydride mixture (1:1, by volume) at 80°C for 30 min.
- Concentrate the sample to dryness under a stream of air.
- Separate the compounds by gas–liquid chromatography, using a Hewlett-Packard apparatus (model 5890) and an Ultra-1 capillary column (0.2 mm × 25 m) with methyl-silicone (OV-1) phase and a temperature gradient from 180 to 290°C, using nitrogen as the carrier gas. Acetylated (*S*)- and (*R*)-2-butylglycosides of the corresponding amino sugars prepared under similar conditions should be used as standards.

(d) The absolute configuration of neutral monosaccharides

The absolute configuration of neutral monosaccharides is determined as follows:

- Hydrolyse a preparation of teichoic acid (1 mg dry weight) with 2 M trifluoroacetic acid for 2 h at 120°C.
- Evaporate the acid under a stream of air.
- Add toluene to, and remove from, the residue to eliminate traces of trifluoroacetic acid.
- Add anhydrous trifluoroacetic acid (15 μl) and (+)-octan-2-ol (0.2 ml), and heat the mixture for 16 h at 120°C.
- Remove the volatiles with a stream of air.
- Acetylate the residue with pyridine:acetic anhydride mixture (1:1, by volume) for 1 h at 100°C.
- Concentrate the sample to dryness in a stream of air.
- Analyse the residue by gas–liquid chromatography, using a Hewlett-Packard 5890A gas chromatograph and an Ultra-1 capillary column (0.2 mm × 25 m) with methyl-silicone (OV-1) phase and a temperature gradient from 180 to 290°C, using nitrogen as the carrier gas. Glycosides of authentic monosaccharides are prepared from (+)-octan-2-ol and (±)-octan-2-ol used as the standards.

(e) The absolute configuration of mannitol

The absolute configuration of mannitol is determined as follows:

- Hydrolyse a preparation of teichoic acid (3 mg dry weight) with 48% hydrofluoric acid (~0.03 ml) at 4°C for 16 h.
- Concentrate the sample to dryness with a stream of air.
- Isolate mannitol on a column (90 × 1.5 cm) of TSK HW-40S gel by elution with 1% acetic acid, using a Knauer differential refractometer for monitoring.
- Measure the optical rotation with a polarimeter in 0.033 M borax at 20°C.

E. NMR Spectroscopy

NMR analysis is used to establish the complete polymer structure, including the configuration of the glycoside centres, the localization of lateral substitutions and the position of phosphodiester bonds.

The structure of the polymer is established using ^1H, ^{13}C, and ^{31}P NMR spectroscopy. The NMR spectra contain almost all the necessary information concerning the primary structure of the polymer in terms of chemical shifts of the corresponding atoms and coupling constants $^nJ_{^1H,^1H}$, $^nJ_{^1H,^{13}C}$, and $^nJ_{^1H,^{31}P}$. To obtain the above-mentioned NMR parameters, it is necessary to complete the full assignment of the signals in the spectra. $^1H/^1H$ 2D COSY, TOCSY, and NOESY (or ROESY) spectra are used for the assignment in the ^1H NMR spectra (for an example, see Figure 3). The protons of a sugar or a polyol residue form a locked spin system. Within the system, the COSY spectrum shows correlation peaks for pairs of protons having a detectable (usually more than 1 Hz) coupling

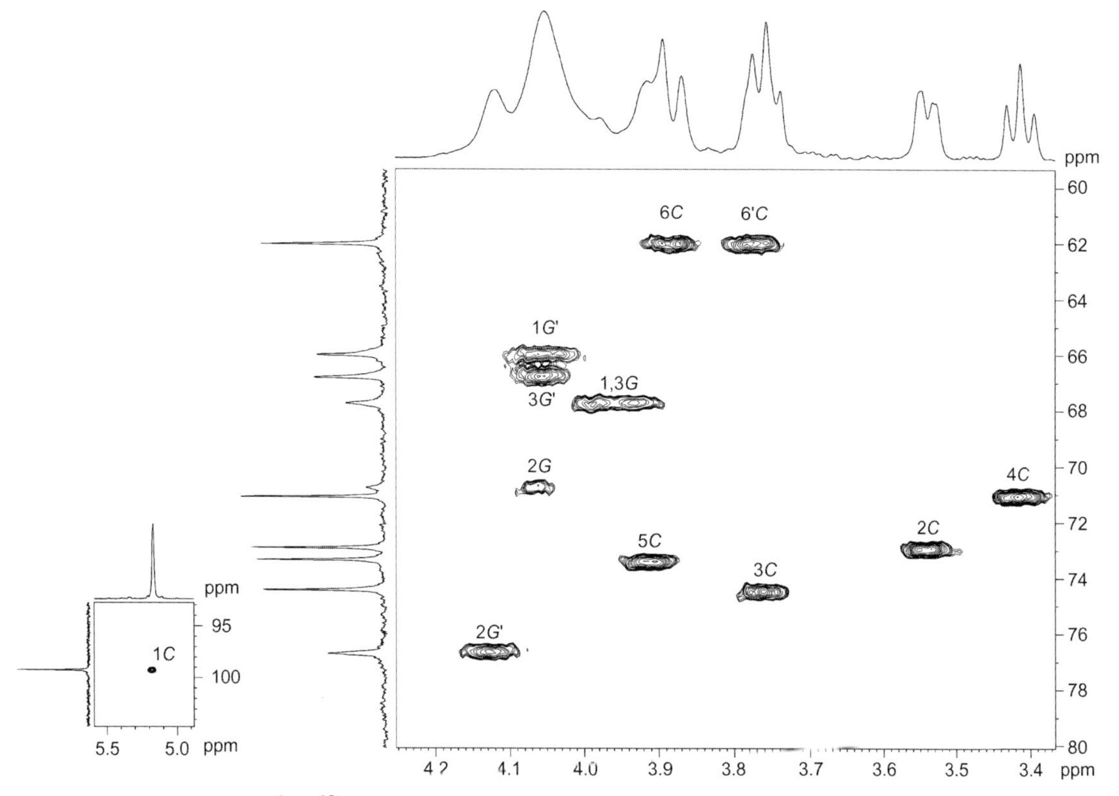

Figure 3. 2D $^1H/^{13}C$ HSQC spectrum of 1,3-poly(glycerol phosphate) with α-D-glucopyranose from cell wall of *Bacillus subtilis* subsp. *subtilis* VKM B-501^T. The corresponding parts of the ^1H and ^{13}C NMR spectra are shown along the horizontal and vertical axes, respectively. Arabic numerals designate atomic numbers in the residues designated by letters: ***C***, α-D-glucopyranose residue; ***G'***, glycerol residues substituted by α-D-glucopyranosyl residues at position 2; ***G***, unsubstituted glycerol residues (Shashkov *et al.*, 2009).

constant. The TOCSY spectrum contains correlation peaks for each proton of the locked spin system with all other protons of the system. The NOESY (or ROESY) spectrum reveals the spatial closeness of the protons (in particular, within the residues). Additionally, correlation peaks of all these 2D spectra contain information about coupling constants, at least in terms of 'small' (<2 Hz), 'medium' (2–6 Hz) or 'large' (7–12 Hz) values. Each sugar or polyol residue is characterized by its individual set of chemical shifts and coupling constants. Thus, full assignment in the ^{1}H NMR spectrum of the polymer reveals the monomer composition of the chain and opens the possibility for signal assignment in the ^{13}C and ^{31}P NMR spectra of the polymer. For this purpose, heteronuclear $^{1}H/^{13}C$ HSQC, HSQC-TOCSY, and HMBC spectra, as well as $^{1}H/^{31}P$ HMQC or HMBC spectra, are used (for an example, see Figure 4). Correlations in the $^{1}H/^{31}P$ spectra are important for the determination of phosphodiester linkages in teichoic acids and in poly(glycosyl phosphates). Analysis of 2D $^{1}H/^{13}C$ NMR spectra confirms the monomer composition of the polymers and reveals positions of substitution from a comparison of ^{13}C chemical shifts of the

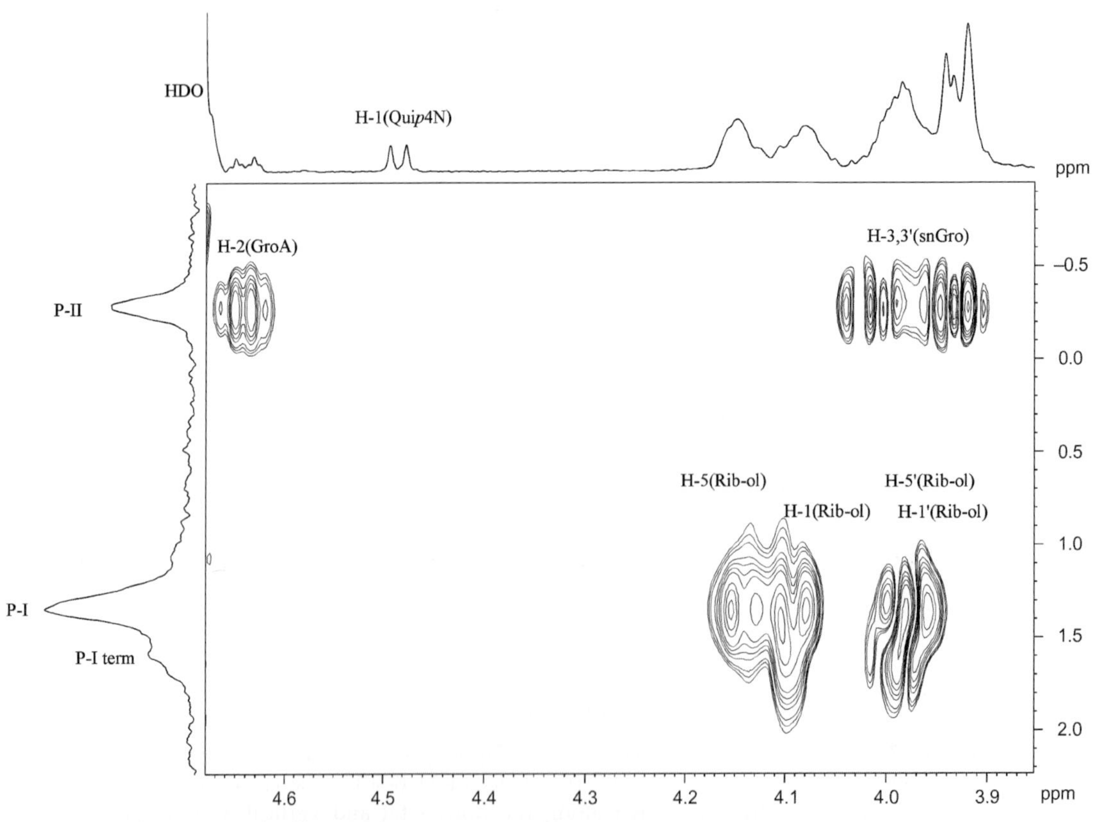

Figure 4. 2D $^{1}H,^{31}P$ HMBC spectrum of the predominant cell wall teichoic acid (P-I)—1,5-poly (ribitol phosphate) and a minor one (P-II)—poly(acylglycosyl polyol phosphate) of *Bacillus subtilis* VKM B-762. The corresponding parts of the ^{1}H and ^{31}P NMR spectra are displayed along the horizontal and vertical axes, respectively (Shashkov *et al.*, 2011).

residues in the polymer with those of the corresponding non-substituted sugars or polyols. Both $^1H/^1H$ ROESY and $^1H/^{13}C$ HMBC spectra are used for the determination of the sequence of the residues linked by glycosidic bonds (for an example, see Figure 5). Inter-residue correlation peaks for anomeric protons of glycosylating residues and nearest (in space) protons of glycosylated residues are used for the sequencing of the residues in the chain from analysis of the ROESY spectra. The anomeric proton of the glycosylating pyranose and the carbon atom of the glycosylated residue ('trans-glycosidic carbon') separated by three single bonds usually have detectable values (2–5 Hz) of the $^3J_{^1H,^{13}C}$ coupling constant. A similar situation occurs for the proton at the trans-glycosidic carbon and the anomeric carbon of the glycosylating pyranose. HMBC spectra contain

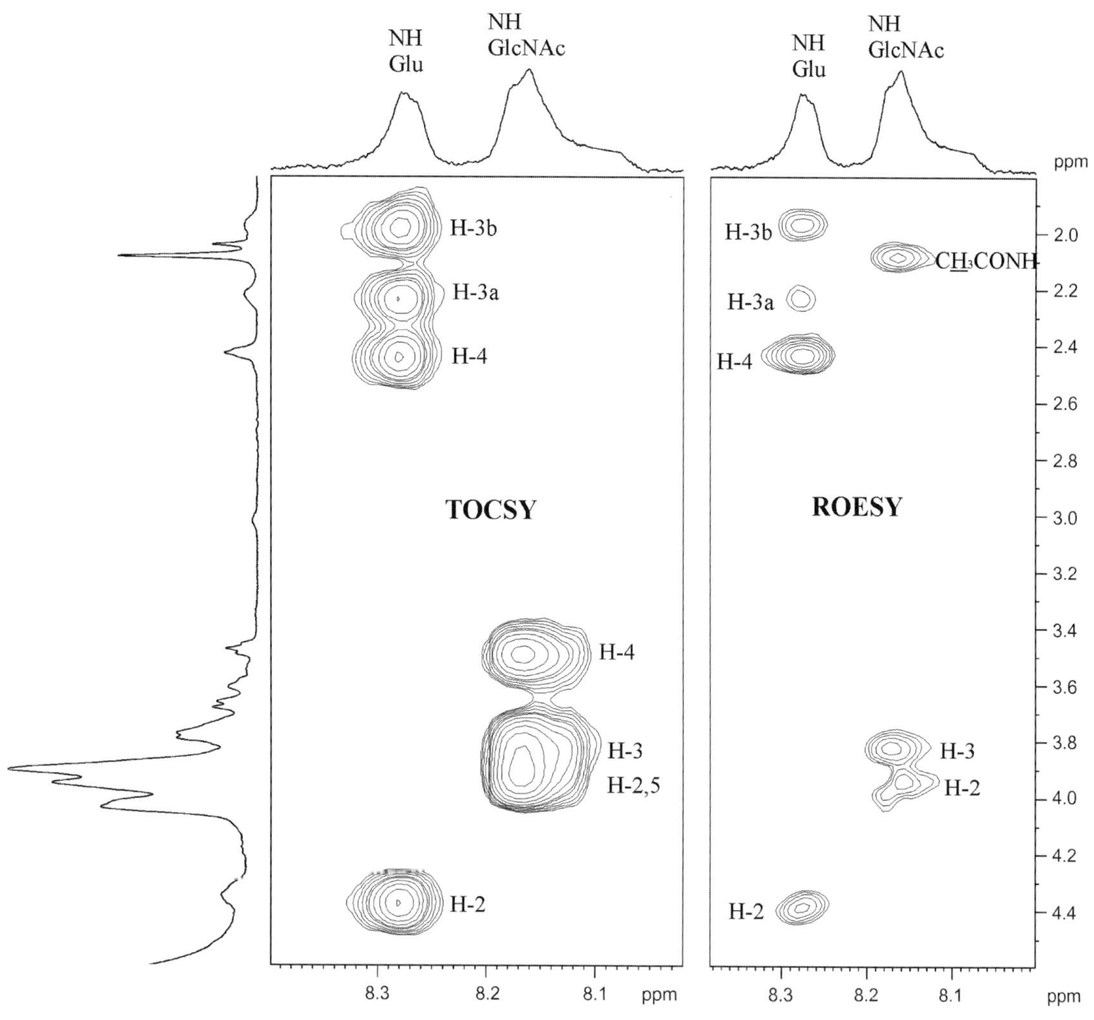

Figure 5. Parts of TOCSY and ROESY spectra of cell wall teichoic acid of *Streptomyces longispororuber* VKM Ac-1735^T. The localization of L-glutamic acid and *N*-acetyl-α-glucosamine at O2 of glycerol are shown (Shashkov *et al.*, 2006).

correlation peaks due to these coupling constants, demonstrating connectivity of the residues. Thus, the primary structure of polymers, excluding the absolute configuration of sugar and polyol residues, can be established by NMR spectroscopy.

NMR spectroscopy does not provide the information to conclude whether monomers belong to different polymers or one heterogeneous polymer. This point can be addressed by applying chromatographic and/or electrophoretic methods; unfortunately, this cannot be successfully resolved in all cases. DOSY ^{1}H NMR (diffusion-ordered spectroscopy) offers an alternative method. In the case of a mixture of polymers differing noticeably in the diffusion coefficients, DOSY makes it possible to distinguish their ^{1}H signals. If the diffusion coefficients are close, the problem still remains.

♦♦♦♦♦♦ V. CONCLUSION

The information provided in this chapter suggests that CWTAs can be reliable taxonomic tool to support the phylogenetic delineation of bacterial taxa (the order Actinomycetales and the genus *Bacillus*) and enable an improved identification of organisms at different taxonomic levels. Structures of CWTAs seem to be especially valuable for the differentiation and description of species when the traditional differentiating properties are variable or indistinct. Further comparative studies of CWTAs in organisms of different groups and the development of simple and efficient methods for the identification of these polymers will be of great importance in the polyphasic approach to bacterial taxonomy, which represents the state-of-the-art approach in the unambiguous circumscription of taxa, and will enable improved strain identification. Thus, it is topical to establish a database comprising such parameters as the taxonomic affiliation of an organism, the presence of CWTAs, and their composition and structure(s). Unlike the existing database (http://www.glyco.ac.ru/bcsdb3/), which lists the structures of all the known bacterial glycopolymers, the database-to-be will be confined to teichoic acids and their importance in the taxonomy of bacteria.

References

Anderson, A. S. and Wellington, E. M. H. (2001). The taxonomy of *Streptomyces* and related genera. *Int. J. Syst. Evol. Microbiol.* **51**, 797–814.

Archibald, A. R. (1974). The structure, biosynthesis and function of teichoic acid. *Adv. Microb. Physiol.* **11**, 53–95.

Archibald, A. R. (1972). Teichoic acids. In: *Methods in Carbohydrate Chemistry* (R. L. Whistler and J. N. Bemiller., eds.), Vol. 6, pp. 162–172. Academic Press, London.

Baddiley, J. (1972). Teichoic acids in cell walls and membranes of bacteria. *Essays Biochem.* **8**, 35–77.

Bergström, N., Jansson, P. E., Kilian, M. and Sørensen, U. B. S. (2003). A unique variant of streptococcal group *O*-antigen (C-polysaccharide) that lacks phosphocholine. *Eur. J. Biochem.* **270**, 2157–2162.

Deak, T. (2011). A survey of current taxonomy of common foodborne bacteria. *Acta Alimentaria* **40**, 95–116.
Endl, J., Seidle, P. H., Fiedler, F. and Schleifer, K. H. (1983). Chemical composition and structure of cell wall teichoic acids of staphylococci. *Arch. Microbiol.* **135**, 215–223.
Euzéby, J. (2011) List of bacterial names with standing in nomenclature, http://www.bacterio.cict.fr/.
Evtushenko, L. I., Bueva , O. V., Stupar, O. V., Shashkov, A. S., Potekhina, N. V., and Streshinskaya, G. M. (2007). *Kroppenstedtia citrinispora* gen. nov., sp. nov., a novel member of the family *Streptomycetaceae*. In: *Proceedings of the Eleventh International Conference on Culture Collections ICCC 11. Connections between Collections* (E. Stackebrandt, M. Wozniczka, V. Weihs, and J. Sikorski, eds.), p. 254. Goslar, Germany.
Fiedler, F. (1988). Biochemistry of the cell surface of *Listeria* strains: a locating general view. *Infection.* **16**, 92–97.
Gavrish, E. Yu., Krauzova, V. I., Potekhina, N. V., Karasev, S. G., Plotnikova, E. G., Altyntseva, O. V., Korosteleva, L. A. and Evtushenko, L. I. (2004). Three new species of brevibacteria, *Brevibacterium antiquum* sp. nov., *Brevibacterium aurantiacum* sp. nov., and *Brevibacterium permense* sp. nov. *Microbiology (Moscow)* **73**, 176–183.
Grund, E. and Kroppenstedt, R. M. (1990). Chemotaxonomy and numerical taxonomy of the genus *Nocardiopsis* Meyer 1976. *Int. J. Syst. Bacteriol.* **40**, 5–11.
Holt, J. G., Krieg, N. R., Sneath, P. H. A., Staley, J. T. and Williams, S. T. (eds.), (1997). *Bergey's Manual of Determinative Bacteriology* 9th ed. Williams & Wilkins, Baltimore.
Kozlova, Yu. I., Streshinskaya, G. M., Shashkov, A. S., Evtushenko, L. I. and Naumova, I. B. (1999). Anionic carbohydrate-containing polymers of cell walls in two streptoverticille genospecies. *Biochemistry (Moscow)* **64**, 671–677.
Kozlova, Yu. I., Streshinskaya, G. M., Shashkov, A. S., Evtushenko, L. I., Gavrish, E. Yu. and Naumova, I. B. (2000). Structure of carbohydrate-containing cell wall polymers in several representatives of the order *Actinomycetales*. *Biochemistry (Moscow)* **65**, 1432–1439.
Kroppenstedt, R. M. and Evtushenko, L. I. (2006). The family *Nocardiopsaceae*. In: *The Prokaryotes. A Handbook on the Biology of Bacteria* (M. Dworkin, S. Falkow, E. Rosenberg, K. -H. Schleifer and E. Stackebrandt, Eds.), 3rd Ed., Vol. 3, pp. 754–795. Springer-Verlag, New York.
Kumar, Y. and Goodfellow, M. (2008). Five new members of the *Streptomyces violaceusniger* 16S rRNA gene clade: *Streptomyces castelarensis* sp. nov., comb. nov., *Streptomyces himastatinicus* sp. nov., *Streptomyces mordarskii* sp. nov. and *Streptomyces ruanii* sp. nov. *Int. J. Syst. Evol. Microbiol.* **58**, 1369–1378.
Labeda, D. P. (1992). DNA-DNA hybridization in the systematics of *Streptomyces*. *Gene* **115**, 249–253.
Labeda, D. P. (1996). DNA-relatedness among verticil-forming *Streptomyces* species (formerly *Streptoverticillium* species). *Int. J. Syst. Bacteriol.* **46**, 699–703.
Labeda, D. P. (1998). DNA relatedness among the *Streptomyces fulvissimus* and *Streptomyces griseoviridis* phenotypic cluster groups. *Int. J. Syst. Bacteriol.* **48**, 829–832.
Labeda, D. P. and Lyons, A. J. (1991). Deoxyribonucleic-acid relatedness among species of the *Streptomyces cyaneus* cluster. *Syst. Appl. Microbiol.* **14**, 158–164.
Naumova, I. B. (1988). The teichoic acids of actinomycetes. *Microbiol. Sci.* **5**, 275–279.
Naumova, I. B. and Shashkov, A. S. (1997). Anionic polymers in cell walls of gram-positive bacteria. *Biochemistry (Moscow)* **62**, 809–840.
Naumova, I. B., Shashkov, A. S., Tul'skaya, E. M., Streshinskaya, G. M., Kozlova, Yu. I., Potekhina, N. V., Evtushenko, L. I. and Stackebrandt, E. (2001). Cell wall teichoic

acids: structural diversity, species specificity in the genus *Nocardiopsis*, and chemotaxonomic perspective. *FEMS Microbiol. Rev.* **25**, 269–283.

Neuhaus, F. C. and Baddiley, J. (2003). A continuum of anionic charge: structures and functions of D-alanyl-techoicacids in gram-positive bacteria. *Microbiol. Mol. Biol. Rev.* **67**, 686–723.

Potekhina, N. V. (2006). Teichoic acid of actinomycete and other gram-positive bacteria *Uspekhi Biol. Khim. (Moscow)* **46**, 225–278.

Potekhina, N. V., Tul'skaya, E. M., Shashkov, A. S., Taran, V. V., Evtushenko, L. I. and Naumova, I. B. (1998). Species specificity of teichoic acids in the actinomycete genus *Glycomyces*. *Microbiology (Moscow)* **67**, 330–334.

Potekhina, N. V., Shashkov, A. S., Evtushenko, L. I., Gavrish, E. Y., Senchenkova, S. N., Usov, A. I., Naumova, I. B. and Stackebrandt, E. (2003a). A novel mannitol teichoic acid with side phosphate groups of *Brevibacterium* sp. VKM Ac-2118. *Eur. J. Biochem.* **270**, 4420–4425.

Potekhina, N. V., Shashkov, A. S., Evtushenko, L. I., Senchenkova, S. N. and Naumova, I. B. (2003b). The mannitol teichoic acid from the cell wall of *Brevibacterium permense* VKM Ac-2280. *Carbohydr. Res.* **338**, 2745–2749.

Potekhina, N. V., Shashkov, A. S., Evtushenko, L. I. and Naumova, I. B. (2003c). Teichoic acids in the cell walls of *Microbispora mesophila* Ac-1953^T and *Thermobifida fusca* Ac-1952^T. *Microbiology (Moscow)* **72**, 157–161.

Potekhina, N. V., Evtushenko, L. I., Senchenkova, S. N., Shashkov, A. S. and Naumova, I. B. (2004). Structures of cell wall teichoic acids of *Brevibacterium iodinum* VKM Ac-2106^T. *Biochemistry (Moscow)* **69**, 1353–1359.

Potekhina, N. V., Streshinskaya, G. M., Tul'skaya, E. M., Kozlova, Yu. I., Senchenkova, S. N. and Shashkov, A. S. (2011). Phosphate-containing cell wall polymers of bacilli. *Biochemistry (Moscow)* **76**, 745–754.

Rooney, A. P., Price, N. P. J., Ehrhardt, C., Swezey, J. L. and Bannan, J. D. (2009). Phylogeny and molecular taxonomy of the *Bacillus subtilis* species complex and description of *Bacillus subtilis* subsp. *inaquosorum* subsp. nov. *Int. J. Syst. Evol. Microbiol.* **59**, 2429–2436.

Schubert, K. and Fiedler, F. (2001). Structural investigations on the cell surface of *Erysipelothrix rhusiopathiae*. *Syst. Appl. Microbiol.* **24**, 26–30.

Schubert, K., Reiml, D., Accolas, J.-P. and Fiedler, F. (1993). A novel type of meso-diaminopimelic acid-based peptidoglycan and novel poly(erythritol phosphate) teichoic acids in cell walls of two coryneform isolates from the surface flora of French cooked cheeses. *Arch. Microbiol.* **160**, 222–228.

Shashkov, A. S., Tul'skaya, E. M., Grachev, A. A., Evtushenko, L. I., Bueva, O. A. and Naumova, I. B. (1998). Structure of teichoic acid of the cell wall of *Streptomyces sparsogenes* VKM Ac-1744^T. *Biochemistry (Moscow)* **63**, 1098–1103.

Shashkov, A. S., Potekhina, N. V., Naumova, I. B., Evtushenko, L. I. and Widmalm, G. (1999a). Cell wall teichoic acids of *Actinomadura viridis* VKM Ac-1315^T. *Eur. J. Biochem.* **262**, 688–695.

Shashkov, A. S., Tul'skaya, E. M., Evtushenko, L. I. and Naumova, I. B. (1999b). A teichoic acid of *Nocardioides albus* VKM Ac-805 cell wall. *Biochemistry (Moscow)* **64**, 1305–1309.

Shashkov, A. S., Tul'skaya, E. M., Evtushenko, L. I., Gratchev, A. A. and Naumova, I. B. (2000). Structure of a teichoic acid of *Nocardioides luteus* VKM Ac-1246^T cell wall. *Biochemistry (Moscow)* **65**, 509–514.

Shashkov, A. S., Streshinskaya, G. M., Evtushenko, L. I. and Naumova, I. B. (2001). NMR-based identification of cell wall anionic polymers of *Spirilliplanes yamanashiensis* VKM Ac-1993^T. *Carbohydr. Res.* **336**, 237–242.

Shashkov, A. S., Potekhina, N. V., Evtushenko, L. I. and Naumova, I. B. (2004). Cell wall teichoic acids of two *Brevibacterium* strains. *Biochemistry (Moscow)* **69**, 658–664.

Shashkov, A. S., Streshinskaya, G. M., Senchenkova, S. N., Kozlova, Y. I., Alferova, I. V., Terekhova, L. P. and Evtushenko, L. I. (2006). Cell wall teichoic acids of streptomycetes of the phenetic cluster *'Streptomyces fulvissimus'*. *Carbohydr. Res.* **341**, 796–802.

Shashkov, A. S., Potekhina, N. V., Senchenkova, S. N. and Kudryashova, E. B. (2009). Anionic polymers of the cell wall of Bacillus subtilis subsp. subtilis VKM B-501T. *Biochemistry (Moscow)* **74**, 745–754.

Shashkov, A. S., Streshinskaya, G. M., Kozlova, Y. I., Senchenkova, S. N., Arbatsky, N. P. and Kudryashova, E. B. (2011). A novel type of teichoic acid from the cell wall of Bacillus subtilis VKM B-762. *Carbohydr. Res.* **346**, 1173–1177.

Stackebrandt, E., Koch, C., Gvozdiak, O. and Schumann, P. (1995). Taxonomic dissection of the genus *Micrococcus*: *Kocuria* gen. nov., *Nesterenkonia* gen. nov., *Kytococcus* gen. nov., *Dermacoccus* gen. nov., and *Micrococcus* Cohn 1872 gen. emend. *Int. J. Syst. Bacteriol.* **45**, 682–692.

Streshinskaya, G. M., Shashkov, A. S., Usov, A. I., Evtushenko, L. I. and Naumova, I. B. (2002). Cell wall teichoic acids of Actinomycetes of three genera of the order *Actinomycetales*. *Biochemistry (Moscow)* **67**, 778–785.

Streshinskaya, G. M., Kozlova, Yu. I., Shashkov, A. S., Evtushenko, L. I. and Naumova, I. B. (2003). The cell wall teichoic acids of Streptomycetes from the *'Streptomyces cyaneus'* cluster. *Microbiology (Moscow)* **72**, 455–460.

Streshinskaya, G. M., Kozlova, Yu. I., Alferova, I. V., Shashkov, A. S. and Evtushenko, L. I. (2005). Cell wall teichoic acids from *Streptomyces daghestanicus* VKM Ac-1722^{T} and *Streptomyces murinus* INA-00524^{T}. *Microbiology (Moscow)* **74**, 40–45.

Streshinskaya, G. M., Shashkov, A. S., Senchenkova, S. N., Bueva, O. V., Stupar, O. S. and Evtushenko, L. I. (2007). A novel teichoic acid from the cell wall of *Streptomyces* sp. VKM Ac-2275. *Carbohydr. Res.* **342**, 659–664.

Swoboda, J. G., Campbell, J., Meredith, T. C. and Walker, S. (2010). Wall teichoic acid function, biosynthesis, and inhibition. *ChemBioChem.* **11**, 35–45.

Takeuchi, M. and Yokota, A. (1989). Cell-wall polysaccharides in coryneform bacteria. *J. Gen. Appl. Microbiol.* **35**, 233–252.

Tomita, S., Furihata, K., Nukada, T., Satoh, E., Uchimura, T. and Okada, S. (2009). Structures of two monomeric units of teichoic acid prepared from the cell wall of *Lactobacillus plantarum* NRIC 1068. *Biosci. Biotechnol. Biochem.* **73**, 530–535.

Tul'skaya, E. M., Shashkov, A. S., Evtushenko, L. I., Bueva, O. V. and Naumova, I. B. (1997). Structural identity of teichoic acids from actinomycete species *Streptomyces hygroscopicus*. *Biochemistry (Moscow)* **62**, 289–293.

Tul'skaya, E. M., Krausova, V. I., Gavrish, E. Yu., Shashkov, A. S., Evtushenko, L. I., and Naumova, I. B. (2003). Cell wall teichoic acid composition is a chemical marker of the mycelium-forming *Nocardioides* species and the proposal of *Nocardioides prauserii* sp. nov. In: *1st FEMS Congress of European Microbiologists. Ljubljana, Slovenia. (Abstract book)*, p. 357. Elsevier, The Netherlands.

Tul'skaya, E. M., Shashkov, A. S., Bueva, O. V. and Evtushenko, L. I. (2007a). Anionic carbohydrate-containing cell wall polymers of *Streptomyces melanosporofaciens* and related species. *Microbiology (Moscow)* **76**, 39–44.

Tul'skaya, E. M., Streshinskaya, G. M., Kozlova, Yu. I., Shashkov, A. S. and Evtushenko, L. I. (2007b). Species and strain specific features of cell wall teichoic acids in *Nocardiopsis*. In: *Proceedings of the Eleventh International Conference on Culture Collections ICCC 11. Connections between Collections* (E. Stackebrandt, M. Wozniczka, V. Weihs, and J. Sikorski, eds.), p. 272. Goslar, Germany.

Uchikawa, K., Sekikawa, I. and Azuma, I. (1986). Structural studies on teichoic acids in cell walls of several serotypes of *Listeria monocytogenes*. *J. Biochem.* **99**, 315–327.

Weidenmaier, C. and Peschel, A. (2008). Teichoic acids and related cell-wall glycopolymers in Gram-positive physiology and host interactions. *Nat. Rev. Microbiol.* **6**, 276–287.

Williams, S. T., Goodfellow, M., Alderson, G., Wellington, F. M. H., Sneath, P. H. A. and Sackin, M. J. (1983). Numerical classification of *Streptomyces* and related genera. *J. Gen. Microbiol.* **129**, 1743–1813.

Zhang, Z. S., Wang, Y. and Ruan, J. S. (1998). Reclassification of *Thermomonospora* and *Microtetraspora*. *Int. J. Syst. Bacteriol.* **48**, 411–422.

7 The Identification of Polar Lipids in Prokaryotes

Milton S. da Costa[1], Luciana Albuquerque[2], M. Fernanda Nobre[1] and Robin Wait[3]

[1] *Department of Life Sciences, University of Coimbra, Coimbra, Portugal;* [2] *Center for Neuroscience and Cell Biology, University of Coimbra, Coimbra, Portugal;* [3] *Kennedy Institute of Rheumatology, Imperial College, London, United Kingdom*

♦♦

CONTENTS

♦♦♦♦♦♦ I. INTRODUCTION

All prokaryotes contain polar lipids in their membranes, and these have an enormous variety of structures, most of which have not been examined in any detail. In some cases, particularly in deep branching lineages of the bacteria, very rare polar lipids are found, most of which are unknown. For comparative taxonomic purposes, it is not always necessary to identify the structure of these lipids by nuclear magnetic resonance (NMR) and mass spectrometry (MS), since it is sufficient to show the native lipids on two-dimensional thin-layer chromatography (2D-TLC) and identify some properties of the lipids on the basis of staining. The most common polar lipids of bacteria are phospholipids, glycolipids and glycophospholipids, aminolipids and sulfur-containing lipids. The Alpha-, Beta- and Gammaproteobacteria, among other bacterial groups, generally possess three major phospholipids, which can be visualized by 2D-TLC; these are phosphatidylglycerol, phosphatidylethanolamine and

METHODS IN MICROBIOLOGY, VOLUME 38
0580-9517 DOI 10.1016/B978-0-12-387730-7.00007-3

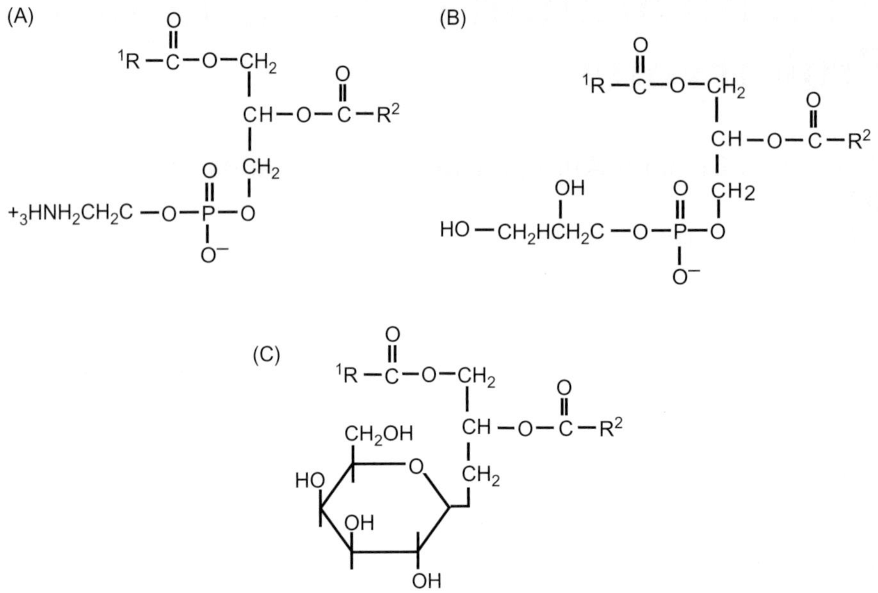

Figure 1. Examples of common polar lipids. (A) phosphatidylethanolamine; (B) phosphatidylglycerol; and (C) galactosylglycerolipid.

diphosphatidylglycerol, but other prominent phospholipids, such as phosphatidylcholine, may also be encountered in some groups (Albuquerque *et al.*, 2002, 2006, 2008, 2010b; Alves *et al.*, 2003). Yet, other groups of bacteria have a complex mixture of polar lipids, most of which have not been identified (Figure 1). The bacteria of the Firmicutes and the Actinobacteria are good examples of organisms that have complex mixtures of polar lipids, most of which have not been studied in any depth (Albuquerque *et al.*, 2011a, 2011b).

The majority of the polar lipids of bacteria have a glycerol backbone to which fatty acids are attached through ester linkages. Alkyl glycerol ethers are also known among the bacteria, but are relatively rare or present in small amounts. Organisms such as *Aquifex pyrophilus* (Huber *et al.*, 1992; Jahnke *et al.*, 2001) possess glycerol monoethers and glycerol diethers. Many other bacteria also generally possess small amounts of glycerol monoethers (Langworthy and Pond, 1986). Rare polar lipids with a long-chain 1,2-diol core are present in *Thermomicrobium roseum*, some species of the phylum Chloroflexi and the glycolipids of the species of genera *Thermus* and *Meiothermus*, but are not known in other bacteria (Balkwill *et al.*, 2004; Ferreira *et al.*, 1999; Pond *et al.*, 1986; Pond and Langworthy, 1987; Wait *et al.*, 1997) (Figure 2).

The diversity of archaeal polar lipids is apparently much smaller than that of the bacteria because the number of species described is smaller and because the polar lipids of many species have not been examined. The analysis of the polar lipids, as required by the minimum standards for the description of the members of the Halobacteriales, leads to a reasonable knowledge of their membrane components (Oren *et al.*, 1997), but is rather scanty in other euryachaeota and

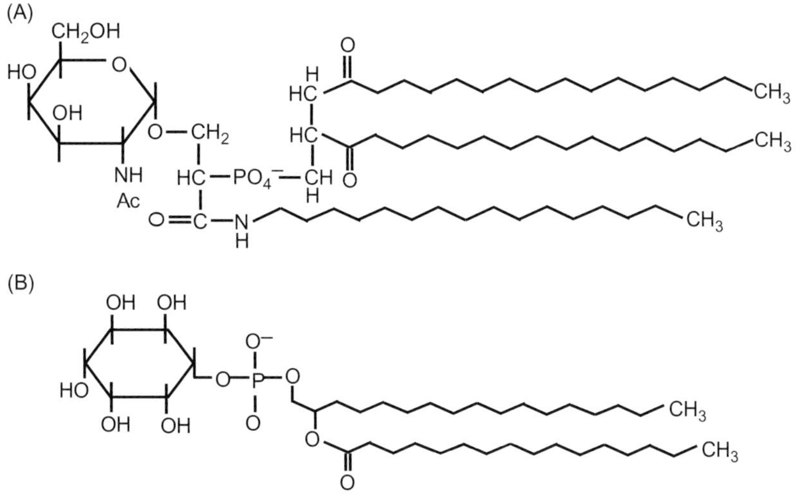

Figure 2. Complex and rare phospholipids from bacteria. (A) The 2′-*O*-(1,2-diacyl-*sn*-glycero-3-phospho)-3′-*O*-(α-*N*-acetylglucosaminyl)-*N*-D-glyceroyl alkylamine (GL-6) of *Deinococcus radiodurans* (Anderson and Hansen, 1985; Huang and Anderson, 1989; 1991). (B) The diol-based phosphatidylinositol of *Thermomicrobium roseum* (Pond et al., 1986; Pond and Langworthy, 1987).

crenarchaeota. However, some 2,3-di-*O*-isopranyl sn-glycerol diether (archaeol)- and ditetraterpenediyl glycerol tetraether (caldarchaeol)-derived phospholipids and glycolipids appear throughout this domain, namely of phosphatidylserine, phosphatidylglycerol, phosphatidylglycerolphosphate methyl ester, phosphatidylethanolamine, phosphatidylinositol, phosphatidylglycerol sulfate, several sulfated glycolipids, glucosaminylphosphatidylinositol and glucosylphosphatidylinositol, among others (Ferrante *et al.*, 1987; Koga *et al.*, 1993; Koga and Morii, 2005; Morii *et al.*, 1999; Nishihara *et al.*, 1992; Trincone *et al.*, 1993). Archaea also possess very rare polar lipids like those containing gulose and mannose in *Thermoplasma acidophilum* (Shimada *et al.*, 2002) (Figure 3).

Polar lipid patterns can sometimes change drastically for apparently minor reasons. The first four species of the genus *Meiothermus* had a polar lipid pattern composed of a major phospholipid, designated PL-2, and two glycolipids designated GL-1a and GL-1b that migrated very closely to each other on one-dimensional TLC. Glycolipid GL-1a was composed of one molecular species that contained an iso-2-OH fatty acid on the oligosaccharide chain, whereas GL-2a had an iso-3-OH fatty acid or a saturated iso-branched fatty acid linked to the oligosaccharide group (Ferreira *et al.*, 1999). Two species named *M. granaticius* and *M. rufus* do not synthesize iso-2-OH fatty acids and, therefore, lacked glycolipid GL-1a (Albuquerque *et al.*, 2009, 2010a; see also Figure 6).

Recent recommendations on the characterization of prokaryotes make no mention of using 2D-TLC for the analysis of the polar lipid pattern of bacteria or archaea (Tindall *et al.*, 2010). However, most journals like to see the polar lipid pattern being presented as 2D-TLC instead of one-dimensional TLC. In this respect, it is advisable that 2D-TLC be used for polar lipid analysis, although in some cases one-dimensional TLC is perfectly capable of separating the polar lipids.

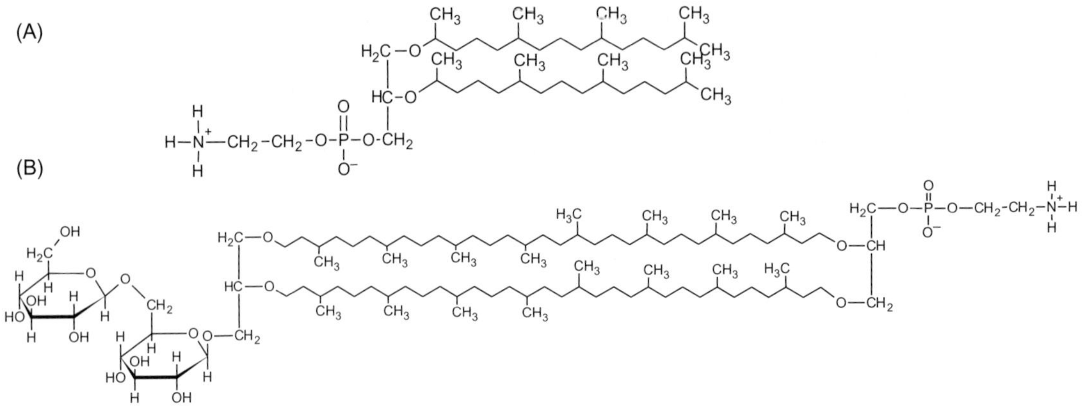

Figure 3. Examples of archaeal polar lipids. (A) Archaetidylphosphoethanolamine and (B) Gentiobiosyl caldarchaetidylphosphoethanolamine.

♦♦♦♦♦♦ II. GROWTH OF ORGANISMS AND EXTRACTION OF POLAR LIPIDS

Several methods to extract polar lipids have been described, but the method of Bligh and Dyer (1959) is the most commonly used method and the one that we have found to be suitable for all prokaryotes.

If possible, the organisms should be grown in liquid medium so that cells from the same phase of growth are always recovered to obtain reproducible results from replicate experiments. Individual polar lipid levels may change with the phase of growth and under different growth conditions, and it is, therefore, important to maintain the culture conditions as similar as possible. Normally the organisms are grown at the optimum growth temperature and harvested for polar lipid analysis during mid-exponential or late exponential phase by placing the culture flasks on ice. Some organisms that are difficult to cultivate in liquid medium can be cultivated on solid medium. The cells recovered from liquid media are washed twice using an appropriate buffer (e.g. 10 mM phosphate buffer at pH 7.5), followed by centrifugation at 4°C. However, appropriate levels of NaCl for halophiles and other buffer systems should be used to take into account the optimum pH for the growth of acidophilic and alkaliphilic organisms.

About 1.0–4.0 g wet weight of cells is used for extraction of polar lipids, but lesser amounts can be used for chemotaxonomic analysis of the polar lipid profiles of prokaryotes by TLC. All glassware should be washed with chloroform: methanol (1:1, by volume) to remove hydrophobic compounds.

A. Precautions

The quantities of cells and volumes of solvents given in the following methodology are for large-scale extraction of polar lipids, but these can be decreased as long as the proportions of the solvents for extraction are maintained. All procedures involving the use of chloroform, other solvents and spray reagents to detect lipids separated by thin-layer chromatography (TLC) must be performed in a fume hood. All the spray reagents are hazardous and care must be taken not to inhale them.

Method 1

- Resuspend the cell pellets in 10 ml of the appropriate washing buffer in large glass centrifuge tubes (90 ml). Add 25 ml of methanol to the suspension and stir magnetically for 10 min. Add 12.5 ml of chloroform and stir for 15 min. Centrifuge the suspension at 2000 × *g* for 15 min in a swinging-bucket centrifuge, at room temperature, until the cells sediment completely. Transfer the extract to a 300 ml Erlenmeyer flask fitted with a sintered-glass stopper. The cell pellet is resuspended in 8 ml of water, stirred and re-extracted using the same volumes of chloroform and methanol. Repeat the centrifugation and transfer the extract to the Erlenmeyer flask. If phase separation occurs at this stage, add methanol dropwise with slow stirring until a one-phase mixture is obtained.
- Add 25 ml of chloroform and 50 ml of 0.5 M KCl so that the final proportion of chloroform:methanol:water is (2:2:1.8, by volume). Shake mixture vigorously to form an emulsion.
- Distribute the emulsion in glass centrifuge tubes and centrifuge at 2000 × *g* for about 5 min to obtain phase separation.
- For routine TLC separation of polar lipids, the lower phase is removed once and transferred to round-bottom flasks with sintered-glass joints. Care should be taken to avoid removing the interface and upper phase. For quantitative analysis of the polar lipids, the upper phase is transferred to the same Erlenmeyer, 50 ml of chloroform are added, the extraction is repeated and the lower phase is removed and added to the first extract.
- Evaporate the lipid extract to dryness with a rotary evaporator under low vacuum at 30–35°C.
- Resuspend the lipid extract in a small volume of chloroform:methanol (1:1, by volume). The extract should be removed from the round-bottom evaporation flask to a 35 ml Teflon-coated screw capped-tube. For quantitative analysis, repeat at least twice with the solvent mixture to recover as much of the extract as possible.
- Store the extract at −20°C under N_2 until utilized.

Method 2

The method described above is very useful to obtain large quantities of polar lipids for several types of analysis. But for small amounts of lipids, there is a variant of the above method that is quicker and uses much less cell material and solvents. One of the drawbacks of the method described by Ross and Grant (1985) is that carotenoids are not removed and can interfere with some analysis.

- Wet cell paste is resuspended in about 1.0 ml of ultrapure water in a brown glass tube with a Teflon-lined screw cap and freeze-dried (100–150 mg) to which 2.75 ml of methanol:NaCl (10 ml, 0.3% wt/vol, aqueous NaCl in 100 ml of methanol) is added to the cells and stirred for 1 min. The mixture is then heated to 100°C for 5 min.
- This mixture is cooled after which 0.75 ml of 0.3% aqueous NaCl is added, followed by 1.25 ml of chloroform. This mixture is stirred with a Teflon-coated

magnetic bar for 2 h at room temperature and then centrifuged for 2 min to remove the cell debris. The supernatant is removed with a glass pipettes to a brown glass tube with a Teflon-lined screw cap and 2 ml of chloroform:methanol:aqueous 0.3% NaCl (5:10:4, by volume) is added to the cell pellet, stirred for 15 minutes and centrifuged again. The two supernatants are pooled.

- Aqueous 0.3% NaCl (0.75 ml) is added to the pooled extract, followed by 1.75 ml of chloroform. The mixture is shaken by hand (3–5 min) and then centrifuged for 3–4 min. Two phases form and the upper phase is removed and discarded. The lower phase, containing the lipids, should be dried under a stream of nitrogen.
- The lipids should be dissolved in chloroform:methanol (1:1, by volume) before carrying out TLC.

♦♦♦♦♦♦ III. PREPARATIVE THIN-LAYER CHROMATOGRAPHY

Sometimes it is necessary to remove pigments and other lipid components from the sample before analysis of the polar lipids by TLC because they may interfere with the migration or the visualization of the polar components on the plates.

- Dry silica gel 60 (Merck 5745, 20 cm × 20 cm, 2 mm thickness) plates for 30 min at 120°C and cool in a desiccator containing dry silica gel until used.
- Add acetone to a chromatography tank in which one of the sides has been lined with filter paper to aid saturation of the atmosphere inside the tank and wait about 1 h before placing the silica Gel 60 plate in the tank. The solvent should be below the band (origin) where the lipid extract will be applied (Figure 4).
- Evaporate the polar lipid extract to dryness under a stream of N_2. Resuspend the extract in a small volume (about 250 μl) of chloroform:methanol (1:1, by volume). Apply all of the extract slowly on the silica gel G plate along a line about 2 cm from the lower edge of the plate. A small amount of chloroform: methanol can be added to the tube to further remove the extract completely and then applied to the silica gel 60 plate. More than one sample can be applied to the same plate, but a space is left between the two sample applications. It is important not to apply the sample closer than about 1.5 cm from the vertical edges of the plate.
- Place the silica Gel 60 plate in the chromatography tank with the upper edge leaning against the inside of the tank and wait until the solvent front reaches about three quarters of the height of the plate.
- Remove the plate from the tank and allow the solvent to evaporate at room temperature or blow dry with cold air from a hair dryer to speed up drying.
- Scrape the silica gel of the area surrounding the origin (about 1 cm above and below) to recover the polar lipids that do not migrate in this solvent. Place the silica on smooth non-absorbent paper, cover with another piece of paper and grind the silica to a fine powder by pressing with finger tips. Place the powdered silica into a glass centrifuge tube. Alternatively, add the silica to a glass centrifuge tube and grind the silica to a fine powder with a glass rod. Add a clean magnetic stirring bar to the centrifuge tube. Add 40 ml

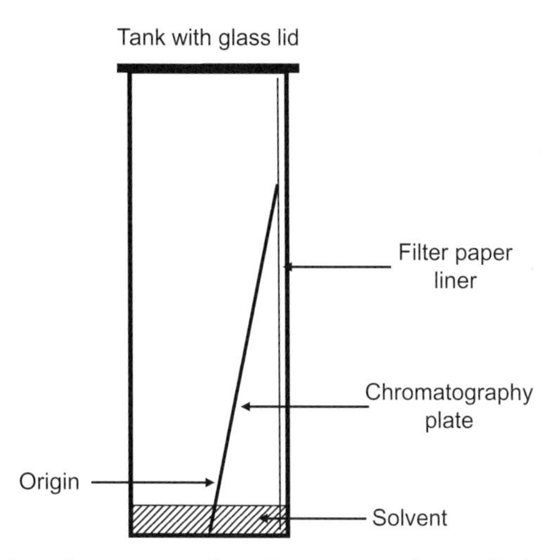

Figure 4. Diagram showing the set up of a chromatography tank for separation of the polar lipids.

of chloroform:methanol:water (45:45:10, by volume), stir slowly on a magnetic stirrer for about 10 min and centrifuge for 5 min in a swing-bucket centrifuge.

- Remove the extract to a 300 ml Erlenmeyer flask fitted with a sintered-glass stopper. Repeat the extraction of the silica and transfer to the Erlenmeyer flask.
- Add 36 ml of 0.5 M KCl to obtain a chloroform:methanol:water mixture (2:2:1.8, by volume). Shake the tube vigorously to obtain an emulsion.
- Distribute the emulsion in glass centrifuge tubes and centrifuge at $2000 \times g$ for about 5 min to obtain phase separation.
- For routine TLC separation of polar lipids, the lower phase is removed once and transferred to round-bottom flasks with sintered-glass joints. Care should be taken to avoid removing the interface and upper phase. For quantitative analysis of the polar lipids, the upper phase is transferred to the same Erlenmeyer, 50 ml of chloroform are added, the extraction is repeated and the lower phase is removed and added to the first extract.
- Evaporate the lipid extract to dryness with a rotary evaporator under low vacuum at 30–35°C.
- Resuspend the lipid extract in a small volume of chloroform:methanol (1:1, by volume). The extract should be removed from the round-bottom evaporation flask to a 35 ml Teflon-coated screw capped-tube. For quantitative analysis, repeat at least twice with the solvent mixture to recover as much of the extract as possible.
- Evaporate to dryness under a stream of N_2. Store at −20°C until needed.

♦♦♦♦♦♦ IV. THIN-LAYER CHROMATOGRAPHY

Silica Gel 60 plates (Merck No. 5626, 0.25 mm thickness, 10 cm × 20 cm) are most often used. Cut the plates with a diamond knife to obtain plates

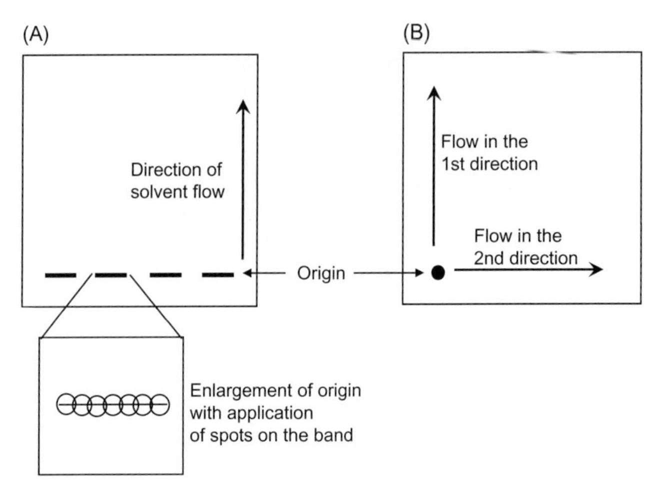

Figure 5. TLC plates showing application of samples and direction of flow of solvents for A, one-dimensional TLC and B, two-dimensional TLC. The enlargement shows application of small spots along a band (origin) for one-dimensional TLC.

that are 10 cm × 10 cm, but smaller plates can also be used. Other silica gel plates are available, but for most purposes, silica gel 60 plates are generally used.

There are many solvent systems that can be used to separate polar lipids by TLC and that are thoroughly described ranging from neutral solvent systems to mixtures of solvents with neutral or basic pH (Zweig and Sherma, 1972; Christie, 1982; Ratledge and Wilkinson, 1989). We generally use an acidic mixture that is adequate for one-dimensional TLC, and neutral and acidic solvent systems for 2D-TLC.

♦♦♦♦♦♦ V. ONE-DIMENSIONAL TLC

One-dimensional TLC is generally sufficient for the separation of the polar lipids of many organisms without recourse to 2D-TLC (Figure 5). Better separation and definition of the lipid bands are obtained if the extract is placed along a line at the origin, instead of a spot. Spotting can lead to overloading of the sample and to vertical smear of the lipids that can mask some of the components (Figure 6).

- Dry silica gel plates for 30 min at 120°C and cool in a desiccator containing dry silica gel until used.
- Add a solvent mixture composed of chloroform:acetic acid (glacial): methanol:water (80:15:12:4, by volume) to a chromatography tank with one side lined with filter paper and allow the atmosphere to saturate with the solvent vapours for about 1 h before introducing the chromatographic plates.

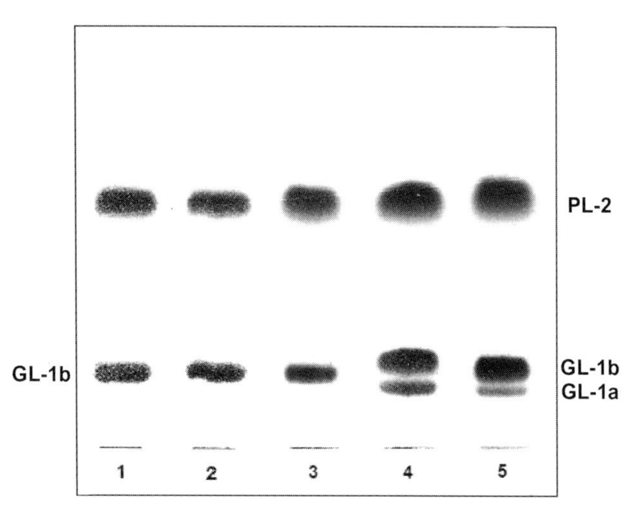

Figure 6. Monodimensional thin-layer chromatography of polar lipids of 1. *M. granaticius* AF-49, 2. *M. granaticius* AF-68^T, 3. *M. rufus* CAL-4^T, 4. *M. ruber* DSM 1279^T and 5. *M. cerbereus* GY-1^T at the optimum growth temperature. The lipids were stained by spraying with 5% molybdophosphoric acid in ethanol followed by heating at 160°C. GL-1a, glycolipid 1a; GL-1b, glycolipid 1b; PL-2, phospholipid 2.

- Several samples can be applied on the same plate about 1 cm from the lower edge. Apply the extracts on the silica Gel 60 plate using a microsyringe or a capillary tube along a band about 1.0 cm in length with about 0.5 cm space between applications (Figures 5A and 6). The lines of application should be traced with a pencil before applying the extract.
- Place the silica Gel 60 plate vertically in the chromatography tank with the upper edge leaning against the inside of the tank, place lid on the tank and leave until the solvent front reaches the upper edge of the plate.
- Remove the plate from the tank and allow the solvent to evaporate at room temperature or blow dry with cold air from a hair dryer to speed up drying.

♦♦♦♦♦♦ VI. TWO-DIMENSIONAL TLC

Two dimensional TLC may be necessary to resolve complex mixtures of polar lipids or to separate components that migrate close to each other on one-dimensional TLC. Overloading the silica at the application spot is not a problem in 2D-TLC because of the fact that individual polar lipids will become separated from each other during the second chromatographic step (Figure 7).

- Prepare two chromatography tanks: (1) a tank containing a solvent mixture composed of chloroform:methanol:water (65:25:4, by volume) and (2) a tank containing a solvent mixture of chloroform:acetic acid (glacial):methanol: water (80:15:12:4, by volume), and allow the atmosphere of the tanks to saturate for about 1 h.

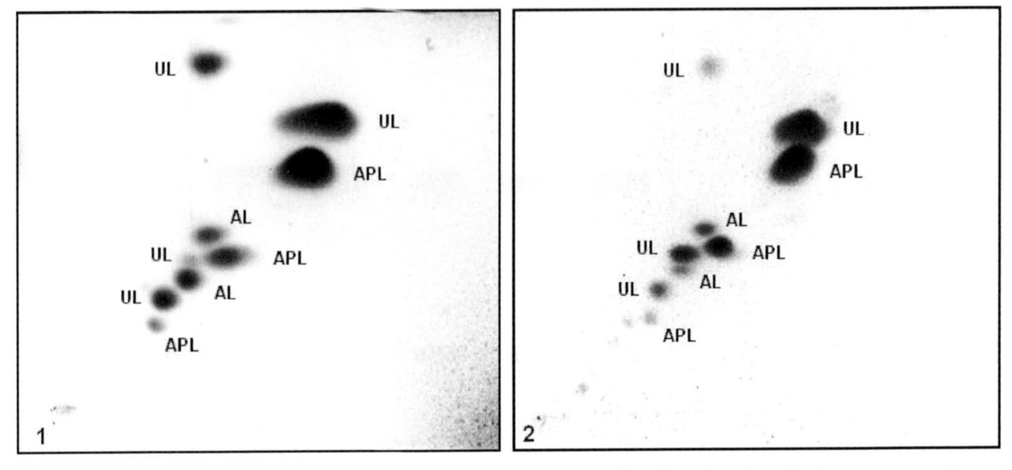

Figure 7. Two-dimensional thin-layer chromatography of polar lipids of the type strains of 1. *Hydrotalea sandarakina* and 2. *Hydrotalea flava* at the optimum growth temperature. The lipids were stained by spraying with 5% molybdophosphoric acid in ethanol followed by heating at 160°C. APL, aminophospholipid; AL, aminolipid; UL, unknown lipid.

- Dry several silica gel G plates at 120°C for 30 min and cool in a desiccator containing dry silica gel until used.
- Apply the sample slowly on one spot about 1.5 cm from the lower edge and the left-hand edge of a 10 cm × 10 cm plate as shown in Figures 5B and 7.
- The silica Gel 60 plate is placed in the first chromatographic tank (1) until the solvent reaches the upper edge. Remove the plate and allow it to dry with the aid of a cold air stream from a hair dryer.
- Rotate the plate 90° counterclockwise and place in the second tank (2). Allow the solvent to reach the upper edge of the plate. Remove and dry the plate.

♦♦♦♦♦♦ VII. DETECTION OF POLAR LIPIDS ON TLC PLATES

The detection reagents described here are generally sufficient to obtain a polar lipid profile from many bacteria and archaea, but other reagents can be used for a more detailed presumptive identification of polar lipid types (Christie, 1982; Ratledge and Wilkinson, 1989; Kates, 1993). It is important to remember that none of the spray reagents are completely specific for any type of polar lipid. Standard mixtures of commercially available polar lipids or lipid extracts of prokaryotes with well-defined polar lipid profiles can be used to test the efficiency and specificity of the spray reagents.

The reagents for the detection of the polar lipids should be sprayed at a distance of 20–30 cm. Spray the reagents evenly and avoid soaking of the plate until the reagent drips from the plate, because this will not improve the detection of polar lipids and the silica may detach from the plate.

A. Phospholipids – Molybdenum Blue Reagent

This reagent is commercially available from Sigma–Aldrich, but preferably should be made in the laboratory for consistently better results:

Solution A: Add 10.0 g molybdenum oxide (MoO_3) to 250 ml of H_2SO_4 (12.5 M). Boil gently in a fume hood until all the MoO_3 dissolves. Allow the solution to cool and store for a few days at room temperature before use.
Solution B: Add 400 mg of molybdenum powder (Mo) to 112.5 ml of solution A. Boil gently in a fume hood for about 15 min to dissolve the Mo. Allow to cool and decant into a glass bottle. Store this solution at room temperature until use.
Spray reagent: Add slowly 20 ml of solution A, 20 ml of solution B and 60–80 ml of water until the reagent becomes green-gray in colour. Store the reagent at room temperature in the dark. If the colour of this reagent changes to blue, water should be added dropwise until the reagent becomes green-gray again. A green-grey-coloured spray reagent is more effective in staining phospholipids than one with blue colour.

Blue spots appear almost immediately after spraying. The plate can be sprayed second time to enhance the blue-staining phospholipids. The reagent is very specific for phosphorus-containing lipids.

B. Lipids with Free Amino Groups – Ninhydrin

- Add 200 mg ninhydrin to 100 ml of 1-butanol:acetic acid (glacial) (95:5, by volume). Store the reagent at room temperature in the dark.
- Spray and place the silica gel 60 plate in an oven at 120°C for about 5 min. The lipids with free amino groups stain pink. The same plate sprayed with the ninhydrin reagent can be used to stain glycolipids with the α-naphthol reagent, after marking the aminolipids with a pencil.

C. Glycolipids – α-Naphthol-Sulfuric Acid

- Prepare a 15% solution of α-naphthol in absolute ethanol.
- Add 10.5 ml of the above solution to a mixture containing 6.5 ml of H_2SO_4 (95–97%), 40.5 ml of absolute ethanol and 4 ml water. Store for a few days before using.
- Spray the silica gel 60 plate and place in an oven heated at 120°C for 5 min. The glycolipids appear as purple spots.

D. Glycolipids – Diphenylamine

- Dilute 20 ml of a 10% solution of diphenylamine in absolute ethanol in 100 ml of HCl (32%) and 80 ml of glacial acetic acid. Store the reagent at room temperature in the dark.

- Spray lightly and cover the plate with a clean glass plate and heat at 110 °C until spots appear (about 40 min). The glycolipids stain blue.

E. Choline-Containing Phospholipids – Dragendorff's reagent

Solution A: Dissolve 17 g basic bismuth nitrate ($Bi_5O(OH)_9(NO_3)_4$) in 20% aqueous glacial acetic acid. Store at 4°C.
Solution B: Dissolve 40 g potassium iodide in 100 ml of water. Store at 4 °C.
Spray reagent: Immediately before use mix 4 parts of solution A and 1 part of solution B with 14 parts of water.

Choline-containing lipids appear orange or red-orange after spraying.

F. Total Lipids – Molybdophosphoric Acid

- Prepare the reagent by adding 5 g of molybdophosphoric acid in 100 ml of absolute ethanol. Store the reagent at room temperature in the dark.
- After spraying, place the silica gel 60 plate in an oven at 160°C for about 20 min. The lipids stain grey on a yellow background. The yellow background can be removed by placing the plate in a small covered glass tank with a few drops of 25% ammonia for a few seconds.

G. Sulphur Containing Lipids – Azure A-Sulfuric Acid

- Dissolve Azure A to saturation in H_2SO_4 at room temperature with slow magnetic stirring.
- Spray the reagent evenly on the plate.
- Immerse the plate in a small tank containing 0.04 M H_2SO_4:methanol (3:1, by volume) to remove excess reagent. The sulfolipids and sulfonolipids stain blue.

♦♦♦♦♦♦ VIII. CHARACTERIZATION OF POLAR LIPIDS BY MASS SPECTROMETRY

Mass spectrometry (MS) is one of the most sensitive and powerful techniques available for biomolecular structure determination, and has played a key role in studies of prokaryote lipids, both in characterization of intact lipids and in identification of their constituent fatty acids, long-chain diols and monosaccharides.

The requirement for thermal vapourization generally precludes analysis of intact glycoconjugates by electron ionization (EI) and GC/MS. Such compounds only became routinely amenable to direct MS characterization with the development of so-called soft ionization methods, in which sample molecules are converted to gas phase ions without the need for heating. These methods include fast atom bombardment (FAB), matrix-assisted laser desorption ionization (MALDI) and electrospray ionization (ESI).

Lipid compositions and monosaccharide sequences of a family of glycosylated diacyl glycerols from *Thermus* and *Meiothermus* have been determined, using a range of mass spectrometric approaches (Carreto *et al.*, 1996; Wait *et al.*, 1997). Tandem MS of acetates and deuteroacetate derivatives established the stereochemistry of the terminal residues, as variously as glucopyranose, galactopyranose, galactofuranose or ribopyranose. The presence of hydroxylated fatty acids in several glycolipids was confirmed, and it was shown that these are exclusively amide-linked to hexosamine in the glycan head-group, and are never ester-linked to glycerol (Carreto *et al.*, 1996; Wait *et al.*, 1997). Several novel glycolipids have been characterized, in which similar tri- and tetraglycosyl head-groups were found to be linked to 16-methyl-1,2-heptadecanediol or 15-methyl-1,2-heptadecanediol, rather than to glycerol (Carreto *et al.*, 1996; Wait *et al.*, 1997; Ferreira *et al.*, 1999).

More recently, the structure of the major acylglycerol glycolipid from *Thermus oshimai*, NTU-063 was established as β-Glc*p*-(1-6)-β-Glc*p*-(1-6)-β-Glc*pN*Acyl-(1-2)-α-Glc*p*(1-1)glycerol diester by a combination of NMR spectroscopy, methylation analysis and MS (Lu *et al.*, 2004), which supports the original proposal and supplies the missing anomeric configuration and linkage information (Wait *et al.*, 1997).

The analytical strategy entails molecular mass measurement in positive- and negative-ion modes of the underivatized TLC-purified glycolipids, which enables determination of their composition in terms of sugar, glycerol/diol and acyl units. The glycolipids are then acetylated and the mass spectrometric analysis is repeated after acetylation and deuteroacetylation of two aliquots of sample. Comparison of the masses of the acetylated and deuteroacetylated derivatives enables counting the number of hydroxyl groups derivatized, which is helpful for determining the true molecular mass, as it is not always obvious if the observed molecular ions are protonated or are cationized by alkali metal ions. Acetylated glycolipids undergo simple and predictable fragmentation (Dell, 1990), via pathways producing predominantly non-reducing terminal carbenium (B_n-type) ions (Domon and Costello, 1988). For example, the FAB spectrum of the peracetylated GL-1 from *Thermus* strain SPS-11 had an abundant B_1 ion; diagnostic of a terminal hexose, at *m/z* 331; further B ions at *m/z* 619 (B_2), 1,116 (B_3) and 1,404 (B_4) established the sequence of the polar head-group as hexose-hexose-(*N*-acyl)hexosamine-hexose. Glycosidic cleavage is particularly favoured at hexosamine, resulting in prominent B_3 fragments, which incorporate the *N*-acylated hexosamine residue that is diagnostic of the *N*-acyl substituents. The major B_3 ion at *m/z* 1,116 indicates that the hexosamine is *N*-acylated by iso-17:0, and other B_3 fragments at *m/z* 1,088 and 1,102 indicate that a subpopulation is acylated instead by iC15:0 or iC16:0. The ratio of the abundances of *m/z* 1,102 and *m/z* 1,088 indicated that approximately 54% of the fatty acids amide-linked to glucosamine were C17 fatty acids and about 40% were C15 fatty acids, the balance being C16 fatty acids.

Ions at *m/z* 523 and 551 originate by cleavage of the sugar–glycerol bond and identify the ester-linked substituents; *m/z* 523 corresponds to acylation with two C15 fatty acids, whereas *m/z* 551 indicates substitution with one C15 and one C17 acid. Comparison of the relative intensities of the *N*-acylated B_3 fragments

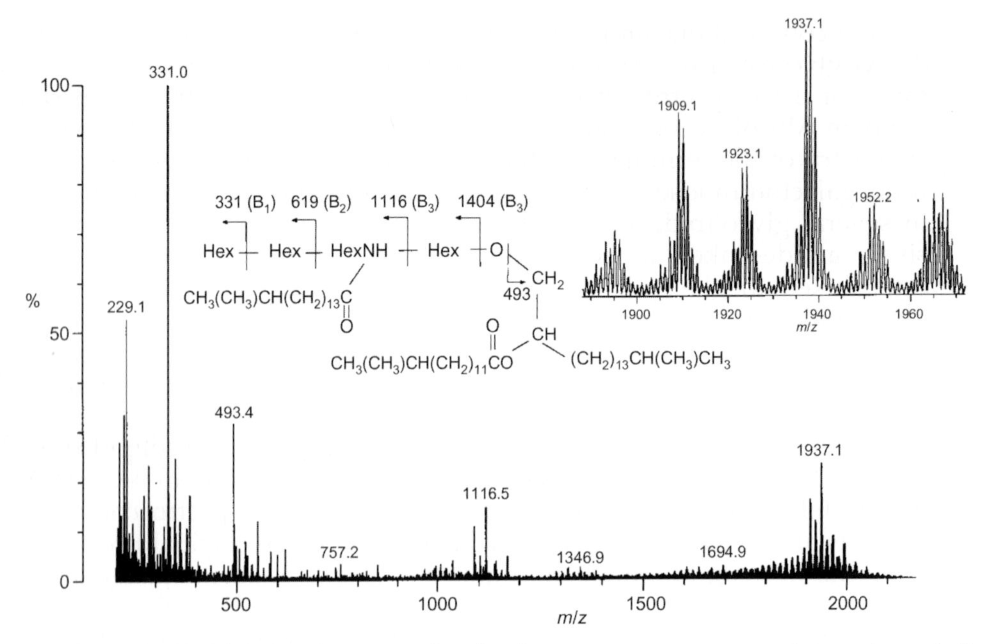

Figure 8. Positive ion fast atom bombardment mass spectrum of tetraglycosyl lipids from *Thermus filiformis*. The major sodium cationized molecule at *m*/*z* 1937.1 originates from a diol-linked glycolipid in which the *N*-acyl substituent is iso17:0 and the *O*-acyl group is iso15:0. The sequence of the glycan chain is defined by B-type carbenium ions at *m*/*z* 331, 619 and 1116 (see text). The fragment at *m*/*z* 493 represents loss of the glycan head-group with charge retention on the lipid moiety and is diagnostic of a C18 diol *O*-acylated with a C15 fatty acid.

and the acylglycerol ions enables an approximate estimate of the distributions of fatty acids at these locations. The ratio of the relative intensities of the acylglycerol fragments at *m*/*z* 551 and 523 was about 1.0/0.9, suggesting that of the total *O*-linked fatty acids, slightly more than 25%, were C17 fatty acids, whereas about 70% were C15 fatty acids.

The mass spectra of glycolipids that are linked to long-chain diols rather than to glycerol exhibit similar fragmentation of the sugar chains, but acylglycerol fragments are absent and are replaced by additional diagnostic fragments that define the diol moiety. Figure 8 shows the mass spectrum of the peracetyl derivative of the major diol-linked glycolipid from *Thermus filiformis*. The fragment at *m*/*z* 493 represents loss of the glycan head-group with charge retention on the lipid moiety and is diagnostic of a C18 diol *O*-acylated with a C15 fatty acid. As before, the sugar sequence of the glycan chain is defined by B-type carbenium ions at *m*/*z* 331, 619, and 1116 (Figure 8).

A. Acid-Catalysed Peracetylation of Intact Polar Lipids (Dell, 1990)

- Cautiously add two volumes of trifluoroacetic anhydride (TFAA) to one volume of acetic acid in a screw-capped tube (danger exothermic reaction; fume hood; and eye protection required). Allow to cool before use, but do not store; make fresh as required.

- Thoroughly dry the samples (5–50 μg in 5 ml screw-capped tubes) by vacuum centrifugation, then add 100 μl of the TFAA/acetic acid mixture, vortex and leave at room temperature for 10 min.
- Remove the reagent by vacuum centrifugation, dissolve the residue in 1 ml of chloroform and desalt by washing three times with equal volumes of water; discard the water washes.
- Evaporate the chloroform in a vacuum centrifuge and dissolve the derivatives in 10 μl chloroform:methanol (1:1, by volume) prior to FAB MS.

Deuteroacetates are prepared identically, except that the acetic acid in the reaction mixture is replaced by d_4-acetic acid.

♦♦♦♦♦♦ IX. FINAL COMMENTS

Even after the advent of the post-genomic era, it is clear that the polar lipid composition of prokaryotes continues to give us information that may not be apparent from the genomic analysis, since so many polar lipids are unknown and the polar lipid patterns can only be known from the analysis of the polar lipids from biomass of organisms that are grown in the laboratory or from biomass obtained directly from the environment. We are of the opinion that polar lipid composition should be included in the description of organisms, since by doing this for all novel taxa new and interesting polar lipids will be found that will add to our knowledge of the diversity of these cell components.

References

Albuquerque, L., Santos, J., Travassos, P., Nobre, M. F., Rainey, F. A., Wait, R., Empadinhas, N., Silva, M. T. and da Costa, M. S. (2002). *Albidovulum inexpectatum* gen. nov., sp. nov., a nonphotosynthetic and slightly thermophilic bacterium from a marine hot spring that is very closely related to members of the photosynthetic genus *Rhodovulum*. *Appl. Environ. Microbiol.* **68**, 4266–4273.

Albuquerque, L., Tiago, I., Veríssimo, A. and da Costa, M. S. (2006). *Tepidimonas thermarum* sp. nov., a new slightly thermophilic betaproteobacterium isolated from the Elisenquelle in Aachen and emended description of the genus *Tepidimonas*. *Syst. Appl. Microbiol.* **29**, 450–456.

Albuquerque, L., Rainey, F. A., Nobre, M. F. and da Costa, M. S. (2008). *Elioraea tepidiphila* gen. nov., sp. nov., a slightly thermophilic member of the *Alphaproteobacteria*. *Int. J. Syst. Bacteriol.* **58**, 773–778.

Albuquerque, L., Fereira, C., Tomaz, D., Tiago, I., Veríssimo, A., da Costa, M. S. and Nobre, M. F. (2009). *Meiothermus rufus* sp. nov., a new slightly thermophilic red-pigmented species and emended description of the genus *Meiothermus*. *Syst. Appl. Microbiol.* **32**, 306–313.

Albuquerque, L., Rainey, F. A., Nobre, M. F. and da Costa, M. S. (2010a). *Meiothermus granaticius* sp. nov., a new slightly thermophilic red-pigmented species from the Azores. *Syst. Appl. Microbiol.* **33**, 243–246.

Albuquerque, L., Rainey, F., Pena, A., Tiago, I., Veríssimo, A., Nobre, M. F. and da Costa, M. S. (2010b). *Tepidamorphus gemmatus* gen. nov., sp. nov., a slightly thermophilic member of the *Alphaproteobacteria*. *Syst. Appl. Microbiol.* **33**, 60–65.

Albuquerque, L., França, L., Rainey, F. A., Schumann, P., Nobre, M. F., and da Costa, M. S. (2011a). *Gaiella occulta* gen. nov., sp. nov., a novel representative of a deep branching phylogenetic lineage within the class *Actinobacteria* and proposal of *Gaiellaceae* fam. nov. and *Gaiellales* ord. nov. Syst. Appl. Microbiol. doi:10.1016/j.syapm.2011.07.001.

Albuquerque, L., Rainey, F. A., Nobre, M. F., and da Costa, M. S. (2011b). *Hydrotalea sandarakina* sp. nov., isolated from a hot spring runoff and emended description of the genus *Hydrotalea* and the species *Hydrotalea flava*. Int. J. Syst. Evol. Microbiol. doi:10.1099/ijs.0.034496-0.

Alves, M. P., Rainey, F. A., Nobre, M. F. and da Costa, M. S. (2003). *Thermomonas hydrothermalis* sp. nov., a new slightly thermophilic γ-proteobacterium isolated from a hot spring in central Portugal. *Syst. Appl. Microbiol.* **26**, 70–75.

Anderson, R. and Hansen, K. (1985). Structure of a novel phosphoglycolipid from *Deinococcus radiodurans*. *J. Biol. Chem.* **260**, 12219–12223.

Balkwill, D. L., Kieft, T. L., Tsukuda, T., Kostandarithes, H. M., Onstott, T. C., Macnaughton, S., Bownas, J. and Fredrickson, J. K. (2004). Identification of iron-reducing *Thermus* strains as *Thermus scotoductus*. *Extremophiles* **8**, 37–44.

Bligh, E. G. and Dyer, W. J. (1959). A rapid method of total lipid extraction and purification. *Can. J. Biochem. Physiol.* **37**, 911–917.

Carreto, L., Wait, R., Nobre, M. F. and da Costa, M. S. (1996). Determination of the structure of a novel glycolipid from *Thermus aquaticus* and demonstration that hydroxy fatty acids are amide linked to glycolipids in *Thermus* spp. *J. Bacteriol.* **178**, 6479–6486.

Christie, W. W. (1982). *Lipid Analysis*. 2nd ed. Pergamon Press, Oxford.

Dell, A. (1990). Preparation and desorption mass spectrometry of permethyl and peracetyl derivatives of oligosaccharides. *Methods Enzymol.* **193**, 647–660.

Domon, B. and Costello, C. E. (1988). A systematic nomenclature for carbohydrate fragmentations in FAB MS/MS spectra of glycoconjugates. *Glycoconj. J.* **5**, 397–409.

Ferrante, G., Ekiel, I. and Sprott, G. D. (1987). Structures of diether lipids of *Methanospirillum hungatei* containing novel head groups *N,N*-dimethylamino and *N,N,N*-trimethylaminopentanetetrol. *Biochim. Biophys. Acta* **921**, 281–291.

Ferreira, A. M., Wait, R., Nobre, M. F. and da Costa, M. S. (1999). Characterization of glycolipids from *Meiothermus* spp. *Microbiology* **145**, 1191–1199.

Huang, Y. and Anderson, R. (1989). Structure of a novel glucosamine-containing phosphoglycolipid from *Deinococcus radiodurans*. *J. Biol. Chem.* **264**, 18667–18672.

Huang, Y. and Anderson, R. (1991). Phosphatdylglycerolalkylamine, a novel phophoglycolipid precursor in *Deinococcus radiodurans*. *J. Bacteriol.* **173**, 457–462.

Huber, R., Wilharm, T., Huber, D., Trincone, A., Burggraf, S., König, H., Rachel, R., Rockinger, I., Fricke, H. and Stetter, K. O. (1992). *Aquifex pyrophilus* gen. nov., sp. nov., represents a novel group of marine hyperthermophilic hydrogen-oxidizing bacteria. *Syst. Appl. Microbiol.* **15**, 340–351.

Jahnke, L. L., Eder, W., Huber, R., Hope, J. M., Hinrichs, K.-U., Hayes, J. M., Des Marais, D. J., Cady, S. L. and Summons, R. E. (2001). Signature lipids and stable carbon isotope analyses of Octopus Spring hyperthermophilic communities compared with those of *Aquificales* representatives. *Appl. Environ. Microbiol.* **67**, 5179–5189.

Kates, M. (1993). Membrane lipids of Archaea. In: *The Biochemistry of Archaea (Archaebacteria)* (M. Kates, D. J. Kushner and A. T. Matheson, eds.), pp. 261–295. Elsevier, Amsterdam.

Koga, Y. and Morii, H. (2005). Recent advances in structural research on ether lipids from archaea including comparative and physiological aspects. *Biosci. Biotechnol. Biochem.* **69**, 2019–2034.

Koga, Y., Nishihara, M., Morii, H. and Akagawa-Matsushita, M. (1993). Ether polar lipids of methanogenic bacteria: structures, comparative aspects, and biosynthesis. *Microbiol. Rev.* **57**, 164–182.

Langworthy, T. A. and Pond, J. L. (1986). Membranes and lipids of thermophiles. In: *Thermophiles: General, Molecular and Applied Microbiology* (T. D. Brock, ed.), pp. 107–135. John Wiley and Sons, New York.

Lu, T. L., Chen, C. S., Yang, F. L., Fung, J. M., Chen, M. Y., Tsay, S. S., Li, J., Zou, W. and Wu, S. H. (2004). Structure of a major glycolipid from *Thermus oshimai* NTU-063. *Carbohyd. Res.* **339**, 2593–2598.

Morii, H., Yagi, H., Akutsu, H., Nomura, N., Sako, Y. and Koga, Y. (1999). A novel phosphoglycolipid archaetidyl(glucosyl)inositol with two sesterterpenyl chains from the aerobic hyperthermophilic archaeon *Aeropyrum pernix* K1. *Biochim. Biophys. Acta.* **1436**, 426–436.

Nishihara, M., Utagawa, M., Akutsu, H. and Koga, Y. (1992). Archaea contain a novel diether phosphoglycolipid with a polar head group identical to the conserved core of eucaryal glycosyl phosphatidylinositol. *J. Biol. Chem.* **267**, 12432–12435.

Oren, A., Ventosa, A. and Grant, W. D. (1997). Proposed minimal standards for description of new taxa in the order *Halobacteriales*. *Int. J. Syst. Evol. Microbiol.* **47**, 233–238.

Pond, J. P. and Langworthy, T. A. (1987). Effect of growth temperature on the long-chain diols and fatty acids of *Thermomicrobium roseum*. *J. Bacteriol.* **169**, 1328–1330.

Pond, J. P., Langworthy, T. A. and Holzer, G. (1986). Long-chain diols: a new class of membrane lipids from a thermophilic bacterium. *Science* **231**, 1134–1136.

Ratledge, C. and Wilkinson, S. G. (eds.), (1989). *Microbial Lipids, Vols. I and II*, Academic Press, London.

Ross, H. N. M. and Grant, W. D. (1985). Lipids in archaebacterial taxonomy. In: *Chemical Methods in Bacterial Systematics* (M. Goodfellow, ed.), pp. 290–291. Academic Press, New York.

Shimada, H., Nemoto, N., Shida, Y., Oshima, T. and Akihiko Yamagishi, A. (2002). Complete polar lipid composition of *Thermoplasma acidophilum* HO-62 determined by high-performance liquid chromatography with evaporative light-scattering detection. *J. Bacteriol.* **184**, 556–563.

Tindall, B. J., Rosselló-Móra, R., Busse, H. J., Ludwig, W. and Kämpfer, P. (2010). Notes on the characterization of prokaryote strains for taxonomic purposes. *Int. J. Syst. Evol. Microbiol.* **60**, 249–266.

Trincone, A., Trivellone, E., Nicolaus, B., Lama, L., Pagnotta, E., Grant, W. D. and Gambacorta, A. (1993). The glycolipid of *Halobacterium trapanicum*. *Biochim. Biophys. Acta* **1210**, 35–40.

Wait, R., Carreto, L., Nobre, M. F., Ferreira, A. M. and da Costa, M. S. (1997). Characterization of novel long-chain 1,2-diols in *Thermus* species and demonstration that *Thermus* strains contain both glycerol-linked and diol-linked glycolipids. *J. Bacteriol.* **179**, 6154–6162.

Zweig, G. and Sherma, J. (eds.), (1972). *Handbook of Chromatography, Vol. II*, CRC Press, Cleveland.

8 The Identification of Fatty Acids in Bacteria

Milton S. da Costa[1], Luciana Albuquerque[2], M. Fernanda Nobre[1] and Robin Wait[3]

[1] Department of Life Sciences, University of Coimbra, Coimbra, Portugal; [2] Center for Neuroscience and Cell Biology, University of Coimbra, Coimbra, Portugal; [3] Kennedy Institute of Rheumatology, Imperial College, London, United Kingdom

CONTENTS

I. INTRODUCTION

The determination of the fatty acyl composition is increasingly becoming a common application for microbiologists interested in the taxonomy of prokaryotes since it is easy to perform and can resolve many phylogenetic relationships. Fatty acids can also be used as a chemotaxonomic parameter to distinguish many closely related species; in fact, the determination of fatty acid analysis, along with polar lipid composition and respiratory quinone analysis, is recommended for publications where new prokaryote taxa are described (Tindall *et al.*, 2010). Sometimes, in fact, fatty acid analysis is the basis for the classification of some organisms that have phenotypic characteristics that do not clearly differentiate them from closely related species such as in the case of most *Legionella* spp. (Diogo *et al.*, 1999). Fatty acid composition has also proved an easy and sure way to identify bacteria.

Bacteria have an enormous variety of fatty acyl chains that include straight-chain saturated and unsaturated fatty acids, iso- and anteiso-branched fatty acids, internally branched fatty acids, hydroxy fatty acids, cyclopropane fatty acids, ω-cyclic fatty acids, dicarboxylic fatty acids, ladderane fatty acids, among others (Figure 1). Polyunsaturated fatty acids are rare, but have been encountered most frequently in bacteria that grow at low temperatures and cyanobacteria

METHODS IN MICROBIOLOGY, VOLUME 38
0580-9517 DOI 10.1016/B978-0-12-387730-7.00008-5

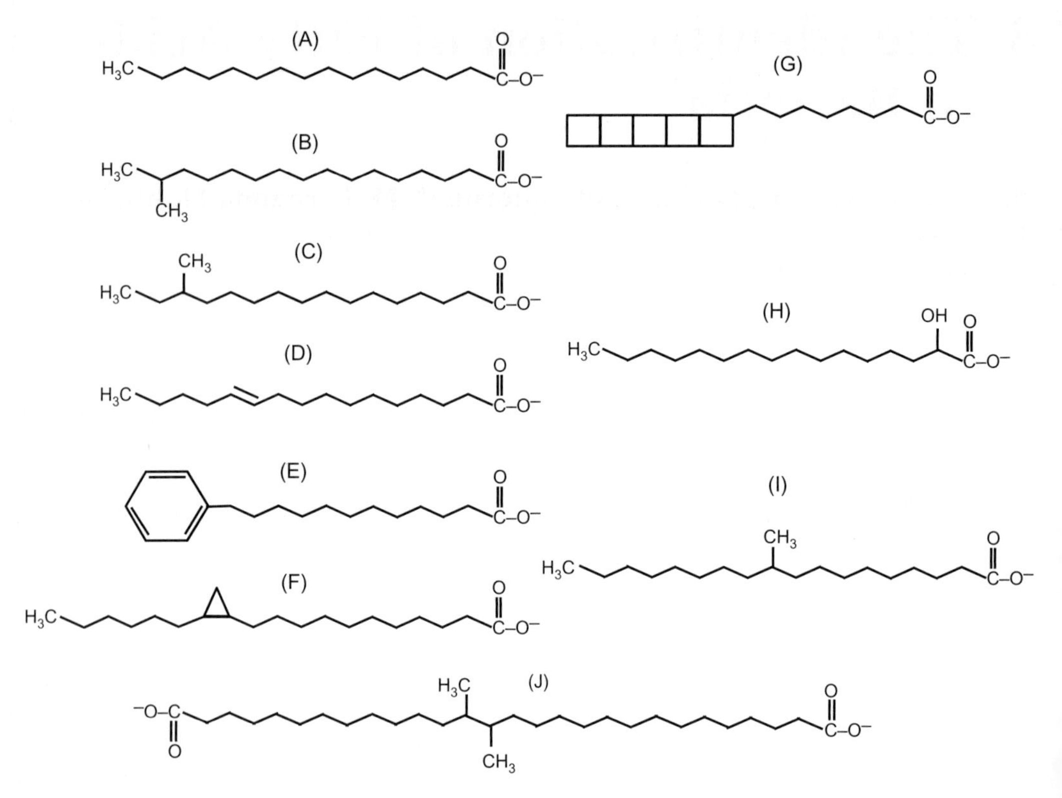

Figure 1. Examples of fatty acids found in bacteria. (A) saturated straight chain; (B) iso-branched fatty acid; (C) anteiso-branched fatty acid; (D) unsaturated straight-chain fatty acid; (E) cyclohexyl fatty acid; (F) cyclopropane fatty acid; (G) ladderane fatty acid; (H) 2-OH fatty acid fatty acid; (I) internally branched fatty acid; (J) dicarboxylic fatty acid.

(Russel and Nichols, 1999; Singh *et al.*, 2002; Yano *et al.*, 1997). Fatty acyl chains such as fatty alcohols are also generally found in small amounts. Some fatty acid species are common across large groups of bacteria, such as those of the Alpha-, Beta- and Gammaproteobacteria, but others seem to be restricted to only a few small groups of organisms such as the cyclohexane and cycloheptane fatty acids of some of the species of the genus *Alicyclobacillus*, *Sulfobacillus* and *Curtobacterium*, unusual the internally branched fatty acids in the species of *Rubrobacter* and *Gaiella occulta* (Albuquerque *et al.*, 2011a; Carreto *et al.*, 1996). The dicarboxylic fatty acids, known as diabolic fatty acids, are present in the bacteria of the order Thermotogales and in *Thermoanaerobacter ethanolicus* (de Rosa *et al.*, 1988, 1989; Jung *et al.*, 1994; Sinninghe Damsté *et al.*, 2007). Perhaps the most unexpected fatty acyl chains are the so-called ladderane fatty acids of anammox bacteria (Hopmans *et al.*, 2006; Rattray *et al.*, 2010; Sinninghe Damsté *et al.*, 2005).

In some cases, fatty acid composition appears not to be very useful in the systematics of some groups of bacteria. For example, only the strains of two of the recognized species of the genus *Thermus*, namely *T. oshimai* and *T. scotoductus*, have uniform fatty acid compositions that are helpful in the identification and classification of these organisms, while the strains assigned to the species

T. thermophilus and to the species *T. filiformis* have extremely variable fatty acid compositions. The type strain of *T. filiformis* possesses 3-OH fatty acids, while the other strains assigned to this species do not (Nobre *et al.*, 1996). Of course, the presence/absence of hydroxyl fatty acids in these organisms is related to the composition of the polar head-group of the glycolipids, but still there is a taxonomic perspective to these observations that cannot be overlooked (Carreto *et al.*, 1996, Ferreira *et al.*, 1999; Wait *et al.*, 1997). On the other hand, the species of the related genus *Meiothermus* appear to have fairly stable fatty acid compositions, although two species, namely *M. granaticius* and *M. rufus*, only have 3-OH fatty acids, while the other species possess 2-OH and 3-OH fatty acids (Albuquerque *et al.*, 2009, 2010; Nobre *et al.*, 1996).

The taxonomic significance of fatty acids is obscure in some groups of bacteria, namely the species of the genus *Deinococcus*, where some species possess large quantities of iso- and anteiso-fatty acids, while others possess straight-chain fatty acids. These species are dispersed within the genus with no apparent taxonomic significance when compared to the 16S rRNA gene sequence-based phylogeny (Callegan *et al.*, 2008; Rainey *et al.*, 2005).

The most important factor behind reproducible fatty acid composition is the extreme care that must be taken to standardize the growth conditions of the organisms being examined, since the levels of the acyl chains vary with the growth temperature, the phase of growth and the medium composition, among other growth conditions. It is not uncommon to obtain differences in the levels of individual fatty acids of the same strain from two different laboratories that seem to use or claim to use identical media and growth conditions. For this reason, it is currently recommended to compare the fatty acid composition of new isolates with that of previously described closely related species growing the new organism alongside the reference type strains under identical culture conditions. Normally, growth conditions and sources of media or of medium components will affect the fatty acid composition leading to differences that can alter our perception of the taxonomy of the organisms we are dealing with. It is even sometimes important to use the same batch or lot of medium to grow the organisms concurrently to compare the fatty acid composition of closely related strains or species, since slight differences in the growth condition can lead to changes of their relative proportions of fatty acids. For example, during a study of the fatty acid composition of 250 strains of the known species of the genus *Legionella* at the time (Diogo *et al.*, 1999), we always used the same reference strains to check variations of the relative proportions of the fatty acids, due to changes in growth conditions over a period of several months. Microbiologists usually determine the fatty acid composition of bacteria of the same genus at the one optimum growth temperature because organisms of the same genus usually have similar growth temperature ranges. However, sometimes two or more strains of the same genus have different optimum growth temperatures, and it is necessary to compare the fatty acid composition at a common growth temperature, otherwise a different perception of the taxonomy of the organisms will be obtained. Growth at one common temperature will ensure that the fatty acids can be used to discriminate among different species (Callegan *et al.*, 2008; Rainey *et al.*, 2003, 2005). For example, the recently described *Hydrotalea sandarakina*, belonging to the phylum Bacteroidetes, has a growth temperature range between 25°C and 52.5°C and an optimum growth temperature around 45°C, while the only other species of the same genus, *Hydrotalea flava* (Kämpfer *et al.*, 2011), has a growth temperature range of 20–37°C. To ensure that both species could be distinguished from each on the basis of the fatty acid composition, they were grown at a common growth temperature of 30°C as well as at their respective optimum growth temperatures (Albuquerque *et al.*, 2011b).

Fatty acid composition is assessed by gas chromatography (GC) and by comparison of the peak retention times of samples with those of known standards. Mass spectrometry can then be used to identify those fatty acids that cannot be identified by GC alone or fatty acids that we suspect are not those identified by automated systems. The Sherlock® Microbial Identification Systems (MIS) (http://www.midi-inc.com) is used by many microbial taxonomists to identify a large variety of fatty acids from microorganisms. This system provides a highly standardized methodology that is primarily directed to the identification of fatty acids and other acyl compounds from pathogenic bacteria and yeast. However, it is used very successfully to determine the fatty acid composition of many different microorganisms. Identification of acyl chains is based on the equivalent chain length (ECL), which refers to a linear extrapolation of each peak's retention time between two straight-chain saturated fatty acid methyl ester reference peaks. The MIS software compares the ECL of each peak in the sample with the expected ECL of fatty acyl compounds in the database. Some peaks are not identified and may correspond to novel fatty acids. Other peaks may be labelled "Summed Feature" because the ECL value corresponds to one fatty acid that cannot be separated from another fatty acid under the standard chromatographic conditions. The relative concentration of possibly two or more fatty acids is given as one value. It is sometimes possible to tentatively identify a particular fatty acid within a "Summed Feature" by comparing these fatty acids with other fatty acids identified by the database from the sample. For example, in *Meiothermus* spp., one peak labelled as "Summed Feature" may contain two fatty acids, namely 15 iso 2-OH and 16:1 $\omega 7c$. As the organism possesses other 2-OH fatty acids, but no detectable monounsaturated straight-chain fatty acids, it is very likely that this "Summed Feature" corresponds to 15 iso 2-OH fatty acid alone (Nobre *et al.*, 1996).

The MIS produces several databases for different types of microorganisms, such as anaerobic bacteria, aerobic bacteria, yeasts, among others. The TSBA database methodology is one of the most frequently used methodologies of MIS for aerobic and facultatively anaerobic bacteria grown at 28°C on Tryptic Soy Agar (BBL 11849) for 24 ± 1 h. Under these growth conditions, many aerobic and facultative bacteria can be identified using the databases, primarily the TSBA database, provided by MIS. But these growth conditions can be modified to suit the identification of fatty acids from most bacteria. Additional databases can be constructed by the investigator, to aid in the identification of certain groups of bacteria, such as *Legionella* spp. (Diogo *et al.*, 1999), because without them many strains cannot be identified. Moreover, some organisms have rare fatty acids that are not in any of the MIDI databases because these strains may represent new organisms that have not been included in the manufacturer's own databases or do not conform to the growth conditions of the MIS. It is not necessary, for comparative purposes, to identify strains by fatty acid analysis; it is merely necessary to identify the majority of the acyl compounds and compare them to closely related organisms grown under the same conditions.

The Sherlock® MIS cannot be used with alternate chromatographs or columns than those provided or recommended by the manufacturer, because of the interdependency of the system. The manufacturer also provides a standard mixture of fatty acids that is automatically injected to calibrate the process at the onset and after every 11 sample injections. However, some modifications of cultivation, extraction and methylation are possible and can be very useful in determining the fatty acid composition of bacteria. Moreover, fatty acyl compounds not identified by MIS can be identified by comparison with fatty acid methyl esters from other bacteria where they have been identified. New unknown fatty acids, however, will have to be identified by mass spectroscopy (MS). In some cases, minor unknown fatty acids (less than 0.5%) are not normally relevant for characterization of the organisms, but each author should be aware of the importance of all fatty acids detected by GC. For example, a new fatty acid was identified by MS that was not found in the MIS database, because it was deemed important for the characterization of a new organism, *Albidovulum inexpectatum* (Albuquerque *et al.*, 2002). This fatty acid was identified by MS as the very rare 11-methyl-11,12-octadecenoic acid (11-Me 18:1[11]=11 met-18:1ω7). It is, however, possible to build large databases for several groups of microorganisms that can be used to tentatively identify new isolates of the species. It has been shown that, under controlled conditions, extremely reproducible results can be obtained. For example, we routinely use fatty acid analysis to identify *Legionella* spp. from environmental samples (Diogo *et al.*, 1999).

Sometimes the MIS software misidentifies fatty acids. This is a problem that each microbiologist must be aware of and must take steps to correct. For example, the major fatty acid (51% of the total) of an organism, *Gaiella occulta* (Albuquerque *et al.*, 2011a) was recently identified by the MIS software as 16:1 2-OH (ECL 17.045). The organism belongs to the Actinobacteria where hydroxyl fatty acids are minor components or are not detected; moreover, there were two other unknown fatty acids with ECLs of 16.050 and 15.056, indicating that the three fatty acids belonged to the same family of fatty acids. To investigate the possibility that these fatty acids were not hydroxylated, *O*-trimethylsilyl derivatives of the fatty acid extract were obtained and GC was performed as recommended by the MIDI system using standard methods. It is well known that hydroxy fatty acids loose the hydroxyl group upon trimethylmethylsilylation and have a different retention time, while trimethylsilylated non-hydroxy fatty acids retain the same RT as before derivatization. The misidentified major fatty acid and the two other unknown fatty acids were clearly not hydroxyl fatty acids because there was no alteration in the retention time. These were identified by GC/MS as a new family of internally branched iso-fatty acids (Albuquerque *et al.*, 2011a).

The methods for the growth and the Anaerobic Bacteria Libraries, named MOORE for the identification anaerobic bacteria, are based on the results obtained by Moore *et al.* (1994). These methods recommend the growth of the organisms in PYG liquid media and use the same acid methanolysis method used throughout the MIDI system for other organisms. Many bacteria and especially anaerobic species frequently possess ether-linked lipids, such as plasmogens, that produce dimethyl acetals (DMAs) during acid hydrolysis. DMAs elute with fatty acids and some elute with the same retention time as hydroxylated fatty acids, leading to erroneous identification of these fatty acids as DMAs (Helander and Haikara, 1995; Kämpfer *et al.*, 2000; Strömpl *et al.*, 2000). However, one should be aware that either the DMAs or the OH fatty acids may have been misidentified and the appropriate derivatizations and GC/MS performed to identify the acyl chains correctly.

This partially automated MIDI system relies on a saponification step followed by acid methanolysis to produce of the methyl esters (Kuykendall *et al.*, 1988), which is reliable for most bacteria. GC is performed with a Hewlett-Packard chromatograph equipped with an automated injector and a 5% phenyl methyl silicone capillary column (0.2 mm by 25 m), and the fatty acid methyl esters are detected with a flame ionization detector (FID). All reagents should be stored at room temperature in dark glass bottles with Teflon-lined screw caps.

♦♦♦♦♦♦ II. CULTIVATION AND HARVESTING OF CELLS

The cells can be grown on solid media, as recommended by the manufacturer, or they can be grown in liquid medium. The cultivation time depends on the organism, the medium used and the temperature, among other growth conditions. As the fatty acid composition can change dramatically with culture conditions, it is very important to always use the same conditions and to harvest at the same growth phase. The MIS recommends removing cells from a specific quadrant of the culture plates. Cells, about 40 mg, are scraped from solid media with a bacteriological loop (10 μl loop) and placed in a 12 cm × 100 cm Teflon-lined screw cap culture tube. Liquid cultures are harvested by refrigerated centrifugation and washed once with an appropriate buffer such as phosphate buffered saline (PBS). The cell pellets are transferred to a Teflon screw-capped glass tube (12 mm × 100 mm) with a bacteriological loop.

♦♦♦♦♦♦ III. PREPARATION OF FATTY ACID METHYL ESTERS

A. Saponification

- Mix 45 g NaOH (reagent grade), 150 ml methanol (HPLC grade) and 150 ml ultrapure water until the NaOH pellets are completely dissolved.
- Add 1 ml to the Teflon glass tube containing fresh cell mass, vortex for about 10 s and then heat for 5 min in a boiling water bath. Vortex again and incubate in the boiling water bath for 25 min. Cool in water at room temperature.

B. Methylation

- Add 325 ml of titrated 6.0 N HCl to 225 ml of methanol (HPLC grade).
- Add 2 ml to the tube and vortex for about 10 s and heat in a water bath at 80°C for 10 min. Cool tube in water to room temperature.

C. Extraction

Reagent A: Add 200 ml methyl-tert-butyl ether (HPLC grade) to 200 ml hexane (HPLC grade).
Reagent B: Dissolve 10.8 g NaOH (reagent grade) in 900 ml of ultrapure water.

Add 1.25 ml of reagent A to each tube and mix end-over-end with a rocking motion on a laboratory rotator for 10 min. Remove the lower phase with a Pasteur pipette and discard. Keep the upper phase that is washed with 3 ml of reagent B. Mix end-over-end with a rocking motion for 5 min, add a few drops of a saturated solution of NaCl and transfer about two-thirds of the upper phase containing the fatty acid methyl esters to an auto-sampler vial and seal. Continue or store at −20°C under a N_2 atmosphere.

D. Modifications

We have omitted the saponification step to recover larger relative proportions of 1,2-long-chain diols from *Thermomicrobium roseum* and some *Thermus* spp., which appear not to be recovered or are recovered in low relative proportions following the MIS extraction and methylation protocol and modified the methylation step by incubating the samples at 100°C for 30 min (Carreto *et al.*, 1996; Wait *et al.*, 1997). This modification does not affect the recovery of other fatty acyl methyl esters from bacteria of the genera *Thermus*, *Meiothermus*, *Deinococcus* and *Truepera*. However, cyclopropane and other acyl compounds may be degraded. It is, therefore, important to compare the fatty acid profiles derived from modified extraction methods with those obtained from the standard method described by the manufacturer.

♦♦♦♦♦♦ IV. MASS SPECTROMETRY OF FATTY ACID AND DIOL DERIVATIVES

In a mass spectrometric experiment, sample molecules are ionized and the masses of their gas phase ions are determined. If sufficient excess energy is deposited, or if the ions are additionally excited, e.g., by collision with neutral gas, then a proportion will dissociate in a structure-specific fashion, producing a fragmentation pattern, from which the mass and structure of the original molecule can be deduced. Low molecular mass compounds such as fatty acid methyl esters can be ionized by bombardment with electrons emitted from a heated filament (electron ionization, EI). This is relatively energetic, and considerable fragmentation usually occurs. Since ionization occurs in the vapour phase, EI MS is easily interfaced to GC, and a combined GC/MS instrument provides detailed structural information on each component of a mixture as it elutes from the chromatograph. Most commercial GC/MS instruments are suitable for fatty acid analysis, whether of magnetic sector, quadrupole, ion trap or time of flight design. The chromatographic columns, injection methods and oven condition conditions needed are similar to those described above, though some differences in retention behaviour and peak shape may be noticed as a result of interfacing to the mass spectrometer.

The mass spectra of methyl esters of saturated fatty acids have easily recognizable molecular ions (typically about 20% of the intensity of the base peak), which define the number of carbon atoms. A series of weaker, regularly spaced fragments 14 *m/z* units apart of composition $C_nH_{2n-1}O_2$ (at *m/z* 73, 87, 101, etc.) originate from cleavage of the carbon skeleton and, together with the well-known McLafferty rearrangement product at *m/z* 74 (McLafferty and Tureck, 1993), afford information on the chain structure. Elimination of CH_2CH_3 and the methoxy group CH_3O from the carboxyl terminus results in fragments at *m/z* M-29 and M-31. The location of alkyl branches within the chain can usually be inferred from the fragmentation pattern because cleavage adjacent to the branch will produce a stable secondary carbocation, though information on branch positions is more reliably obtained from the spectra of such as picolinyl esters or 2-alkenyl-4,4-dimethyloxazoline (DMOX) derivatives. Terminal methyl groups are not easily recognized from the spectra of methyl esters. The spectra of iso-branched and straight-chain methyl esters are virtually indistinguishable, but the former can be recognized by their shorter retention times. Anteiso-branched FAME can be identified because the fragment at M-29 is stronger than that at M-31, whereas the converse is true for iso and normal compounds. However, picolinyl or DMOX derivatives provide a better method for unambiguous branch location.

Unsaturation and the presence of carbocyclic rings decrease the mass of the molecular ion by two *m/z* units per ring or bond compared to the corresponding saturated acid. The overall appearance of the spectra of ω-cyclohexane ring-containing FAME, such as those present in *Alicyclobacillus*, is similar to that of saturated compounds, except for the presence of an ion at *m/z* 83, which originates from cleavage of the cyclohexane ring. The size and locations of carbocyclic rings are best confirmed from the spectra of picolinyl or DMOX derivatives.

The fragmentation of unsaturated methyl esters is usually more extensive than that of the corresponding saturated compound, but it is not normally possible to deduce the locations of unsaturation because extensive double bond migration occurs under EI conditions. A number of derivatives are available that permit unambiguous localization of double bond positions by MS. The main strategies are either to derivatize the double bonds themselves, so that the resulting spectra contain fragments diagnostic of their positions, or to modify the carboxyl group so as to stabilize the positive charge and prevent bond migration.

bis-Methylthio (dimethyl disulfide) formation represents an example of the former approach (Francis, 1981). These derivatives are prepared from FAME by a simple one-step reaction, have good chromatographic properties and are easily interpretable mass spectra.

For example, GC/MS analysis of the FAME of a major component with an ECL of 17.83 from *A. inexpectatum* was consistent with an octadecenoic acid. Synthesis of its corresponding *bis*(methylthio) derivative confirmed this (M^+ at *m/z* 390) and fragment ions at *m/z* 145 and *m/z* 245, resulting from cleavage at the original site of unsaturation, enabled localization of the double bond to the

Δ11 position (Albuquerque *et al.*, 2002). A further advantage is that methylthio addition is stereospecific, so chromatographically resolvable *erythro* and *threo* products result from *E* and *Z* isomers of fatty acids, respectively.

The alternative carboxyl derivatization strategy enables location of features such as alkyl branches and rings, as well as double bonds in the same experiment; however, preparation of the derivatives is a little more complex because FAMEs have to be de-esterified first.

Picolinyl esters are easy to prepare (Christie and Stephanov, 1987; Harvey, 1992) and their mass spectra are extremely simple to interpret. However, they elute at substantially higher temperatures than methyl esters and, because of the ring nitrogen, are prone to peak tailing, which degrades chromatographic resolution resulting in peak overlap, which may complicate spectral interpretation.

2-Alkenyl-4,4-dimethyloxazoline (DMOX) derivatives yield informative mass spectra from which the locations of features such as rings, branches and olefinic bonds are readily deduced (Zhang *et al.*, 1988), since the charge stabilization conferred by the heterocyclic ring largely suppresses double bond migration.

These derivatives have been used to establish the structure of a novel FAME from *A. inexpectatum*, which had an ECL 18.08 and a molecular ion at *m*/*z* 310. Its DMOX derivative had a molecular ion at *m*/*z* 349 and a spectrum characterized by a regularly spaced even mass ion series (*m*/*z* $126+(14)_n$) originating by cleavage, without rearrangement, at every carbon atom of the alkyl chain. This regular 14 *m*/*z* spacing was interrupted between *m*/*z* 264 and 224, suggesting that the compound was 11-methyl-11,12-octadecenoic acid (Albuquerque *et al.*, 2002).

A. Identification of Hydroxylated Fatty Acids

3-Hydroxy acids are present in some *Thermus* species, whereas 2-hydroxy acids are characteristic of *Meiothermus*, though they may be accompanied by 3-hydroxy acids in some cases. Whenever hydroxy acids are present, they *N*-acylate the hexosamine residue in position two of the sugar chain and are never glycerol-linked (Carreto *et al.*, 1996; Ferreira *et al.*, 1999; Wait *et al.*, 1997).

The presence of a hydroxyl group provides an additional site for ionization, so the mass spectra of hydroxylated fatty acids usually contain fragments diagnostic of the location of the substituent. Methyl esters of 2-hydoxy acids undergo cleavage between carbons 1 and 2, resulting in an intense ion at *m*/*z* M-59. If the hydroxyl group is on carbon 3, a molecular ion is not usually detectable in EI spectra, and the major process observed is cleavage on the alkyl side of the hydroxyl-substituted carbon giving an ion at *m*/*z* 103, normally the base peak in the spectrum. *O*-trimethylsilyl derivatives are useful for the characterization of hydroxylated FAME, since they usually exhibit improved chromatographic behaviour, with sharper and more symmetrical peaks. Since the derivatives exhibit increased retention times on non-polar columns, hydroxylated acids can be recognized even without mass spectrometric analysis. Molecular ions are usually weak or absent in the spectra of *O*-trimethylsilyl ethers, but the molecular mass may usually be deduced from a peak at *m*/*z* M-15.

The mass spectrum of the *O*-TMS ether of the major hydroxylated acid from *Thermus aquaticus*, for example, had a prominent M-15 peak at *m*/*z* 357 and an abundant α-cleavage fragment at *m/z* 175 (the silylated form of the *m/z* 103 fragment in underivatized hydroxyl FAME) consistent with a 3-hydroxylated 17 carbon compound. It had a retention time shorter than authentic methyl 3-OH heptadecanoate, but identical to 3-OH-15-methylhexadecanoate.

The major hydroxyl acid in *M. ruber* had a molecular ion at *m/z* 300 (underivatized), which shifted to *m/z* 372 after trimethylsilylation. The loss of 59 *m/z* units from the molecular ion (representing a facile cleavage between C-1 and C-2, with charge retention on the substituent-bearing fragment) identified it as a 2-hydroxy 17:0 FAME. Its retention behaviour was consistent with an iso-branched alkyl chain, and this was supported by the observation of traces of a later-eluting 2-hydroxy acid in the position expected for the anteiso-branched isomer.

B. Preparation of Bis(methylthio) Derivatives

- After GC and GC/MS analyses of fatty acid methyl esters, dry down the residual sample in a round-bottomed tube with a PTFE-faced screw cap and redissolve in 100 μl of hexane.
- Add 100 μl of dimethydisulfide and two drops of a 6% (wt/vol) solution of iodine in diethyl ether.
- Allow to react at room temperature overnight.
- Shake with 0.5 ml 5% aqueous sodium thiosulfate to remove iodine and recover the derivatives by extracting (twice) with an equal volume of hexane.

C. Preparation of 2-Alkenyl-4,4-dimethyloxazoline Derivatives

- Demethylate the fatty acid methyl esters by treatment with 0.5 ml of 1 M sodium hydroxide in 50% aqueous methanol at 60°C overnight.
- Cool, reduce the pH to below 2 by addition of 0.5 ml of 1 M HCl and recover the free fatty acids by extraction with hexane:chloroform (4:1, by vol.).
- Dry the extracts, redissolve in 50 μl chloroform, mix with 100 μl of 2-methyl-2-amino propanol and heat for 3 h at 285°C in a sealed tube.
- Cool to room temperature, add two volumes of chloroform and wash with 1 ml of distilled water, made alkaline with a few drops of 1 M sodium hydroxide solution.
- Remove the water layer and wash the chloroform solution of DMOX derivatives twice more with 1 ml of water.
- Remove the chloroform layer to a clean tube and dry with a vacuum centrifuge.

D. Preparation of O-trimethylsilyl Ethers

- Dry the samples in a suitable reaction tube using a vacuum centrifuge.
- Add 100 μl of bis(trimethylsilyl)trifluoroacetamide and allow to react for 30 min at 60°C (heating block or GC oven).
- Remove the reagent with a vacuum centrifuge and dissolve the derivatives in trimethylpentane or hexane for GC/MS.

E. Mass Spectrometry of Long-Chain Diols

Moderate amounts of two late eluting components with ECLs of 19.060 (major peak) and 19.160 (minor peak) have been detected in FAME extracts of some *Thermus* strains (Carreto *et al.*, 1996). On trimethysilylation, the retention times of these components increased to ECLs 19.832 and 19.925, respectively, suggesting the presence of hydroxyl groups.

Abundant protonated molecules were observed at *m/z* 431 in the isobutane chemical ionization spectra of the TMS derivatives of both compounds. An abundant fragment at *m/z* 327 was assigned as an α-cleavage ion, which would be consistent with the presence of an *O*-trimethylsilylated alkyl chain of at least 17 carbon atoms ($CH_3-(CH_2)_{15}-CHOSi(CH_3)_3=327$).

The spectra do not resemble those of hydroxylated fatty acids, but are consistent with octadecanediols, the masses of which would be 286 underivatized or 430 after trimethylsilylation. Detailed characterization of the structure of the alkyl chains was achieved by mass spectrometry of their acetate derivatives.

Molecular ions are not observed, the highest mass signal observed being *m/z* 310, attributable to elimination of acetic acid. The structure of the alkyl chain is defined by an important group of ions at odd mass number (*m/z* 295, 281, 267 and 253), which originate by cleavage, without rearrangement, of the terminal region of the alkyl backbone. The absence of the fragment at *m/z* 281, which can only be formed by cleavage of two bonds, is diagnostic of an iso branch. Conversely, the fragment at *m/z* 267 is absent from anteiso-branched octadecanediols.

Deuteroacetylation provides useful confirmation since the ions at *m/z* 310, 295, 281, 267 and 253 will be shifted by three *m/z* units, showing they contained a single acetate group. The fragment at *m/z* 284 will be absent from the spectra of the iso-branched octadecanediols and that at *m/z* 270 from the anteiso isomer. Fragments not containing acetate groups, such as those at *m/z* 268, 250 and 222 will be unaffected by deuteroacetylation.

The long-chain diols from *T. roseum* lipids exhibit a greater degree of structural diversity, with both internally branched and unbranched examples. Most of the high-mass ions in the spectra of acetate derivatives of the straight-chain compounds are odd electron products, attributable to the loss of one or more of the acetate groups (either as acetic acid [M-60] or as ketene [M-42]). The main exception is an even-electron fragment at M-73, which originates by cleavage between C-1 and C-2. The spectra of the internally branched diacetates are similar, except for the presence of additional even-electron fragment ions, probably originating by a charge-site remote mechanism, which are diagnostic of methyl branch location. In 13-methyl compounds, these were observed at *m/z* 253 and 267. In the acetate derivative of 15-methyl-1,2 henicosanediol by contrast, the analogous ions are shifted to *m/z* 281 and 295.

F. Base Catalysed Acetylation of Long-Chain Diols

- Dry 1 ml of pyridine (Aldrich, HPLC grade) by addition of a spatula full of P_2O_5, vortex and stand for an hour, then pellet the P_2O_5 with a bench-top centrifuge and aspirate off the pyridine in a fume hood.

- Dissolve the sample (5–50 μg) in 100 μl of the dried pyridine, and add 100 μl of acetic anhydride and vortex mix.
- Leave at room temperature for 24 h and then remove the reagents by vacuum centrifugation.
- Dissolve the derivatives in chloroform and desalt by washing three times with an equal volume of water; discard the washes.

Deuteroacetates are prepared identically except for the use of d_6-acetic anhydride in place of acetic anhydride.

Final Comments

Prokaryote acyl chains provide us with a grand view of chemical diversity that makes up the hydrophobic part of cell membranes. They are also important in taxonomy, and fatty acid composition should be included in all new descriptions of novel taxa. It is important to include the type strains of the most closely related species in such studies and comparisons. Despite pitfalls common to all phenotypic and chemotaxonomic analyses, these parameters provide us with a better view of the relationships of organisms and the uniqueness of the individual organisms we are trying to describe.

References

Albuquerque, L., Santos, J., Travassos, P., Nobre, M. F., Rainey, F. A., Wait, R., Empadinhas, N., Silva, M. T. and da Costa, M. S. (2002). *Albidovulum inexpectatum* gen. nov., sp. nov., a nonphotosynthetic and slightly thermophilic bacterium from a marine hot spring that is very closely related to members of the photosynthetic genus *Rhodovulum*. *Appl. Environ. Microbiol.* **68**, 4266–4273.

Albuquerque, L., Fereira, C., Tomaz, D., Tiago, I., Veríssimo, A., da Costa, M. S. and Nobre, M. F. (2009). *Meiothermus rufus* sp. nov., a new slightly thermophilic red-pigmented species and emended description of the genus *Meiothermus*. *Syst. Appl. Microbiol.* **32**, 306–313.

Albuquerque, L., Rainey, F. A., Nobre, M. F. and da Costa, M. S. (2010). *Meiothermus granaticius* sp. nov., a new slightly thermophilic red-pigmented species from the Azores. *Syst. Appl. Microbiol.* **33**, 243–246.

Albuquerque, L., França, L., Rainey, F. A., Schumann, P., Nobre, M. F., and da Costa, M. S. (2011a). *Gaiella occulta* gen. nov., sp. nov., a novel representative of a branching phylogenetic lineage within the class *Actinobacteria* and proposal of *Gaiellaceae* fam. nov. and *Gaiellales* ord. nov. Syst. Appl. Microbiol. doi:10.1016/j.syapm.2011.07.001.

Albuquerque, L., Rainey, F. A., Nobre, M. F., and da Costa, M. S. (2011b). *Hydrotalea sandarakina* sp. nov., isolated from a hot spring runoff and emended description of the genus *Hydrotalea* and the species *Hydrotalea flava*. Int. J. Syst. Evol. Microbiol. doi:10.1099/ijs.0.034496-0.

Callegan, R. P., Nobre, M. F., McTernan, P. M., Battista, J. R., González, R. N., McKay, C. P., da Costa, M. S. and Rainey, F. A. (2008). Description of four novel psychrophilic, ionizing radiation-sensitive *Deinococcus* species from alpine environments. *Int. J. Syst. Evol. Microbiol.* **58**, 1252–1258.

Carreto, L., Wait, R., Nobre, M. F. and da Costa, M. S. (1996). Determination of the structure of a novel glycolipid from *Thermus aquaticus* and demonstration that

hydroxy fatty acids are amide linked to glycolipids in *Thermus* spp. *J. Bacteriol.* **178**, 6479–6486.
Christie, W. W. and Stephanov, K. (1987). Separation of picolinyl derivatives by high performance liquid chromatography for identification by mass spectrometry. *J. Chromatogr.* **392**, 259–265.
De Rosa, M., Gambacorta, A., Huber, R., Lanzotti, V., Nicolaus, B., Stetter, K. O. and Trincone, A. (1988). A new 15,16-dimethyl-30-glyceryloxytriacontanoic acid from lipids of *Thermotoga maritima*. *J. Chem. Soc. Chem. Commun.* **19**, 1300–1301.
De Rosa, M., Gambacorta, A., Huber, R., Lanzotti, V., Nicolaus, B., Stetter, K. O. and Trincone, A. (1989). Lipid Structures in *Termotoga maritima*. In: *Microbiology of Extreme Environments and Its Potential for Biotechnology* (M. S. da Costa, J. C. Duarte and R. A. Williams, eds.), pp. 167–173. Elsevier, London.
Diogo, A., Veríssimo, A., Nobre, M. F. and da Costa, M. S. (1999). Usefulness of fatty acid composition for differentiation of *Legionella* species. *J. Clin. Microbiol.* **37**, 2248–2254.
Ferreira, A. M., Wait, R., Nobre, M. F. and da Costa, M. S. (1999). Characterization of glycolipids from *Meiothermus* spp. *Microbiology* **145**, 1191–1199.
Francis, G. W. (1981). Alkylthiolation for the determination of double bond positions in unsaturated fatty acid esters. *Chem. Phys. Lipids.* **29**, 369–374.
Harvey, D. (1992). Mass spectrometry of picolinyl and other nitrogen-containing derivatives of lipids. In: *Advances in Lipid Methodology-1* (W. W. Christie, ed.), pp. 19–80. The Oily Press, Ayr.
Helander, I. M. and Haikara, A. (1995). Cellular fatty acyl and alkenyl residues in *Megasphaera* and *Pectinatus* species: contrasting profiles and detection of beer spoilage. *Microbiology* **141**, 1131–1137.
Hopmans, E. C., Kienhuis, M. V., Rattray, J. E., Jaeschke, A., Schouten, S. and Sinninghe Damsté, J. (2006). Improved analysis of ladderane lipids in biomass and sediments using high-performance liquid chromatography/atmospheric pressure chemical ionization tandem mass spectrometry. *Rapid Commun. Mass Spectrom.* **20**, 2099–2103.
Jung, S., Zeikus, J. G. and Hollingsworth, R. I. (1994). A new family of very long chain α,ω-dicarboxylic acids is a major structural fatty acyl component of the membrane lipids of *Thermoanaerobacter ethanolicus* 39E. *J. Lipid Res.* **35**, 1057–1065.
Kämpfer, P., Rainey, F. A., Andersson, M. A., Murmiaho Lassila, E. L., Ulrych, U., Busse, H.-J., Weiss, N. and Salkinoja-Salonen, M. (2000). *Frigoribacterium faeni* gen. nov., sp. nov., a novel psychrophilic genus of the family *Microbacteriaceae*. *Int. J. Syst. Evol. Microbiol.* **50**, 355–363.
Kämpfer, P., Lodders, N. and Falsen, E. (2011). *Hydrotalea flava* gen. nov., a new species of the phylum *Bacteroidetes* and allocation of the genera *Chitinophaga*, *Sediminibacterium*, *Lacibacter*, *Flavihumibacter*, *Flavisolibacter*, *Niabella*, *Niastella*, *Segetibacter*, *Parasegetibacter*, *Terrimonas*, *Ferruginibacter*, *Filimonas* and *Hydrotalea* to the family *Chitinophagaceae* fam. nov. *Int. J. Syst. Bacteriol.* **61**, 518–523.
Kuykendall, L. D., Roy, M. A., O'Neill, J. J. and Devine, T. E. (1988). Fatty acids, antibiotic resistance, and deoxyribonucleic acid homology groups of *Bradyrhizobium japonicum*. *Int. J. Syst. Bacteriol.* **38**, 358–361.
McLafferty, F. W. and Tureck, F. (1993). *Interpretation of Mass Spectra*. University Science Books, Mill Valley.
Moore, L. V. H., Bourne, D. M. and Moore, W. E. C. (1994). Comparative distribution and taxonomic value of cellular fatty acids in thirty-three genera of anaerobic Gram-negative bacilli. *Int. J. Syst. Microbiol.* **44**, 338–347.
Nobre, M. F., Carreto, L., Wait, R., Tenreiro, S., Fernandes, O., Sharp, R. J. and da Costa, M. S. (1996). Fatty acid composition of the species of the genera *Thermus* and *Meiothermus*. *Syst. Appl. Microbiol.* **19**, 303–311.

Rainey, F. A., Silva, J., Nobre, M. F., Silva, M. T. and da Costa, M. S. (2003). *Porphyrobacter cryptus* sp. nov., a novel slightly thermophilic, aerobic, bacteriochlorophyll *a*-containing species. *Int. J. Syst. Evol. Microbiol.* **53**, 35–41.

Rainey, F. A., Ray, K., Ferreira, M., Gatz, B. Z., Nobre, M. F., Bagaley, D., Rash, B. A., Park, M. J., Earl, A. M., Shank, N. C., Small, A. M., Henk, M. C., Battista, J. R., Kämpfer, P. and da Costa, M. S. (2005). Extensive diversity of ionizing-radiation-resistant bacteria recovered from Sonoran desert soil and description of nine new species of the genus *Deinococcus* obtained from a single soil sample. *Int. J. Syst. Evol. Microbiol.* **71**, 5225–5235.

Rattray, J. E., van de Vossenberg, J., Jaeschke, A., Hopmans, E. C., Wakeham, S. G., Lavik, G., Kuypers, M. M. M., Strous, M., Jetten, M. S. M., Schouten, S. and Sinninghe Damsté, J. S. (2010). Impact of temperature on ladderane lipid distribution in anammox bacteria. *Appl. Environ. Microbiol.* **76**, 1596–1603.

Russel, N. J. and Nichols, D. S. (1999). Polyunsaturated fatty acids in marine bacteria – a dogma rewritten. *Microbiology.* **145**, 767–779.

Singh, S. C., Sinha, R. P. and Häder, D. P. (2002). Role of lipids and fatty acids in stress tolerance in Cyanobacteria. *Acta Protozool.* **41**, 297–308.

Sinninghe Damsté, J. S., Rijpstra, W. I., Geenevasen, J. A., Strous, M. and Jetten, M. S. (2005). Structural identification of ladderane and other membrane lipids of planctomycetes capable of anaerobic ammonium oxidation (anammox). *FEBS J.* **272**, 4270–4283.

Sinninghe Damsté, J. S., Rijpstra, W. I. C., Hopmans, E. C., Schouten, S., Balk, M. and Stams, A. J. (2007). Structural characterization of diabolic acid-based tetraester, tetraether and mixed ether/ester, membrane-spanning lipids of bacteria from the order Thermotogales. *Arch. Microbiol.* **188**, 629–641.

Strömpl, C., Tindall, B. J., Lünsdorf, H., Wong, T.-Y., Moore, E. R. B. and Hippe, H. (2000). Reclassification of *Clostridium quercicolum* as *Dendrosporobacter quercicolus* gen. nov. comb nov. *Int. J. Syst. Evol. Microbiol.* **50**, 101–106.

Tindall, B. J., Rosselló-Móra, R., Busse, H. J., Ludwig, W. and Kämpfer, P. (2010). Notes on the characterization of prokaryote strains for taxonomic purposes. *Int. J. Syst. Evol. Microbiol.* **60**, 249–266.

Wait, R., Carreto, L., Nobre, M. F., Ferreira, A. M. and da Costa, M. S. (1997). Characterization of novel long-chain 1,2-diols in *Thermus* species and demonstration that *Thermus* strains contain both glycerol-linked and diol-linked glycolipids. *J. Bacteriol.* **179**, 6154–6162.

Yano, Y., Nakayama, A. and Yoshida, K. (1997). Distribution of polyunsaturated fatty acids in bacteria present in intestines of deep-sea fish and shallow-sea poikilothermic animals. *Int. J. Syst. Evol. Microbiol.* **63**, 2572–2577.

Zhang, J. Y., Yu, Q. T., Liu, B. N. and Huang, Z. H. (1988). Chemical modification in mass spectrometry. IV 2-alkenyl-4,4-dimethyloxazolines as derivatives for the double bond location of long-chain olefinic acids. *Biomed. Environ. Mass Spectrom.* **15**, 33–44.

9 The Extraction and Identification of Respiratory Lipoquinones of Prokaryotes and Their Use in Taxonomy

Milton S. da Costa[1], Luciana Albuquerque[2], M. Fernanda Nobre[1] and Robin Wait[3]

[1] Department of Life Sciences, University of Coimbra, Coimbra, Portugal; [2] Center for Neuroscience and Cell Biology, University of Coimbra, Coimbra, Portugal; [3] Kennedy Institute of Rheumatology, Imperial College, London, United Kingdom

CONTENTS

I. INTRODUCTION

Isoprenoid quinones are constituents of prokaryote cell membranes and are found in the vast majority of the aerobic or anaerobic organisms that have been examined, where they play important functions in electron transport. Ubiquinones (2,3-dimethoxy-5-methyl-1,4-benzoquinone with a polyprenyl side chain of varying length, abbreviated U-*n*, where *n* denotes the number of isoprenyl units) are widely distributed in bacteria and eukaryotes, while the chemically related plastoquinones have been identified in plants, algae and cyanobacteria. The menaquinones (2-methyl-3-phytyl-1,4-naphthoquinone, with a polyprenyl side chain of varying length, abbreviated MK-*n*) are also widely distributed in prokaryotes. Saturation or hydrogenation of the menaquinones' polyprenyl side chain has been reported in many actinobacteria, such as *Corynebacterium* spp., *Mycobacterium* spp., *Actinomyces* spp., *Microcella* spp., and *Arthrobacter* spp. as well as many other bacteria and archaea (Collins and Jones, 1981). These saturations are indicated by the abbreviation MK-$n(H_n)$ referring to the number of hydrogens in the isoprenoid side chain, such as MK-8(H_2). Fully

METHODS IN MICROBIOLOGY, VOLUME 38
0580-9517 DOI 10.1016/B978-0-12-387730-7.00009-7

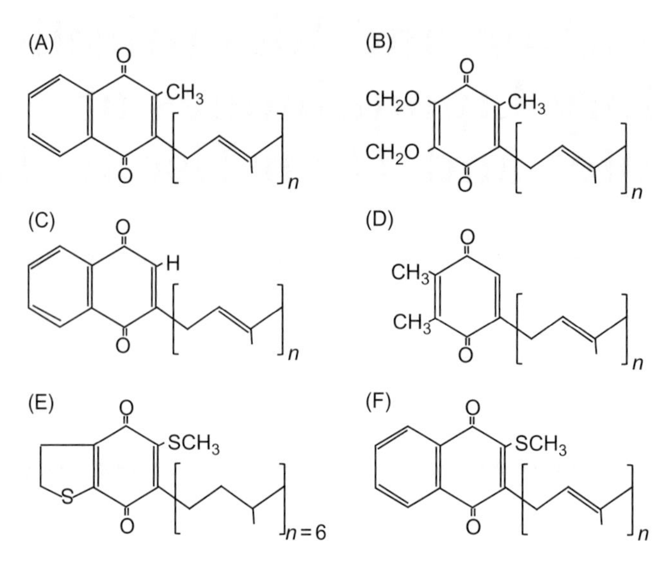

Figure 1. Structures of some isoprenoid quinones. (A) menaquinone; (B) ubiquinone; (C) demethylmenaquinone; (D) plastoquinone; (E) caldariellaquinone; (F) methionaquinone.

saturated menaquinones have been found in archaea such as *Pyrobaculum islandicum* (Tindall, 1989). Demethylmenaquinones (DMK) are not uncommon in some bacteria such as members of the family *Pasteurellaceae* (class Gammaproteobacteria) where these isoprenoid quinones are present in addition to menaquinones and ubiquinones (Kroppenstedt and Mannheim, 1989; Carlone *et al.*, 1988). It is interesting to note that all known genera of the suborder Micrococcinae possess menaquinones, some of which are partially saturated; however, the three species of the genus *Demequina* possess DMK-9(H_2), which can be used as a marker for this genus (Matsumoto *et al.*, 2010; Schumann *et al.*, 2009). Figure 1 shows the structures of representative respiratory quinones of prokaryotes.

Isoprenoid quinones have not been examined in most archaea described and the data on the distribution of these chemotaxonomic markers are limited. However, there is a reasonable amount of information on the members of the *Halobacteriaceae*, menaquinone 8 (MK-8) and MK-8(H_2) are found in many of these organisms (although the two species of the genus *Halorhabdus* possess only MK-8(H_2) (Antunes *et al.*, 2008). *Halococcus dombrowskii* and *Halococcus salifodinae* possess trace levels of MK-7(H_2) in addition to MK-8 and MK-8(H_2) (Stan-Lotter *et al.*, 2002). Rare quinones have been encountered in some archaea and bacteria, for example the archaea of the genera *Sulfolobus*, *Acidianus* and *Stygiolobus* possess a sulphur-containing isoprenoid quinone, designated caldariellaquinone (Collins and Jones, 1981; De Rosa *et al.*, 1977; Plumb *et al.*, 2007; Segerer *et al.*, 1986, 1991; Thurl *et al.*, 1986), whereas some members of the order *Thermoplasmatales* such as *Thermoplasma acidophilum* possess unique isoprenoid quinones such as thermoplasmaquinone-7 (TPQ-7), which is a derivative of methylmenaquinone as well as MK-7 (Shimada *et al.*, 2001), but other species of this order possess MK-7 (Itoh *et al.*, 2007). Some hyperthermophilic archaea such as *Thermococcus celer* and

Desulfurococcus mobilis appear to be devoid of lipoquinones (Thurl *et al.*, 1986), but *Pyrobaculum islandicum* has a fully saturated menaquinones designated MK-6 (H_{12} (Tindall, 1989). A sulfur-containing 2-methylthio-3-VI,VII-tetrahydromultipreny17-l,4-naphthoquinone, designated methionaquinone, is found in the thermophilic bacterium *Hydrogenobacter thermophilus* (Kawasumi *et al.*, 1984; Ishii *et al.*, 1987). Therefore, care should be taken to account for these and other rare quinones in taxonomic studies (Collins and Jones, 1981; Tindall, 1989).

The respiratory quinones have great taxonomic significance because the type of quinone present in an organism and the isoprenoid chain length reflect, in many cases, the phylogenetic affiliation of bacterium. For example, the major respiratory quinone of the bacteria of the phylum *Deinococcus/Thermus* is MK-8, although minor amount of MK-7 and MK-9 may also be detected (Battista and Rainey, 2001). Many known members of the classes Alphaproteobacteria, Gammaproteobacteria and Betaproteobacteria have ubiquinones with variable isoprene chain lengths (Tindall, 2005), but menaquinones and demethylmenaquinones can also be found in some groups. It should be noted that the presence of menaquinones and ubiquinones is, in part, determined by the growth characteristics in some organisms. For example, *Escherichia coli* produces primarily U-8 under aerobic growth but MK-8 under anaerobic growth on nitrate (van Hellemond and Tielens, 1994). The species of the phylum Bacteroidetes (formerly known as *Cytophaga/Flavobacterium/Bacteroides* CFB group), on the other hand, possess menaquinones, generally MK-7 and MK-6 (Albuquerque *et al.*, 2011b, 2011c; Collins and Jones, 1981; Hudson *et al.*, 1989).

Most organisms possess mixtures of either menaquinones or ubiquinones with different polyprenyl lengths, where one is generally the major respiratory quinone. For example, the species of the genus *Legionella* possess ubiquinones with polyprenyl chains, which, depending on the species, range between U-10 and U-14, one of which is the major quinone of the species; however, minor components, namely U-7, U-8, U-9 and U-15, have also been identified in some species (Wait, 1988). There are also cases where two quinones (menaquinones or ubiquinones) are present in about the same concentration; in *Microcella putealis*, MK-12 and MK-13 are present in about the same proportion, there being also very low levels of MK-11 (Tiago *et al.*, 2005). The order *Rubrobacterales*, comprising the species of the genus *Rubrobacter*, *Solirubrobacterales* comprising the species of the genera *Solirubrobacter*, *Patulibacter* and *Conexibacter* and the order *Thermoleophilales* comprising the species of the genus *Thermoleophylum* form three deep branching lineages of the class Actinobacteria (Albuquerque *et al.*, 2011a). Each of these orders comprises organisms that have different major respiratory menaquinones; however, the species of the genera *Solirubrobacter*, *Patulibacter* and *Conexibacter* have menaquinone variants based on MK-7, namely MK-7(H4), MK-7(H2) and DMK-7 (Albuquerque *et al.*, 2011a; Gundlapally *et al.*, 2009). This example serves to support the view that we cannot extrapolate the type of respiratory quinones from one group to another and that the respiratory quinones should be examined for each species.

Isoprenoid quinones are rather delicate cell components that are easily degraded, so all extraction steps must be performed in subdued light and preferably in brown glass tubes to prevent photo-oxidation; the samples should always be flushed with N_2 before storage at $-20°C$. It is also important to note that all glassware and Teflon-coated screw-capped tubes should be washed thoroughly with hexane to remove hydrophobic impurities before extraction of respiratory lipoquinones. Disposable plastic material such as automatic pipette tips and tubes should be avoided when working with solvents and solvent system because that may leach organic substances into the sample or plastics that are dissolved in the solvent system used, thus contributing to the interferences in identification and detection of isoprenoid quinones.

♦♦♦♦♦♦ II. GROWTH OF ORGANISMS AND EXTRACTION OF QUINONES

- When possible, grow organisms on solid media, although liquid medium can be used to grow organisms that grow poorly on solid media. Normally 100 mg of cell dry weight is sufficient for the analysis of lipoquinone composition, but it may be necessary to obtain larger quantities of cells when respiratory quinones are present in very small quantities. The biomass should be stored under N_2 atmosphere at $-20°C$ or lower.
- Resuspend the wet cell paste in about 1.0 ml of ultrapure water in a brown glass tube with a Teflon-lined screw cap and freeze-dry. Extended storage can be considered safe under these specific conditions, and we generally store standard respiratory quinones for over a year at $-70°C$.
- Add 3.0 ml of hexane:methanol (1:2, by volume) to lyophilized cells, place a small magnetic bar in the tube and flush with N_2. The suspension is stirred on a magnetic stirrer for about 30 min. The samples are placed on ice for 30 min, until phase separation takes place.
- Add ice-cold hexane (3.0 ml) to the samples, followed by 2.0 ml of 0.3% NaCl under a N_2 atmosphere slowly flushed into the tube.
- Centrifuge at $2000 \times g$ for 5 min in a swinging bucket centrifuge at room temperature for phase separation.
- Remove the upper phase with a Pasteur pipette into a 5 ml brown glass tube with Teflon-coated screw cap.
- Evaporate to dryness under a N_2 atmosphere and store, if necessary, at $-20°C$ or lower.

♦♦♦♦♦♦ III. SEPARATION AND PARTIAL PURIFICATION OF QUINONES

- Prepare a solvent composed of hexane:diethyl ether (85:15, by volume).
- Add the solvent to a chromatography tank (internal dimensions 20 cm [height] by 20 cm [length] by 8 cm [width]), with one side lined with filter paper and wait about 60 min for good vapour saturation of the atmosphere.

Smaller tanks can be used for smaller plates. The solvent level in the tank should be lower than the location of application (origin) of the samples.

- Resuspend the sample in about 0.4 ml diethyl ether. Apply the extract along a thin line about 1.5 cm from the lower edge of a 10 cm × 10 cm plastic sheet coated with Silica Gel 60 F_{254} (Merck No. 5735) under subdued light. The amount of extract applied to the silica gel depends on the concentration of quinones in the samples; we normally start with about 0.1 ml. Two separate samples can be applied on the same thin-layer chromatography (TLC) plate. Note that silica gel sheets are generally available in sizes of 20 cm × 20 cm or 20 cm × 10 cm and should be cut with scissors to the appropriate dimensions.
- Place the TLC plate in the tank and allow the solvent to ascend to the upper edge of the plate. Remove the silica gel-coated plate and allow the solvent to evaporate. This last step can be speeded up with cold air from a hair dryer.
- Observe the plate under UV light (254 nm) for only a few seconds and mark the contours of the dark bands with a pencil. Protect hands and face from the UV radiation. The menaquinones migrate further from the origin (R_F of about 0.7), while the ubiquinones migrate nearer to the origin (R_F of about 0.4). This step leads to the presumptive identification of the respiratory quinones as ubiquinones or menaquinones and their partial purification, since the extracts contain other hydrophobic substances that can interfere with the final HPLC step. It is noteworthy that in rare cases menaquinones and ubiquinones can be present in the same sample, so both bands should be extracted.
- Scrape the silica that contains the band with the quinone using a spatula onto non-absorbent paper. Place a piece of paper above the silica and press with finger tips to grind it into a fine powder. The silica powder is poured in a Pasteur pipette that has been previously plugged with hexane-washed glass wool so that this plug retains most of the silica powder.
- Elute the samples twice from the silica with about 0.5 ml hexane:methanol (1:2, by volume) into a 1.5 ml brown glass Teflon-coated screw-cap tube and place it on ice. Add 0.3 ml of cold hexane and four drops of 0.3% NaCl. Mix thoroughly, allow the phases to separate and transfer the upper phase into a new 1.5 ml brown glass tube. Care must be taken not to remove the interface or the lower phase of the sample. The purified sample can be stored at −20°C under a N_2 atmosphere.
- The TLC-purified sample is filtered through a hydrophobic membrane syringe filter (0.2 μm pore size and 0.3 mm diameter [provide brand, part number and supplier]) to remove silica particles that could block the HPLC column. Evaporate under N_2 until dry.
- Dissolve the sample in about 0.2 ml of methanol:heptane (10:2, by volume).

♦♦♦♦♦♦ IV. HIGH-PERFORMANCE LIQUID CHROMATOGRAPHY

- Inject the sample into a 20 μl capacity loop (a loop refers to the small tubing of various volumes where the sample is injected) using an appropriate

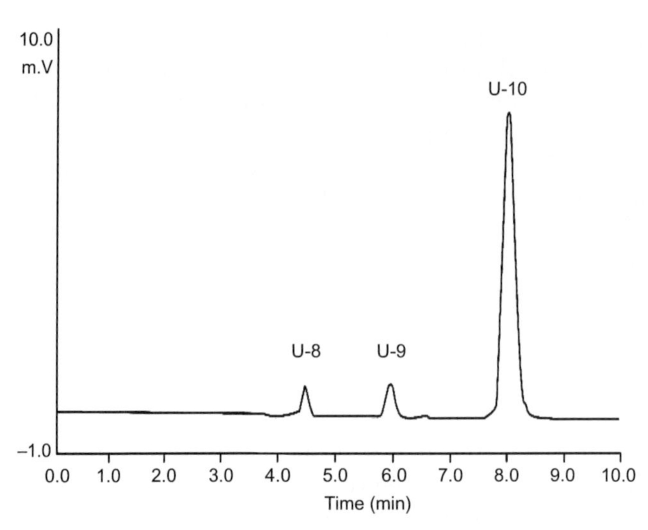

Figure 2. Ubiquinone (U) profile on HPLC of the type strain of *Sphingomonas adhaesiva* (DSM 7418) grown aerobically under optimum conditions.

syringe. The rest of the quinone extract can be stored at −20°C under a N_2 atmosphere, or it can be evaporated to dryness and dissolved in a small volume to inject a more concentrated extract, if necessary.

- Separate the quinones on a reverse phase ODS 2 (25 cm by 4 mm internal diameter) column equilibrated with degassed methanol:heptane (10:2, by volume). Degassing of the eluent is achieved by closing the top of a side arm flask with a rubber stopper and applying vacuum from a pump connected to a side arm with stirring for about 30 min. The samples are eluted with this solvent system at a rate of 2.0 min/ml at 37°C and detected at 269 nm.
- Other solvent systems for elution of lipoquinones are also used. One commonly used elution system is composed of methanol:1-chlorobutane (80:20, by volume) or (70:30, by volume), but methanol:heptane (10:2, by volume) is less corrosive for the HPLC column and tubing and has the same capacity to separate quinones (Wait, 1988).

The affinity of the quinones for the columns is primarily determined by the nature of isoprene unit at carbon 6, which is non-polar and strongly absorbed by the packing of the column, so that longer chains tend to be more tightly bound. The isoprene quinones will therefore be eluted from the column in order of the chain length with shorter chain quinones' elution before those with longer chains.

Isoprenoid quinones should be tentatively identified by comparison of their retention times with those of external standards eluted before or after the sample (Figures 2 and 3). Some commercial ubiquinones are available, but it is preferable to grow bacteria with known respiratory quinones and use these for comparison with the retention time to identify those from new strains. Short-chain quinones are rarer (less than six isoprenyl units) than those with longer chains,

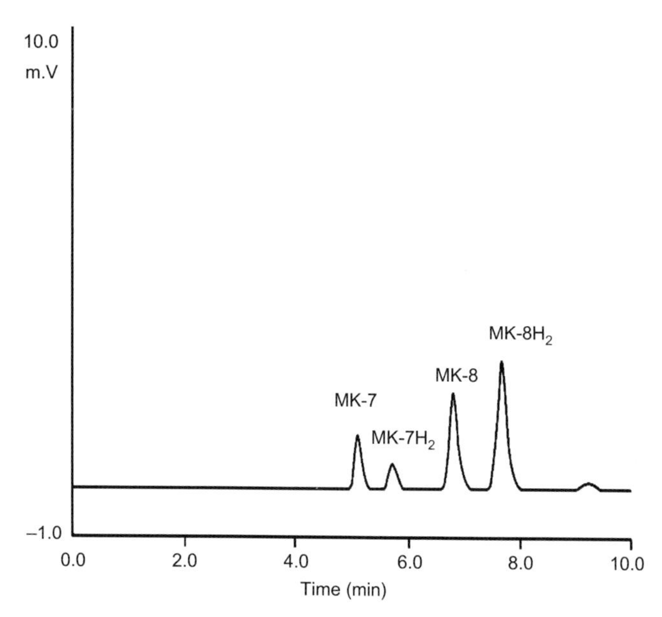

Figure 3. Menaquinone (MK) profile of the type strain of *Micrococcus luteus* (DSM 4698) grown aerobically under optimum conditions.

but some bacteria are known to have short-chain quinones (menaquinones and ubiquinones) that can be used as standards. In some cases, we mix bacterial samples of known quinone composition with those that were being previously identified, as internal standards, to ensure that the peaks have identical retention times. If the quinones cannot be identified by comparison with known standards, then mass spectroscopy must be used for their identification (Wait, 1988). Respiratory quinones may, as stated above, be present in very small amounts and the cell mass used to extract them may have to be increased substantially as was recently found with the new actinobacterium, *Microcella putealis*, where 300 mg of cell dry weight had to be used to identify small levels of menaquinones ot ubiquinones (Tiago *et al.*, 2005).

♦♦♦♦♦♦ V. IDENTIFICATION OF ISOPRENOID QUINONES BY MASS SPECTROMETRY

Definitive identification of quinones requires the use of physicochemical methods such as nuclear magnetic resonance or mass spectrometry. Mass spectra may be obtained from quinones purified as described above by TLC or HPLC using electron ionization in the positive-ion mode and a direct insertion probe. Rapid heating of the probe is necessary to favour evaporation over thermal degradation. Electron ionization mass spectra (EI MS) of quinones usually display abundant molecular ions and intense low-mass ions, which are diagnostic of the nature of the quinone nucleus. Ubiquinones, for example, have prominent peaks at *m/z* 197 and 235, derived from their 2,3-dimethoxy-1,4-benzoquinone group, whereas menaquinones have peaks at *m/z* 225 and 187 (Wait, 1988).

Because isoprenoid quinones are thermally labile, the sensitivity obtainable using direct probe EI MS is often poor, and the experiments are technically demanding. It is therefore preferable to employ soft ionization mass spectrometric methods such as fast atom bombardment or electrospray. Fast atom bombardment in the negative mode using 3-nitrobenzyl alcohol as matrix gives rise to abundant semiquinone molecular anion radicals $[M]^{\cdot -}$, revealing the mass of each quinone molecular species present. The relative intensities of these molecular anions reflect the concentration of the individual quinones in the mixture and generally agree with compositions determined by HPLC (Dennis *et al.*, 1993). More detailed structural information may be obtained by collisional fragmentation and tandem mass spectrometry in the positive-ion mode. The fragmentation pathways operating are generally similar to those observed in EI mass spectra.

The purpose of taxonomy should be to seek to describe taxa and their relationship to other taxa, but it should seek to add to our knowledge of the diversity of metabolic pathways and chemical structures. The analysis of isoprenoid quinones certainly helps in adding to our knowledge of biodiversity of bacteria and archaea, but it is also important for confirmation of taxonomic assignment of new taxa.

References

Albuquerque, L., França, L., Rainey, F. A., Schumann, P., Nobre, M. F., and da Costa, M. S. (2011a). *Gaiella occulta* gen. nov., sp. nov., a novel representative of a deep branching phylogenetic lineage within the class *Actinobacteria* and proposal of *Gaiellaceae* fam. nov. and *Gaiellales* ord. nov. *Syst. Appl. Microbiol.* (in press).

Albuquerque, L., Rainey, F. A., Nobre, M. F., and da Costa, M. S. (2011b). *Hydrotalea sandarakina* sp. nov., isolated from a hot spring runoff and emended description of the genus *Hydrotalea* and the species *Hydrotalea flava*. *Int. J. Syst. Evol. Microbiol.* (in press).

Albuquerque, L., Rainey, F. A., Nobre, M. F., and da Costa, M. S. (2011c). *Schleiferia thermophila* gen. nov., sp. nov., a slightly thermophilic bacteria of the phylum *Bacteroidetes* and the proposal of *Schleiferiaceae* fam. nov. *Int. J. Syst. Evol. Microbiol.* (in press). doi:10.1099/ijs.0.028852-0.

Antunes, A., Taborda, M., Huber, R., Moissl, C., Nobre, M. F. and da Costa, M. S. (2008). *Halorhabdus tiamatea* sp. nov., a non-pigmented, extremely halophilic archaeon from a deep-sea, hypersaline anoxic basin of the Red Sea, and emended description of the genus *Halorhabdus*. *Int. J. Syst. Evol. Microbiol.* **58**, 215–220.

Battista, J. R. and Rainey, F. A. (2001). The family *Deinococcaceae*. In: *Bergey's Manual of Systematic Bacteriology 2nd ed.* (D. R. Boone and R. W. Castenholz, eds.), Vol. 1, pp. 395–396. Springer, New York.

Carlone, G. M., Schalla, W. O., Moss, C. W., Ashley, D. L., Douglas, M. F., Holler, J. S. and Plikaytis, B. D. (1988). *Haemophilus ducreyi* isoprenoid quinone content and structure determination. *Int. J. Syst. Bacteriol.* **38**, 249–253.

Collins, M. and Jones, D. (1981). Distribution of isoprenoid quinone structural types in bacteria and their taxonomic implications. *Microbiol. Rev.* **45**, 316–354.

Dennis, P. J., Brenner, D. J., Thacker, W. L., Wait, R., Vesey, G., Steigerwalt, A. G. and Benson, R. F. (1993). Five new *Legionella* species isolated from water. *Int. J. Syst. Microbiol.* **43**, 329–337.
De Rosa, M., De Rosa, S., Gambacorta, G., Minale, L., Thomson, R. H. and Worthington, R. D. (1977). Caldariellaquinone, a unique benzo-b-thiopen-4,7-quinone from *Caldariella acidophila*, an extremely thermophilic and acidophilic bacterium. *J. Chem. Soc. Perkin Trans.* **1**, 653–657.
Gundlapally, S., Reddy, N. and Garcia-Pichel, F. (2009). Description of *Patulibacter americanus* sp. nov., isolated from biological soil crusts, emended description of genus *Patulibacter* Takahashi *et al.*, 2006 and proposal of *Solirubrobacterales* ord. nov. and *Thermoleophilales* ord. nov. *Int. J. Syst. Evol. Microbiol.* **59**, 87–94.
Hudson, J. A., Schofield, K. M., Morgan, H. W. and Daniel, R. M. (1989). *Thermonema lapsum* gen. nov., sp. nov., a thermophilic gliding bacterium. *Int. J. Syst. Bacteriol.* **39**, 485–487.
Ishii, M., Kawasumi, T., Igarashi, Y., Komada, T. and Minoda, Y. (1987). 2-Methylthio-1,4-naphtoquinone, a new quinone from an extremely thermophilic hydrogen bacterium. *Agric. Biol. Chem.* **47**, 167–169.
Itoh, T., Yoshikawa, N. and Takashina, T. (2007). *Thermogymnomonas acidicola* gen. nov., sp. nov., a novel thermoacidophilic, cell wall-less archaeon in the order *Thermoplasmatales*, isolated from a solfataric soil in Hakone, Japan. *Int. J. Syst. Evol. Microbiol.* **57**, 2557–2561.
Kawasumi, T., Igarashi, Y., Kodama, T. and Minoda, Y. (1984). *Hydrogenobacter thermophilus* gen. nov., sp. nov., an extremely thermophilic, aerobic, hydrogen-oxidizing bacterium. *Int. J. Syst. Bacteriol.* **34**, 5–10.
Kroppenstedt, R. M. and Mannheim, W. (1989). Lipoquinones in members of the family *Pasteurellaceae. Int. J. Syst. Bacteriol.* **39**, 304–308.
Matsumoto, A., Nakai, K., Morisaki, K., Omura, S. and Takahashi, Y. (2010). *Demequina salsinemoris* sp. nov., isolated on agar media supplemented with ascorbic acid or rutin. *Int. J. Syst. Evol. Microbiol.* **60**, 1206–1209.
Plumb, J. J., Haddad, C. M., Gibson, J. A. E. and Franzmann, P. D. (2007). *Acidianus sulfidivorans* sp. nov., an extremely acidophilic, thermophilic archaeon isolated from a solfatara on Lihir Island, Papua New Guinea, and emendation of the genus description. *Int. J. Syst. Evol. Microbiol.* **57**, 1418–1423.
Schumann, P., Kämpfer, P., Busse, H.-J. and Evtushenko, L. I. (2009). Proposed minimal standards for describing new genera and species of the suborder *Micrococcineae. Int. J. Syst. Evol. Bacteriol.* **59**, 1823–1849.
Segerer, A., Neuner, A., Kristjansson, J. K. and Stetter, K. O. (1986). *Acidianus infernus* gen. nov., sp. nov., and *Acidianus brierleyi* comb. nov.: facultatively aerobic, extremely acidophilic thermophilic sulfur-metabolizing archaebacteria. *Int. J. Syst. Bacteriol.* **36**, 559–564.
Segerer, A. H., Trincone, A., Gahrtz, M. and Stetter, K. O. (1991). *Stygiolobus azoricus* gen. nov., sp. nov. represents a novel genus of anaerobic, extremely thermoacidophilic archaebacteria of the order *Sulfolobales. Int. J. Syst. Bacteriol.* **41**, 495–501.
Shimada, H., Shida, Y., Nemoto, N., Oshima, T. and Yamagishi, A. (2001). Quinone profiles of *Thermoplasma acidophilum* HQ-62. *J. Bacteriol.* **183**, 1462–1465.
Stan-Lotter, H., Pfaffenhuemer, M., Legat, A., Busse, H. J., Radax, C. and Gruber, C. (2002). *Halococcus dombrowskii* sp. nov., an archaeal isolate from a Permian alpine salt deposit. *Int. J. Syst. Evol. Microbiol.* **52**, 1807–1814.
Thurl, S., Witke, W., Buhrow, I. and Schäfer, W. (1986). Quinones from archaebacteria, II. Different types of quinones from sulphur-dependent archaebacteria. *Biol. Chem. Hoppe-Seyler* **367**, 191–198.

Tiago, I., Pires, C., Mendes, V., Morais, P. V., da Costa, M. and Veríssimo, A. (2005). *Microcella putealis* gen. nov., sp. nov., a Gram-positive alkaliphilic bacterium isolated from a nonsaline alkaline groundwater. *Syst. Appl. Microbiol.* **28**, 479–487.

Tindall, B. J. (1989). Fully saturated menaquinones in the archaebacterium *Pyrobaculum islandicum*. *FEMS Microbiol. Lett.* **60**, 251–254.

Tindall, B. J. (2005). Respiratory lipoquinones as biomarkers. *In*: *Molecular Microbial Ecology Manual*, (A. Akkermans, F. de Bruijn and D. van Elsas, Eds.), 2nd ed. Kluwer Academic, Dordrecht. Section 4.1.5, Supplement 1.

Van Hellemond, J. J. and Tielens, A. G. M. (1994). Expression and functional properties of fumarate reductase. *Biochem. J.* **304**, 321–331.

Wait, R. (1988). Confirmation of the identity of *Legionella* by whole cell fatty-acid and isoprenoid quinone profiles. *In*: *A Laboratory Manual for Legionella* (T. G. Harrison and A. G. Taylor, eds.), pp. 69–101. John Wiley and Sons, Chichester.

10 Detection and Characterization of Mycolic Acids and Their Use in Taxonomy and Classification

Atteyet-Alla Fetouh Yassin
Institut für Medizinische Mikrobiologie und Immunologie der Universität Bonn, Bonn, Germany
E-mail: atteyet-alla.yassin@ukb.uni-bonn.de

♦♦

CONTENTS

♦♦♦♦♦♦ I. INTRODUCTION

A. History

The presence of a large amount of wax-like substance in the lipid fraction of the tubercle bacilli was known over 100 years ago. Attention was first drawn to this unsaponifiable wax-like substance by Aronson in 1898, and all subsequent investigators have corroborated Aronson's observation. It was referred to as one of the higher alcohols by Bulloch and Macleod (1904). The chemical composition and properties of this alcohol were studied by Tamura (1913), who described it as a monohydric alcohol of the formula $C_{29}H_{56}O$ and gave it the name mykol. Mykol was acid-fast and thought to be responsible for the acid fastness of the bacilli. The pioneering studies coordinated by R.J. Anderson at Yale in the 1930s clearly indicated the unusual lipophilic nature of this fraction (Anderson, 1929; Anderson *et al.*, 1937). Stodola *et al.* (1938) first proposed to call the ether-soluble, unsaponifiable, high molecular weight hydroxy acid of the human tubercle bacillus 'mycolic acid'. The initial structural features of mycolic acids, as elucidated by Stodola *et al.* (1938), are as follows: pyrolysis at 300°C yielded hexacosanoic acid ($C_{26}H_{52}O_2$) and an unidentified long-chain component, which together showed a rough empirical formula of $C_{88}H_{172}O_4$ or $C_{88}H_{176}O_4$. The

METHODS IN MICROBIOLOGY, VOLUME 38
0580-9517 DOI 10.1016/B978-0-12-387730-7.00010-3

essential structural details of mycolic acids were determined by Asselineau and Lederer in the 1950s who defined mycolic acids as high molecular weight β-hydroxy fatty acids with a long α side chain (Asselineau, 1950; Asselineau and Lederer, 1950).

Similar types of α-branched-β-hydroxylated fatty acids but of smaller molecular weight were isolated from organisms other than mycobacteria. Lederer *et al.*, (1952) reported on the isolation of 2-tetradecyl-3-hydroxy octadecanoic acid, corynemycolic acid ($C_{32}H_{64}O_3$), from *Corynebacterium diphtheriae*. Soon thereafter, a mycolic acid with medium chain length ($C_{50}H_{96}O_3$), named nocardic acid, was isolated from *Nocardia asteroides* (Michel *et al.*, 1960). Later, the name nocardomycolic acid (by analogy to corynemycolic acid) was proposed for nocardic acid Asselinneau, J. (1966). The bacterial lipids. Hermann, Paris and San Francisco. Nocardomycolic acid was subsequently isolated from *Nocardia brasiliensis* and other species of *Nocardia* (Lanéelle *et al.*, 1965). The distribution of mycolic acids in species of the genera *Corynebacterium*, *Mycobacterium* and *Nocardia* led Barksdale (1970) to designate this group of bacteria as the CMN-group. The assignment of bacteria to the CMN-group withstood rigorous numerical, chemotaxonomic and phylogenetic analysis and now it is clear that the CMN-group, from a phylogenetic standpoint, forms a clade within the lineage comprising members of the Actinobacteria. This clade embraces the genera *Corynebacterium* (Lehmann and Neumann, 1896), *Dietzia* (Rainey *et al.*, 1995), *Gordonia* (Tsukamura, 1971; Stackebrandt *et al.*, 1988), *Millisia* (Soddell *et al.*, 2006), *Mycobacterium* (Lehmann and Neumann, 1896), *Nocardia* (Trevisan, 1889), *Rhodococcus* (Goodfellow and Alderson, 1977), *Segniliparus* (Butler *et al.*, 2005), *Skermania* (Chun *et al.*, 1997), *Smaragdicoccus* (Adachi *et al.*, 2007), *Tsukamurella* (Collins *et al.*, 1988) and *Williamsia* (Kämpfer *et al.*, 1999).

Research during the past 40 years has consolidated knowledge of the structure of lipids of *Mycobacterium tuberculosis* and related bacteria. The development of modern tools of analytical chemistry allowed the exact nature and the complexity to become more apparent. The introduction of thin-layer chromatography (TLC) of mycolic acid esters enabled the true complexity of the natural mixtures to be clearly assessed. The application of infrared spectroscopy (IR) and nuclear magnetic resonance (NMR) allowed the essential functional groups to be recognized, and mass spectrometry (MS) provided accurate molecular weights and structural information (Gastambide-Odier *et al.*, 1964; Etémadi, 1967; Minnikin and Polgar, 1966, 1967; Laval *et al.*, 2001; Watanabe *et al.*, 2001). What was once described as a single component that was liberated upon saponification of the 'wax' isolated from *M. tuberculosis* is now recognized as a broad family of over 500 intimately related chemical structures (Barry *et al.*, 1998; Minnikin *et al.*, 2002). It is now known that mycolic acids are not only found in *M. tuberculosis* but are also present in all mycobacteria and related genera of actinobacteria. This staggering number of individual mycolic acid components, which are quantitatively the major components of the cell wall, in a single bacterium is a reflection of the biological role and importance of these molecules. Mycolic acids are currently employed as chemotaxonomic markers, and their analysis has been applied in the identification of mycobacteria and related genera of the actinobacteria.

B. Chemical Structure

Mycolic acids are long-chain 2-alkyl-branched, 3-hydroxy fatty acids. The production of mycolic acids is a genuine synapomorphy unique to taxa, traditionally called actinomycetes (Embley and Stackebrandt, 1994), and now referred to as the actinobacteria. The length and structural complexity of mycolic are differentiating characters at the genus level (Table 1). Mycolic acids found in mycobacteria contain between 60 and 90 carbon atoms and usually occur as a complex mixture of structurally related molecules that differ from one another by the nature of the chemical groups at the so-called 'proximal' and 'distal' positions (relative to the carboxyl group) of their main chain, called the meromycolate chain. The least polar, called α-mycolate, is ubiquitous in mycobacteria; it possesses between 70 and 90 carbons and contains two unsaturations: either two *cis* cyclopropyl groups or one or two *cis*/*trans* double bonds at the proximal and distal positions (Kaneda *et al.*, 1988; Laval *et al.*, 2001). The slightly more polar and lower molecular weight α′-mycolates (60–68 carbon atoms) occur in some mycobacterial species (Minnikin *et al.*, 1982b). More polar mycolates are also produced by most mycobacteria (Minnikin *et al.*, 1984b) and these contain an oxygenated group at the distal position. The oxygenated group may be a methoxy, a keto, an epoxy or an ester group. Representative structures of the seven known subclasses of mycolic acids found in mycobacteria are shown in Figure 1.

In contrast to the *Mycobacterium* species, the mycolic acids of related genera of actinobacteria have lower molecular weight (C_{22}–C_{78}), possess no oxygen functions other than the 3-hydroxy unit, and contain variable numbers of double bonds (Figure 2). The mycolic acids from *Tsukamurella* species have a carbon skeleton range of C_{62}–C_{78} and contain 1–6 double bonds (Collins and Jones, 1982; Tomiysa and Yano, 1984; Collins *et al.*, 1988). The genus *Gordonia* produces

Table 1. Variations in size and structure of mycolic acid within the mycolic acid containing genera of the order *Actinomycetales*

Genus	Total number of carbons	Number of double bonds	Pyrolysis products
Corynebacterium	22–36	0–2	8–16
Dietzia	34–38	0–1	14–18
Gordonia	52–66	1–4	16–18
Millisia	44–52	ND[a]	16–18
Mycobacterium	60–90	2	22–26
Nocardia	46–58	1–3	14–18
Rhodococcus	30–56	0–4	12–18
Segniliparus	ND	ND	22
Skermania	58–64	2–6	16–20
Smaragdicoccus	43–49	ND	ND
Tsukamurella	62–78	1–6	20–22
Williamsia	50–56	ND	16–18

[a]ND = no data available.

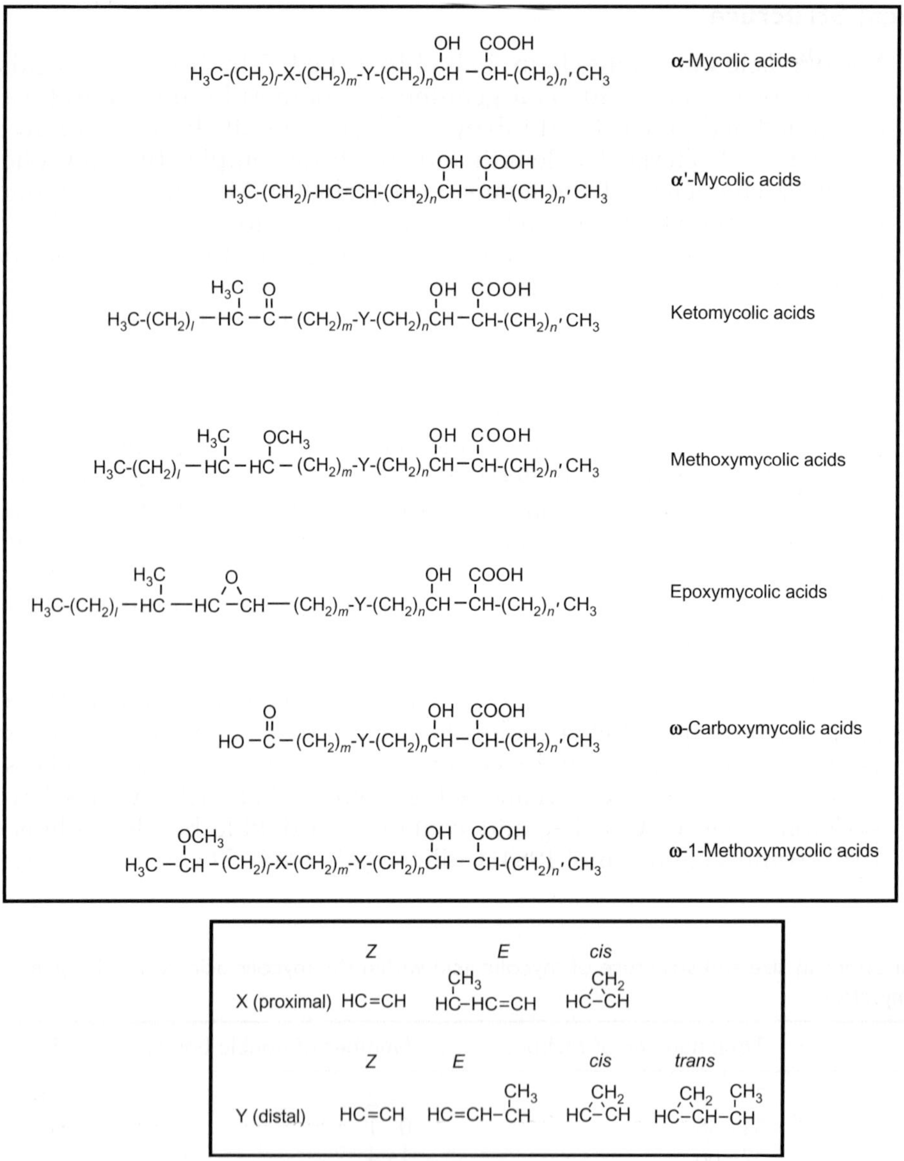

OH COOH
$H_3C-(CH_2)_l-X-(CH_2)_m-Y-(CH_2)_n CH-CH-(CH_2)_{n'} CH_3$ α-Mycolic acids

OH COOH
$H_3C-(CH_2)_l-HC=CH-(CH_2)_n CH-CH-(CH_2)_{n'} CH_3$ α'-Mycolic acids

H_3C O OH COOH
$H_3C-(CH_2)_l-HC-C-(CH_2)_m-Y-(CH_2)_n CH-CH-(CH_2)_{n'} CH_3$ Ketomycolic acids

H_3C OCH_3 OH COOH
$H_3C-(CH_2)_l-HC-HC-(CH_2)_m-Y-(CH_2)_n CH-CH-(CH_2)_{n'} CH_3$ Methoxymycolic acids

H_3C O OH COOH
$H_3C-(CH_2)_l-HC-HC-CH-(CH_2)_m-Y-(CH_2)_n CH-CH-(CH_2)_{n'} CH_3$ Epoxymycolic acids

O OH COOH
$HO-C-(CH_2)_m-Y-(CH_2)_n CH-CH-(CH_2)_{n'} CH_3$ ω-Carboxymycolic acids

OCH_3 OH COOH
$H_3C-CH-(CH_2)_l-X-(CH_2)_m-Y-(CH_2)_n CH-CH-(CH_2)_{n'} CH_3$ ω-1-Methoxymycolic acids

	Z	*E*	*cis*	
X (proximal)	HC=CH	CH_3 HC-HC=CH	CH_2 HC-CH	
	Z	*E*	*cis*	*trans*
Y (distal)	HC=CH	CH_3 HC=CH-CH	CH_2 HC-CH	CH_2 CH_3 HC-CH-CH

Figure 1. General structures of the major mycolic acids found in mycobacteria. The values of $l = n = 15$, 17, or 19; those of m depend on the nature of X and Y. When a methyl branch present or absent in both X and Y, $m = 12$, 14, or 16; when a methyl branch is present in X or Y, $m = 13$, 15, or 17.

a series of mycolic acids with total chain lengths of approximately 52–66 carbon atoms with 1–4 double bonds (Alshamaony *et al.*, 1976a). The total size range of the mycolic acids of the genus *Williamsia* is C_{50}–C_{56} (Kämpfer *et al.*, 1999). The mycolic acids from *Nocardia* species have an average carbon chain length between C_{46} and C_{58} and contain 0–3 double bonds (Alshamaony *et al.*, 1976b). The mycolic acids of *Skermania* species are similar in chain length to those of

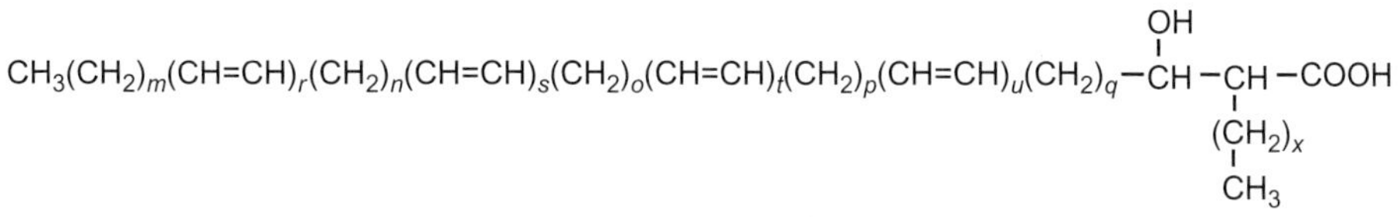

$$
\begin{array}{l}
\qquad\qquad\qquad\qquad\qquad\qquad\qquad\qquad\qquad\qquad\qquad\qquad\qquad\qquad\qquad\quad OH \\
CH_3(CH_2)_m(CH{=}CH)_r(CH_2)_n(CH{=}CH)_s(CH_2)_o(CH{=}CH)_t(CH_2)_p(CH{=}CH)_u(CH_2)_q{-}CH{-}CH{-}COOH \\
\qquad\qquad\qquad\qquad\qquad\qquad\qquad\qquad\qquad\qquad\qquad\qquad\qquad\qquad\qquad\qquad\quad (CH_2)_x \\
\qquad\qquad\qquad\qquad\qquad\qquad\qquad\qquad\qquad\qquad\qquad\qquad\qquad\qquad\qquad\qquad\quad CH_3
\end{array}
$$

Figure 2. General structure of mycolic acids found in *Corynebacterium, Dietzia, Gordonia, Nocardia, Rhodococcus* and other related genera of the order *Actinomycetales*. The value of $r+s+t+u$ lies between 0 and 4, values of m, n, o, p and q are all greater than 1 and the value of x lies between 5 and 19.

Gordonia species but are a little longer than those characteristic of the *Nocardia* species, having a chain length range between C_{58} and C_{64}, and they contain 2–6 double bonds (Blackall *et al.*, 1989; Chun *et al.*, 1997). The mycolic acids from *Rhodococcus* have an overall size of 30–56 carbon atoms and contain 0–4 double bonds (Alshamaony *et al.*, 1976b; Collins *et al.*, 1982; Barton *et al.*, 1989). The mycolic acids present in *Millisia* have 44–52 carbons (Soddell *et al.*, 2006). *Dietzia* possesses short-chain mycolic acids with 34–38 carbon atoms and 0–1 double bonds (Rainey *et al.*, 1995; Nishiuchi *et al.*, 2000). Finally, the mycolic acids from *Corynebacterium* are the smallest in size with an overall chain length of 22–36 carbon atoms and 0–2 double bonds (Collins *et al.*, 1982).

C. Biological Functions

Mycolic acids are characteristic lipid components of the cell envelope of mycobacteria and related actinobacteria, constituting approximately 40–60% by weight of the cell envelope (Goren and Brennan, 1979; Minnikin, 1982). Although they are present in the free lipids of the cell envelope associated mainly with glucose and trehalose forming the mycolyl glycolipids outermost monolayer, the majority are esterified to form the terminal [5-mycoloyl-β-Araf-(1→2)-5-mycoloyl-α-Araf(1→)] units of arabinogalactan attached to the peptidoglycan of the cell envelope (McNeil *et al.*, 1991). Mycobacteria and related mycolic acid containing bacteria possess many unique characteristics that are directly attributed to the presence of mycolic acids in their cell wall (Minnikin, 1982; Barry *et al.*, 1998; Daffé and Draper, 1998). The most important characteristics conferred by this molecule include properties such as resistance to chemical injury, low permeability to hydrophobic antibiotic substances (Nikaido, 1994), resistance to dehydration and the ability to persist and thrive within the hostile environment of the macrophage phagolysosome (Armstrong and Hart, 1971; Barry and Mdluli, 1996). Another familiar property, related to the physical organization of mycolic acids in the cell wall, is the ability of mycolic acid containing bacteria to retain coloration with carbol fuchsin (acid-fast staining) following acid treatment of intact stained cells (Barksdale and Kim, 1977). Mycolic acids, either in the form of cord factor (trehalose 6,6′-dimycolate) or other mycolyl glycolipids, play crucial roles in host immune response. It has been reported that mycolic acids are presented by the CD1 system (CD1b) to αβTCR+ T cells making these lipid antigens active players in the immune recognition of this pathogen (Beckman *et al.*, 1994; Moody *et al.*, 1999).

♦♦♦♦♦♦ II. PREPARATION OF MYCOLIC ACIDS FOR ANALYSIS

Classical methods of mycolic acid analysis involve extraction of the mycolic acid from the bacterial cells followed by derivatization, separation by various chromatography techniques including TLC, gas chromatography (GC) (Guerrant *et al.*, 1981; Lambert *et al.*, 1986) and high-performance liquid chromatography (HPLC). In addition, a wide variety of spectroscopic methods such as mass spectrometry (MS) (Alshamaony *et al.*, 1976a, 1976b; Minnikin *et al.*, 1982a, 1982b), gas chromatography–mass spectrometry (GC–MS) (Yano *et al.*, 1978; Kaneda *et al.*, 1986) and matrix-assisted laser desorption ionization time of flight (MALDI-TOF) mass spectrometry (Laval *et al.*, 2001) have been used in the determination of mycolic acid structure.

A. Cultivation of Bacteria

- Grow cultures of *Corynebacterium, Dietzia, Gordonia, Millisia, Nocardia, Rhodococcus, Skermania, Segniliparus, Tsukamurella, Williamsia* species and rapidly growing *Mycobacterium* species in shake flasks of Brain Heart Infusion (BD) for 5–7 days at 37°C (*Williamsia* species at 27°C).
- After checking for culture purity at maximum growth stage kill the cells with formaldehyde (1%, v/v).
- Harvest cells by centrifugation, wash with distilled water and lyophilize.

In addition to the above conventional culture method, a simple method suitable for routine analyses of both pathogenic and non-pathogenic organisms was developed, based on acid methanolysis and/or saponification of wet bacterial cell mass:

- Transfer a loopful of bacterial wet cells (5–10 mg) (growing, e.g. on Columbia agar) in a biosafety cabinet type II to a screw-capped tube containing the reagents necessary for either acid or alkaline methanolysis (see Sections III.A.1a or III.A.1b).
- Grow slow-growing mycobacteria such as *M. africanum, M. avium, M. bovis, M. celatum, M. haemophilum, M. kansasii, M. szulgai, M. tuberculosis, M. ulcerans, M. xenopi* on Loewenstein or Sauton medium as surface pellicles at 37°C for a period of time until maximum biomass is obtained.
- Using a loop, collect cells by scraping the colonies from the surface of the medium.

B. Precipitation Techniques

Precipitation of mycolic acids in organic solvents is an appropriate method for use in diagnostic laboratories to distinguish between mycolic acid containing bacteria. The mycolic acids from *Mycobacterium* species are precipitated from solution

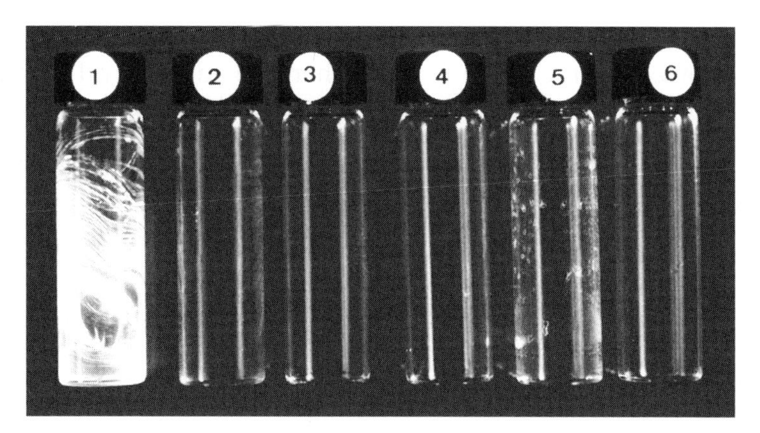

Figure 3. Precipitation of mycolic acid methyl esters extracted with acid methanolysis from petroleum ether (b.p. 60–80°C) solutions after evaporation under a stream of nitrogen at room temperature. (1) *Mycobacterium* (note white precipitate in tube); (2) *Tsukamurella*; (3) *Gordonia*; (4) *Nocardia*; (5) *Rhodococcus*; (6) *Corynebacterium*. Note absence of precipitate in the tubes 2–6.

by the addition of an equal volume of ether (Kanetsuna and Bartoli, 1972) or a double volume of ethanol (Hecht and Causey, 1976), whereas the mycolic acids from the remaining mycolic acid containing taxa remain in solution. A similar differentiation can be made by precipitating the mycolic acid methyl esters in acetonitrile/toluene (Hamid *et al.*, 1993). While all mycolic acid methyl esters are soluble in acetonitrile/toluene (1:2, v/v), the mycolic acid methyl esters of *Mycobacterium* species are insoluble in acetonitrile/toluene (3:2, v/v). A simpler test based on the precipitation of mycolic acid methyl esters extracted from whole cells using the acid methanolysis procedure was designed by Yassin (unpublished data). In this test, mycobacterial mycolates dissolved in petroleum ether form, after evaporation under a stream of nitrogen at room temperature, a white precipitate on the wall of the test tube, while those from the genera *Corynebacterium*, *Dietzia*, *Gordonia*, *Millisia*, *Nocardia*, *Rhodococcus*, *Tsukamurella* and *Williamsia* leave an oily residue after evaporation (Figure 3).

I. Precipitation of mycolic acids from an ethereal solution by ethanol (Kanetsuna and Bartoli, 1972)

- Treat biomass (2 g wet weight) in a closed tube with 5 ml of 2.5% (w/v) KOH in methanol–benzene (1:1, v/v) for one day at 37°C with occasional shaking.
- Add 3 ml of 1 N HCl to the mixture, followed by the addition of 5 ml of ether and mix well.
- Centrifuge at 2000 rpm for 5 min then transfer the clear ethereal layer to a new tube.
- Repeat the ether extraction three or four more times and evaporate the combined ether extracts.
- Extract the residue again with ether three times.
- Evaporate the final ether extract to 1 ml.
- Add 1 ml ethanol and keep the mixture at 0–4°C for 2 h.

- Collect the white precipitate by centrifugation at 200 × *g* for 5 min.
- Redissolve the precipitate in 1 ml ether and add 1–2 ml ethanol.
- Keep the mixture at 0–4°C to reprecipitate the mycolic acid.
- Collect the white precipitate by centrifugation at 200 × *g* for 5 min.
- Dry the precipitate under a stream of nitrogen, weigh and keep for further analysis.

Mycobacteria give large amounts of white precipitates, whereas other mycolic acid containing bacteria give no precipitate or in some cases, a sparse precipitate.

2. Precipitation of mycolic acids by acetonitrile (Hamid *et al.*, 1993)

- Place dry organisms (up to 50 mg) in a Teflon-coated screw-capped tube.
- Treat with 2 ml of 5% (w/v) aqueous tetrabutylammonium hydroxide.
- Heat the mixture at 100°C overnight.
- Cool the mixture at room temperature, then separate the phases by centrifugation at 2000 rpm for 2–3 min.
- Transfer the supernatant to a new Teflon-coated screw-capped tube containing 2 ml dichloromethane and 25 μl iodomethane.
- Shake the mixture for 30 min, then centrifuge at 2000 rpm for 2–3 min.
- Discard the upper layer.
- The lower layer, which contains the mycolic acid methyl ester, to dryness under nitrogen at 37°C.
- Dissolve the crude mycolic acid methyl ester in 0.5 ml dichloromethane.
- Transfer to weighed, 1.5 ml polypropylene microcentrifuge tubes and evaporate to dryness under a stream of nitrogen.
- Dissolve the residue in acetonitrile:toluene (0.1:0.2, v/v).
- Add 0.2 ml of acetonitrile, whereupon a cloudy precipitate appears in the solution.
- Mix the contents of the tubes gently and keep at 4°C for 1 h.
- Collect the precipitate by centrifugation at 2000 rpm for 2–3 min.
- Dry the precipitate under a gentle stream of nitrogen, weigh and keep at 4°C for further analysis.

3. Precipitation of mycolic acids from petroleum ether (Yassin, unpublished)

- The method is exactly the same as described in Section III.A.1.a.
- After evaporation of the petroleum ether extract under a stream of nitrogen, mycolic acids from *Mycobacterium* species leave a waxy residue on the tube, whereas mycolic acids from other bacteria leave an oily residue (Figure 3).

♦♦♦♦♦♦ III. APPLICATION OF THIN-LAYER CHROMATOGRAPHY IN MYCOLIC ACID ANALYSIS

TLC has been used to separate mycolic acids as their methyl esters (Minnikin *et al.*, 1980), and was often used as a preparative step in purifying mycolic acids before separation by GC (Minnikin, 1993; Yassin *et al.*, 1993). Both one- and two-dimensional TLC methodology, employing a variety of solvents, have been used in the field of chemotaxonomy (Goodfellow *et al.*, 1976; Keddie and Cure, 1977; Minnikin *et al.*, 1975, 1980, 1984c; Minnikin and Goodfellow, 1980; Daffé *et al.*, 1983; Dobson *et al.*, 1984; Rainey *et al.*, 1995; Soddell *et al.*, 2006). Analysis of mycolic acids by TLC is a valuable tool in the classification of various mycobacterial species and related genera of actinobacteria. It is a relatively simple and cost-effective method for routine screening of *Mycobacterium* species in the mycobacteriology laboratory with a potential for use in developing countries. Comprehensive catalogues of TLC profiles showing the classes of mycolates of various species have been compiled (Minnikin *et al.*, 1984c, 1985; Yassin *et al.*, 1993).

A. Preparation of Cellular Mycolic Acid Methyl Esters and Silyl Ethers

1. Methanolysis of whole cells

Dry biomass or a loopful of wet bacterial cell mass can be degraded by both acid methanolysis (Minnikin *et al.*, 1980) and alkaline methanolysis (Minnikin *et al.*, 1982a, 1984b). The use of alkaline methanolysis has an advantage over acid methanolysis as degradation of mycolates containing cyclopropane rings is not observed, and therefore clearer patterns are obtained. However, acid methanolysis remains of value in confirming the presence of epoxymycolates (Minnikin *et al.*, 1982a), which have a chromatographic behaviour very similar to that of ketomycolates. Furthermore, acid methanolysis is preferred to alkaline methanolysis in order to avoid the possibility of alkaline isomerization of mycolates.

(a) Acid methanolysis (Minnikin et al.*, 1980)*

It is necessary to include a co-solvent such as toluene in order to dissolve the mycolic acid esters, however, if excess toluene is present, two layers are not always produced on addition of light petroleum. Removal of any traces of acid by passing the petroleum ether extract through a small column of ammonium hydrogen carbonate [$(NH_4)HCO_3$, Merck Cat. no. 101131] is a convenient way of producing a stable methanolysate for analysis.

- Place dry organisms (up to 50 mg) in a Teflon-coated screw-capped tube.
- Add 3 ml of dry methanol:toluene:sulfuric acid (30:15:1, v/v).
- Incubate the tube at 85°C in an oven or heating block overnight (16–18 h).
- After cooling to room temperature, add 3 ml petroleum ether (b.p. 60–80°C), and shake vigorously.
- Centrifuge for 5 min at 3000 rpm to separate the phases.
- Prepare a small column (*ca.* 2 cm) of dry ammonium hydrogen carbonate in a short cotton wool-plugged Pasteur pipette, prewashed with diethyl ether (2 ml).
- Apply the upper petroleum ether layer and collect the eluent in a small vial.

- Add a further portion (2 ml) of petroleum ether to the remaining reaction mixture, and apply the upper layer to the column after shaking and centrifugation.
- Combine the eluents, and evaporate to dryness under a stream of nitrogen at room temperature.

(b) Alkaline methanolysis (Minnikin* et al., *1984a)

- Treat biomass (50 mg) with 20% methanolic tetramethylammonium hydroxide (0.5 ml), toluene (0.5 ml) and methanol (1.0 ml) at 37°C overnight.
- After centrifugation at 2000 rpm for 5 min, remove the supernatant to a Teflon-coated screw-capped tube.
- Wash the cell pellet with 1 ml of methanol:toluene (2:1, v/v), centrifuge, and combine the supernatant with that from the first step.
- Mark the level of the solution on the tube and then add 3 ml dimethylformamide.
- Reduce the volume of the liquid to the level marked on the tube using a stream of nitrogen at room temperature.
- Add 1 ml of 10% solution of iodomethane in dimethylformamide and mix the contents of the tube thoroughly for 1 h on a tube rotator or by mechanical shaking.
- Extract the mixture with 4 ml petroleum ether (b.p. 60–80°C).
- After centrifugation at 2000 rpm for 5 min, transfer the upper layer to a new tube.
- Wash the lower layer with more petroleum ether (2 ml × 2 ml), and reduce the combined upper layers to approximately 1 ml in a stream of nitrogen at room temperature.
- Extract the residue (composed mainly of dimethylformamide) with petroleum ether (1 ml, followed by 2 ml × 0.5 ml).
- Evaporate the combined upper layers to dryness under a stream of nitrogen at room temperature.

2. *t*-Butyldimethylsilyl (TBDMS) ethers of mycolic acid methyl esters (Minnikin and Patel, 1979; Datta et al., 1987)

To chromatographically distinguish similar methoxy and α′-mycolates, mycolic acid methyl esters are converted to *t*-butyldimethylsilyl (TBDMS) ethers following a scaled-down procedure of Corey and Venkateswarlu (1972):

- Dissolve total alkaline methanolysates in 0.15 ml toluene.
- Add 0.15 ml solution prepared by dissolving *t*-butyldimethylsilyl chloride (0.15 g) and imidazole (0.17 g) in dimethylformamide (1 ml).
- Heat the mixture overnight in an oven or heating block at 75°C.
- Cool the mixture to room temperature and extract with 0.3 ml petroleum ether (b.p. 60–80°C).
- Separate the phases by centrifugation at 3000 rpm for 5 min.
- Apply the upper layer onto a 2 cm column of diethyl ether-washed neutral alumina in a cotton wool-plugged Pasteur pipette.

- Collect the eluent in a separate vial.
- Wash the lower layer again with two successive portions of petroleum ether (0.3 ml), and pass these also through the alumina column.
- Elute material from the column with two 0.5 ml portions of diethyl ether.
- Evaporate the combined eluents to dryness under a stream of nitrogen at room temperature, and analyse by one-dimensional TLC, using petroleum ether:toluene (70:30, v/v) as developing solvent.

3. Trimethylsilyl (TMS) ethers of mycolic acid methyl esters (Yano et al., 1972)

- Dry mycolic acid methyl esters, previously purified by preparative TLC, under a stream of nitrogen.
- Add 50 μl pyridine and 100 μl *N,O*-bistrimethylsilyltrifluoroacetamide (BSTFA).
- Incubate the mixture at 70°C for 20 min.
- After completion of the reaction, cool the mixture to room temperature and extract the resultant trimethylsilyl ethers with 2 ml of hexane.
- Dry the extract at room temperature under a stream of nitrogen.
- Dissolve the dried extract in 100 μl *n*-hexane and inject into the gas chromatograph.

B. Development of TLC Plates

Silica gel 60 F254 aluminium sheets (Merck, No. 5554) are generally used. Small sheets (10 cm × 10 cm or 5 cm × 5 cm) are made by cutting them from large sheets (20 cm × 20 cm).

1. One-dimensional TLC

For routine analyses of organisms suspected to contain mycolic acids, methanolysates should be screened using one-dimensional TLC. Triple development in petroleum ether (60–80°C):acetone (95:5, v/v) increases the resolution of the mycolic ester detection. Allow the plate to dry between each development with the aid of a cold air stream from a hair dryer. This simple system is sufficient if only a single component, corresponding to a mycolic acid ester, is observed but is inadequate for a true understanding of the complexity of the mixtures of long-chain compounds in methanolysates of *Mycobacterium* species.

The mycolic acid spots are revealed as follows:

- Evenly spray the plates at a distance of 20–30 cm in a fume hood with a 10% ethanolic solution of molybdophosphoric acid. The reagent is prepared by adding 10 g of molybdophosphoric acid to 100 ml of absolute ethanol, and should be stored at room temperature in the dark.
- After air drying, place the chromatographic plate in an oven at 150°C for about 5 min. The lipids stain grey on a yellow background.

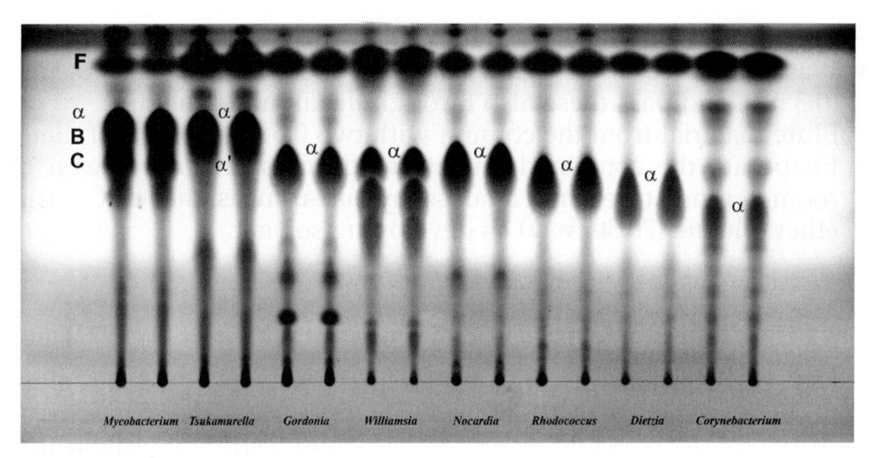

Figure 4. One-dimensional TLC of whole-cell acid methanolysates of selected genera of mycolic acid containing *Actinobacteria* after triple development in petroleum ether–acetone (95:5, v/v). *Abbreviations*: α,α′=α-mycolates, α′-mycolates; B=methoxymycolate; C=ketomycolate; F=non-hydroxylated fatty acids.

The patterns obtained following one-dimensional TLC of methanolysates of typical strains belonging to the genera *Dietzia, Gordonia, Mycobacterium, Nocardia, Rhodococcus, Tsukamurella* and *Williamsia* are shown in Figure 4. The components with Rf values of 0.1–0.5 usually correspond to methyl esters of mycolic acids, whereas those with higher Rf values (0.8–1.0) are attributable to the methyl esters of non-hydroxylated fatty acids. Thus, whole-cell methanolysates from *Mycobacterium* species (with the exceptions of *Mycobacterium fallax* and *Mycobacterium triviale* that produce only α-mycolates) give multi-spot patterns that correspond to the different types of mycolates found in mycobacteria. Methanolysates from species of the genera *Tsukamurella* and *Segniliparus* give mycolic acid patterns consisting of two spots corresponding to α- and α′-mycolates (Figure 4), whereas methanolysates from species of the genera *Corynebacterium, Dietzia, Gordonia, Millisia, Nocardia, Rhodococcus, Skermania* and *Williamsia* exhibit simple patterns consisting of a single spot corresponding to α-mycolates. Notably, the migration (chromatographic) behaviour of α-mycolic acids from the different taxa reflects differences in chain length and structure. The Rf values increase in proportion to the total carbon number of the mycolic acids but decrease with increasing polarity as shown in Figure 4. Therefore, the mycolates from the representative strains of *Mycobacterium* have a relatively high mobility in comparison with mycolates from other taxa. The mycolic acid esters from *Tsukamurella* species are characteristically slightly more mobile than those of the genus *Segniliparus*, which in turn run a little faster than *Gordonia* species, which in turn run faster than mycolates from *Williamsia* species, which run little faster than those of *Nocardia* species, which run little faster than those from species of the genus *Rhodococcus*, which in turn run little faster than mycolic acid esters from *Dietzia* species. Mycolic acids of *Corynebacterium* species demonstrate the lowest Rf value of the known mycolic acids (Figure 4). Actinobacteria and other taxa that do not contain any mycolic acids give a single spot of high Rf value on TLC, which corresponds to non-hydroxylated fatty acids.

2. Two-dimensional TLC

Two-dimensional TLC is necessary to resolve complex mixtures of mycolic acids as is the case of *Mycobacterium*, *Segniliparus* and *Tsukamurella* species. The approach allows the true complexity of natural mycolic acid mixtures to be established and provides an efficient procedure for the recognition of characteristic patterns of the general types of mycolic acids. Accordingly, mycolic acid esters from *Mycobacterium* species can be resolved by two-dimensional TLC into several patterns based on the presence or absence of different functional groups (keto, methoxy, ω-carboxy, epoxy) in the longer carbon chain of the mycolic acid molecule.

- Apply the samples on the lower-right corners of 10 cm × 10 cm Merck 5554 TLC silica gel F254 aluminium sheets.
- Apply triple development in the first direction with petroleum ether (b.p. 60–80°C):acetone (95:5, v/v). Allow the plate to dry between each development.
- After the third development, rotate the plate 90° counter-clockwise and develop by a single development with toluene:acetone (97:3, v/v) in the second direction.
- Reveal the mycolic acid spots by spraying with an ethanolic 10% solution of molybdophosphoric acid as described above.

Extensive systematic investigations of methanolysates from representative strains of *Mycobacterium* species indicate that 12 distinct mycolic acid patterns are commonly encountered. Accordingly, members of the genus *Mycobacterium* can be grouped in 12 aggregate groups as follows (Table 2):

Aggregate group 1: The mycolic acid pattern of this aggregate group represents the simplest one found in *Mycobacterium* species. It consists of only one spot that corresponds to α-mycolate (Figure 5A).

Aggregate group 2: The mycolic acid pattern of members of this group is composed of approximately equal amounts of a diunsaturated α-mycolate and a lower molecular weight α′-mycolate (Figure 5B).

Aggregate group 3: The mycolic acid pattern contains α-, α′- and methoxymycolates (Figure 5C).

Aggregate group 4: The mycolic acid pattern is composed of α-, α′- and ketomycolates (Figure 5D).

Aggregate group 5: The mycolic acid pattern consists of α-, α′-, methoxy and ketomycolates (Figure 5E).

Table 2. Aggregate groups of *Mycobacterium* species according to mycolic acid pattern with reference to mycolic acid pyrolytic products of the α-mycolates isolated from the different species

Aggregate group	Mycolic acids composition	Organism	Pyrolytic cleavage products of α-mycolate
1	α	*M. brumae*	$C_{22:0}$
		M. fallax	$C_{22:0}$,[a] $C_{24:0}$[a]
		M. triviale	$C_{24:0}$
2	α, α′	*M. abscessus*	$C_{24:0}$
		M. chelonae	$C_{24:0}$
3	α, α′, methoxy	*M. agri*	$C_{24:0}$
4	α, α′, keto	*M. malmoense*	$C_{24:0}$, $C_{26:0}$[a]
		M. simiae	$C_{24:0}$
5	α, α′, methoxy, keto	*M. thermoresistibile*	$C_{22:0}$, $C_{24:0}$
6	α, keto	*M. bovis*	$C_{26:0}$
		M. bovis BCG	$C_{26:0}$
		M. leprae	$C_{22:0}$, $C_{24:0}$
7	α, methoxy, keto	*M. africanum*	$C_{26:0}$
		M. asiaticum	$C_{24:0}$
		M. gastri	$C_{22:0}$, $C_{24:0}$[a]
		M. gordonae	$C_{22:0}$,[a] $C_{24:0}$[a]
		M. haemophilum	$C_{22:0}$,[a] $C_{24:0}$[a]
		M. kansasii	$C_{24:0}$
		M. intermedium	$C_{22:0}$,[a] $C_{24:0}$,[a] $C_{26:0}$[a]
		M. marinum	$C_{24:0}$
		M. microti	$C_{24:0}$, $C_{26:0}$[a]
		M. tuberculosis	$C_{24:0}$, $C_{26:0}$[a]
		M. szulgai	$C_{24:0}$
		M. ulcerans	$C_{22:0}$,[a] $C_{24:0}$[a]
8	α, α′, epoxy	*M. chitae*	$C_{22:0}$, $C_{24:0}$[a]
		M. confluentis	$C_{22:0}$,[a] $C_{24:0}$[a]
		M. farcinogenes	$C_{24:0}$
		M. fortuitum	$C_{24:0}$
		M. peregrinum	$C_{24:0}$
		M. porcinum	$C_{24:0}$
		M. senegalense	$C_{24:0}$
		M. smegmatis	$C_{24:0}$
9	α, keto, wax esters[b]	*M. aichiense*	$C_{22:0}$, $C_{24:0}$
		M. avium subsp. *avium*	$C_{24:0}$
		M. avium subsp. *paratuberculosis*	$C_{24:0}$
		M. aurum	$C_{22:0}$
		M. celatum	$C_{24:0}$,[a] $C_{26:0}$[a]
		M. conspicuum	ND[c]
		M. chlorophenolicum	ND
		M. diernhoferi	$C_{22:0}$
		M. flavescens	$C_{22:0}$, $C_{24:0}$[a]
		M. gadium	$C_{22:0}$,[a] $C_{24:0}$[a]
		M. hibernae	$C_{22:0}$,[a] $C_{24:0}$[a]
		M. intracellulare	$C_{22:0}$, $C_{24:0}$[a]

(Continued)

Table 2. (Continued)

Aggregate group	Mycolic acids composition	Organism	Pyrolytic cleavage products of α-mycolate
		M. interjectum	$C_{24:0}$,[a] $C_{26:0}$[a]
		M. lepraemurium	$C_{24:0}$
		M. mucogenicum	$C_{22:0}$,[a] $C_{24:0}$[a]
		M. nonchromogenicum	$C_{22:0}$, $C_{24:0}$[a]
		M. neoaurum	$C_{22:0}$
		M. nonchromogenicum	$C_{22:0}$, $C_{24:0}$[a]
		M. phlei	$C_{24:0}$
		M. rhodesiae	$C_{22:0}$,[a] $C_{24:0}$[a]
		M. scrofulaceum	$C_{22:0}$, $C_{24:0}$[a]
		M. sphagni	$C_{22:0}$, $C_{24:0}$[a]
		M. terrae	$C_{24:0}$
		M. tokaiense	$C_{22:0}$, $C_{24:0}$[a]
		M. xenopi	$C_{26:0}$
10	α, α′, keto, wax esters[b]	*M. austroafricanum*	$C_{22:0}$
		M. chubuense	$C_{22:0}$,[a] $C_{24:0}$[a]
		M. duvalii	$C_{22:0}$,[a] $C_{24:0}$[a]
		M. gilvum	$C_{22:0}$
		M. komossense	$C_{24:0}$
		M. obuense	$C_{22:0}$,[a] $C_{24:0}$[a]
		M. parafortuitum	$C_{22:0}$
		M. pulveris	$C_{22:0}$, $C_{24:0}$[a]
		M. shimoidei	$C_{24:0}$
		M. vaccae	$C_{22:0}$,[a] $C_{24:0}$[a]
11	α, α′, wax esters[b]	*M. cookii*	$C_{22:0}$,[a] $C_{24:0}$[a]
12	α-, ω-1 methoxy	*M. alvei*	$C_{22:0}$,[a] $C_{24:0}$[a]

[a]Main pyrolytic cleavage products. In the case of two fatty acids this means that both are released during pyrolysis in approximately the same amounts.
[b]Wax ester mycolates are composed of ω-carboxymycolates esterified to 2-eicosanol and homolog. The latter two compounds are released as hydrolysis products from wax ester mycolates during acid and alkaline methanolysis.
[c]ND=no data available.

Aggregate group 6: Members of this group possess a mycolic acid pattern that consists of α- and ketomycolates (Figure 5F).

Aggregate group 7: The mycolic acid pattern of this aggregate group composed of α-, methoxy and ketomycolates (Figure 5G).

Aggregate group 8: The mycolic acid pattern recorded for alkaline methanolysates shows the presence of α-, α′- and epoxymycolates (Figure 5H). On acid methanolysis the epoxymycolates are converted to the polar components I, J, N and O.

Aggregate group 9: The mycolic acid pattern consists of α-, keto and wax ester mycolates. The latter compounds consist of ω-carboxymycolates and long-chain alcohols (E) homologous with 2-eicosanol (Figure 5I).

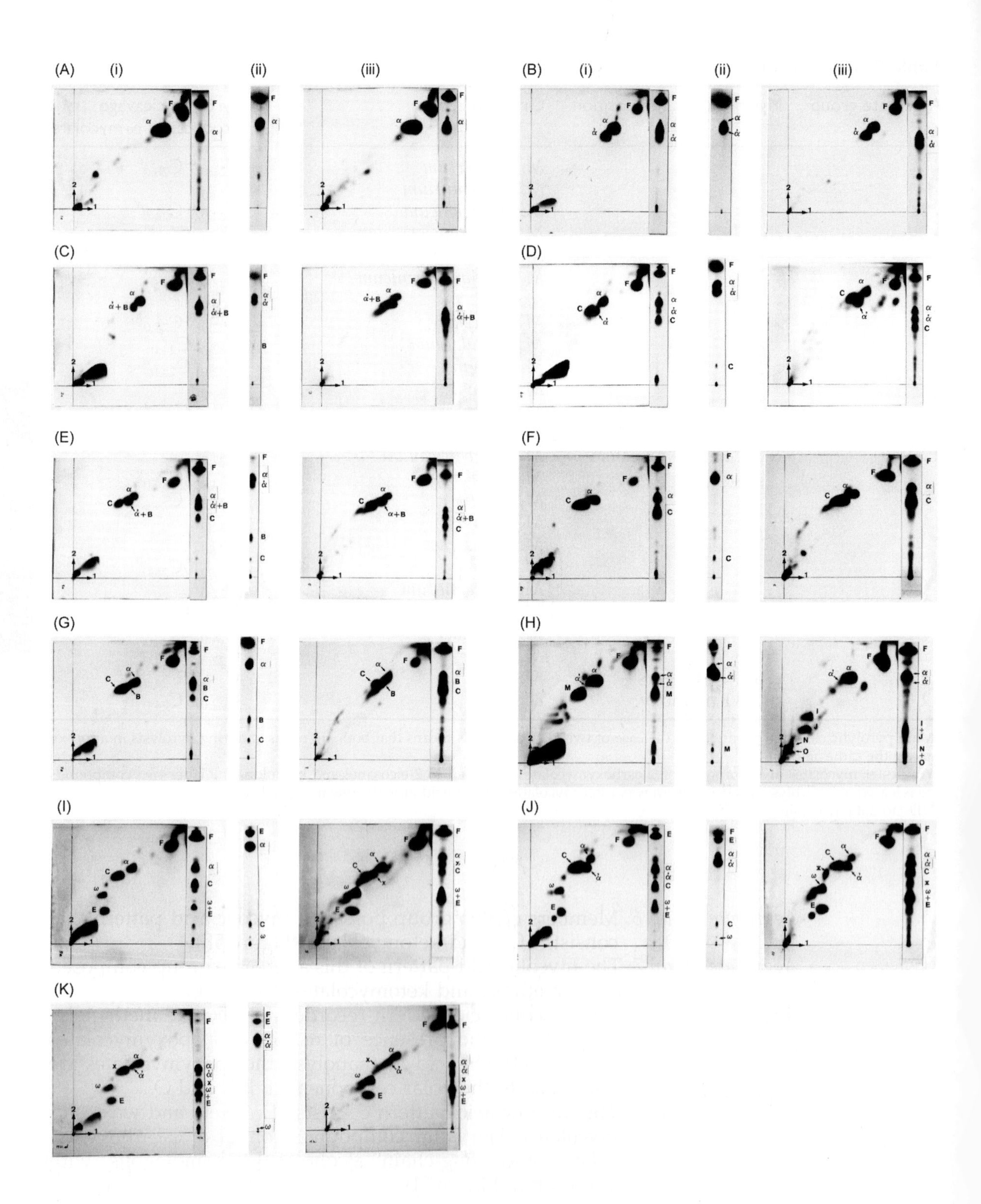

Aggregate group 10: The mycolic acid pattern is composed of α-, α′-, keto and wax ester mycolates. The latter is composed of ω-carboxy-mycolates and long-chain alcohols (E) homologous with 2-eicosanol (Figure 5J).

Aggregate group 11: The mycolic acid pattern consists of α-, α′-, ω-carboxymycolates and long-chain alcohols (E) homologous with 2-eicosanol (Figure 5K).

Aggregate group 12: The mycolic acid pattern of this group consists of two spots, the higher one corresponds to α-mycolate and the lower one corresponds to ω-1 methoxymycolates.

It is noteworthy to mention that certain *Nocardia* isolates occasionally contain substantial proportions of characteristic long-chain alcohols (nocardols) structurally related to mycolic acids. The presence of nocardols in *Nocardia* strains may lead to misinterpretation of one-dimensional TLC of methanolysates, since in addition to the single spot recognizable as a mycolic ester, another component having similar TLC properties is observed. In this case, two-dimensional TLC was shown to be of assistance in recognizing the presence of nocardols since they migrate in the second solvent with a similar mobility to that of the non-hydroxylated fatty acid methyl esters.

3. Preparative TLC

Preparative one-dimensional TLC can be used to purify mycolic acids:

- Apply extracts produced with acid methanolysis as a band using 20 cm × 20 cm pieces of Merck 5554 TLC silica gel F254 aluminium sheets.
- Develop the plates with petroleum ether (b.p. 60–80°C):acetone (95:5, v/v) to about 10 cm.
- Dry the plates in a fume hood at room temperature.
- Develop to 20 cm, and dry the plates in a fume hood at room temperature.
- Reveal the components by spraying the dried plates with 0.2% solution of 2′,7′-dichlorofluorescein in ethanol. After drying at room temperature, locate the separated components by viewing under UV light (366 nm).
- Scrape the mycolic acid methyl esters' spots off the plates, and elute them from the silica gel with 2 ml of chloroform:methanol (2:1, v/v).
- Remove the silica gel by centrifugation at 3000 rpm for 5 min.
- Remove the dye by washing the supernatant in *ca.* 10 ml of a solution made of 0.5 M ammonia.

Figure 5. One- and two-dimensional TLC of whole-cell alkaline (i) and acid methanolysates (iii) and one-dimensional TLC of TBDMS ethers of alkaline methanolysates (ii) of aggregate group 1, 2, 3, 4, 5, 6, 7, 8, 9, 10 and 11 mycobacteria (panels A–K, respectively). For panels (i) and (iii), triple development with petroleum ether–acetone (95:5, v/v) in the first direction was followed by single development with toluene–acetone (97:3, v/v) in the second direction. For the TBDMS ethers (panels ii), a single development was used with petroleum ether–toluene (70:30, v/v). *Abbreviations*: B = methoxymycolate; C = ketomycolate; E = 2-eicosanol; F = non-hydroxylated fatty acids; I, J, N and O = components derived from epoxymycolates by acid methanolysis; M = epoxymycolate; ω = ω-carboxymycolate; α, = α-mycolates; α′ = α′-mycolates.

- Wash the chloroform-based layer with water.
- Pass the washed chloroform layer through diethyl ether-washed anhydrous sodium sulfate in a cotton wool-plugged Pasteur pipette (*ca.* 4 cm).
- Collect the washed-eluent, in a separate vial and evaporate to dryness under a stream of nitrogen at room temperature.

Purified mycolates can be subjected to pyrolysis gas chromatography or be further derivatized as trimethylsilyl ethers.

♦♦♦♦♦♦ IV. OTHER METHODS OF MYCOLIC ACID ANALYSIS

A. High-Performance Liquid Chromatography (HPLC)

HPLC is a suitable technique for separation, quantification and analysis of fatty acids, including mycolic acids. However, the detection of underivatized acids is neither sensitive nor selective because these compounds generally do not contain suitable chromophores. To increase the sensitivity and selectivity of detection, a number of UV-absorbing or fluorescent derivatives have been used. The most frequently used derivatives are phenaycl and substituted phenacyl esters. They are easily prepared in quantitative yields and are stable, and so permit the determination of lipids even at nanogram levels (Wood and Lee, 1983). Phenacylbromide (2-bromoacetophenone) (PB) and other PBs are among the reagents often used for fatty acids analysis (Hanis *et al.*, 1988). A broad choice of stationary phases can be used for separation, such as reversed-phase (Engelmann *et al.*, 1988; Chen and Anderson, 1992) or silver ion modes (Christie and Breckenridge, 1989). The eluates are detected with UV detectors operating at 254 nm.

HPLC characterization of mycolic acids has been in use for several years and is gradually finding its way into clinical laboratories as a useful and rapid method for identification of bacteria at the species level. *p*-Bromophenacyl derivatives of mycolic acids have also been used to generate characteristic profiles of mycolates for many *Mycobacterium* species and these have been used diagnostically (Butler *et al.*, 1986, 1991, 1999; Butler and Guthertz, 2001; Glickman *et al.*, 1994). These methods have become so highly developed that the Centers for Disease Control (CDC) in 1996 published a book entitled '*Standardized Method for HPLC Identification of Mycobacteria*' to facilitate the use in the clinical laboratory setting of mycolic acid HPLC patterns as a rapid diagnostic tool for speciating mycobacteria (Centers for Disease Control, U.S. Department of Health and Human Services, Public Health Service).

1. Sample preparation

The sample preparation protocol involves the following:

(a) Saponification of mycolic acid (Butler and Kilburn, 1988)

- Place dry organisms (up to 10 mg) or a loopful of cells in a Teflon-coated screw-capped tube.

- Add 2 ml of saponification reagent (25% KOH in 50% ethanol).
- Mix the suspension vigorously and saponify the cells by heating in an oven or a heating block at 85°C for 18 h (or alternative by autoclaving for 1 h at 121°C).
- Cool to room temperature and add 1.5 ml of 50% HCl to acidify the mycolic acids.
- Add 2 ml chloroform and shake vigorously.
- Separate the phases by centrifugation at 3000 rpm for 5 min.
- Transfer the lower organic layer with a Pasteur pipette to a new Teflon-coated screw-capped tube and evaporate to dryness.

(b) Preparation of* p*-bromophenacyl esters (PBPA) of mycolic acid (Durst* et al.*, 1975)

- To the dry mycolic acid from the previous step add the following: 0.5 ml chloroform, 2 mg potassium carbonate, 25 μl *p*-bromophenacyl bromide in acetonitrile. (A kit containing the individual components is available from Alltech Associates, Inc., Product No. 18036.)
- Heat in an oven or a heating block at 85°C for 45 min.
- Cool to room temperature and clarify the derivative by filtration through 0.45 μm pore nylon 66 membrane filter.
- Evaporate to dryness and resuspend in 100 μl of chloroform for analysis.

2. Instrumentation

- Use a HPLC equipped with a variable UV detector (set at 260 nm) and a reverse-phase C-18 cartridge column (4.6 mm × 7.5 cm) with 3 μm particle size.

3. Conditions of HPLC analysis (*Butler* et al., *1991*)

- *Solvent A*: 98% methanol
- *Solvent B*: 2% methylene chloride
- Maintain the solvent flow rate constant (2.5 ml/min).
- Equilibrate the column with a mixture of A:B (98:2, v/v).
- During the first minute following injection change the solvent mixture, using a linear gradient, to 80% A and 20% B.
- Use a linear gradient for the next 9 min to change the solvent mixture to 35% A and 65% B.
- Return the gradient during the next 0.5 min to 98% A and 2% B.
- Finally, equilibrate the column for 4.5 min with a solvent mixture of 98% A and 2% B.
- For identification, match the chromatogram of the unknown strain with those of reference ones analysed under the same conditions.

B. Gas–Liquid Chromatography (GLC)

GLC has been used to characterize microorganisms through analysis of cellular structural components such as mycolic acids. Mycolic acid analysis using GLC is performed after its conversion to methyl esters, otherwise mycolic acid could not be amenable to analysis by gas–liquid chromatography. There are two independent methods for mycolic acid analysis by GLC. Pyrolysis gas chromatography involves the thermal fragmentation of mycolic acid methyl esters in an inert atmosphere at elevated temperature (300°C and higher). This technique takes advantage of the fact that cleavage of the carbon chain of the mycolic acid molecule between the alkyl-branched alpha carbon and the hydroxyl-substituted beta carbon forms short-chain fatty acid methyl esters and long-chain meroaldehydes (Stodola *et al.*, 1938; Etémadi, 1967). The long-chain meroaldehydes cannot be detected because they do not elute from the GLC column, but the short-chain fatty acid methyl esters can be detected and identified by comparison of their chain length with those of standard fatty acids. The detected fatty acid methyl esters are useful for differentiating bacteria belonging to different genera (Lechevalier *et al.*, 1971, 1973).

For further characterization of mycolic acids using GLC, mycolic acid methyl esters are converted to trimethylsilyl (TMS)-derivatives. Although, mycolic acid methyl esters decompose by pyrolysis at 300°C or higher temperatures, the TMS-methyl esters of mycolic acids are stable under this analytical condition. The gas chromatographic profile of the TMS derivatives of methyl mycolates is characteristic for each species of bacteria (Yano and Saito, 1972; Yano *et al.*, 1972, 1978). However, the gas chromatographic separation of the TMS derivatives of the total mycolic acid methyl esters is not sufficient for identification of mycolic acid's molecular species (Figure 6).

- Prepare mycolic acid methyl esters using acid methanolysis as described in Section III.A.1.a.
- Separate and purify α-mycolic acid methyl esters using preparative TLC as described in Section III.B.3.
- Prepare TMS derivatives of mycolic acid methyl esters as described in Section III.A.3.
- Separate the TMS derivatives of methyl mycolate using a gas chromatograph (Shimadzu GC-14A, equipped with a flame-ionization detector and coupled to a Shimadzu CR5A integrator for recording and processing chromatograms).
- The GC conditions are as follows:
  - Capillary column (12 m × 0.25 mm internal diameter) coated with a chemically bound 0.10 μm SIM-DIST CB phase (Chrompack 99927).
  - Carrier gas nitrogen, column flow rate approximately 6 ml/min, column inlet pressure 1 kg/cm^2.
  - Hydrogen flow rate to the detector 30 ml/min.
  - Oxygen flow rate to the detector 60 ml/min.
  - Split ratio 1:50.
  - Injector temperature 397°C.
  - Detector temperature 397°C.
  - Column temperature programming from 290°C (hold for 4 min) to 370°C (hold for 40 min) at a rate of 2°C/min.

For identification, the mycolic acid profiles of unknown *Mycobacterium* species are compared with representative mycolic acid profiles of known *Mycobacterium* species (Figure 6).

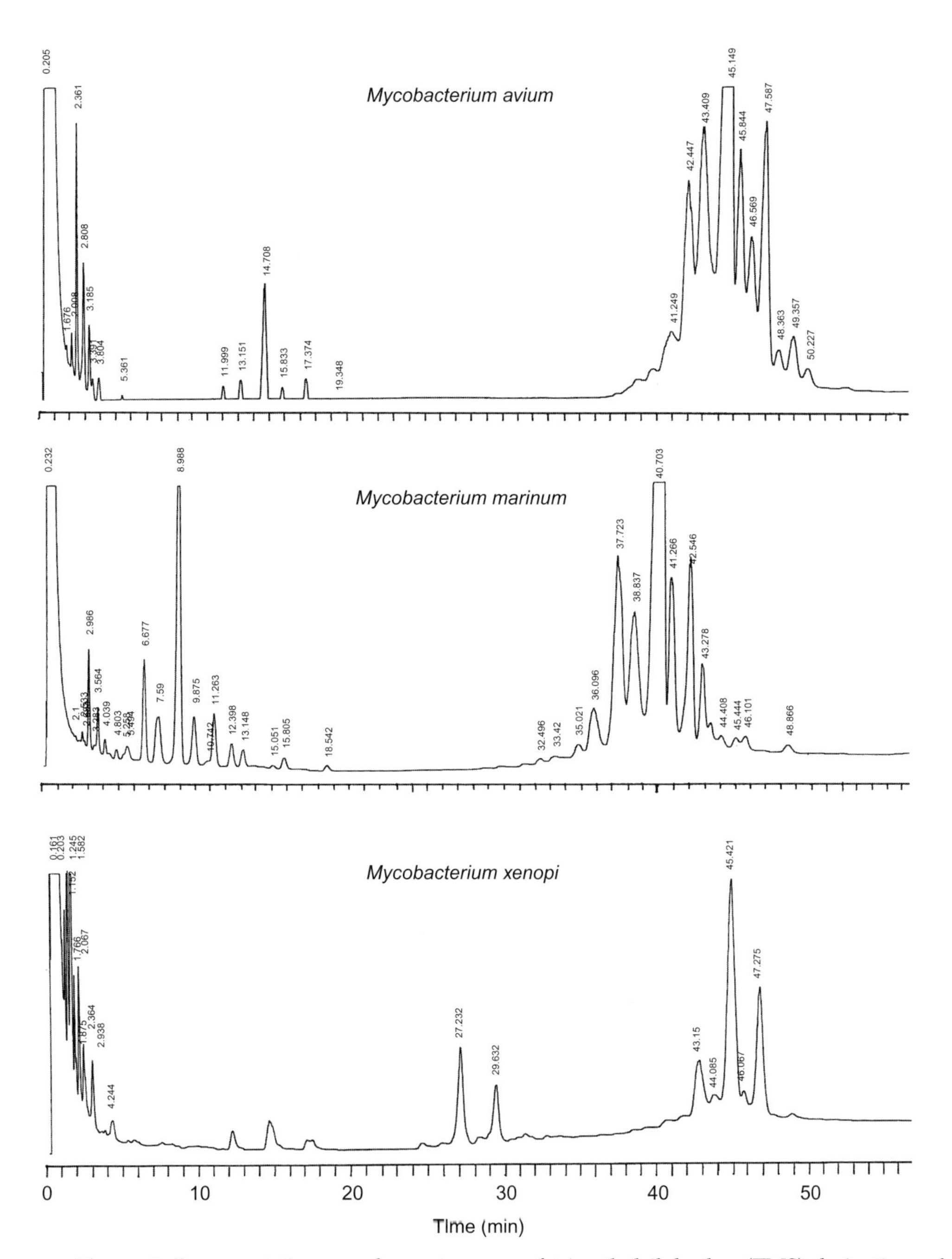

Figure 6. Representative gas chromatograms of trimethylsilyl ether (TMS) derivatives of methyl α-mycolates from *M. avium*, *M. marinum* and *M. xenopi*. The peaks eluted late in the chromatograms correspond to the different molecular species of α-mycolates. Gas chromatography conditions as described in the text.

C. Mass Spectrometry (MS)

Mass spectrometry has been used as a method for characterizing bacteria on the basis of mycolic acids analysis. Electron impact (EI) mass spectrometry, gas chromatography/mass spectrometry, fast atom bombardment mass spectrometry, electrospray ionization (ESI) mass spectrometry and MALDI-TOF have all been employed. EI mass spectrometry has been frequently used for the structural elucidation of mycolic acids. Although this technique provides important structural information on the overall size of mycolates, the degree of unsaturation, the nature of both the α- and the meromycolate chains, the fragmentation patterns are complicated so that, in some cases, the interpretation of spectra is difficult (Alshamaony *et al.*, 1976a, 1976b; Minnikin, 1982). However, the application of MS in conjunction with gas chromatography–mass spectrometry (GC–MS) for the analysis of the TMS-ether derivatives of methyl mycolates offers a good separation of mycolic acids according to the total number of carbons and double bonds. These characteristics of mycolic acids are well known as essential criteria for the chemotaxonomy of genera (Collins *et al.*, 1982; Yano *et al.*, 1978, 1990; Kaneda *et al.*, 1986; Baba *et al.*, 1997; Nishiuchi *et al.*, 1999, 2000). A highly complex mixture of mycolic acids can clearly be separated, and the individual molecular species can be identified satisfactorily. Even the acids of the same carbon numbers, which are detected as a single peak on gas chromatograms, were shown to consist of several isomers having different branch structures. Moreover, this technique allows the detection of minor components of molecular species and reveals the differences of the mycolic acid profiles among different taxa (Baba *et al.*, 1997; Nishiuchi *et al.*, 1999, 2000).

MALDI-TOF mass spectrometry is a more reliable procedure for determining the chain lengths of mycolic acid methyl esters and recognizing the presence of particular types of unsaturation and additional homologous series for representative mycolates (Laval *et al.*, 2001; Watanabe *et al.*, 2001). It is a sensitive analytical method requiring less than 10 pmol of mycolates to obtain well-resolved mass spectra, and has marked advantages in direct analysis of mycolic acids as no pretreatment of these compounds is required (Laval *et al.*, 2001; Watanabe *et al.*, 2001; Fujita *et al.*, 2005a, 2005b). In this technique mycolic acid methyl esters and 2,5-dihydroxybenzoic acid (2,5-DHB) as matrix are dissolved in chloroform:methanol (2:1, v/v) at low concentrations, then aliquots of *ca.* 5 μl of both mycolic acid methyl esters and the matrix are mixed and allowed to crystallize at room temperature before exposure to laser. MALDI-TOF provides a characteristic fingerprint of the whole mycolate content from a given bacterium such as *Mycobacterium* species and related genera, thus facilitating the identification of these taxa. Interestingly, this technique gives precise structure information of the individual mycolic acid molecules that can be used not only for chemotaxonomy but also for virulence analysis of *Mycobacterium* species and related taxa (Dubnau *et al.*, 2000; Ueda *et al.*, 2001; Fujita *et al.*, 2005a, 2005b). It has been reported that replacement of a cyclopropane ring by a double bond in the α-mycolate esterifying trehalose totally abolishes the formation of cords (aggregation to form serpentine structures) typical of virulent mycobacteria and related taxa and thus affects the virulence of the *Mycobacterium* strain (Dubnau *et al.*, 2000; Glickman *et al.*, 2000; Ueda *et al.*, 2001).

D. Application of Pyrolysis Gas Chromatography

A characteristic property of mycolic acids and esters is their facile cleavage at high temperatures (300°C) between the alkyl-branched alpha carbon and the hydroxyl-substituted beta carbon to yield straight-chain fatty acid esters that can be analysed by GC (Etémadi, 1967), and long-chain meroaldehydes (Stodola *et al.*, 1938; Asselineau and Lederer, 1950), as shown in Figure 7 (Etémadi, 1964).

Pyrolysis gas chromatography of methyl mycolates provides information on the structure of the released fatty acid methyl esters such as saturation or chain-branching. Studies on the nature of the long-chain fatty acid esters released by pyrolysis have been found to be of great value in the differentiation of mycolic acid containing bacteria (Lechevalier *et al.*, 1971, 1973). Systematic investigations of mycolic acid cleavage products from *Mycobacterium* species revealed that all species usually release $C_{22:0}$, $C_{24:0}$ and/or $C_{26:0}$ completely saturated fatty acid methyl esters as characteristic thermal cleavage products (Table 1). The relative amount of $C_{22:0}$, $C_{24:0}$ or $C_{26:0}$ fragments resulting from cleavage can be used to distinguish between the different *Mycobacterium* species (Table 1). In contrast, the mycolic acids from all other mycolic acid containing actinobacteria cleave on pyrolysis to yield fatty acids methyl esters with smaller chain lengths. Mycolic acids from *Tsukamurella* species release, on pyrolysis, $C_{20:0}$ and $C_{22:0}$ straight-chain saturated and unsaturated fatty acid esters, while mycolic acids from *Segniliparus* species produce only $C_{22:0}$ fatty acid ester. Examination of the pyrogram of mycolic acid methyl esters from *Skermania* revealed the presence of straight-chain saturated and unsaturated fatty acids methyl esters with chain lengths varying from $C_{16:0}$ to $C_{20:0}$ with $C_{18:1}$ as the main pyrolysis product; lesser proportions of $C_{16:0}$, $C_{16:1}$ and $C_{20:0}$ were recorded. The mycolic acid methyl esters from *Gordonia*, *Millisia* and *Williamsia* species usually cleave on pyrolysis to release $C_{16:0}$ and $C_{18:0}$ saturated fatty acid esters. Mycolic acids from *Nocardia* species split on pyrolysis to yield fatty acid methyl esters having

Figure 7. Proposed pyrolytic fragmentation of mycolic acid methyl esters.

chain lengths $C_{14:0}$, $C_{16:0}$ and $C_{18:0}$. In the majority of *Nocardia* species $C_{14:0}$ and $C_{16:0}$ constitute the main pyrolytic cleavage products, and in some other species $C_{18:0}$ is absent. Mycolic acids from some *Nocardia* species also produce monounsaturated $C_{16:1}$ and $C_{18:1}$ fatty acids. The mycolic acids from members of the genus *Rhodococcus* generally cleaves into $C_{12:0}$, $C_{14:0}$, $C_{16:0}$ and $C_{18:0}$ fatty acid methyl esters. In contrast to the aforementioned taxa, mycolic acids from species of the genus *Dietzia* have their own characteristic cleavage products. *Dietzia* mycolates cleave on pyrolysis to release a series of saturated fatty acids with odd-numbered and even-numbered carbons, $C_{14:0}$, $C_{15:0}$, $C_{16:0}$, $C_{17:0}$ and $C_{18:0}$ in addition to $C_{18:1}$. The mycolates from *Corynebacterium* strains are particularly distinctive, having shorter chain lengths, and on pyrolysis release $C_{6:0}$, $C_{8:0}$, $C_{10:0}$, $C_{12:0}$, $C_{13:0}$, $C_{14:0}$, $C_{15:0}$ and $C_{16:0}$ saturated fatty acid methyl esters. Monounsaturated fatty acids, e.g. $C_{14:1}$, $C_{16:1}$ are detected in some *Corynebacterium* strains.

It is noteworthy to mention that the carbon chain lengths of the pyrolytic cleavage products among the three genera, *Dietzia*, *Nocardia* and *Rhodococcus*, overlap, so that the chain lengths of mycolic acids from isolated strains cannot be used as a satisfactory index for identification at the generic level. However, a distinction can be made between the genera *Nocardia* and *Rhodococcus* on the basis of the meroaldehyde moiety formed during pyrolysis. The meroaldehydes cleaved from *Nocardia* mycolates are retained on the column of the gas chromatograph because of their large size, whereas those released from *Rhodococcus* mycolates, being smaller, are eluted. Moreover, the three genera *Gordonia*, *Millisia* and *Williamsia* cannot be distinguished from each other on the basis of the mycolic acids cleavage products. On the other hand, the carbon chain lengths of the mycolic acid cleavage products are a quite useful marker to distinguish the genera *Mycobacterium*, *Tsukamurella* and *Segniliparus* from each other and from other mycolic acid containing genera of the order *Actinomycetales*.

Since the composition of the pyrolysis products depends on the specific conditions of pyrolysis, e.g. temperature, duration, sample size, carrier gas velocity, stationary phase, column parameters, and so on, investigation of mycolic acid methyl esters by this technique requires strict standardization of the pyrolysis conditions. Many examples of pyrolysis design have been recorded in the literature that often impedes the comparison of the results.

The current method is based on the use of a Shimadzu GC-14A gas chromatograph equipped with a flame-ionization detector. The chromatograph is coupled to a Shimadzu CR5A integrator for recording and processing chromatograms. The conditions are as follows:

- Capillary column (25 m $\times$ 0.25 mm internal diameter) coated with a chemically bound 0.25 μm-CP-CIL 5CB phase (Chrompack 07441).
- Carrier gas nitrogen, column flow rate approximately 6 ml/min, column inlet pressure 1 kg/cm^2.
- Hydrogen flow rate to the detector 30 ml/min.
- Oxygen flow rate to the detector 60 ml/min.
- Split ratio 1:50.
- Injector temperature 350°C.

- Detector temperature 300°C.
- Column temperature programming from 190°C (hold for 5 min) to 280°C (hold for 10 min) at a rate of 2 °C/min.
- Sample size is 1 μl.
- The identity of individual fatty acid esters is established by comparison of the retention times with those of standard straight-chain saturated and mono-unsaturated esters.

♦♦♦♦♦♦ V. CONCLUSION

Mycolic acids are useful chemotaxonomic markers for the delineation of taxa. A wide array of chemical methods are available for mycolic acid analysis. In particular, TLC is a relatively simple, cost-effective and reliable method to distinguish mycobacteria from other mycolic acid containing taxa and for routine screening of *Mycobacterium* species in mycobacteriology laboratories with a potential for use in developing countries. Notwithstanding, modern methods such as GC–MS permit fingerprinting of the whole mycolate and provide detailed structural information on mycolic acid molecular species, which could be used as models to illustrate differences of strains within a bacterial species. HPLC offers an alternative approach to mycolic acid analysis. A major advantage of HPLC is that mycolic acid separation is carried out at ambient temperature. It is a rapid method that is widely used for the identification of *Mycobacterium* species. However, the interpretation of the generated mycolic acid fingerprint requires expertise with mycolic acid pattern recognition practices.

Mycolic acid data increased confidence in the accuracy of species names allocated to different genera. Mycolic acid profiles are a stable feature that is independent of the culture medium and age of the culture, and are shared by all strains of the same species. Therefore, a taxonomy based on mycolic acid composition would be more consistent with a taxonomy based on genetic characteristics. The mandatory requirements for mycolic acid data needed for the assignment of a bacterial strain to a novel genus are (i) determination of the mycolic acid pattern by one- and two-dimensional TLC; (ii) determination of the chain length of the α-branch cleaved during pyrolysis gas chromatography; (iii) determination of the carbon chain lengths of the meromycolic residues; and (iv) determination of the numbers of double bonds in the meromycolate chain. To secure a coherent framework for the classification of novel mycolata strains to a new species it is proposed, as a minimum, to determine the mycolic acid pattern by TLC and to determine mycolic acid cleavage products by pyrolysis gas chromatography.

References

Adachi, K., Katsuta, A., Matsuda, S., Peng, X., Misawa, N., Shizuri, Y., Kroppenstedt, R. M., Yokota, A. and Kasai, H. (2007). *Smaragdicoccus niigatensis* gen. nov., sp. nov., a novel member of the suborder *Corynebacterineae*. *Int. J. Syst. Evol. Microbiol.* **57**, 297–301.

Alshamaony, L., Goodfellow, M., Minnikin, D. E. and Mordarska, H. (1976a). Free mycolic acids as criteria in the classification of *Gordona* and the "*rhodochrous*" complex. *J. Gen. Microbiol.* **92**, 183–187.

Alshamaony, L., Goodfellow, M. and Minnikin, D. E. (1976b). Free mycolic acids as criteria in the classification of *Nocardia* and the "*rhodochrous*" complex. *J. Gen. Microbiol.* **92**, 188–199.

Anderson, R. J. (1929). The chemistry of the lipoids of tubercle bacilli. Concerning the so-called tubercle bacilli wax. Analysis of the purified wax. *J. Biol. Chem.* **83**, 505–522.

Anderson, R. J., Reeves, R. E. and Stodola, F. H. (1937). The chemistry of the lipids of tubercle bacilli: LI. Concerning the family bound lipids of the human tubercle bacillus. *J. Biol. Chem.* **121**, 649–668.

Armstrong, J. A. and Hart, P. D. (1971). Response of cultured macrophages to *Mycobacterium tuberculosis*, with observations on fusion of lysosomes with phagosomes. *J. Exp. Med.* **134**, 713–740.

Aronson, H. (1898). Zur Biologie der Tuberkelbacillen. *Ber. Klin. Wochenschr.* **35**, 484–486.

Asselineau, J. (1950). Sur la structure chemique des acides mycoliques isolés de *Mycobacterium tuberculosis* var. *hominis*; déterminations de la position de l'hydroxyle. *C. R. Hebd. Séances Acad. Sci.* **230**, 1620–1622.

Asselineau, J. and Lederer, E. (1950). Structure of mycolic acids of mycobacteria. *Nature* **166**, 782–783.

Asselineau, J. (1966). *The bacterial lipids*. Hermann, Paris and San Francisco.

Baba, T., Nishiuchi, Y. and Yano, I. (1997). Composition of mycolic acid molecular species as a criterion in nocardial classification. *Int. J. Syst. Bacteriol.* **47**, 795–801.

Barksdale, L. (1970). *Corynebacterium diphtheriae* and its relatives. *Bacteriol. Rev.* **34**, 378–422.

Barksdale, L. and Kim, K.-S. (1977). *Mycobacterium*. *Bacteriol. Rev.* **41**, 217–372.

Barry, C. E., III and Mdluli, K. (1996). Drug sensitivity and environmental adaptation of mycobacterial cell wall components. *Trends Microbiol.* **4**, 275–281.

Barry, C. E., III, Lee, R. E., Mdluli, K., Sampson, A. E., Schoeder, B. G., Slayden, R. A. and Yuan, Y. (1998). Mycolic acids: structure, biosynthesis and physiological functions. *Prog. Lipid Res.* **37**, 143–179.

Barton, M. D., Goodfellow, M. and Minnikin, D. E. (1989). Lipid composition in the classification of *Rhodococcus equi*. *Zentralbl. Bakteriol.* **272**, 154–170.

Beckman, E. M., Porcelli, S. A., Morita, C. T., Behar, S. M., Furlong, S. T. and Brenner, M. B. (1994). Recognition of a lipid antigen by CD1-restricted $\alpha\beta$+ T cell. *Nature* **372**, 691–694.

Blackall, L. L., Parlett, J. H., Hayward, A. C., Minnikin, D. E., Greenfield, P. F. and Harbers, A. (1989). *Nocardia pinensis* sp. nov., an actinobacterium found in activated sludge foams in Australia. *J. Gen. Microbiol.* **135**, 1547–1558.

Bulloch, W. and Macleod, J. J. R. (1904). The chemical constitution of tubercle bacillus. *J. Hyg.* **4**, 1–10.

Butler, W. R. and Guthertz, L. S. (2001). Mycolic acid analysis by high-performance liquid chromatography for identification of *Mycobacterium* species. *Clin. Microbiol. Rev.* **14**, 704–726.

Butler, W. R., Ahearn, D. G. and Kilburn, J. O. (1986). High-performance liquid chromatography of mycolic acids as a tool in the identification of *Corynebacterium*, *Nocardia*, *Rhodococcus*, and *Mycobacterium* species. *J. Clin. Microbiol.* **23**, 182–185.

Butler, W. R. and Kilburn, J. O. (1988). Identification of major slowly growing pathogenic mycobacteria and *Mycobacterium gordonae* by high-performance liquid chromatography of their mycolic acids. *J. Clin. Microbiol.* **26**, 50–53.

Butler, W. R., Jost, K. C., Jr and Kilburn, J. O. (1991). Identification of mycobacteria by high-performance liquid chromatography. *J. Clin. Microbiol.* **29**, 2468–2472.

Butler, W. R., Floyd, M. M., Silcox, V., Cage, G., Desmond, E., Duffey, P. S., Guthertz, L. S., Gross, W. M., Jost, K. C., Jr, Ramos, L. S., Thibert, L. and Warren, N. (1999). *Mycolic Acid Pattern Standards for HPLC Identification of Mycobacteria*. Centers for Disease Control and Prevention, Atlanta, GA.

Butler, W. R., Floyd, M. M., Brown, J. M., Toney, S. R., Daneshvar, M. I., Cooksey, R. C., Carr, J., Steigerwalt, A. G. and Charles, N. (2005). Novel mycolic acid-containing bacteria in the family *Segniliparaceae* fam. nov., including the genus *Segniliparus* gen. nov., with descriptions of *Segniliparus rotundus* sp. nov. and *Segniliparus rugosus* sp. nov.. *Int. J. Syst. Evol. Microbiol.* **55**, 1615–1624.

Chen, H. and Anderson, R. E. (1992). Quantitation of phenacyl esters of retinal fatty acids by high-performance liquid chromatography. *J. Chromatogr.* **578**, 124–129.

Christie, W. W. and Breckenridge, G. H. McG. (1989). Separation of cis and trans isomers of unsaturated fatty acids by high-performance liquid chromatography in the silver ion mode. *J. Chromatogr.* **469**, 261–269.

Chun, J., Blackall, L. L., Kang, S.-O., Hah, Y. C. and Goodfellow, M. (1997). A proposal to reclassify *Nocardia pinensis* Blackall *et al.* as *Skermania piniformis* gen. nov., comb. nov.. *Int. J. Syst. Bacteriol.* **47**, 127–131.

Collins, M. D. and Jones, D. (1982). Lipid composition of *Corynebacterium paurometabolum* (Steinhaus). *FEMS Microbiol. Lett.* **13**, 13–16.

Collins, M. D., Goodfellow, M. and Minnikin, D. E. (1982). A survey of the structures of mycolic acids in *Corynebacterium* and related taxa. *J. Gen. Microbiol.* **128**, 129–149.

Collins, M. D., Smida, J., Dorsch, M. and Stackebrandt, E. (1988). *Tsukamurella* gen. nov. harbouring *Corynebacterium paurometabolum* and *Rhodococcus aurantiacus*. *Int. J. Syst. Bacteriol.* **38**, 385–391.

Corey, E. J. and Venkateswarlu, A. (1972). Protection of hydroxyl groups as *tert*-butyldimethylsilyl derivatives. *J. Am. Chem. Soc.* **94**, 6190–6191.

Daffé, M. and Draper, P. (1998). The envelope layers of mycobacteria with reference to their pathogenicity. *Adv. Microbiol. Physiol.* **39**, 131–203.

Daffé, M., Lanéelle, M. A., Asselineau, C., Lévy-Frébault, V. and David, H. (1983). Intéret taxonomique des acides gras des mycobactéries: proposition d'une méthode d'analyse. *Ann. Microbiol.* **134B**, 241–256.

Datta, A. K., Katoch, V. M., Katoch, K., Sharma, V. D. and Shivannavar, C. T. (1987). Appearance of a methoxy mycolate-like component by the acid methanolysis of *Mycobacterium leprae*. *Int. J. Lepr.* **55**, 680–684.

Dobson, G., Minnikin, D. E., Parlett, J. H., Goodfellow, M., Ridell, M. and Magnusson, M. (1984). Systematic analysis of complex mycobacterial lipids. In: *Chemical Methods in Bacterial Systematics* (M. Goodfellow and D. E. Minnikin, eds.), Academic Press, London.

Dubnau, E., Chan, J., Raynaud, C., Mohan, V. P., Lanéelle, M. A., Yu, K., Quemard, A., Smith, I. and Daffé, M. (2000). Oxygenated mycolic acids are necessary for virulence of *Mycobacterium tuberculosis* in mice. *Mol. Microbiol.* **36**, 630–637.

Durst, H. D., Milano, M., Kikta, E. J., Connelly, S. A. and Grushka, E. (1975). Phenacyl esters of fatty acids via crown ether catalysts for enhanced ultraviolet detection in liquid chromatography. *Anal. Chem.* **47**, 1797–1801.

Embley, T. M. and Stackebrandt, E. (1994). The molecular phylogeny and systematic of the actinomycetes. *Annu. Rev. Microbiol.* **48**, 257–289.

Engelmann, G. J., Esmans, E. L. and Alderweireldt, F. C. (1988). Rapid method for analysis of red blood cell fatty acids by reversed-phase high-performance liquid chromatography. *J. Chromatogr.* **432**, 29–36.

Etémadi, A. H. (1964). Techniques microanalytiques d'étude de structure d'esters α-ramifiés, β-hydroxylés. Chromatographie en phase vapeur et spectrométrie de masse. *Bull. Soc. Chim. Fr.*, 1535–1541.

Etémadi, A. H. (1967). The use of pyrolysis gas chromatography and mass spectroscopy in the study of the structure of mycolic acids. *J. Gas Chromatogr.* **5**, 447–456.

Fujita, Y., Naka, T., Doi, T. and Yano, I. (2005a). Direct molecular mass determination of trehalose monomycolate from 11 species of mycobacteria by MALDI-TOF mass spectrometry. *Microbiology* **151**, 1443–1452.

Fujita, Y., Naka, T., McNeil, M. R. and Yano, I. (2005b). Intact molecular characterization of cord factor (trehalose 6,6′-dimycolate) from nine species of mycobacteria by MALDI-TOF mass spectrometry. *Microbiology* **151**, 3403–3416.

Gastambide-Odier, M., Delauményy, J. M. and Lederer, E. (1964). Demonstration of propane rings in various mycolic acids from human and bovine strains of *Mycobacterium tuberculosis. C. R. Hebd. Séances Acad. Sci.* **259**, 3404–3407.

Glickman, M. S., Cox, J. S. and Jacobs, W. R., Jr. (2000). A novel mycolic acid cyclopropane synthetase is required for cording, persistence, and virulence of *Mycobacterium tuberculosis. Mol. Cell* **5**, 717–727.

Glickman, S. E., Kilburn, J. O., Butler, W. R. and Ramos, L. S. (1994). Rapid identification of mycolic acid patterns of mycobacteria by high-performance liquid chromatography using pattern recognition software and a *Mycobacterium* library. *J. Clin. Microbiol.* **32**, 740–745.

Goodfellow, M. and Alderson, G. (1977). The actinomycete-genus *Rhodococcus*: a home for the '*rhodochrous*' complex. *J. Gen. Microbiol.* **100**, 99–122.

Goodfellow, M., Collins, M. D. and Minnikin, D. E. (1976). Thin-layer chromatographic analysis of mycolic acid and other long-chain components in whole-organism methanolysates of coryneform and related taxa. *J. Gen. Microbiol.* **96**, 351–358.

Goren, M. B. and Brennan, P. J. (1979). Mycobacterial lipids: chemistry and biologic activities. In: *Tuberculosis* (G. P. Youmans, ed.), pp. 63–193. W. B. Saunders Co., Philadelphia.

Guerrant, G. O., Lambert, M. A. and Moss, C. W. (1981). Gas-chromatographic analysis of mycolic acid cleavage products in mycobacteria. *J. Clin. Microbiol.* **13**, 899–907.

Hamid, M., Minnikin, D. E. and Goodfellow, M. (1993). A simple chemical test to distinguish mycobacteria from other mycolic-acid-containing actinomycetes. *J. Gen. Microbiol.* **139**, 2203–2213.

Hanis, T., Smrz, M., Klir, P., Macek, K., Klima, J., Base, J. and Deyl, Z. (1988). Determination of fatty acids as phenacyl esters in rat adipose tissue and blood vessel walls by high-performance liquid chromatography. *J. Chromatogr.* **452**, 443–457.

Hecht, T. and Causey, A. (1976). Rapid method for the detection and identification of mycolic acids in aerobic actinomycetes and related bacteria. *J. Clin. Microbiol.* **4**, 284–287.

Kämpfer, P., Andersson, M. A., Rainey, F. A., Kroppenstedt, R. M. and Salkinoja-Salonen, M. (1999). *Williamsia muralis* gen. nov., sp. nov., isolated from the indoor environment of a children's day care centre. *Int. J. Syst. Bacteriol.* **49**, 681–687.

Kaneda, K., Naito, S., Imaizumi, S., Yano, I., Mizuno, S., Tomiyasu, I., Baba, T., Kusunose, E. and Kusunose, M. (1986). Determination of molecular species composition of C80 or longer-chain alpha-mycolic acids in *Mycobaterium* spp. by gas chromatography-mass spectrometry and mass chromatography. *J. Clin. Microbiol.* **24**, 1060–1070.

Kaneda, K., Imaizumi, S., Mizuno, S., Baba, T., Tsukamura, M. and Yano, I. (1988). Structure and molecular species composition of three homologous series of α-mycolic acids from *Mycobacterium* spp. *J. Gen. Microbiol.* **134**, 2213–2229.

Kanetsuna, F. and Bartoli, A. (1972). A simple chemical method to differentiate *Mycobacterium* from *Nocardia*. *J. Gen. Microbiol.* **70**, 209–212.

Keddie, R. M. and Cure, G. L. (1977). The cell wall composition and distribution of free mycolic acids in named strains of coryneform bacteria and in isolates from various natural sources. *J. Appl. Bacteriol.* **42**, 229–252.

Lanéelle, M.-A., Asselineau, J. and Castelnuovo, G. (1965). Études sur les mycobactéries et les nocardiae. IV. Composition des lipides de *Mycobacterium rhodochrous, M. pellegrino* sp., et de quelques souches de nocardiae. *Ann. Inst. Pasteur (Paris)* **108**, 69–82.

Lambert, M. A., Moss, C. W., Silcox, V. A. and Good, R. C. (1986). Analysis of mycolic acid cleavage products and cellular fatty acids of *Mycobacterium* species by capillary gas chromatography. *J. Clin. Microbiol.* **23**, 731–736.

Laval, F., Lanéelle, M.-A., Déon, C., Monsarrat, B. and Daffé, M. (2001). Accurate molecular mass determination of mycolic acids by MALDI-TOF mass spectrometry. *Anal. Chem.* **73**, 4537–4544.

Lechevalier, M. P., Horan, A. C. and Lechevalier, H. (1971). Lipid composition in the classification of nocardiae and mycobacteria. *J. Bacteriol.* **105**, 313–318.

Lechevalier, M. P., Lechevalier, H. and Horan, A. C. (1973). Chemical characteristics and classification of nocardiae. *Can. J. Microbiol.* **19**, 965–972.

Lederer, E., Pudles, J., Barbezat, S. and Trillat, J. J. (1952). Sur la constitution chimique de l'acide coryno-mycolique de bacille diphthérique. *Bull. Soc. Chim. Fr.* **19**, 93–95.

Lehmann, K. B. and Neumann, R. (1896). *Atlas und Grundriss der Bakteriologie und Lehrbuch der speciellen bakteriologischen Diagnostik.* 1st ed. and J. F. Lehmann, München.

McNeil, M. R., Daffé, M. and Brennan, P. J. (1991). Locations of the mycolyl ester substituents in the cell wall of mycobacteria. *J. Biol. Chem.* **266**, 13217–13223.

Michel, G., Bordet, C. and Lederer, E. (1960). Isolement d'un nouvel acide mycolique: l'acide nocardique, à partir d'une souche de *Nocardia asteroides. Comp. Rend. Acad. Sci.* **250**, 3518–3520.

Minnikin, D. E. (1982). Lipids: complex lipids, their chemistry, biosynthesis and roles. In: *The Biology of the Mycobacteria* (C. Ratledge and J. L. Stanford, eds.), pp. 95–184. Academic Press, London.

Minnikin, D. E. (1993). Mycolic acids. In: *CRC Handbook of Chromatography: Analysis of Lipids* (K. D. Mukherjee and N. Weber, eds.), pp. 339–348. CRC Press, Boca Raton.

Minnikin, D. E. and Goodfellow, M. (1980). Lipid composition in the classification and identification of acid-fast bacteria. In: *Microbiological Classification and Identification* (M. Goodfellow and R. G. Board, eds.), pp. 189–256. Academic Press, London.

Minnikin, D. E. and Patel, D. V. (1979). Butyldimethylsilyl ethers of iodine-catalysed solvolysis products of long-chain epoxides. *Chem. Phys. Lipid* **23**, 173–178.

Minnikin, D. E. and Polgar, N. (1966). Stereochemical studies on the mycolic acids. *Chem. Commun.*, 648–649.

Minnikin, D. E. and Polgar, N. (1967). The methoxymycolic and ketomycolic acids from human tubercle bacilli. *Chem. Commun.*, 1172–1174.

Minnikin, D. E., Alshamaony, L. and Goodfellow, M. (1975). Differentiation of *Mycobacterium, Nocardia* and related taxa by thin-layer chromatographic analysis of whole-organism methanolysates. *J. Gen. Microbiol.* **88**, 200–204.

Minnikin, D. E., Hutchinson, I. G., Caldicott, A. B. and Goodfellow, M. (1980). Thin-layer chromatography of methanolysates of mycolic acid containing bacteria. *J. Chromatogr.* **188**, 221–233.

Minnikin, D. E., Minnikin, S. M. and Goodfellow, M. (1982a). The oxygenated mycolic acids of *Mycobacterium fortuitum, M. farcinogenes* and *M. senegalense. Biochim. Biophys. Acta* **712**, 616–620.

Minnikin, D. E., Minnikin, S. M., Goodfellow, M. and Stanford, J. L. (1982b). The mycolic acids of *Mycobacterium chelonei. J. Gen. Microbiol.* **128**, 817–822.

Minnikin, D. E., Minnikin, S. M., O'Donnell, A. G. and Goodfellow, M. (1984a). Extraction of mycobacterial mycolic acids and other long-chain compounds by an alkaline methanolysis procedure. *J. Microbiol. Meth.* **2**, 243–249.

Minnikin, D. E., Minnikin, S. M., Hutchinson, I. G., Goodfellow, M. and Grange, J. M. (1984b). Mycolic acid patterns of representative strains of *Mycobacterium fortuitum*, *"Mycobacterium peregrinum"* and *Mycobacterium smegmatis*. *J. Gen. Microbiol.* **130**, 363–367.

Minnikin, D. E., Minnikin, S. M., Parlett, J. H., Goodfellow, M. and Magnusson, M. (1984c). Mycolic acid pattern of some species of *Mycobacterium*. *Arch. Microbiol.* **139**, 225–231.

Minnikin, D. E., Minnikin, S. M., Parlett, J. H. and Goodfellow, M. (1985). Mycolic acid patterns of some rapidly-growing species of *Mycobacterium*. *Zentrabl. Bakteriol. Hyg. A* **259**, 446–460.

Minnikin, D. E., Kremer, L., Dover, L. G. and Besra, G. S. (2002). The methyl-branched fortifications of *Mycobacterium tuberculosis*. *Chem. Biol.* **9**, 545–553.

Moody, B. D., Besra, G. S., Wilson, I. A. and Porcelli, S. A. (1999). The molecular basis of CD1-mediated presentation of lipid antigens. *Immunol. Rev.* **172**, 285–296.

Nikaido, H. (1994). Prevention of drug access to bacterial targets: permeability barriers and active efflux. *Science* **264**, 382–388.

Nishiuchi, Y., Baba, T., Hotta, H. H. and Yano, I. (1999). Mycolic acid analysis in *Nocardia* species. The mycolic acid compositions of *Nocardia asteroides*, *N. farcinica*, and *N. nova*. *J. Microbiol. Meth.* **37**, 111–122.

Nishiuchi, Y., Baba, T. and Yano, I. (2000). Mycolic acids from *Rhodococcus*, *Gordonia*, and *Dietzia*. *J. Microbiol. Meth.* **40**, 1–9.

Rainey, F. A., Klatte, S., Kroppenstedt, R. M. and Stackebrandt, E. (1995). *Dietzia*, a new genus including *Dietzia maris* comb. nov., formerly *Rhodococcus maris*. *Int. J. Syst. Bacteriol.* **45**, 32–36.

Soddell, J. A., Stainsby, F. M., Eales, K. L., Kroppenstedt, R. M., Seviour, R. J. and Goodfellow, M. (2006). *Millisia brevis* gen. nov., sp. nov., an actinomycete isolated from activated sludge foam. *Int. J. Syst. Evol. Microbiol.* **56**, 739–744.

Stackebrandt, E., Smida, J. and Collins, M. D. (1988). Evidence of phylogenetic heterogeneity within the genus *Rhodococcus*: revival of the genus *Gordona* (Tsukamura). *J. Gen. Appl. Microbiol.* **34**, 341–348.

Stodola, F. H., Lesuk, A. and Anderson, R. J. (1938). The chemistry of the lipids of tubercle bacilli. LIV. The isolation and properties of mycolic acid. *J. Biol. Chem.* **126**, 505–513.

Tamura, S. (1913). Zur Chemie der Bakterien. I. Mitteilung. *Hoppe-Seyler's Zeitschr. Physiol. Chem.* **87**, 85–114.

Tomiyasu, I. and Yano, I. (1984). Separation and analysis of novel polyunsaturated mycolic acids from a psychrophilic, acid-fast bacterium *Gordona aurantiaca*. *Eur. J. Biochem.* **139**, 173–180.

Trevisan, V. (1889). I *Generi e le Specie delle Battieriacee*. Zanaboni and Gabuzzi, Milano.

Tsukamura, M. (1971). Proposal of a new genus, *Gordona*, for slightly acid-fast organisms occurring in sputa of patients with pulmonary disease and in soil. *J. Gen. Microbiol.* **68**, 15–26.

Ueda, S., Fujiwara, N., Naka, T., Sakaguchi, I., Ozeki, Y., Yano, I., Kasama, T. and Kobayashi, K. (2001). Structure–activity relationship of mycoloyl glycolipids derived from *Rhodococcus* sp. 4306. *Microb. Pathog.* **30**, 91–99.

Watanabe, M., Aoyagi, Y., Ridell, M. and Minnikin, D. E. (2001). Separation and characterization of individual mycolic acids in representative mycobacteria. *Microbiology* **147**, 1825–1837.

Wood, R. and Lee, T. (1983). High-performance liquid chromatography of fatty acids: quantitative analysis of saturated, monoenoic, polyenoic and geometrical isomers. *J. Chromatogr.* **254**, 237–246.

Yano, I. and Saito, K. (1972). Gas chromatographic and mass spectrometric analysis of molecular species of corynomycolic acids from *Corynebacterium ulcerans*. *FEBS Lett.* **23**, 352–356.

Yano, I., Saito, K., Furukawa, Y. and Kusunose, M. (1972). Structural analysis of molecular species of nocardomycolic acids of *Nocardia erythropolis* by the combined system of gas chromatography and mass spectrometry. *FEBS Lett.* **21**, 215–219.

Yano, I., Kageyama, K., Ohno, Y., Masui, M., Kusunose, M., Kusunose, E. and Akimori, N. (1978). Separation and analysis of molecular species of mycolic acids in *Nocardia* and related taxa by gas chromatography mass spectrometry. *Biomed. Mass Spectrom.* **5**, 14–24.

Yano, I., Imaeda, T. and Tsukamura, M. (1990). Characterization of *Nocardia nova*. *Int. J. Syst. Bacteriol.* **40**, 170–174.

Yassin, A. F., Binder, C. and Schaal, K. P. (1993). Identification of mycobacterial isolates by thin-layer and capillary gas-liquid chromatography under diagnostic routine conditions. *Zentralbl. Bakteriol.* **278**, 34–48.

11 Polyamines

Hans-Jürgen Busse
Institut für Bakteriologie, Mykologie und Hygiene, Department für Pathobiologie, Veterinärmedizinische Universität Wien, Wien, Austria

♦♦

CONTENTS

♦♦♦♦♦♦ I. INTRODUCTION

Polyamines in a wide concentration range are found in the majority of prokaryotes. In some extremely halophilic *Archaea* these cellular components are either absent or found in only trace amounts (Hamana *et al.*, 1985). Knowledge on polyamines in *Escherichia coli* and several eukaryotic microorganisms has been excellently reviewed by Tabor and Tabor (1985). More recent studies have indicated that polyamines may specifically and unspecifically interact with DNA in stabilizing its structure and conformation and affecting expression of certain genes (Feuerstein *et al.*, 1991; Eraso and Kaplan, 2009). Furthermore, polyamines enable bacteria to react to osmotic shock that leads to an intercellular increase of inorganic cations, which is compensated by excretion of the polycationic polyamines (Tkachenko *et al.*, 1997; Schiller *et al.*, 2000). In agreement with this finding are the observations that some halophilic species, *Vibrio alginolyticus*, *Vibrio parahaemolyticus*, *Halomonas marina* (formerly *Deleya marina*) and *Rhodothalassium salexigens*, have intracellular polyamine content that is inversely related to the salinity of the growth medium (Yamamoto *et al.*, 1986; Auling *et al.*, 1991a; Hamana *et al.*, 2001).

In one of the first reports dealing with polyamine patterns in different bacteria, Yamamoto *et al.* (1979) showed that three *Vibrio* and three *Beneckea* species exhibit the presence of the unusual triamine *sym*-norspermidine (Table 1). Independently from these results, Baumann *et al.* (1980) proposed reclassification of all *Beneckea* species in the genus *Vibrio*. Later, Yamamoto *et al.* (1983)

METHODS IN MICROBIOLOGY, VOLUME 38
0580-9517 DOI 10.1016/B978-0-12-387730-7.00011-5

Table 1. Polyamines most commonly detected in bacteria

Structure	Name	Abbreviations
$H_2N-(CH_2)_3-NH_2$	1,3-Diaminopropane	DAP, C3
$H_2N-(CH_2)_4-NH_2$	Putrescine	PUT, C4
$H_2N-CH_2-CHOH-(CH_2)_2-NH_2$	2-Hydroxyputrescine	HPUT, OH–C4
$H_2N-(CH_2)_5-NH_2$	Cadaverine	CAD, C5
$H_2N-(CH_2)_3-NH-(CH_2)_3-NH_2$	*sym*-Norspermidine	NSPD, C3C3
$H_2N-(CH_2)_3-NH-(CH_2)_4-NH_2$	Spermidine	SPD, C3C4
$H_2N-(CH_2)_4-NH-(CH_2)_4-NH_2$	*sym*-Homospermidine	HSPD, C4C4
$H_2N-(CH_2)_3-NH-(CH_2)_4-NH-(CH_2)_3-NH_2$	Spermine	SPM, C3C4C3

showed that also other *Vibrio* species are characterized by the presence of *sym*-norspermidine, whereas this polyamine was lacking in representatives of related genera, including *Aeromonas* and *Plesiomonas* and two *Vibrio* species. These two species, *Vibrio fischeri* and *Vibrio costicola*, were later reclassified as species of the genera *Aliivibrio* and *Salinivibrio*, respectively (Urbanczyk *et al.*, 2007; Huang *et al.*, 2000). By examination of *Vibrio* strains and species, Yamamoto *et al.* (1991) substantiated the finding that the presence of *sym*-norspermidine is a common trait of the genus.

Scherer and Kneifel (1983) reported on the polyamines in methanogenic *Archaea* and showed that the predominant polyamine in two *Methanococcus* species is spermidine, whereas the polyamine patterns of *Methanococcus mazei* and *Methanosarcina barkeri* contain predominantly putrescine and significant amounts of *sym*-homospermidine (Table 1). In contrast, other methanogenic taxa were shown to contain at least tenfold lower polyamine concentrations. Results obtained from additional methanogenic archaea published by Kneifel *et al.* (1986) were in accordance with these previous observations. These authors also confirmed previous observations reported by Hamana *et al.* (1985) that polyamines are not detectable or are present only in trace amounts in extremely halophilic *Archaea*.

In a comprehensive study covering numerous representatives of the classes Alphaproteobacteria, Betaproteobacteria and Gammaproteobacteria, Busse and Auling (1988) showed that within these taxa (formerly designated subclasses) several different polyamine patterns can be found that correspond well with the phylogenetic relationships of the examined bacterial species. It was demonstrated that the polyamine patterns of different isolates then classified as *Pseudomonas* species reflected their distribution over the three classes. Furthermore, it was shown that representatives of the class Betaproteobacteria contain the unusual diamine 2-hydroxyputrescine.

Usually when polyamine data are reported, the concentration of each compound is given in μmol/g dry weight or wet weight of biomass. In most cases, this information is less important for classification. More important is the identification of the predominant compounds in the polyamine patterns, but in certain cases minor components may be of significance as well. For instance, representatives of the genus *Pasteurella* sensu stricto share the presence of minor amounts of *sym*-norspermidine that is absent in species of closely related genera of the family *Pasteurellaceae* (Busse *et al.*, 1997).

♦♦♦♦♦♦ II. ANALYSIS OF POLYAMINES

A. Production of Biomass and Extraction of Polyamines

Busse and Auling (1988) showed that the composition of the growth medium has little influence on the qualitative polyamine composition. The standard medium used for cultivation is PYE broth (0.3% peptone from casein, 0.3% yeast extract, pH 7.2), a cheap, easy-to-prepare growth medium on which a wide range of bacteria are able to grow. If required, the strength of the medium can be easily adapted by increasing medium strength 3.3-fold ($3.3 \times$ PYE, containing similar amounts of nutrients like trypticase soy broth) or reducing the strength to a tenth ($0.1 \times$ PYE, useful for cultivation of oligotrophic bacteria), supplemented with NaCl for halophilic organisms or with specific carbon sources. Furthermore, Busse and Auling (1988) showed that the overall polyamine content and amounts of different polyamines depend on the physiological age at which the biomass is harvested, with high polyamine contents in the early–mid exponential growth phase and sometimes greatly reduced amounts at the stationary growth phase. Considering this observation, Busse and Auling (1988) recommended the standardization of the physiological age of the biomass and the extraction of polyamines from biomass harvested from late exponential growth phase as determined by measurement of the optical density of the culture.

Polyamines are extracted using perchloric acid, and the polyamine patterns are usually quantitatively analysed by high-performance liquid chromatography (Scherer and Kneifel, 1983; Hamana *et al.*, 1985). In addition, a qualitative analysis using thin-layer chromatography has been published (Nigam *et al.*, 2010).

Extraction and analysis of polyamines described below are based on the method reported by Scherer and Kneifel (1983) and slightly modified by Busse and Auling (1988).

- Harvest the cells at approximately 70% of the maximum optical density of the liquid culture, to standardize the physiological age of the biomass to be subjected to polyamine extraction.
- Harvest the cells by centrifugation and wash once with 0.9% NaCl solution.
- Lyophilize the biomass and transfer 40 mg to a tube.
- Add internal standard 1,8-diaminooctane (1 μmol/40 mg biomass) and 1 ml perchloric acid ($HClO_4$ [0.2 M]).
- Close the tube tightly and incubate at 100°C for 30 min with thorough shaking after 15 min.
- Remove cell debris by centrifugation.
- Mix 200 μl of the supernatant with 300 μl Na_2CO_3 (100 mg/ml) and 800 μl 5-dimethylamino-naphthalene-1-sulfonyl chloride (dansyl chloride; 7.5 mg/ml in acetone).
- Incubate for 20 min at 60°C.

Polyamines

- Add 100 μl of proline solution (50 mg/ml) and continue incubation at 60°C for another 10 min, in order to bind excess dansyl chloride.
- Cool the reaction mixture to 4°C.
- Add 100 μl toluene and extract the polyamines into the toluene phase by shaking.
- Inject 10 μl from the upper toluene phase into the high-performance liquid chromatograhy (HPLC).

B. High-Performance Liquid Chromatograhy for Analysis of Polyamine Patterns

For quantitative analysis of dansylated polyamines, the HPLC must be equipped with a gradient system, a fluorescence detector, a column oven and a 250 mm × 4.6 mm reversed phase column ODS RP18 (5-μm particles). The adjustments of the fluorescence detector are excitation wavelength, 312–450 nm with a filter at 360 nm; emitted light is passed through a cut-off filter at 450 nm. Separation of the polyamine extract is carried out at a column temperature of 40°C and a linear gradient of acetonitrile: H_2O (40:60 to 85:15) is applied for 35 min, followed by 15 min elution with the 85:15 mixture. With this system, the polyamines 2-hydroxyputrescine (1,4-diamino-2-butanol), 1,3-diaminopropane, putrescine (1,4-diaminobutane), cadaverine (1,5-diaminopentane), *sym*-norspermidine (N-(3-aminopropyl)propane-1,3-diamine), spermidine N-(3-aminopropyl)butane-1,4-diamine, *sym*-homospermidine N-(4-aminobutyl)butane-1,4-diamine and spermine (N,N′-bis(3-aminopropyl)butane-1,4-diamine) can be reliably separated and identified. Amounts of different polyamines are calculated relative to the area under the internal standard peak 1,8-diaminooctane. To determine the amounts for several polyamines, correction factors have to be applied (Busse and Auling, 1988): 2.8 for 2-hydroxyputrescine; 1.8 for 1,3-diaminopropane; 1.4 for putrescine; 1.2 for cadaverine, 1.0 for *sym*-norspermidine, spermidine and *sym*-homospermidine; and 0.8 for spermine.

Except 2-hydroxyputrescine and *sym*-homospermidine, all polyamines are commercially available. They can be dansylated as described above and the derivatized polyamines can be applied to calibrate the HPLC analysis system and to allow analysis of polyamine patterns of unknown samples based on retention times. Retention times of 2-hydroxyputrescine and *sym*-homospermidine can be identified by extraction of polyamines from bacterial species that have been described to contain one or the other polyamine as a major compound in the polyamine pattern. 2-Hydroxyputrescine is a major compound in the polyamine patterns of *Burkholderia cepacia*, *Achromobacter denitrificans* and *Comamonas testosteroni*, whereas *sym*-homospermidine is predominant in *Rhizobium leguminosarum*, *Brevundimonas diminuta* and *Sphingomonas paucimobilis* (Busse and Auling, 1988). Chromatograms of polyamine patterns of representatives of each, an alphaproteobacterium, a betaproteobacterium and a gammaproteobacterium, obtained after HPLC analysis according to the method described above are reported by Busse *et al.* (1989). Retention times are dependent on the HPLC apparatus applied. However, 2-hydroxyputrescine is usually released from the column 12–13 min earlier than the internal standard 1,8-diaminooctane; putrescine is released 7–8 min earlier than the internal standard; 1,3-diaminopropane is released 1 min earlier than putrescine; putrescine is released 1.7 min earlier than cadaverine; spermidine is released approximately 3.5 min later than the internal standard and *sym*-norspermidine and *sym*-homospermidine are released 1 min earlier and 1 min later, respectively, than spermidine; and spermine is released approximately 11 min later than the internal standard.

♦♦♦♦♦♦ III. POLYAMINES IN BACTERIA

The following information could be of use to researchers in the interpretation of data they obtain using the methodologies described above.

A. Proteobacteria

Within the phylum *Proteobacteria*, all polyamines listed in Table 1 have been found and numerous different polyamine patterns or combinations have been reported. Alphaproteobacteria species usually contain predominantly triamines, either spermidine (SPD) and/or *sym*-homospermidine (HSPD), and often, in addition, high amounts of putrescine (PUT). In contrast, taxa of the class Betaproteobacteria are rather homogeneous in containing 2-hydroxyputrescine (HPUT), putrescine and rarely significant amounts of spermidine. However, few species of this class have been reported to lack HPUT. A higher variability is found in the class Gammaproteobacteria where the majority of taxa have the major polyamine SPD, but numerous different polyamine patterns have been reported consisting of the characteristic PUT and SPD in the genus *Pseudomonas* and its nearest relatives, 1,3-diaminopropane (DAP) in *Acinetobacter* and several taxa within the family *Pasterurellaceae*, cadaverine (CAD) in *Stenotrophomonas* or the presence of *sym*-norspermidine (NSPD) in *Pasteurella* sensu stricto and *Vibrio*. So far, data on the classes Deltaproteobacteria and Epsilonproteobacteria are rather limited; SPD or HSPD is predominant in representatives of the Deltaproteobacteria, but high amounts of CAD may be detected in some taxa. Taxa of the Epsilonproteobacteria predominantly contain the polyamine SPD.

1. Alphaproteobacteria

Within the taxa of the class Alphaproteobacteria, seven different polyamine patterns have been reported (Table 2). These polyamine patterns are characterized by either a single predominant compound (SPD or HSPD), a pair of major polyamines (SPD/HSPD, SPD/PUT, HSPD/PUT and SPD/SPM) or the presence of three major polyamines (PUT/SPD/HSPD). Generally these polyamine patterns are genus-, family- or order-specific, but within the genus *Rhizobium* two different polyamine patterns have been reported.

A polyamine pattern with the single major polyamine SPD has been detected in genera of the families *Acetobacteriaceae*, *Beijerinckiaceae*, *Rhizobiaceae*, *Rhodobacteraceae*, *Hyphomonadaceae*, *Erythrobacteraceae*, *Sphingomonadaceae* and *Rhodospirillaceae*. HSPD is the only major polyamine in taxa of the families *Beijerinckiaceae*, *Rhizobiaceae*,

Table 2. Polyamine patterns within representatives of the Alphaproteobacteria. Only major or diagnostic polyamines are listed. Taxonomic affiliation according to Garrity *et al.* (2007) and/or Yarza *et al.* (2008)

Taxonomic affiliation	Major polyamine(s)	Reference(s)
Acetobacteriaceae		
Acteobacter, Asaia, Craurococcus, Gluconobacter, Gluconacetobacter, Kozakia, Paracraurococcus, Roseomonas, Saccharibacter	SPD, (PUT)[a]	Hamana *et al.*, 2003b (2006a, 2006b) and Sánchez-Porro *et al.* (2009)
Rhodopila	SPD, HSPD	Hamana and Takeuchi (1998)
Zavarzinia	PUT, HSPD	Auling *et al.* (1988) and Hamana *et al.* (2003b)
Brucellaceae		
Brucella, Ochrobactrum, Mycoplana ramosa[b]	PUT, SPD, HSPD	Kämpfer *et al.* (2007a, b, 2008); Scholz *et al.* (2008, 2010); Huber *et al.* (2010)
Pseudochrobactrum, Paenochrobactrum	PUT, SPD	Kämpfer *et al.* (2006, 2007b, 2009a, 2010a)
Beijerinckiaceae		Hamana *et al.* (2006b)
Beijerinckia	HSPD	Hamana *et al.* (1988)
Methylocapsa	SPD	
Rhizobiaceae		
Rhizobium, Carbophilus, Ensifer	HSPD, (PUT)	Auling *et al.* (1988); Busse and Auling (1988); and Hamana *et al.* (2003b)
Rhizobium radiobacter, Rhizobium undicola, Rhizobium rubi, Rhizobium vitis	SPD, HSPD[b]	Busse and Auling (1988) and Hamana *et al.* (2003b)
Mycoplana dimorpha[b]	SPD, HSPD	
Kaistia, Blastobacter capsulatus	SPD, (PUT)	Hamana *et al.* (2003b)
Phyllobacteriaceae		
Aminobacter, Phyllobacterium	HSPD, PUT, (SPD)	Busse and Auling (1988); Auling *et al.* (1991a); and Hamana and Kishimoto (1996)
Pseudaminobacter	HSPD, PUT, SPD	Kämpfer *et al.* (1999) and Hamana *et al.* (2006b)
Mesorhizobium	HSPD, (PUT, CAD)	Hamana and Takeuchi (1998)
Bradyrhizobiaceae		
Bradyrhizobium, Nitrobacter, Oligotropha, Rhodopseudomonas	HSPD, (PUT, SPD)	Auling *et al.* (1988, 1991a); Busse and Auling (1988); and Hamana *et al.* (2003b)
Rhodobacteraceae		
Catellibacterium, Jannaschia, Paracoccus, Pelagibaca, Pseudorhodobacter, Pseudovibrio, Rhodobacter, Rhodovulum, Roseibium, Roseibacterium, Roseivarius, Yangia	PUT, SPD, (CAD)	Busse and Auling (1988); Hamana and Matsuzaki (1990); Hamana and Takeuchi (1998); and Hamana *et al.* (2003b, 2006b)
Roseobacter, Pseudovibrio denitrificans	SPD	Hamana and Takeuchi (1998)

(*Continued*)

Table 2. (Continued)

Taxonomic affiliation	Major polyamine(s)	Reference(s)
Hyphomonadaceae		
Hyphomonas	SPD	Hamana and Takeuchi (1998)
Xanthobacteraceae		
Ancylobacter, Starkeya, Xanthobacter	HSPD, (PUT, SPD)	Busse and Auling (1988); Hamana and Takeuchi (1998); and Hamana *et al.* (2003b)
Sphingomonadaceae		
Sphingomonas, Sphingosinicella	HSPD	Busse and Auling (1988); Busse *et al.* (1999); Takeuchi *et al.* (2001); Hamana *et al.* (2003b), 2006b; and Geueke *et al.* (2007)
Sphingobium, Novosphingobium, Sphingopyxis, Blastomonas	SPD	Busse and Auling (1988); Hamana and Takeuchi (1998); Busse *et al.* (1999); Takeuchi *et al.* (2001); and Hamana *et al.* (2003b, 2006b)
Erythrobacteraceae		
Altererythrobacter, Erythrobacter, Erythromicrobium, Porphyrobacter	SPD	Hamana *et al.* (2006b) and Kumar *et al.* (2008)
Methylobacteriaceae		
Methylobacterium	HSPD, (PUT)	Busse and Auling (1988) and Auling *et al.* (1991a)
Rhodobiaceae		
Rhodobium	SPD, SPM	Urdiain *et al.* (2008)
Caulobacteraceae		
Asticcacaulus, Phenylobacterium, Brevundimonas	HSPD, (SPD)	Busse and Auling (1988); Hamana *et al.* (2006b) and Segers *et al.* (1994)
Brevundimonas nasdae, Brevundimonas vesicularis	SPD	Segers *et al.* (1994); and Hamana *et al.* (2006b)
Rhodospirillaceae		
Rhodospirillum, Terasakiella	PUT, SPD	Busse and Auling (1988) and Hamana et al. (2006b)
Azospirillum	HSPD, PUT, (SPD)	Busse and Auling (1988)
Defluviicoccus, Hoeflea	SPD	Hamana *et al.* (2006b)
Tistrella	CAD, SPD, HSPD	Hamana *et al.* (2006b)

[a]Polyamine shown in brackets are present in some but not in all representatives of the corresponding group.
[b]Diagnostic minor comound.

Phyllobacteriaceae, *Bradyrhizobiaceae*, *Xanthobacteraceae*, *Sphingomonadaceae*, *Methylobacteriaceae* and *Caulobacteriaceae*. The major polyamines SPD/HSPD are present in *Rhodopila* (*Acetobacteraceae*), *Mycoplana dimorpha* (*Rhizobiaceae*) and *Caulobacteraceae*. The pair SPD/PUT characterizes some taxa of the *Acetobacteraceae, Brucellaceae, Rhodobacteraceae* and *Rhodospirillaceae*. The two major polyamines HSPD/PUT are present in *Zavarzinia* (*Acetobacteraceae*), *Rhizobiaceae, Phyllobacteriaceae, Bradyrhizobiaceae, Xanthobacteraceae, Methylobacteriaceae* and *Rhodospirillaceae*. So far, the genus *Rhodobium* is the only taxon within the *Alphaproteobacteria* that exhibits a polyamine pattern composed of the two major polyamines SPD/SPM. The polyamine pattern PUT/SPD/HSPD has been reported for several species of the genera of the family *Xanthobacteraceae* as well as species of the families *Phyllobacteriaceae*, *Bradyrhizobiaceae*, *Brucellaceae* and *Rhodospirillaceae*.

Originally, *Zavarzinia compransoris* (formerly *Pseudomonas compransoris*) was phylogenetically placed within the class Alphaproteobacteria but without assignment to any line of descent within this class (Auling *et al.*, 1988). Garrity *et al.* (2007) placed *Z. compransoris* in the family *Acetobacteraceae*. The majority of taxa within this family contain SPD as the major polyamine and HSPD is lacking. In contrast, the polyamine pattern of *Z. compransoris* exhibits a polyamine pattern with PUT and HSPD as the major components. This observation may suggest that *Z. compransoris* is misplaced and future examinations may result in its assignment to another family. The same might be true for *Rhodophila globiformis* that contains HSPD and PUT as major polyamines.

2. Betaproteobacteria

Polyamine patterns within the class Betaproteobacteria are rather homogenous (Busse and Auling, 1988; Auling *et al.*, 1991a; Hamana and Takeuchi, 1998, Hamana *et al.*, 2000a, 2006b). The majority of taxa within this class contain HPUT, which has so far been reported outside the *Betaproteobacteria* only for selected species of the gammaproteobacterial genus *Shewanella* and a *Colwellia* species (Table 3). HPUT may be present as a major compound or only in minor amounts. Furthermore, all species of this class contain PUT as a major polyamine. SPD and CAD may be present as major compounds, in moderate or minor amounts or even absent. The presence of SPD or CAD as a major compound does not indicate close phylogenetic relationships but appears to be randomly distributed over several higher taxa within the class Betaproteobacteria, possibly reflecting certain physiological properties that are related to the presence of these polyamines. SPM, if detected, is only present in minor amounts. However, a few taxa have been reported to lack the *Betaproteobacteria*-specific HPUT. These species include *Spirillum volutans*, *Oxalobacter formigenes*, *Eikenella corrodens*, *Kingella denitrificans*, *Kingella kingae*, *Kingella oralis*, *Vitreoscilla beggiatoides* and *Vitreoscilla stercoraria*. *S. volutans* and *O. formigenes* both are located on separate lines of descent within the Betaproteobacteria. The genera *Eikenella*, *Kingella* and *Vitreoscilla* form a separate branch within the family *Neisseriaceae* together with species of *Neisseria*, *Bergeriella*, *Simonsiella*, *Conchiformibius*, *Uruburuella*, *Alysiella* and *Stenoxybacter*. Other species of the family *Neisseriaceae* like members of the genera *Chromobacterium* and *Aquabacterium* show polyamine patterns with the Betaproteobacteria-specific diamine HPUT. If future analyses of polyamines show patterns lacking HPUT for the close relatives of *Eikenella*, *Kingella* and *Vitreoscilla*, an emended description of the family *Neisseriaceae* might be considered restricting the family to the genera *Neisseria*, *Eikenella*, *Kingella*, *Vitreoscilla*, *Bergeriella*, *Simonsiella*, *Conchiformibius*, *Uruburuella*, *Alysiella* and *Stenoxybacter* based on phylogenetic data and absence of HPUT as a family specific trait.

Table 3. Predominant compounds and diagnostic minor components in the polyamine patterns of representatives of the Gammaproteobacteria. Taxonomic affiliation according to Garrity *et al.* (2007) and/or Yarza *et al.* (2008)

Taxonomic affiliation	Major polyamine	Reference(s)
Pseudoalteromonadaceae		
Pseudoalteromonas	SPD, (SPM)[a]	Hamana *et al.* (2000b, 2006b)
Alteromonadaceae		
Alteromonas	CAD, SPD	Hamana *et al.* (2006b)
Psychromonadaceae		
Psychromonas	CAD, (NSPD, PUT)	Hamana *et al.* (2006b)
Idiomarinaceae		
Idiomarina	SPD	Hamana *et al.* (2006b)
Pseudidiomarina	PUT	Hamana *et al.* (2006b)
Shewanellaceae		
Shewanella	PUT, (CAD)	Hamana *et al.* (2000a, 2006b)
Shewanella gelidimarina	SPD	Hamana *et al.* (2006b)
Shewanella violaceae, S. gaetbuli, S. marinitestina, S. sediminis, S. schlegeliana, S. halifaxensis, S. morhuae	HPUT, (PUT)	Hamana *et al.* (2000a, 2006b)
Colwelliaceae		
Colwellia	Absent	Hamana *et al.* (2006b)
Colwellia maris	HPUT, PUT, SPD	Hamana *et al.* (2006b)
Moritellaceae		
Moritella	CAD	Hamana *et al.* (2006b)
Moritella yayanosii	NSPD	Hamana *et al.* (2006b)
Oceanospirillaceae		
Marinobacterium, Marinomonas, Oceanobacter	SPD, (PUT, CAD)	Hamana *et al.* (2006b)
Ferrimonadaceae		
Ferrimonas	SPD	Hamana *et al.* (2006b)
Vibrionaceae		
Vibrio, Photobacterium	NSPD, SPD	Yamamoto *et al.* (1983, 1991) and Hamana *et al.* (2000a, 2006b)
Halomonadaceae		
Halomonas, Chromohalobacter	SPD	Auling *et al.* (1991a) and Hamana *et al.* (2006b)
Xanthomonadaceae		
Dyella, Stenotrophomonas	CAD, SPD	Auling *et al.* (1991a); Hamana *et al.* (2006b); and Yang *et al.* (1993)
Aquimonas, Pseudoxanthomonas, Rhodanobacter, Xanthomonas, Frateuria, Pseudofulvimonas	SPD	Auling *et al.* (1991a); Yang *et al.* (1993); Hamana and Matsuzaki (1993); Hamana *et al.* (2006b); and Kämpfer *et al.* (2010b)

(Continued)

Table 3. (Continued)

Taxonomic affiliation	Major polyamine	Reference(s)
Pseudomonadaceae		
Pseudomonas, Azotobacter, Azomonas, Azorhizophilus	PUT, SPD, (CAD)	Busse and Auling (1988); Auling *et al.* (1991a); Goris *et al.* (1998); and Hamana *et al.* (2003a, 2006b)
Moraxellaceae		
Acinetobacter, Moraxella caprae, Moraxella lacunata	DAP	Busse and Auling (1988); Auling *et al.* (1991b); Hamana and Matsuzaki (1992); Hamana *et al.* (2003a, 2006b); and Kämpfer *et al.* (1991)
Moraxella, Psychrobacter	SPD	Hamana *et al.* (2006b)
Moraxella boevrei	DAP, SPD	Hamana *et al.* (2003a)
Psychrobacter urativorans	PUT	Hamana *et al.* (2006b)
Ectothiorhodospiraceae		
Ectothiorhodospira, Halorhodospira, Thioalkalivibrio	SPD, (SPM)	Hamana *et al.* (2006b)
Alcanivoraceae		
Kangiella	PUT, CAD, SPD	Hamana *et al.* (2006b)
Enterobacteriaceae		
Enterobacter, Citrobacter, Rahnella	PUT, CAD	Hamana (1996)
Escherichia, Yokenella	PUT, CAD, SPD	Hamana (1996)
Erwinia	(DAP, PUT, CAD, SPD)	Hamana (1996); Hamana *et al.* (2006b); Zherebilo *et al.* (2001)
Proteus, Providencia	PUT, (DAP, CAD, SPD)	Hamana (1996)
Serratia	DAP, PUT, CAD, (SPD)	Hamana (1996)
Tatumella	DAP, PUT, SPD	
Yersinia	PUT, (CAD, SPD)	Hamana (1996) and Sprague *et al.* (2008)
Klebsiella	PUT, CAD, (DAP, SPD)	Hamana (1996)
Raoultella, Hafnia	CAD, (DAP, PUT)	Hamana (1996)
Leminorella	PUT, SPD	Hamana (1996)
Brenneria, Pantoea, Pectobacterium, Dickeya, Morganella	PUT, (DAP, CAD)	Hamana (1996) and Zherebilo *et al.* (2001)
Photorhabdus, Xenorhabdus	PUT, SPD, (CAD)	Hamana *et al.* (2003a)
Thiotrichaceae		
Thiothrix disciformis	PUT, CAD, SPD	Hamana *et al.* (2006b)
Thiothrix flexilis	PUT, CAD	Hamana *et al.* (2006b)
Thiothrix eikelboomii, T. unzii, T. fructosivorans	SPD	Hamana *et al.* (2006b)
Pasteurellaceae		
Pasteurella sensu stricto	PUT, (SPD, DAP, CAD), NSPD[b]	Busse *et al.* (1997)
Actinobacillus sensu stricto, *Aggregatibacter, Bibersteinia, Haemophilus*	DAP	Busse *et al.* (1997)

(Continued)

Table 3. (Continued)

Taxonomic affiliation	Major polyamine	Reference(s)
Gallibacterium	PUT, SPD, (DAP, CAD)	Busse *et al.* (1997) and Bisgaard *et al.* (2009)
Avibacterium, Mannheimia	DAP, SPD	Busse *et al.* (1997)
Aeromonadaceae		
Aeromonas	DAP, PUT, (CAD)	Kämpfer *et al.* (1994)
Chromatiaceae		
Allochromatium, Chromatium, Thiohalocapsa	SPD, (SPM)	Hamana and Takeuchi (1998)
Legionellaceae		
Legionella	PUT, (SPD)	Hamana and Takeuchi (1998)

[a]Polyamine shown in brackets are present in some but not in all representatives of the corresponding group.
[b]Diagnostic minor compound.

3. Gammaproteobacteria

A large variety of polyamine patterns has been observed among the Gammaproteobacteria (Table 3). Major components can be DAP, PUT, CAD and SPD alone or in combination. Additionally, HPUT, HSPD and SPM may be present. The polyamine patterns are rarely homogeneous within families but often genus-specific. SPD, sometimes in combination with another polyamine, is the major polyamine in species of the families *Oceanospirillaceae, Ectothiorhodospiraceae, Chromatiaceae, Halomonadaceae,* certain genera of the *Xanthomonadaceae* and *Moraxellaceae,* the genera *Pseudoalteromonas, Idiomarina, Ferrimonas,* certain species of *Thiothrix* and a single species of the genus *Shewanella.* PUT is the characteristic major polyamine in *Pseudidiomarina, Legionella,* the majority of *Shewanella* species and a single *Psychrobacter* species. Several taxa of the families *Moraxellaceae* and *Pasteurellaceae* contain the single major component DAP. *Moritella* species contain predominantly CAD, but *Moritella yayanosii* is characterized by the major polyamine NSPD. CAD is also the major component in *Psychromonas* species, which may also contain NSPD and PUT. The major polyamines CAD and SPD are characteristic of the genus *Alteromonas,* and also of the family *Xanthomonadaceae* genera, *Stenotrophomonas* and *Dyella,* which clearly distinguishes them from other genera of the family. Several *Shewanella* species (and *Colwellia maris*) contain HPUT, a diamine that is a characteristic trait for the majority of *Betaproteobacteria* but not detected in other bacterial lineages. However, the presence of this diamine does not reflect phylogenetic relationships of these species within the genus *Shewanella.* Species of the family *Pseudomonadaceae* contain the major polyamines PUT and SPD and often also CAD. Species of the genus *Aeromonas* contain DAP and PUT as major components, and significant amounts of CAD may also be present. *Vibrio* species and several *Photobacterium* species contain the rare triamine NSPD as a predominant component. NSPD is also a diagnostic trait for the genus *Pasteurella* sensu stricto. Species of this group also contain the major polyamine PUT; the presence of additional major components DAP and CAD is variable. Other genera within the family *Pasteurellaceae* show a polyamine pattern either with DAP and SPD (*Avibacterium, Mannheimia*), or PUT, SPD and variable for DAP and CAD (*Gallibacterium*).

The family *Enterobacteriaceae* requires separate consideration. Almost all species of the assigned genera contain the major diamine PUT and the majority of them also contain major amounts of DAP and/or CAD and some also contain SPD. Unfortunately, different compositions of polyamine patterns are only rarely reflecting genus affiliations. Recently, we have carried out studies on the correlation between the physiological age and corresponding polyamine patterns of selected *Enterobacteriaceae* species, including *Yersinia enterocolitica, Yersinia ruckeri, Klebsiella pneumoniae, Enterobacter cloacae, Citrobacter freundii* and *Yokenella regenburgei* (Amann, Frischmann and Busse, unpublished results). The results from these studies demonstrated that the physiological age of the biomass from which polyamines were extracted can greatly influence the resulting polyamine pattern. In late exponential phase cells of *K. pneumoniae*, PUT is predominant and DAP is a second major polyamine. In contrast, CAD is predominant in late stationary phase cells and PUT is a second major polyamine. These changes in the polyamine contents are reflected by a twofold reduction of relative amounts of DAP and PUT in late stationary phase cells, whereas the amount of CAD increases 10fold. Applying the same test conditions, DAP is the major polyamine in *E. cloacae* and its relative amount is almost unaffected (reduced from 51% to 45%) by the growth phase of the cells. At the same time, CAD is increased from 2% to 22% and relative amounts of PUT are unaffected as well. In late-exponential phase cells of *Y. regensburgei*, PUT is the major polyamine (54%), whereas in stationary cells PUT and CAD are predominant ($33.5 \pm 1.3\%$). The relative amount of DAP is stable in both late exponential phase and stationary phase cells. These results indicated that part of the PUT proportion is replaced by CAD in stationary phase cells. In *C. freundii*, cells of the two physiological ages show the same polyamine patterns with DAP and PUT predominant and traces of CAD. In *Y. enterocolitica* subsp. *palearctica*, DAP is absent or detectable only in trace amounts. PUT is the major polyamine decreasing from 94% in early exponential phase cells to 84% in stationary phase cells. At the same time, the relative amount of CAD is increasing from 3% to 14%, indicating partial replacement of PUT by CAD. In another *Yersinia* species, *Y. ruckeri*, PUT was a major polyamine in early exponential phase cells, mid exponential phase cells and early stationary phase cells. During the same growth phases, the content of DAP decreased from 48% to 3%, whereas the content of CAD increased from 6% to 58%, demonstrating an almost complete replacement of DAP by CAD. Comparable results were also obtained with additional strains of *Y. ruckeri* demonstrating species specificity.

These data demonstrate that in *Enterobacteriaceae* species, the polyamine pattern is heavily dependent on the physiological age of the cells from which polyamines are extracted. A polyamine pattern obtained from mid-exponential phase cells may be clearly different from that of stationary phase cells of the same strain. Hence, a strain might be classified with different groups depending on the physiological age at which the biomass was harvested. Analysis of

Table 4. Predominant compounds in the polyamine patterns of representatives of the Deltaproteobacteria. Taxonomic affiliation according to Garrity *et al.* (2007) and/or Yarza *et al.* (2008)

Taxonomic affiliation	Major polyamine(s)	Reference(s)
Bacteriovoracaceae		
Bacteriovorax stolpii	PUT	Hamana and Takeuchi (1998)
Bdellovibrio bacteriovorus	PUT, SPD	Hamana and Takeuchi (1998)
Peredibacter starrii	SPD, HSPD	Hamana and Takeuchi (1998)
Myxococcaceae		
Archangium, Corallococcus, Cystobacter, Melittangium, Myxococcus, Stigmatella	HSPD	Hamana and Takeuchi (1998)
Desulfovibrionaceae		
Desulfosarcina cetonica	CAD, HSPD	Hamana *et al.* (2006b)
Desulfobaculum vacuolatum, Desulfovibrio desulfuricans	CAD, SPD	Hamana *et al.* (2006b)
Desulfovibrio, Desulfococcus, Desulfobulbus	SPD	Hamana and Takeuchi (1998) and Hamana *et al.* (2006a)
Desulfuromonadaceae		
Geothermobacter	SPD, SPM	Hamana *et al.* (2006b)
Geopsychrobacter, Pelobacter, Desulfuromusa	SPD	Hamana *et al.* (2006b)
Geobacteraceae		
Geobacter	SPD	Hamana *et al.* (2006a, b)

biomass without indicating precisely the physiological age at which polyamines are extracted is not useful for classification of members of the family *Enterobacteriaceae*. The majority of polyamine data for species of the family *Enterobacteriaceae* listed in Table 3 were obtained from the article of Hamana (1996). In Hamana (1996), it is indicated that biomass was collected from exponential or stationary phase cells, but without any clear indication of the physiological age of the cells analysed. Hence, the data from Hamana (1996) cannot serve for a polyamine-based classification scheme for species of the family *Enterobacteriaceae*. Careful reanalysis of polyamine patterns based on data obtained from biomass standardized for its physiological age needs to be carried out in order to demonstrate the value of this approach for classification of species of the family *Enterobacteriaceae*.

4. Deltaproteobacteria

Polyamine data from the class Deltaproteobacteria are too limited to draw detailed conclusions for all associated families. However, available data indicate that polyamine pattern might be useful for classification within this class (Table 4). Only three representatives of the family *Bacteriovoraceae*, each representing a different genus, have been analysed, and each showed a different polyamine pattern. The genera *Desulfovibrio, Desulfococcus and Desulfobulbus* of the family *Desulfovibrionaceae* show a polyamine pattern with the predominant polyamine being SPD. Other polyamine patterns within this family contained either CAD and HSPD or CAD and SPD, but these traits were only found in one or two species so far analysed. Representatives of both the families *Desulfuromonadaceae* and *Geobacteraceae* contain the major polyamine SPD. High amounts of SPM in *Geothermobacter ehrlichii* might be related to the thermophily of this organism, since the tendency to produce long-chained polyamines appears to be a reaction to growth at higher temperatures (Hamana *et al.*, 1989; Hosoya *et al.*, 2004). Homogeneous polyamine patterns with the single major polyamine being HSPD have been reported for six genera of the family *Myxococcaceae*, suggesting this trait to be family specific.

5. Epsilonproteobacteria

From the class Epsilonproteobacteria, representatives of the families Helicobacteraceae and *Campylobacteraceae*, including *Helicobacter, Campylobacter, Arcobacter, Hydrogenimonas, Sulfospirillum, Sulfurimonas* and *Wolinella*, show rather homogeneous polyamine patterns (Hamana and Takeuchi, 1998; Hamana *et al.*, 2006a, 2006b). No significant differences are observed among species of these genera. All are characterized by the major component SPD, but some differences in the total amount of this polyamine were reported. In contrast, members of the family *Nautiliaceae*, including *Cuminibacter* and *Lebetimonas* species, predominantly contain SPM, but SPD can occur in significant amounts and the most deeply branching species of the family shows only SPD as the major polyamine and hence more closely resemble the species of the families *Helicobacterceae* and *Campylobacteraceae.*

B. Bacteroidetes

Within the phylum Bacteroidetes, the majority of taxa are characterized by a polyamine pattern that contains either HSPD or SPD as the predominant polyamine (Hamana *et al.*, 1995, 2008; Hamana and Nakagawa, 2001a, 2001b; Buczolits *et al.*, 2006; Hosoya and Hamana, 2008; Kämpfer *et al.*, 2009c).

HSPD is the major polyamine of representatives of the families *Flavobacteriaceae, Sphingobacteriaceae, Cyclobacteriaceae* and *Chitinophagaceae.* Species of the genera *Cytophaga* and *Hymenobacter* of the family *Cytophagaceae* also contain the major polyamine HSPD, but within the same family a clade comprising the genera *Arcicella, Dyadobacter, Runella, Spirosoma* and *Flectobacillus* is characterized by the major polyamine being SPD. All taxa of the *Bacteroidales,* including the families *Bacteroidaceae, Rikenellaceae, Porphyromonadaceae* and *Prevotellaceae,* as well as the *Saprospiraceae,* also contain polyamine patterns with SPD as the major compound. Within the family *Flammeovirgaceae,* the species of the genus *Flammerovirga* contain 50-fold lower polyamine content than the majority of other Bacteroidetes species. Two species of the genus *Flammeovirga, F. aprica* and *F. yaeyamenis* contain CAD as the major polyamine, whereas SPD is the major polyamine in *F. kamogawensis* and other species of the family *Flammeovirgaceae.*

C. Actinobacteria

The polyamine contents of numerous actinobacterial species are significantly lower than those of the Proteobacteria and Bacteroidetes. Available polyamine data for several taxa including species of the genera *Actinomyces*, *Arthrobacter*, *Microbacterium*, *Cellulomonas*, *Corynebacterium* and *Tsukamurella* do not establish the usefulness of this approach for the classification of these taxa due to the heterogeneity of polyamine patterns or very low polyamine contents (Hamana, 1995; Altenburger *et al.*, 1997). However, interesting polyamine patterns have been detected in representatives of several genera of the family *Microbacteriaceae*. *Agrococcus jenensis* is characterized by a polyamine pattern with the single major polyamine SPM that is so far unique among actinobacteria. The significance of this finding was later confirmed from polyamine analyses of two other *Agrococcus* species (Wieser *et al.*, 1999; Zlamala *et al.*, 2002). Another specific polyamine pattern (major compounds DAP and CAD) was detected in *Pseudoclavibacter helvolus* (formerly *Brevibacterium helvolum*). The third polyamine pattern within this family is characteristic of the genera *Clavibacter* and *Rathayibacter*. Species of these two genera have relatively high polyamine content and the major polyamines are SPD and SPM. The same polyamine patterns were also detected in *Curtobacterium* species, but in significantly lower concentrations. Major polyamines SPD and SPM also characterize the family *Propionibacteriaceae* (Busse and Schumann, 1999; Kämpfer *et al.*, 2009b). Representatives of the family *Nocardioidaceae* are characterized by the presence of the major polyamine CAD. Major amounts of PUT or SPD may also be present. The three examined representatives of the family *Intrasporangiaceae* all contain PUT as the major polyamine, but in *Terrabacter* PUT is the sole polyamine, while in *Terracoccus* almost equal amounts of PUT, SPD and SPM are detected. In the genus *Intrasporangium* PUT and SPD are predominant. However, further studies of polyamine patterns with additional representatives of the three genera have to be carried out to substantiate their possible use for identification. A characteristic polyamine pattern with PUT and CAD has been shown to be a common trait of species of the genus *Brevibacterium* (Altenburger *et al.*, 1997; Kämpfer *et al.*, 2010c).

♦♦♦♦♦♦ IV. FINAL REMARKS

The above summary of available polyamine data indicates that polyamine patterns are a useful tool for classification of bacteria. The advantage over all other chemotaxonomic approaches is the fact that polyamines are not or only rare constituents of the cell envelope but found dissolved in the cytoplasm or associated to DNA, RNA or ribosomes. Though polyamines' analyses might not provide helpful results in certain taxa, e.g. *Corynebacteriaceae*, *Microbacterium* and *Arthrobacter* species, there are other taxa where polyamines excellently reflect established phylogenetic relationships. A good overview of variations of polyamine patterns is available for the Proteobacteria and the Bacteroidetes. Concerning Actinobacteria, promising results are available, but a major shortcoming is the fact that so far only a very limited number of taxa have been examined for polyamines. Further studies are desirable to fill this gap. Compared to other chemotaxonomic approaches, polyamine patterns are more conserved than fatty acid profiles, and hence they cannot be applied for species differentiation. Usually polyamine patterns are also less conserved than polar lipid profiles, but in some families, e.g. the family *Brucellaceae*, they are of almost the same level of discriminatory power. On the other hand, polyamine patterns are useful for classification at the same taxonomic level as quinone systems or even lower. Inclusion of polyamine patterns for description of novel genera supplements their phenotypic classification and can support the development of a stable taxonomy.

References

Altenburger, P., Kämpfer, P., Akimov, V. N., Lubitz, W. and Busse, H.-J. (1997). Polyamine distribution in actinomycetes with group B peptidoglycan and species of the genera *Brevibacterium, Corynebacterium* and *Tsukamurella*. *Int. J. Syst. Bacteriol.* **47**, 270–277.

Auling, G., Busse, J., Hahn, M., Hennecke, H., Kroppenstedt, R.-M., Probst, A. and Stackebrandt, E. (1988). Phylogenetic heterogeneity and chemotaxonomic properties of certain gram-negative aerobic carboxydobacteria. *Syst. Appl. Microbiol.* **10**, 264–272.

Auling, G., Busse, H.-J., Pilz, F., Webb, L., Kneifel, H. and Claus, D. (1991a). Rapid differentiation, by polyamine analysis, of *Xanthomonas* strains from phytopathogenic pseudomonads and other members of the class *Proteobacteria* interacting with plants. *Int. J. Syst. Bacteriol.* **41**, 223–228.

Auling, G., Pilz, F., Busse, H.-J., Karrasch, S., Streichan, M. and Schön, G. (1991b). Analysis of the polyphosphate accumulating microflora in phosphorus eliminating anaerobic-aerobic activated sludge systems using diaminopropane (DAP) as a biomarker for rapid estimation of *Acinetobacter*. *Appl. Environ. Microbiol.* **57**, 3585–3592.

Baumann, P., Baumann, L., Bang, S. S. and Woolkalis, M. J. (1980). Reevaluation of the taxonomy of *Vibrio, Beneckea*, and *Photobacterium*: abolition of the genus *Beneckea*. *Curr. Microbiol.* **4**, 127–132.

Bisgaard, M., Korczak, B. M., Busse, H.-J., Kuhnert, P., Bojesen, A. M. and Christensen, H. (2009). Classification of the taxon 2 and taxon 3 complex of Bisgaard within *Gallibacterium* and description of *Gallibacterium melopsittaci* sp. nov., *Gallibacterium trehalosifermentans* sp. nov. and *Gallibacterium salpingitidis* sp. nov. *Int. J. Syst. Evol. Microbiol.* **59**, 735–744.

Buczolits, S., Denner, E. B. M., Kämpfer, P. and Busse, H.-J. (2006). Proposal of *Hymenobacter norwichensis* sp. nov., classification of '*Taxeobacter ocellatus*', '*Taxeobacter gelupurpurascens*' and '*Taxeobacter chitinovorans*' as *Hymenobacter ocellatus* sp. nov., *Hymenobacter gelipurpurascens* sp. nov. and *Hymenobacter chitinivorans* sp. nov., respectively, and emended description of the genus *Hymenobacter* Hirsch et al. 1999. *Int. J. Syst. Evol. Microbiol.* **56**, 2071–2078.

Busse, H.-J. and Auling, G. (1988). Polyamine pattern as a chemotaxonomic marker within the *Proteobacteria*. *Syst. Appl. Microbiol.* **11**, 1–8.

Busse, H.-J. and Schumann, P. (1999). Polyamine profiles within genera of the class *Actinobacteria* with LL-diaminopimelic acid in the peptidoglycan. *Int. J. Syst. Bacteriol.* **49**, 179–184.

Busse, H.-J., El-Banna, T. and Auling, G. (1989). Evaluation of different approaches for identification of xenobiotic-degrading pseudomonads. *Appl. Environ. Microbiol.* **55**, 1578–1583.

Busse, H.-J., Bunka, S., Hensel, A. and Lubitz, W. (1997). Discrimination of members of the family *Pasteurellaceae* based on polyamine patterns. *Int. J. Syst. Bacteriol.* **47**, 698–708.
Busse, H.-J., Kämpfer, P. and Denner, E. B. (1999). Chemotaxonomic characterisation of *Sphingomonas*. *J. Ind. Microbiol. Biotechnol.* **23**, 242–251.
Eraso, J. M. and Kaplan, S. (2009). Regulation of gene expression by PrrA in *Rhodobacter sphaeroides* 2.4.1: role of polyamines and DNA topology. *J. Bacteriol.* **191**, 4341–4352.
Feuerstein, B. G., Williams, L. D., Basu, H. S. and Marton, L. J. (1991). Implications and concepts of polyamine-nucleic acid interactions. *J. Cell. Biochem.* **46**, 37–47.
Garrity, G. M., Lilburn, T. G., Cole, J. R., Harrison, S. H., Euzéby, J., and Tindall, B. J. (2007). Taxonomic outline of the bacteria and archaea, Release 7.7. March 6, 2007. Part 3 – The *Bacteria*: Phylum "*Proteobacteria*", class *Alphaproteobacteria*. 52-111. http://www.taxonomicoutline.org/index.php/toba/article/view/179/212.
Geueke, B., Busse, H.-J., Fleischmann, T., Kämpfer, P. and Kohler, H.-P. E. (2007). Description of *Sphingosinicella xenopeptidilytica* sp. nov., a β-peptide-degrading species, and emended descriptions of the genus *Sphingosinicella* and the species *Sphingosinicella microcystinivorans*. *Int. J. Syst. Evol. Microbiol.* **57**, 107–113.
Goris, J., Kersters, K. and De Vos, P. (1998). Polyamine distribution among authentic pseudomonads and *Azotobacteraceae*. *Syst. Appl. Microbiol.* **21**, 285–290.
Hamana, K. (1995). Polyamine distribution patterns in coryneform bacteria and related Gram-positive eubacteria. *Ann. Rep. Coll. Med. Care Technol. Gumma Univ.* **16**, 69–77.
Hamana, K. (1996). Distribution of diaminopropane and acetylspermidine in *Enterobacteriaceae*. *Can. J. Microbiol.* **42**, 107–114.
Hamana, K. and Kishimoto, N. (1996). Polyamine distribution patterns in C_1 compound-utilizing eubacteria and acidophilic eubacteria. *J. Gen. Appl. Microbiol.* **42**, 431–437.
Hamana, K. and Matsuzaki, S. (1990). Polyamines and their biosynthetic activities in nonphytopathogenic marine agrobacteria. *Can. J. Microbiol.* **36**, 567–572.
Hamana, K. and Matsuzaki, S. (1992). Diaminopropane occurs ubiquitously in *Acinetobacter* as the major polyamine. *J. Gen. Appl. Microbiol.* **38**, 191–194.
Hamana, K. and Matsuzaki, S. (1993). Polyamine distribution patterns serve as a phenotypic marker in the chemotaxonomy of the *Proteobacteria*. *Can. J. Microbiol.* **39**, 304–310.
Hamana, K. and Nakagawa, Y. (2001a). Polyamine distribution profiles in the eighteen genera phylogenetically located within the *Flavobacterium-Flexibacter-Cytophaga* complex. *Microbios*, **106**, 7–17.
Hamana, K. and Nakagawa, Y. (2001b). Polyamine distribution profiles in newly validated genera and species within the *Flavobacterium-Flexibacter-Cytophaga-Sphingobacterium* complex. *Microbios*, **106**, 105–116.
Hamana, K. and Takeuchi, M. (1998). Polyamine profiles as chemotaxonomic marker within alpha, beta, gamma, delta and epsilon subclasses of class *Proteobacteria*: distribution of 2-hydroxyputrescine and homospermidine. *Microbiol. Cult. Coll.* **14**, 1–14.
Hamana, K., Kamekura, M., Onishi, H., Akazawa, T. and Matsuzaki, S. (1985). Polyamines in photosynthetic eubacteria and extreme-halophilic archaebacteria. *J. Biochem.* **97**, 1653–1658.
Hamana, K., Matsuzaki, S. and Sakakibara, M. (1988). Distribution of *sym*-homospermidine in eubacteria, cyanobacteria, algae and ferns. *FEMS Microbiol. Lett.* **50**, 11–16.
Hamana, K., Akiba, T., Uchino, F. and Matsuzaki, S. (1989). Distribution of spermine in bacilli and lactic acid bacteria. *Can. J. Microbiol.* **35**, 450–455.
Hamana, K., Nakagawa, Y. and Yamasato, K. (1995). Chemotaxonomic significance of polyamine distribution patterns in the *Flavobacterium-Cytophaga* complex and related genera. *Microbios.* **81**, 135–145.

Hamana, K., Okada, M., Saito, T. and Nogi, Y. (2000a). Polyamine distribution profiles among some members of the gamma subclass of the class *Proteobacteria*. *Microbiol. Cult. Coll.* **16**, 51–61.
Hamana, K., Saito, T. and Okada, M. (2000b). Polyamine profiles within the beta subclass of the class *Proteobacteria*: distribution of 2-hydroxyputrescine. *Microbiol. Cult. Coll.* **16**, 63–69.
Hamana, K., Niitsu, M. and Samejima, K. (2001). Occurrence of aminopropylhomospermidine as the major cellular polyamine in a halophilic, phototrophic alpha proteobacterium, *Rhodothalassium salexigens*. *J. Gen. Appl. Microbiol.* **47**, 99–101.
Hamana, K., Sakamoto, A., Tachiyanagi, S. and Terauchi, E. (2003a). Polyamine profiles of some members of the gamma subclass of the class *Proteobacteria*: Polyamine analysis of twelve recently described genera. *Microbiol. Cult. Coll.* **19**, 3–11.
Hamana, K., Sakamoto, A., Tachiyanagi, S., Terauchi, E. and Takeuchi, M. (2003b). Polyamine profiles of some members of the alpha subclass of the class *Proteobacteria*: polyamine analysis of twenty recently described genera. *Microbiol. Cult. Coll.* **19**, 13–21.
Hamana, K., Saito, T., Okada, M. and Niitsu, M. (2006a). Polyamine distibution profles among some members within the delta- and epsilon-subclasses of *Proteobacteria*. *Microbiol. Cult. Coll.* **20**, 3–8.
Hamana, K., Sato, W., Gouma, K., Yu, J., Ino, Y., Umemura, Y., Mochizuki, C., Takatsuka, K., Kigure, Y., Tanaka, N., Itoh, T. and Yokoto, A. (2006b). Cellular polyamine catalogues of the five classes of the phylum Proteobacteria: distributions of homospermidine within the class Alphaproteobacteria, hydroxyputrescine within the class Betaproteobacteria, norspermidine within the class Gammaproteobacteria, and spermidine within the classes Deltaproteobacteria and Epsilonproteobacteria. *Ann. Gunma Health Sci.* **27**, 1–16.
Hamana, K., Itoh, T., Benno, Y. and Hayashi, H. (2008). Polyamine distribution profiles of new members of the phylum *Bacteroidetes*. *J. Gen. Appl. Microbiol.* **54**, 229–236.
Hosoya, R. and Hamana, K. (2008). Distribution of two triamines, spermidine and homospermidine, and an aromatic amine, 2-phenylethylamine, within the phylum *Bacteroidetes*. *J. Gen. Appl. Microbiol.* **50**, 255–260.
Hosoya, R., Hamana, K., Niitsu, M. and Itoh, T. (2004). Polyamine analysis for chemotaxonomy of thermophilic eubacteria: polyamine distribution profiles within the orders *Aquificales*, *Thermotogales*, *Thermodesulfobacteriales*, *Thermales*, *Thermoanaerobacteriales*, *Clostridiales* and *Bacillales*. *J. Gen. Appl. Microbiol.* **50**, 271–287.
Huang, C. Y., Garcia, J. L., Patel, B. K. C., Cayol, J. L., Baresi, L. and Mah, R. A. (2000). *Salinivibrio costicola* subsp. *vallismortis* subsp. nov., a halotolerant facultative anaerobe from Death Valley, and emended description of *Salinivibrio costicola*. *Int. J. Syst. Evol. Microbiol.* **50**, 615–622.
Huber, B., Scholz, H. C., Kämpfer, P., Falsen, E., Langer, S. and Busse, H.-J. (2010). *Ochrobactrum pituitosum* sp. nov., isolated from an industrial environment. *Int. J. Syst. Evol. Microbiol.* **60**, 321–326.
Kämpfer, P., Bark, K., Busse, H.-J., Auling, G. and Dott, W. (1991). Numerical and chemotaxonomy of polyphosphate accumulating *Acinetobacter* strains with high polyphosphate: AMP Phosphotransferase (PPAT) activity. *Syst. Appl. Microbiol.* **15**, 409–419.
Kämpfer, P., Blasczyk, K. and Auling, G. (1994). Characterization of *Aeromonas* genomic species by using quinone, polyamine, and fatty acid patterns. *Can. J. Microbiol.* **40**, 844–850.
Kämpfer, P., Müller, C., Mau, M., Neef, A., Auling, G., Busse, H.-J., Osborn, A. M. and Stolz, A. (1999). Description of *Pseudaminobacter* gen. nov. with two species,

Pseudaminobacter salicylatoxidans sp. nov and *Pseudaminobacter defluvii* sp. nov. *Int. J. Syst. Bacteriol.* **49**, 887–897.

Kämpfer, P., Rosselló-Mora, R., Scholz, H., Welinder-Olsen, C. and Busse, H.-J. (2006). Description of *Pseudochrobactrum* gen. nov., with the two species *Pseudochrobactrum asaccharolyticum* sp. nov. and *Pseudochrobactrum saccharolyticum* sp. nov. *Int. J. Syst. Evol. Microbiol.* **56**, 1823–1829.

Kämpfer, P., Scholz, H. C., Huber, B., Falsen, E. and Busse, H.-J. (2007a). *Ochrobactrum haematophilum* sp. nov. and *Ochrobactrum pseudogrignonense* sp. nov., isolated from human clinical specimens. *Int. J. Syst. Evol. Microbiol.* **57**, 2513–2518.

Kämpfer, P., Scholz, H., Huber, B., Thummes, K., Busse, H.-J., Maas, E. W. and Falsen, E. (2007b). Description of *Pseudochrobactrum kiredjianiae* sp. nov. *Int. J. Syst. Evol. Microbiol.* **57**, 755–760.

Kämpfer, P., Sessitsch, A., Schloter, M., Huber, B., Busse, H.-J. and Scholz, H. C. (2008). *Ochrobactrum rhizosphaerae* sp. nov. and *Ochrobactrum thiophenivorans* sp. nov., isolated from the environment. *Int. J. Syst. Evol. Microbiol.* **58**, 1426–1431.

Kämpfer, P., Huber, B., Lodders, N., Warfolomeow, I., Busse, H.-J. and Scholz, H. C. (2009a). *Pseudochrobactrum lubricantis* sp. nov., isolated from a metal-working fluid. *Int. J. Syst. Evol. Microbiol.* **59**, 2464–2467.

Kämpfer, P., Lodders, N., Warfolomeow, I. and Busse, H.-J. (2009b). *Tessaracoccus lubricantis* sp. nov., isolated from a metalworking fluid. *Int. J. Syst. Evol. Microbiol.* **59**, 1545–1549.

Kämpfer, P., Lodders, N. and Busse, H.-J. (2009c). *Arcicella rosea* sp. nov., isolated from tap water. *Int. J. Syst. Evol. Microbiol.* **59**, 341–344.

Kämpfer, P., Martin, E., Lodders, N., Jäckel, U., Huber, B. E., Schumann, P., Langer, S., Busse, H.-J. and Scholz, H. (2010a). *Paenochrobactrum gallinarii* gen. nov., sp. nov., isolated from air of a duck barn, and reclassification of *Pseudochrobactrum glaciei* as *Paenochrobactrum glaciei* comb. nov. *Int. J. Syst. Evol. Microbiol.* **60**, 1493–1498.

Kämpfer, P., Martin, E., Lodders, N., Langer, S., Schumann, P., Jäckel, U. and Busse, H.-J. (2010b). *Pseudofulvimonas gallinarii* gen. nov., sp. nov., a new member of the family *Xanthomonadaceae*. *Int. J. Syst. Evol. Microbiol.* **60**, 1427–1431.

Kämpfer, P., Schäfer, J., Lodders, N. and Busse, H.-J. (2010c). *Brevibacterium sandarakinum* sp. nov., isolated from a wall of an indoor environment. *Int. J. Syst. Evol. Microbiol.* **60**, 909–913.

Kneifel, H., Stetter, O., Andereesen, J. R., Wiegel, J., König, H. and Schoberth, S. M. (1986). Distribution of polyamines in representative species of archaebacteria. *Syst. Appl. Microbiol.* **7**, 241–245.

Kumar, N. R., Nair, S., Langer, S., Busse, H.-J. and Kämpfer, P. (2008). *Altererythrobacter indicus* sp. nov., isolated from wild rice (*Porteresia coarctata* Tateoka). *Int. J. Syst. Evol. Microbiol.* **58**, 839–844.

Nigam, A., Jit, S. and Lal, R. (2010). *Sphingomonas histidinilytica* sp. nov., isolated from a hexachlorocyclohexane dumpsite. *Int. J. Syst. Evol. Microbiol.* **60**, 1038–1043.

Sánchez-Porro, C., Gallego, V., Busse, H.-J., Kämpfer, P. and Ventosa, A. (2009). Transfer of *Teichococcus ludipueritiae* and *Muricoccus roseus* to the genus *Roseomonas*, as *Roseomonas ludipueritiae* comb. nov. and *Roseomonas rosea* comb. nov., respectively, and emended description of the genus *Roseomonas*. *Int. J. Syst. Evol. Microbiol.* **59**, 1193–1198.

Scherer, P. and Kneifel, H. (1983). Distribution of polyamines in methanogenic bacteria. *J. Bacteriol.* **154**, 1315–1322.

Scherer, P., Stetter, K. O., Andreesen, J. R., Wiegel, J., König, H. and Schoberth, S. M. (1986). Distribution of polyamines in reprsentative species of Archaebacteria. *Syst. Appl. Microbiol.* **7**, 241–245.

Polyamines

Schiller, D., Kruse, D., Kneifel, H., Kramer, R. and Burkovski, A. (2000). Polyamine transport and role of *potE* in response to osmotic stress in *Escherichia coli*. *J. Bacteriol.* **182**, 6247–6249.

Scholz, H. C., Hubalek, Z., Sedláček, I., Vergnaud, G., Tomaso, H., Al Dahouk, S., Melzer, F., Kämpfer, P., Neubauer, H., Cloeckaert, A., Maquart, M., Zygmunt, M. S., Whatmore, A. M., Falsen, E., Bahn, P., Göllner, C., Pfeffer, M., Huber, B., Busse, H.-J. and Nöckler, K. (2008). *Brucella microti* sp. nov., isolated from the common vole *Microtus arvalis*. *Int. J. Syst. Evol. Microbiol.* **58**, 375–382.

Scholz, H. C., Nöckler, K., Göllner, C., Bahn, P., Vergnaud, G., Tomaso, H., Al Dahouk, S., Kämpfer, P., Cloeckaert, A., Maquart, M., Zygmunt, M. S., Whatmore, A. M., Pfeffer, M., Huber, B., Busse, H.-J. and De, B. K. (2010). *Brucella inopinata* sp. nov., isolated from a breast implant infection. *Int. J. Syst. Evol. Microbiol.* **60**, 801–808.

Segers, P., Vancanneyt, M., Pot, B., Torck, U., Hoste, B., Dewettinck, D., Falsen, E., Kersters, K. and De Vos, P. (1994). Classification of *Pseudomonas diminuta* Leifson and Hugh 1954 and *Pseudomonas vesicularis* Büsing, Döll and Freytag 1953 in *Brevundimonas* gen. nov. as *Brevundimonas diminuta* comb. nov. and *Brevundimonas vesicularis* comb. nov., respectively. *Int. J. Syst. Bacteriol.* **44**, 499–510.

Sprague, L. D., Scholz, H. C., Amann, S., Busse, H.-J. and Neubauer, H. (2008). *Yersinia similis* sp. nov. *Int. J. Syst. Evol. Microbiol.* **58**, 952–958.

Tabor, C. W. and Tabor, H. (1985). Polyamines in microorganisms. *Microbiol. Rev.* **49**, 81–99.

Takeuchi, M., Hamana, K. and Hiraishi, A. (2001). Proposal of the genus *Sphingomonas sensu stricto* and three new genera, *Sphingobium*, *Novosphingobium* and *Sphingopyxis*, on the basis of phylogenetic and chemotaxonomic analyses. *Int. J. Syst. Evol. Microbiol.* **51**, 1405–1417.

Tkachenko, A. G., Salakhetdinova, O. and Pshenichnov, M. R. (1997). Exchange of putrescine and potassium between cells and media as a factor in the adaptation of *Escherichia coli* to hyperosmotic shock. *Mikrobiologia*. **66**, 329–334.

Urbanczyk, H., Ast, J. C., Higgins, M. J., Carson, J. and Dunlap, P. V. (2007). Reclassification of *Vibrio fischeri*, *Vibrio logei*, *Vibrio salmonicida* and *Vibrio wodanis* as *Aliivibrio fischeri* gen. nov., comb. nov., *Aliivibrio logei* comb. nov., *Aliivibrio salmonicida* comb. nov. and *Aliivibrio wodanis* comb. nov. *Int. J. Syst. Evol. Microbiol.* **57**, 2823–2829.

Urdiain, M., López-López, A., Gonzalo, C., Busse, H.-J., Langer, S., Kämpfer, P. and Rosselló-Móra, R. (2008). Reclassification of *Rhodobium marinum* and *Rhodobium pfennigii* as *Afifella marina* gen. nov. comb. nov. and *Afifella pfennigii* comb. nov., a new genus of photoheterotrophic *Alphaproteobacteria* and emended descriptions of *Rhodobium*, *Rhodobium orientis* and *Rhodobium gokarnense*. *Syst. Appl. Microbiol.* **31**, 339–351.

Wieser, M., Schumann, P., Martin, K., Altenburger, P., Burghardt, J., Lubitz, W. and Busse, H.-J. (1999). *Agrococcus citreus* sp. nov., isolated from a medieval wall painting of the chapel of Castle Herberstein (Austria). *Int. J. Syst. Bacteriol.* **49**, 1165–1170.

Yamamoto, S., Shinoda, S. and Makita, M. (1979). Occurence of norspermidine in some species of genera *Vibrio* and *Beneckea*. *Biochem. Biophys. Res. Comm.* **87**, 1102–1108.

Yamamoto, S., Shinoda, S., Kawaguchi, M., Wakamatsu, K. and Makita, M. (1983). Polyamine distribution in *Vibrionaceae*: norspermidine as a general constituent of *Vibrio* species. *Can. J. Microbiol.* **29**, 724–728.

Yamamoto, S., Yoshida, M., Nakao, H., Koyama, M., Hashimoto, Y. and Shinoda, S. (1986). Variations in cellular polyamine compositions and contents of *Vibrio* species during growth in media with various NaCl concentrations. *Chem. Pharm. Bull.* **34**, 3038–3042.

Yamamoto, S., Chowdhury, M. A. R., Kuroda, M., Nakano, T., Koumoto, Y. and Shinoda, S. (1991). Further study on polyamine compositions in *Vibrionaceae*. *Can. J. Microbiol.* **37**, 148–153.

Yang, P., De Vos, P., Kersters, K. and Swings, J. (1993). Polyamine patterns as chemotaxonomic markers for the genus *Xanthomonas*. *Int. J. Syst. Bacteriol.* **43**, 709–714.

Yarza, P., Richter, M., Peplies, J., Euzéby, J., Amann, R., Schleifer, K.-H., Ludwig, W., Glöckner, F. O. and Rosselló-Mora, R. (2008). The All-Species Living Tree Project: a 16S rRNA-based phylogenetic tree of all sequenced type strains. *Syst. Appl. Microbiol.* **31**, 241–250. ftp://ftp.arb-silva.de/living_tree/LTP_release_95/LTD_895.pdf

Zherebilo, O. E., Kucheryava, N., Gvozdyak, R. I., Ziegler, D., Scheibner, M. and Auling, G. (2001). Diversity of polyamine patterns in soft rot pathogens and other plant-associated members of the Enterobacteriaceae. *Syst. Appl. Microbiol.* **24**, 54–62.

Zlamala, C., Schumann, P., Kämpfer, P., Rosselló-Mora, R., Lubitz, W. and Busse, H.-J. (2002). *Agrococcus baldri* sp. nov., isolated from the air of the 'Virgilkapelle' in Vienna. *Int. J. Syst. Evol. Microbiol.* **52**, 1211–1216.

12 Characterization of Pigments of Prokaryotes and Their Use in Taxonomy and Classification

Aharon Oren
Department of Plant and Environmental Sciences, Institute of Life Sciences, and the Moshe Shilo Minerva Center for Marine Biogeochemistry, The Hebrew University of Jerusalem, Jerusalem, Israel

♦♦

CONTENTS

♦♦♦♦♦♦ I. INTRODUCTION

Cultures and colonies of many prokaryotes often display a variety of colours including yellow, red, purple, green, as well as some others. Sometimes the pigmentation is immediately connected with the way of life of the cells. This is obviously the case for phototrophic prokaryotes, oxygenic as well as anoxygenic, that use photons as energy source and therefore possess different pigments to harvest light energy and convert it to biologically available energy – ATP and/or proton gradients across membranes. In many other cases the function of the pigmentation and its advantage to the cells is not directly clear. Carotenoids may act as photoprotectants, but to what extent they indeed provide a selective advantage to the cells that produce them is not always known. Sometimes the pigmentation of the cells is a fixed property of the species, in other cases the degree of pigmentation depends on growth conditions.

As pigmentation of colonies is so easy to assess, statements about the colour of the cells are found in all species descriptions. However, a detailed characterization of the type(s) of pigments present is not always easy. Recent notes on the characterization of prokaryote strains for taxonomic purposes (Tindall *et al.*, 2010) only contain a short statement on this subject: 'Cellular pigments may be

METHODS IN MICROBIOLOGY, VOLUME 38
0580-9517 DOI 10.1016/B978-0-12-387730-7.00012-7

soluble either in water or organic solvents. Typical pigments include carotenoids (which turn blue in the presence of concentrated sulfuric acid), flexirubins (which change colour reversibly under acid and alkaline conditions), (bacterio) chlorophyll [which is soluble in organic solvents, but becomes water soluble following saponification; treatment with acid results in the formation of (bacterio) phaeophytin], melanin and pyocyanin (fluorescence at 360 nm, reversible colour change at acid/alkaline pH; this list is not exhaustive.' Representative chemical structures of these and other pigments encountered in prokaryotes are given in Figures 1–5.

For some groups of prokaryotes the characterization of the pigments produced is an indispensable part of the description of new species. This is true, for example, for the anoxygenic phototrophic bacteria (Imhoff and Caumette, 2004). For many other groups a simple statement about the colour of the colonies or cell pellets suffices. This colouration can vary with growth media used

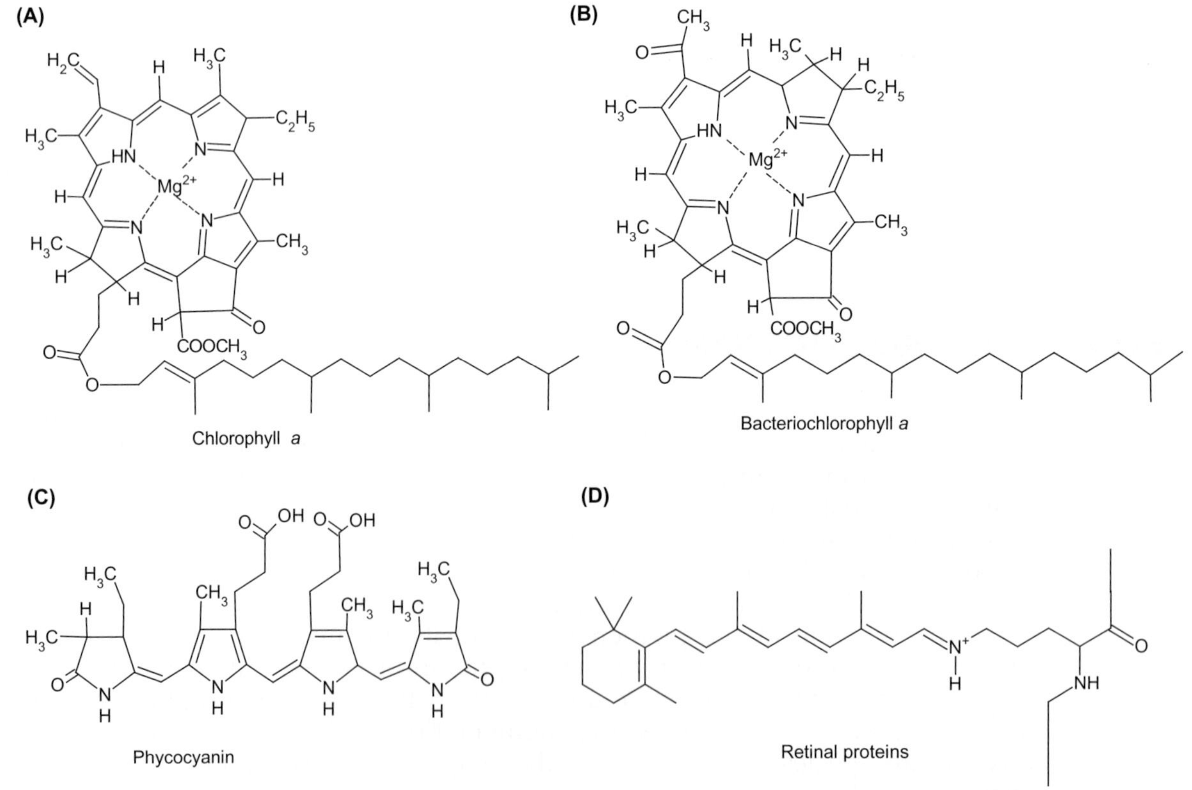

Figure 1. Structures of various pigments involved in light harvesting and photosynthesis in prokaryotes: (A) chlorophyll *a* (cyanobacteria); (B) bacteriochlorophyll *a* (different groups of *Alphaproteobacteria*, *Betaproteobacteria* and *Gammaproteobacteria*); (C) the chromophore of phycocyanin, one of the phycobiliproteins of the cyanobacteria; (D) the retinal chromophore of bacteriorhodopsin (*Halobacteriaceae*), proteorhodopsin (some marine proteobacteria) and xanthorhodopsin (*Salinibacter*).

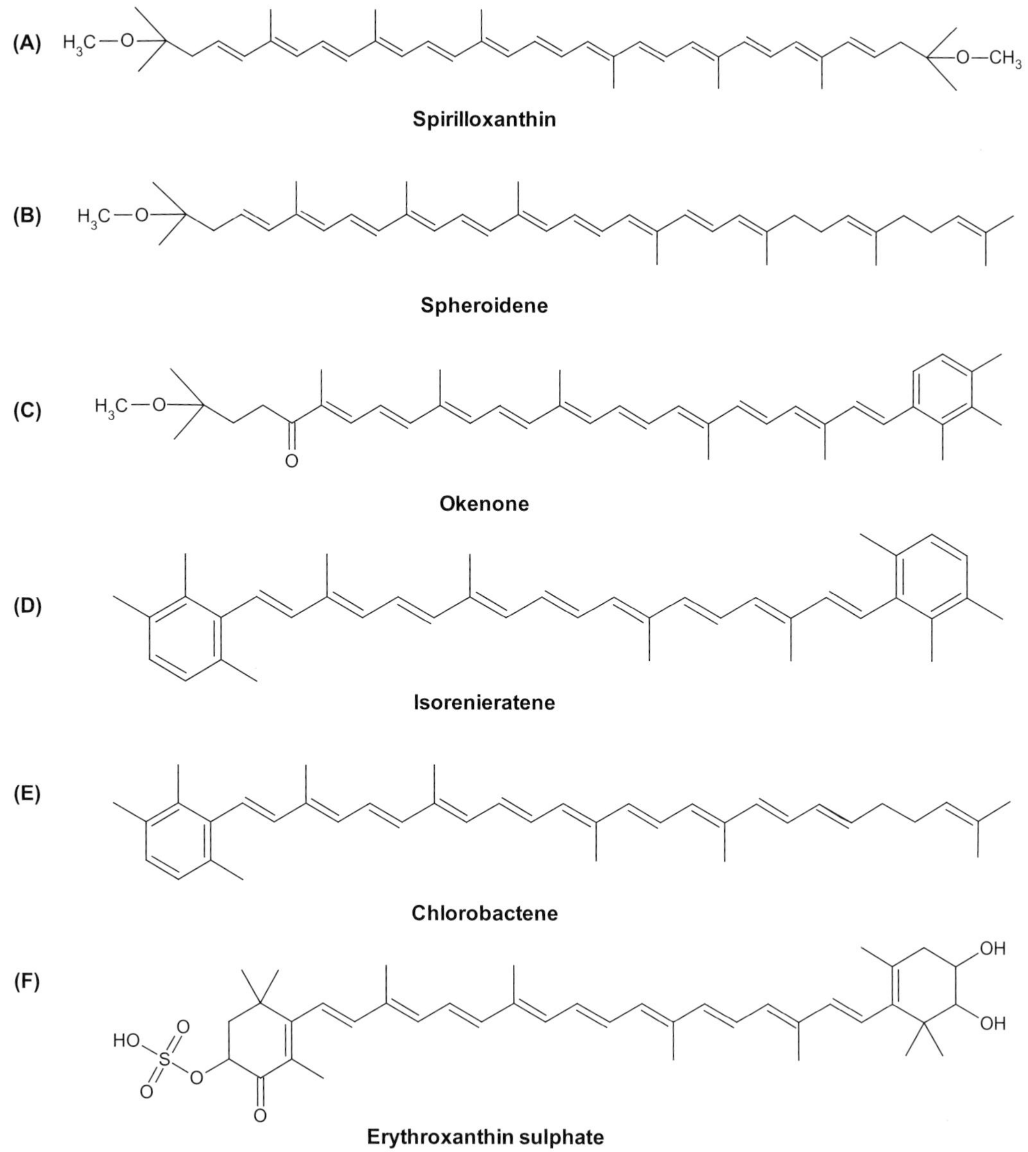

(A) Spirilloxanthin

(B) Spheroidene

(C) Okenone

(D) Isorenieratene

(E) Chlorobactene

(F) Erythroxanthin sulphate

Figure 2. Various carotenoids of anoxygenic phototrophic prokaryotes: (A) spirilloxanthin (*Rhodospirillum rubrum*, *Allochromatium vinosum*); (B) spheroidene of *Rhodobacter sphaeroides*; (C) okenone of *Chromatium okenii*; (D) isorenieratene of *Chlorobium phaeovibrioides*; (E) chlorobactene of *Chlorobium limicola*; (F) erythroxanthin sulphate of *Erythromicrobium ramosum*, *Erythrobacter longus*, *Erythrobacter litoralis*.

as well as cultivation conditions, and these should be stated when such information is included in a species description. However, pigments are useful chemotaxomomic markers, and novel types of prokaryote pigments are regularly discovered in those relatively rare cases in which a more in-depth study was

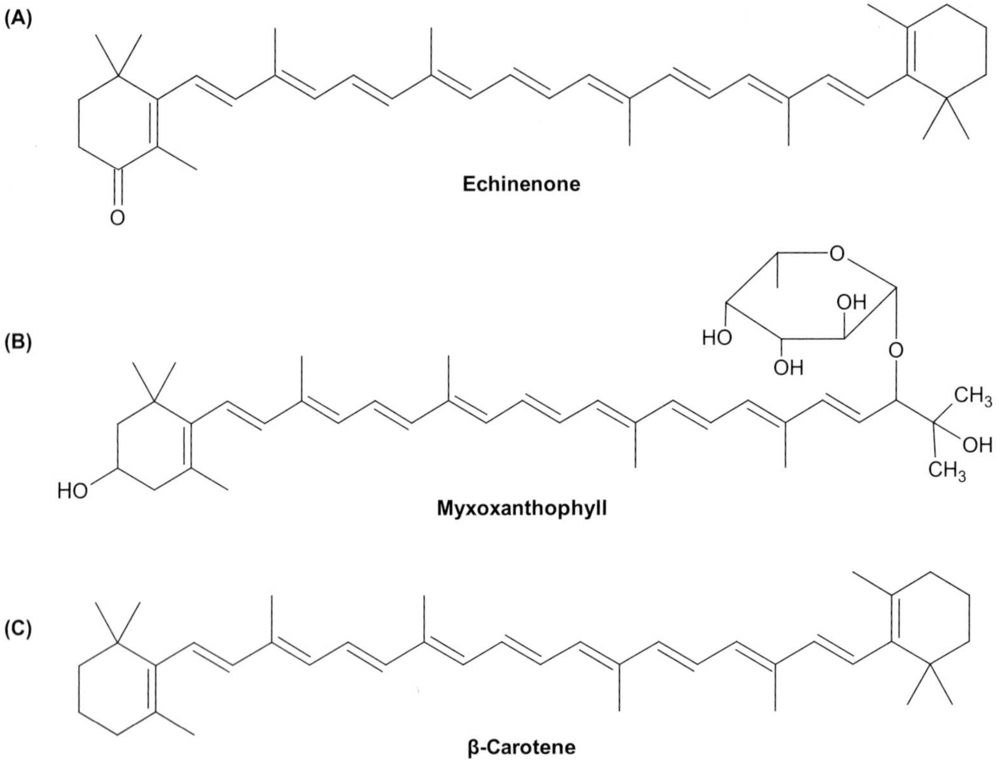

Figure 3. Various carotenoids found in cyanobacteria: (A) echinenone; (B) myxoxanthophyll; (C) β-carotene.

made of the pigment(s) responsible for the colour. Thus, presence of yet unknown types of pigments in newly described taxa can easily be missed. The case studies presented at the end of this chapter provide a few examples.

Because of the great diversity of pigments produced in the prokaryote world, and the fact that they belong to disparate chemical groups, it is not feasible to give a simple, universally applicable protocol that enables the characterization of all the molecules responsible for the colour of the cells within a short time. For novel compounds, mass cultivation may be needed to extract sufficient amounts of the pigments for separation followed by the use of a suite of techniques of organic chemistry that allow the elucidation of the molecular structure: elemental analysis, mass spectrometry, infrared spectroscopy, ^{1}H and ^{13}C nuclear magnetic resonance spectroscopy and so on.

By their very nature, pigments are light-sensitive molecules. When extracted from the cells, the light sensitivity, and often their sensitivity to oxygen as well, may be increased. Therefore, depending on the type(s) of pigments being studied, manipulations should be performed as much as possible at reduced light intensities and purified (bacterio) chlorophylls, carotenoids and so on, should be stored under nitrogen to protect the compounds from chemical oxidation, resulting in the modification of their structure and properties.

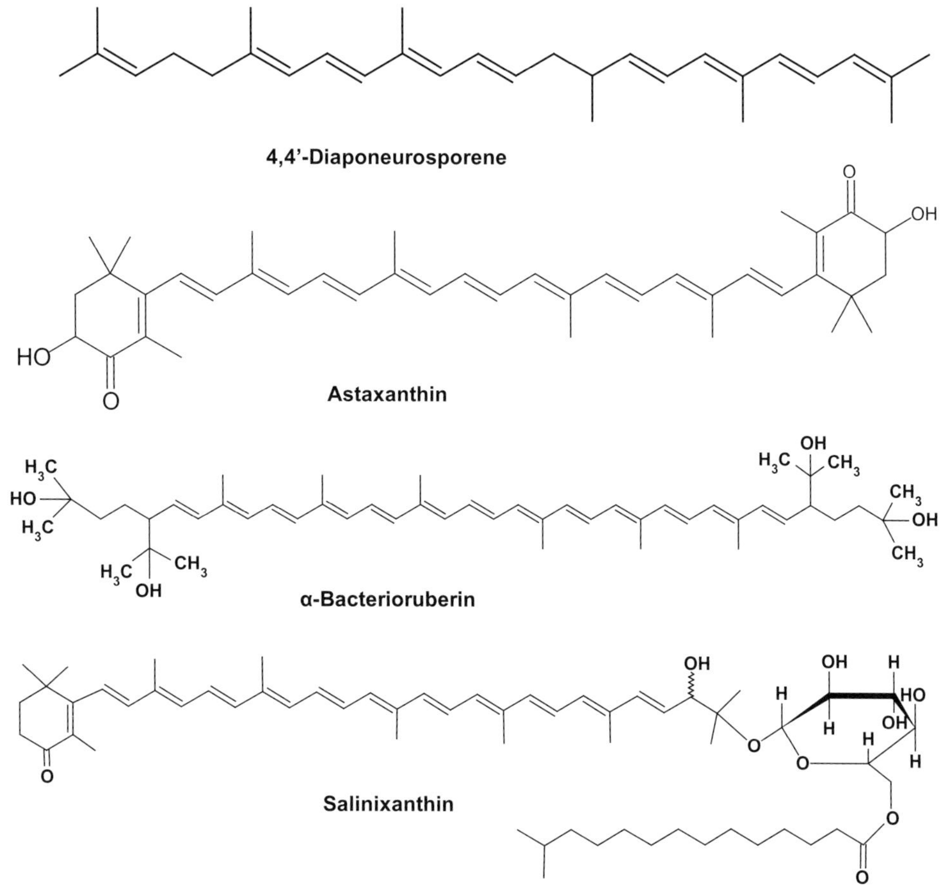

(A)

4,4'-Diaponeurosporene

(B)

Astaxanthin

(C)

α-Bacterioruberin

(D)

Salinixanthin

Figure 4. Various carotenoids of non-phototrophic prokaryotes: (A) 4,4′-diaponeurosporene of *Staphylococcus aureus*; (B) astaxanthin of *Paracoccus carotinifaciens*, *Paracoccus marcusii*; (C) α-bacterioruberin of *Halobacterium* and other members of the *Halobacteriaceae*; (D) salinixanthin of *Salinibacter ruber*.

♦♦♦♦♦♦ II. ABSORPTION SPECTRA – *IN VIVO* AND *IN VIVO*

The first indications of the possible identity of the pigment(s) present in the cells (or in some cases, pigments excreted into the culture medium) are given by the absorption spectrum, both of intact cells and of extracts of the pigments in suitable solvents. Relevant wavelengths used for detection are not only in the visible range (400–700 nm) but also in the infrared (for some photosynthetic bacteria even up to 1050 nm) and ultraviolet ranges down to 300 nm to detect, e.g. mycosporine-like amino acids present in some cyanobacteria (Castenholz and Garcia-Pichel, 2000).

The *in vivo* absorption spectrum of pigmented prokaryotes often differs greatly from the spectrum of the pigments extracted from the same cells. For

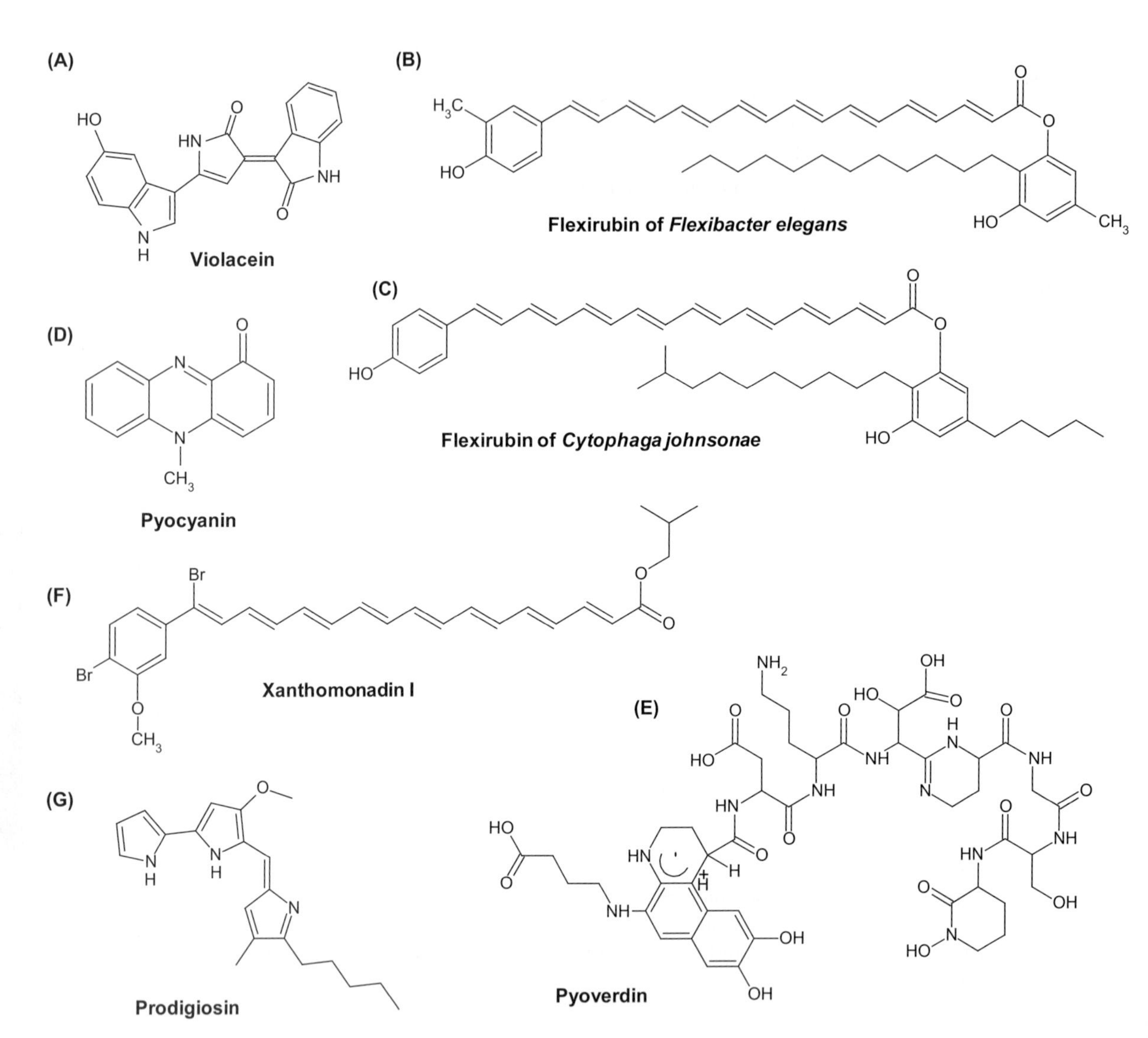

Figure 5. Various pigments of heterotrophic prokaryotes: (A) violacein of *Chromobacterium* spp.; (B) flexirubin of *Flexibacter elegans*; (C) another flexirubin-like pigment of *Flavobacterium johnsonae*; (D) pyocyanin of *Pseudomonas aeruginosa*; (E) pyoverdin of *Pseudomonas aeruginosa*; (F) xanthomonadin I of *Xanthomonas* sp.; (G) prodigiosin of *Serratia marcescens*.

pigments located within lipid membranes, the exact wavelengths of the absorption maxima depend on the chemical environment in which the pigments are found. Extraction of hydrophobic pigments in organic solvents changes their environment and this often causes significant shifts in their absorption maxima. The best-known case is that of different types of bacteriochlorophylls from anoxygenic phototrophic bacteria: extraction of bacteriochlorophyll *a* into organic solvents causes the *in vivo* infrared absorption maximum to shift from

850–910 nm to 775 nm, a red shift of 75–135 nm; for bacteriochlorophyll *b*, this red shift following solvent extraction is even more dramatic (Table 1).

When measuring the absorption spectrum of intact cells in a conventional spectrophotometer, the optical properties of the pigments are generally masked to a large extent by the light scattering caused by the particles in suspension. Because of scattering only part of the light that proceeds in the same direction as the incident light is measured. Moreover, scattering is wavelength-dependent. Thus, normally the absorption curve is superimposed on the high scattering curve. The result is that the absorption peaks of the pigments may hardly be visible. Figure 6 shows an example in which the absorption spectrum of a suspension of *Salinibacter ruber*, a pink-orange pigmented member of the *Bacteroidetes*, was compared with that of cells suspended in a concentrated sucrose solution (see below) and of an extract of hydrophobic pigments in an organic solvent.

To minimize the effect of particle scattering on the shape of the absorption spectrum, different strategies have been proposed. One is by using an integrating sphere setup offered as an accessory for some types of spectrophotometer. Here the sample is placed in the centre of a spherical cavity coated white for high diffusive reflectivity, causing uniform diffusion of the light and abolishing the effect of scattering by the particles in suspension on the amount of light that reaches the detector. This device is seldom, if ever, used in the characterization of new prokaryotes.

Shibata *et al.* (1954) proposed a method to reduce the effect of scattering on the absorption spectrum of intact cells, based on attaching opalescent plates to cuvettes for both the cell suspension and the reference cuvette. Both the sample and the blank cell compartments are provided – on the sides of light transmission – with identical opalescent plates which diffuse uniformly all the light as it leaves the cuvette. The required properties of the opalescent plates are uniform, moderate and constant opalescency over a wide range of wavelengths. Too much opalescency reduces the intensity reaching the detector too much. For measurements in a wide range of wavelengths, from 220 to 1300 nm, filter paper impregnated with paraffin oil was found satisfactory as an opalescent material. For observations in the visible regions, commercial opal glass, one side of which has an opalescent coating, gave a result slightly better than that obtained with oiled paper. For this method it is important that the distances between the cuvettes (sample, blank) and the detector are the same. The method is not in wide use today.

A simpler method to record absorption spectra of intact cells while minimizing light scattering is by suspending the cells in 60% (w/v) sucrose (Biebl and Drews, 1969). The high refractive index of the concentrated sugar solution is not very different from that of the cell material, so that scattering is reduced. This method was included in the recommended standards for the description of new species of anoxygenic phototrophic bacteria (Imhoff and Caumette, 2004): 'An *in vivo* absorption spectrum needs to be published together with the species description. Absorption spectra of whole cells are measured with cell suspensions washed twice in medium or appropriate salt solutions and then suspended in 60% (w/v) sucrose solution. Better results are often achieved by using isolated photosynthetic membranes suspended in buffer. For this

Table 1. Chlorophyll absorption maxima *in vivo* and *in vitro* for the phototrophic prokaryotes

Biological group	Dominant chlorophyll	Ether	Intact cells	Red shift (nm)	Additional *in vivo* absorption maxima (nm)
Cyanobacteria	Chlorophyll *a*	665	680–683	15–18	420
Prochlorococcus (Cyanobacteria)	Divinyl chlorophyll *a*	664 (acetone)	675	11	445
	Divinyl chlorophyll *b*	647 (acetone)	655	8	480
Alphaproteobacteria, *Betaproteobacteria*, *Gammaproteobacteria*	Bacteriochlorophyll *a*	770	830–910	60–135	375, 590, 800–810
Alphaproteobacteria, *Gammaproteobacteria*	Bacteriochlorophyll *b*	790	986–1,035	196–245	400, 605, 835–850
Chlorobiales, 'Chloroflexales'	Bacteriochlorophyll *c*	660	745–755	85–95	457–460
Chlorobiales	Bacteriochlorophyll *d*	650	715–745	65–95	450
Chlorobiales	Bacteriochlorophyll *e*		710–725		460–462
Heliobacteriaceae	Bacteriochlorophyll *g*	763	788–790	25–27	419, 575

Adapted from Bricaud *et al.* (1999), Goericke and Repeta (1992), Kondratieva *et al.* (1992), Overmann and Garcia-Pichel (2006), Pfennig (1978), and other sources.

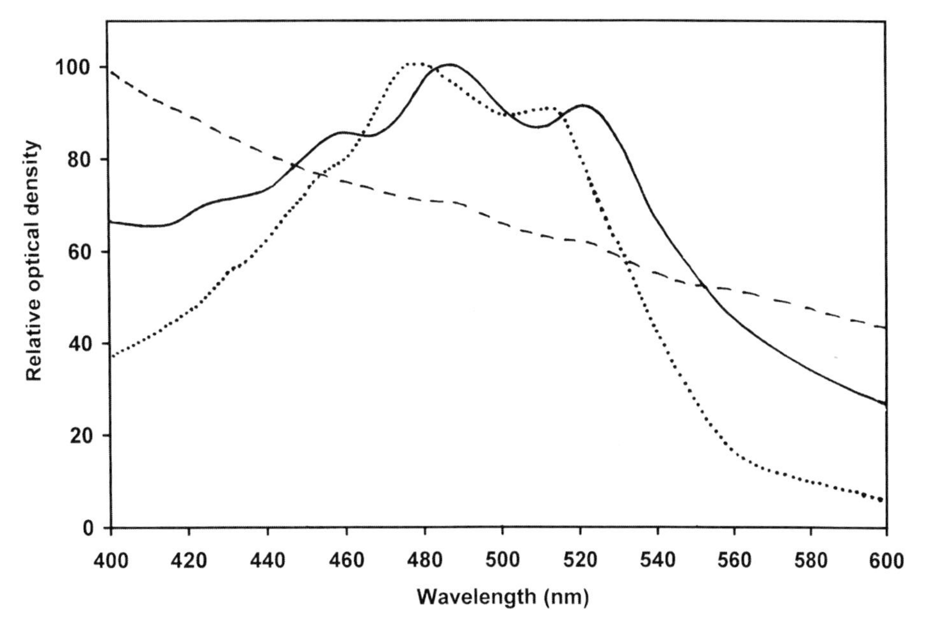

Figure 6. Absorption spectra of *Salinibacter ruber* (DSM 13855^{T}) as measured in a suspension of whole cells (dashed line), in whole cells suspended in 60% sucrose (solid line), and the spectrum of an extract in methanol/acetone (1:1, v/v). Spectra were recorded in an Agilent 8453 UV-Visible Spectrometry System, using respectively growth medium, 60% sucrose, and methanol/acetone (1:1, v/v) as blank.

purpose, it is sufficient to break the cells by ultrasonication or with a French press and to separate whole cells and large cell fragments from the internal membranes at 15,000 × *g*.'

♦♦♦♦♦♦ III. THE MAJOR GROUPS OF PIGMENTS IN THE PROKARYOTE WORLD

Many different classes of pigments are encountered in the prokaryotes. These include magnesium tetrapyrrole pigment chlorophylls and bacteriochlorophylls and the magnesium-lacking phaeophytin forms, the phycobiliproteins of the cyanobacteria, retinal-containing pigments such as bacteriorhodopsin, halorhodopsin and proteorhodopsin, carotenoids, flexirubins and many others.

As stated above, for structure elucidation of a newly discovered pigment it is necessary to purify amounts of the compound sufficient for organic chemistry analysis by mass spectroscopy, NMR, IR spectroscopy and other methods. Knowing the phylogenetic position of the organism together with existing information on the types of pigments found in related taxa is often helpful. Thus, soon after the orange-pink carotenoid of the extremely halophilic *S. ruber*, named salinixanthin, was identified as a novel C_{40}-carotenoid acyl glycoside (Lutnæs *et al.*, 2002) (see Section IV.A), it was ascertained that the phylogenetically related marine thermophile *Rhodothermus marinus* contains very similar carotenoid pigments (Lutnæs *et al.*, 2004).

A. Pigments of Photosynthetic Prokaryotes

1. Chlorophylls and phycobiliproteins of oxygenic photosynthetic prokaryotes

The oxygenic phototrophic prokaryotes all belong to the phylum *Cyanobacteria*. Nearly all species of cyanobacteria contain chlorophyll *a* (Figure 1A) as the only light-harvesting and reaction centre chlorophyll. Chlorophyll *a* has an *in vivo* absorption maximum at 680–683 nm, which shifts to 665 nm in extracts in organic solvents (Table 1). A few species (*Prochloron, Prochlorococcus, Prochlorothrix*) have divinyl derivatives of chlorophyll *a* and chlorophyll *b* (Bricaud *et al.*, 1999; Goericke and Repeta, 1993). *Acaryochloris marina*, anoxygenic phototrophic prokaryote that lives as a symbiont with corals, contains chlorophyll *d* which absorbs at 700–720 nm (Kühl *et al.*, 2005).

Cyanobacteria (with the exception of the above-mentioned types that contain divinyl chlorophylls) all possess different phycobiliproteins as accessory light-harvesting pigments. These are water-soluble proteins with open-chain tetrapyrroles as chromophores (Figure 1C). The main phycobiliproteins are blue phycocyanin (absorption maximum 620–625 nm) and allophycocyanin (650 nm) and the red phycoerythrin (565 nm). Some cyanobacteria modify the relative amounts of the different phycobiliproteins synthesized according to the colour of the light in which they are grown, to maximize the use of the available photons (complementary chromatic adaptation) (Tandeau de Marsac, 1977).

2. Bacteriochlorophylls of anoxygenic photosynthetic prokaryotes

A much greater diversity of chlorophyll derivatives (bacteriochlorophylls; Figure 1B) is found in the different groups of anoxygenic phototrophic bacteria. Chemically they differ in the side groups on the tetrapyrrole ring and the nature of the ester–phytyl (as shown in Figure 1B), phytadienyl, farnesyl or geranylgeraniol. They all have their long-wavelength absorption maximum in the infrared wavelength range. The *in vivo* spectrum depends on the location in the membrane of the bacteriochlorophyll–protein complexes; therefore, the same pigment can have different absorption maxima in different organisms. All bacteriochlorophylls show a very significant red shift in absorption maximum when they are removed from the photosynthetically active membranes by extraction in organic solvents (Figures 7 and 8). Table 1 lists the most common types of bacteriochlorophyll. Bacteriochlorophyll *a*, with absorption peaks *in vivo* at 800–805 nm and around 870 nm, is especially widespread: it is found in many types of photosynthetic Proteobacteria, including anaerobic sulfur oxidizing bacteria (*Chromatium, Ectothiorhodospira*) and others such as *Rhodobacter* and *Rhodospirillum*. Most anoxygenic phototrophs need anaerobic conditions for bacteriochlorophyll production, there are also anoxygenic phototrophs (*Erythrobacter, Roseobacter, Roseovarius, Staleya, Rubriomonas, Roseivivax* and others) that produce bacteriochlorophyll *a* in the presence of oxygen. The

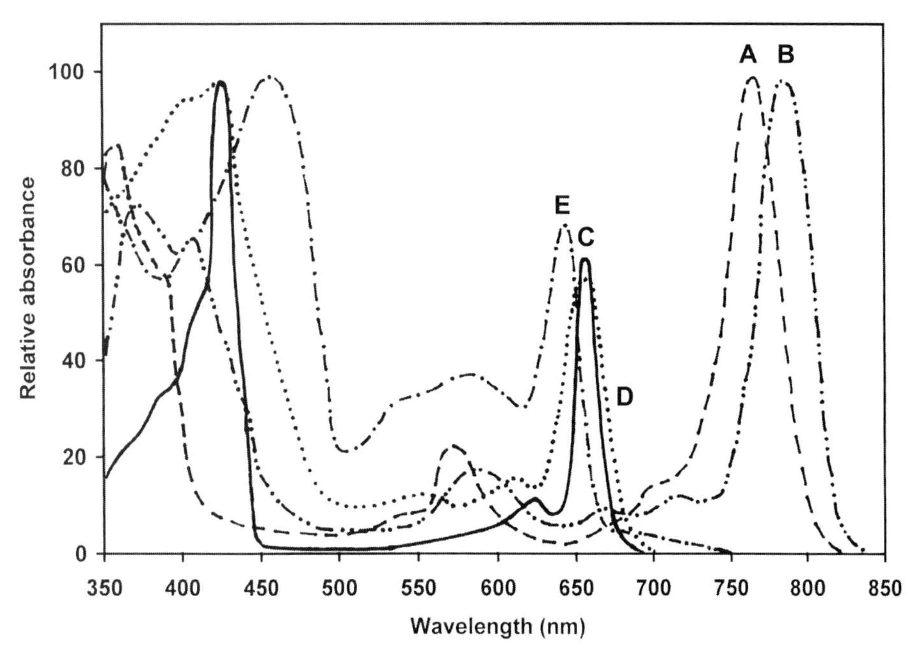

Figure 7. Absorption spectra of different types of bacteriochlorophyll from anoxygenic photosynthetic prokaryotes: (A) bchl *a* of *Rhodobacter sphaeroides*; (B) bchl *b* of *Blastochloris viridis*; (C) bchl *c* of *Chloroflexus aurantiacus*; (D) bchl *d* of *Prosthecochloris vibrioformis*; (E) bchl *e* of *Chlorobium phaeovibrioides* (spectra adapted from Oelze, 1985). Pigments were dissolved in diethyl ether (A–C) or acetone (D and E).

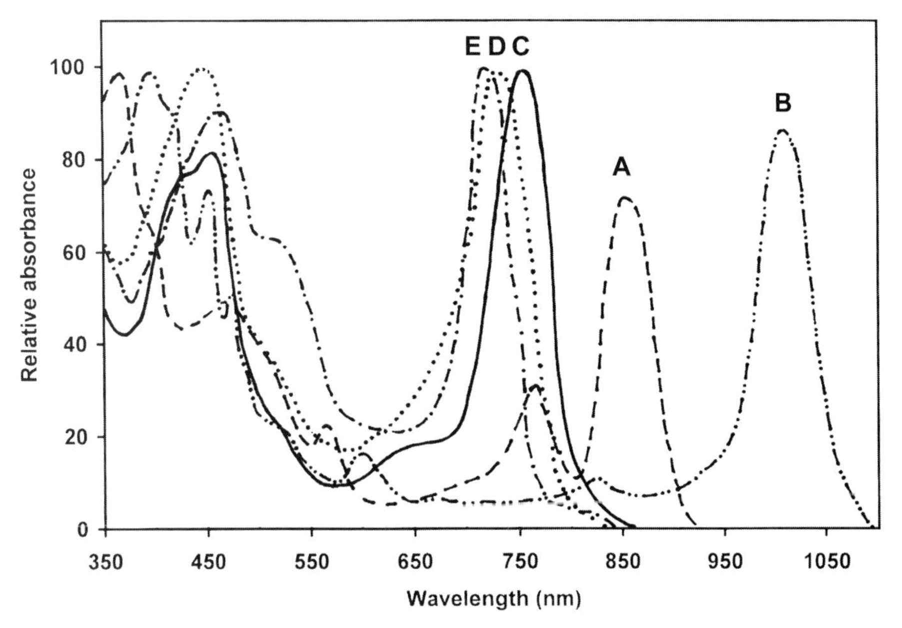

Figure 8. *In vivo* absorption spectra of anoxygenic photosynthetic prokaryotes with different types of bacteriochlorophyll: (A) *Rhodocyclus tenuis* (bchl *a*); (B) *Blastochloris viridis* (bchl *b*); (C) *Chlorobium limicola* (bchl *c*); (D) *Prosthecochloris vibrioformis* (bchl *d*); (E) *Chlorobium phaeovibrioides* (bchl *e*). Spectra adapted from Oelze (1985).

recommended standards for the description of new species of anoxygenic phototrophic bacteria (Imhoff and Caumette, 2004) include a record of the colour of cell suspensions, *in vivo* absorption spectra from 370 to 1020 nm, and information on the type of bacteriochlorophyll present. More detailed information on the analysis of bacteriochlorophylls was given by Oelze (1985).

Extraction and chromatographic separation of bacteriochlorophylls (based on Oelze, 1985; volumes of culture and solvents used for extraction depend on the pigment content of the cells and the amount of pigment needed for analysis).

Note: All chlorophylls and bacteriochlorophylls are sensitive to exposure to light, acids and oxidizing chemicals. All manipulations should therefore be done in the dark or in dim light.

- Collect cells by centrifugation.
- Extract with methanol for 30 min at 4°C; 5 mM Na-ascorbate may be added to prevent oxidation.
- Remove cells and debris by centrifugation.
- Repeat the extraction until sedimented material has become colourless.
- Transfer the extract to a separating funnel.
- Add diethyl ether to the combined methanolic extracts.
- Extract bacteriochlorophyll into the ether layer following the slow addition of water saturated with NaCl.
- Wash the ether three or four times with an identical volume of water (+5 mM Na-ascorbate).
- Dry the extract over anhydrous Na_2SO_4, and evaporate to dryness.
- Dissolve the residue in a small volume of diethyl ether and spot on thin-layer silica gel plates (Merck, no. 105721).
- Develop the plates with chloroform/acetone (92:8, v/v). Bacteriochlorophyll *a* migrates as a blue band about halfway between the origin and the solvent front, while carotenoids and bacteriophaeophytin migrate close to the solvent front.

B. Retinal Proteins

A different class of photoactive pigments used by some prokaryotes to harvest light energy is the retinal-containing membrane-bound proteins such as bacteriorhodopsin found in *Halobacterium* and other Archaea of the family *Halobacteriaceae*, as well as in different groups of Bacteria: xanthorhodopsin of *Salinibacter* (*Bacteroidetes*), proteorhodopsin in some (mostly yet uncultured) marine bacteria (Béjà *et al.*, 2000). The retinal chromophore (Figure 1D) is bound by a Schiff base to a lysine residue of the protein. Upon excitation these pigments cause the formation of a proton gradient over the cell membrane. The exact wavelength of the absorption maximum of retinal proteins depends on the amino acid sequence of the protein. Bacteriorhodopsin of *Halobacterium* species has its absorption peak at 568 nm; *Salinibacter* xanthorhodopsin absorbs light best at 559 nm, and different variants of proteorhodopsin at 490–530 nm (Sabehi *et al.*, 2007). It should be noted that the pink-red colour of *Halobacterium* is primarily caused by C-50 carotenoids: α-bacterioruberin and derivatives (see below). When large amounts of bacteriorhodopsin are present, the colour changes from pink to purple. Other retinal proteins are known with different functions: halorhodopsin of *Halobacterium* is a light-driven inward chloride pump, and other similar proteins act as light sensors.

Another coloured protein that acts as a light sensor is the photoactive yellow protein first discovered in the photosynthetic *Halorhodospira halophila*, an organism that contains bacteriochlorophyll *a* and spirilloxanthin as its main pigments. Photoactive yellow proteins (PYPs) are 14-kDa water-soluble blue-light photoreceptors thought to mediate negative phototactic responses. The chromophore is deprotonated 4-hydroxycinnamic acid covalently bound to the single cysteine in the protein. Maximum absorption is at 446 nm.

C. Carotenoids

Carotenoids are widespread in the prokaryotic world, colouring the organisms in different shades of red, pink, orange or yellow. In photosynthetic prokaryotes they may harvest light and transfer the energy to the chlorophyll or bacteriochlorophyll in the photosynthetic reaction centres; in others cases the carotenoids may provide the cells with protection against photodynamic damage.

At least 500 naturally occurring carotenoids are known (Liaaen-Jensen and Andrewes, 1985). Representative structures of carotenoids of anoxygenic photosynthetic bacteria, of the oxygenic cyanobacteria, and of different non-photosynthetic prokaryotes are shown in Figures 2–4, respectively. Most carotenoids are 40-carbon compounds, formally derived from the basic structure of lycopene with a central chain of 11 conjugated double bonds, modified by hydrogenation, dehydrogenation, cyclization, oxidation, glycosylation or any combination of these. Oxygenated derivatives are termed xanthophylls. Carotenoids with different chain lengths also occur, such as the C-50 bacterioruberin derivatives responsible for most of the colour of the halophilic Archaea of the family *Halobacteriaceae* (Figure 4C) and C-30 carotenoids such as 4-hydroxy-4,4′-diaponeurosporene of *Staphylococcus aureus* (Figure 4A) (Taylor and Davies, 1974).

The diversity of carotenoids is especially great among the anoxygenic phototrophic bacteria, which derive most of their colour from carotenoid pigments (Table 2). More than 100 different carotenoids have been found in the photosynthetic prokaryotes (Schmidt, 1978; Takaichi, 1999). Five main pathways of carotenoid biosynthesis have been identified in the anoxygenic phototrophs: the spirilloxanthin pathway (normal spirilloxanthin, unusual spirilloxanthin, spheroidene and carotenal pathways), the okenone pathway, the isorenieratene pathway (isorenieratene, chlorobactene), the γ- and β-carotene pathway and the diapocarotene pathway. In addition, carotenoid glucosides and carotenoid glucoside fatty acid esters have also been found in some species. The *Rhodospirillaceae* and *Chromatiaceae* use the spirilloxanthin or the okenone pathway, depending on the genus or species. All of the *Ectothiorhodospiraceae* have the spirilloxanthin pathway. Spheroidene is found in a few species only,

Table 2. Carotenogenesis pathways and their distribution within anaerobic photosynthetic bacteria

Carotenogenesis pathway	Major components	Family, representative species
Spirilloxanthin pathway:		
Normal spirilloxanthin	Spirilloxanthin, rhodovibrin, anhydrorhodovibrin	*Rhodospirillaceae* *Chromatiaceae* *Ectothiorhodospiraceae*
Unusual spirilloxanthin	Rhodovibrin, anhydrorhodovibrin, rhodopin, lycopene	*Rhodospirillaceae*, *Chromatiaceae*, *Ectothiorhodospiraceae*
Spheroidene	Spheroidene, neurosporene, OH-spheroidene	*Rhodospirillaceae*
Carotenal	Rhodopinal, rhodopinol, lycopenol, lycopene	*Rhodospirillaceae*, *Chromatiaceae*
Okenone pathway:		
Okenone	Okenone	*Chromatiaceae*
R.g.-Keto-carotenoid[a]	*R.g.*-Keto II, *R.g.*-Keto III	*Rhodospirillaceae*
Isorenieratene pathway:		
Isorenieratene	Isorenieratene	*Chlorobiaceae*
Chlorobactene	Chlorobactene, OH-chlorobactene, γ-carotene	*Chlorobiaceae*
γ- and β-carotene pathway	γ-Carotene, β-carotene	'Chloroflexaceae'
Diapocarotene pathway	Diaponeurosporene	*Heliobacteriaceae*
Additional pathway:		
Glucoside	Rhodopinal glucoside, rhodopin glucoside, methoxyhydroxylycopene glycoside	*Rhodospirillaceae*, *Ectothiorhodospiraceae*
Glucoside ester	Dihydroxylycopene diglycoside ester, dihydroxylycopene diglycoside diester, OH-chlorobactene glucoside ester, OH-γ-carotene glucoside ester	*Ectothiorhodospiraceae*, *Chlorobiaceae*, 'Chloroflexaceae'

[a]R.g.=*Rhodopila globiformis*.
Modified from Takaishi (1999).

e.g. *Rhodobacter sphaeroides*, *Rhodobacter capsulatus*, and *Rubrivivax gelatinosus*. The isorenieratene, the γ- and β-carotene and the diapocarotene pathways are found specifically in the *Chlorobiaceae*, 'Chloroflexaceae' and *Heliobacteriaceae* species, respectively. *Heliobacteriaceae* species have only C-30 carotenes, 4,4′-diapocarotenes. Some *Ectothiorhodospiraceae*, *Chlorobiaceae* and 'Chloroflexaceae' species have carotenoid glucoside esters. Most aerobic photosynthetic bacteria have unusual carotenoids, such as carotenoid sulphates, carotenoic acids and hydroxyl derivatives of β-carotene. The carotenoid content of the cells may vary based on light intensity, age of culture and so on. The recommended standards for the description of new species of anoxygenic phototrophic bacteria (Imhoff and Caumette, 2004) state that 'It should be noted that the specific chemical identity of carotenoids cannot be delineated from absorption spectra and that more sophisticated chemical analyses are required to identify these pigments. Therefore, a qualitative and quantitative detailed pigment analysis is an optional property for the description of a new species.'

Carotenoids can be extracted by organic solvents. There are many different protocols in the literature. Most use a non-polar hydrocarbon (hexane, cyclohexane, isooctane) plus a polar component that is usually acetone or a small chain alcohol. Because of the great variation in properties of carotenoids, there are no simple universal protocols for their separation, purification and characterization. Two examples are presented below.

Extraction without saponification – see Eimhjellen and Liaaen-Jensen (1964).

- To a suspension of cells (e.g. 50 ml) add 2 volumes of acetone.
- Extract for 6 h at 3°C under an atmosphere of nitrogen.
- Remove the cells by filtration, and wash the filters with pure acetone until colourless.
- Transfer the extract to 50 ml petroleum ether in a separatory funnel.
- Re-extract the hypophase with 50 ml portions of petroleum ether until colourless.
- Combine the petroleum ether extracts and wash with water.
- Dry the extract over anhydrous Na_2SO_4.

With saponification – Suzue *et al.* (1967)

- To 100 g (wet weight) of cells add 200 ml 75% methanol, 20 g NaOH and 10 g pyrogallol.
- Heat for 30 min at 80–85°C.
- Add 200 ml distilled water.
- Extract three times with 100 ml hexane.
- Combine the hexane extracts, and wash five times with 300 ml water.
- Dry the extract over anhydrous Na_2SO_4.
- Concentrate to a small volume under reduced pressure.

Most separation protocols are based on thin-layer chromatography on MgO, silica gel or reverse-phase plates (Bolliger and König, 1969; Sadowski and Wójcik, 1983; Singh *et al.*, 1973). For analytical work, reverse-phase HPLC with octadecylsilylated silica is generally used. A typical protocol (Ronen *et al.*, 1999) uses a Spherisorb ODS-2 column (5 μm, 3.2 mm × 250 mm) (Phenomenex®), a sample size of 50 μl of acetone-dissolved pigments, a mobile phase consisting of acetonitrile/H_2O (9:1) (solvent A) and 100% ethyl acetate (solvent B), used in a linear gradient between A and B for 30 min, at a flow of 1 ml/min, and detection of the light absorption peaks in the range of 200–600 nm using a photo diode-array detector. Carotenoids are identified by their characteristic absorption spectra and their typical retention time. Identification of unknown carotenoids is generally first based on ^{1}H NMR and ^{13}C NMR spectra, as well as on mass spectroscopy.

D. Flexirubin Pigments

Many members of the *Bacteroidetes* (genera *Flexibacter*, *Cytophaga*, *Sporocytophaga* and relatives) are coloured yellow-orange to pink-red. Some have carotenoid pigments, but others may instead have additional pigments of the flexirubin group. Flexirubin pigments are aryl polyenes containing a polyenoic acid chromophore terminated by a *p*-hydroxyphenyl group and esterified with a dialkylated resorcinol. The polyene chain is mainly biosynthesized from acetate, the conjugated phenyl as well as the three adjacent carbon atoms along the chain are derived from tyrosine (Reichenbach *et al.*, 1981; Achenbach *et al.*, 1979a). Variations may occur in the length of the polyene chain (6–8 double bonds), in a methyl or chloro substituent in the *meta* position of the conjugated phenyl groups, and in the alkyl substituents of the esterified resorcinol. Around 50 different flexirubin-like pigments are known (Achenbach, 1987; Achenbach *et al.*, 1981; Reichenbach *et al.*, 1974, 1981). Figure 5B and C gives representative examples. Carotenoid pigments are usually produced by members of marine species while flexirubin pigments are more frequently associated with clinical, freshwater or soil organisms (Reichenbach, 1989).

A simple test for the presence of flexirubin pigments is based on the colour shift with alkali. Colonies having a flexirubrin type of pigment exhibit an immediate colour shift from yellow to orange or red, purple or brown when flooded with 20% KOH, and revert to their initial colour when flooded with an acidic solution once the excess of KOH has been removed. The colour change is not absolutely specific for the flexirubrin type of pigment, but it is still helpful when combined with the results of other tests (Bernardet *et al.*, 2002; Fautz and Reichenbach, 1980).

Another test to assess whether flexirubin pigments, carotenoids, or both are present is based on the incorporation of labelled biosynthetic precursors, followed by thin-layer chromatography and autoradiographic analysis of the chromatograms. Tyrosine does not enter carotenoids, mevalonic acid does not incorporate into flexirubins. A control with labelled methionine is added, which should be incorporated into both types of pigments (Fautz and Reichenbach, 1980):

- Grow cells in medium containing 1% peptone from caseine, 0.2% yeast extract, 0.1% $MgSO_4 \cdot 7H_2O$, pH 7.2.
- For radiolabelling inoculate with 20 μl of a dense culture into 2 ml portions of medium in 20 ml tubes.
- Add 0.5 μCi of L-[$^{14}CH_3$]-methionine, L-[U-^{14}C]-tyrosine (e.g. Moravek Biochemicals and Radiochemicals – Cat. No. MC 275) or D,L-[2-^{14}C]-mevalonic acid (DBED salt) (e.g. Moravek Biochemicals and Radiochemicals – Cat. No. MC 411).
- Agitate for 16 h on a rotary shaker at 30°C or a different temperature as appropriate.
- Harvest the cells by centrifugation.
- Extract the cells with 1–2 ml acetone or methanol.
- Dry the extracts in a stream of nitrogen. Dissolve residue in toluene.

- Separate the pigments on 20 cm silica gel plates (KG 60 F254, Merck), with the solvent petroleum ether (b.p. 60–80°C)/toluene/acetone (25:25:16, v/v/v).
- Photograph the chromatogram.
- Cover the chromatogram X ray film, expose for 2–4 weeks at 2°C and develop.

Procedures have been given for the preparative HPLC separation of dimethyl ethers of flexirubin-type pigments on a Nucleosil 10-C_{18} column eluted with acetone/tetrahydrofuran/water (Achenbach *et al.*, 1978, 1979b).

E. Other Pigments

A number of additional pigments have been identified in different groups of prokaryotes that cannot be classified within the classes of pigments discussed above. Table 3 provides a non-exhaustive list:

- Violacein of *Chromobacterium violaceum* (Figure 5A) is a purple pigment (maximum absorption at 570 nm) that has bactericidal, trypanocidal, tumoricidal and antiviral activity. It can be extracted from the cells with ethanol. It is produced only on media containing tryptophan (Rettori and Durán, 1998).
- Bacteria of the genus *Xanthomonas* contain yellow xanthomonadins. Xanthomonadin I (Figure 5F) is an isobutyl derivative of a natural ester in which the esterifying alcohol has not been identified. Xanthomonadins can be separated by TLC on silica gel developed with acetone–chloroform mixtures (Jenkins and Starr, 1982; Starr *et al.*, 1977).
- *Pseudomonas aeruginosa* strains produce two types of soluble pigments, the fluorescent pigment pyoverdin (Figure 5E) and the blue pigment pyocyanin (Figure 5C). The latter is produced abundantly in media of low-iron content and functions in iron metabolism in the bacterium. Pyocyanin (from 'pyocyaneus') refers to 'blue pus', which is a characteristic of suppurative infections caused by *P. aeruginosa*.
- *Serratia marcescens* is well known for its red colonies, caused by prodigiosin (Figure 5G).

Table 3. Various miscellaneous pigments from different prokaryotes

Pigment name	Genera	Absorption maximum (nm)	Illustration
Violacein	*Chromobacterium*	585 (in butanol)	Figure 5A
Xanthomonadin	*Xanthomonas*	444 (in methanol)	Figure 5F
Pyocyanin	*Pseudomonas*		Figure 5D
Pyoverdin	*Pseudomonas*	400 (in water)	Figure 5E
Prodigiosin	*Serratia*	535 (*in vivo*)	Figure 5G
Photoactive yellow protein	*Halorhodospira*, *Rhodobacter*	446	
Mycosporine-glycine	Cyanobacteria	310	

- A variety of UV-absorbing pigments have been identified from different types of cyanobacteria. These include both mycosporine-like amino acids dissolved in the cytoplasm (maximum absorption generally 330–335 nm or 360–365 nm, depending on the chemical structure) and scytonemin located in the extracellular sheath (maximum absorption at 384 nm). Mycosporine-like amino acids are condensation derivatives of a cyclohexenone ring and amino acid or imino alcohol residues; scytonemin is a dimeric indole alkaloid with no related compounds among natural products. Both protect the cells from harmful UV radiation (Castenholz and Garcia-Pichel, 2000).

♦♦♦♦♦♦ IV. CASE STUDIES

To provide recent examples of the elucidation of the nature of the pigments of new isolates of prokaryotes, two case studies are presented here: the finding of a novel carotenoid that acts as a light antenna for a retinal protein proton pump in *S. ruber*, and the identification of astaxanthin in novel isolates of *Paracoccus*.

A. The Pigments of *Salinibacter ruber*

Salinibacter ruber, a red-orange pigmented member of the *Bacteroidetes*, was first isolated from solar salterns in Spain (Antón *et al.*, 2002). The brines of solar saltern crystallizer ponds are coloured pink-red to orange-red, mainly due to the presence of C-50 bacterioruberin carotenoids of halophilic Archaea of the family *Halobacteriaceae* (Figure 4B) and to β-carotene accumulated in massive amounts within the chloroplast of the unicellular green alga *Dunaliella salina*. Colonies of *Salinibacter* are little more orange than the colonies of the Halobacteriaceae on agar plates, suggesting presence of a different pigment. The absorption spectrum (Figure 6) confirmed this. HPLC (reverse-phase, LiChrospher 100 RP-18 column eluted with a gradient from 70% acetone–30% water to 100% acetone) enabled separation of the single carotenoid pigment of *Salinibacter* from bacterioruberin pigments and from β-carotene, and this procedure was also used to directly demonstrate the presence of the novel carotenoid in the biomass of saltern crystallizer ponds (Oren and Rodríguez-Valera, 2001).

The pigment was purified by TLC and preparative HPLC. Lyophilized cells (9.4 g) were submitted to lysis with water (15 ml). The carotenoid was extracted with methanol/acetone (3:7, 3 × 150 ml) to give 2.5 mg of carotenoid. The solution was taken to dryness at reduced pressure and purified by repeated precipitation of white contaminants with cold acetone. The residue was subjected to column chromatography (CC) (silica) with increasing percentage of acetone in hexane. The carotenoid was eluted with acetone/hexane (1:1). The solution was taken almost to dryness (reduced pressure, room temperature) and subjected to preparative TLC (acetone/hexane, 2:3) eluted with methanol.

HPLC was performed on a Hewlett-Packard Series HP 1050 instrument equipped with a diode-array detector with detection wavelengths set to 450 and 480 nm. A Brownlee Spheri-5 RP-18 250 mm × 4.6 mm column was used with two different elution systems: (1) hexane/acetone/methanol/1 M ammonium acetate buffer 0:0:80:20 (0 min)–0:30:70:0 (30 min)–20:50:30:0 (50 min), 1.25 ml/min. (2) Acetone/methanol 0:100 (0–1 min)–15:85 (15 min), 1 ml/min. Silica/calcium carbonate (2:1) was used as stationary phase for preparative thin-layer chromatography. The purified pigment was characterized by a combination of visible spectra, EIMS (Electron Impact Mass Spectrometry), ^{1}H-^{1}H COSY (COrrelated SpectroscopY) and TOCSY (TOtal Correlation SpectroscopY) NMR, CD circular dichroism and chemical reactions. The fatty acid residue bound to the pigment was analysed by GC/MS as methyl ester following acid methanolysis. The structure of the novel pigment, named salinixanthin, was described as a new C_{40}-carotenoid acyl glycoside (Figure 4D) (Lutnæs *et al.*, 2002).

Genes encoding four retinal proteins were identified in the genome of *S. ruber*: a putative bacteriorhodopsin-like proton pump, a putative halorhodopsin-like chloride pump and two putative sensory rhodopsins. The proton pump, named xanthorhodopsin (absorption maximum ~575 nm) was found to be present in the membrane adjacent to salinixanthin, which acts as a light antenna, transferring light energy to the retinal of xanthorhodopsin with a high efficiency (Balashov *et al.*, 2005; Lanyi and Balashov, 2008).

B. Astaxanthin Production by *Paracoccus marcusii* and *Paracoccus carotinifaciens*

Several isolates belonging to the genus *Paracoccus* are coloured brightly orange. One such yellow strain appeared as a contaminant on an agar plate and was subsequently described as a new species, *Paracoccus marcusii* (Harker *et al.*, 1998). A related organism, also brightly orange, was obtained from Japanese soil, and named *Paracoccus carotinifaciens* (Tsubokura *et al.*, 1999).

To characterize the pigments(s) of *P. marcusii*, cells were collected by centrifugation; hydrophobic pigments were extracted with acetone at 25°C for 15 min in the dark, dried under a stream of nitrogen and stored at −20°C until analysis. Analysis included thin-layer chromatography on Silica gel 60 F254 plates, using either diethyl ether or hexane/ethyl acetate 3:2 (v/v) as solvent system and reverse-phase HPLC as described earlier (Fraser *et al.*, 1997):

- Extract cells with 3 volumes of 10% diethyl ether in petroleum ether (40–60°C boiling point).
- Centrifuge at 3000 × *g* for 4 min at 4°C to obtain phase separation.
- Collect the organic phase and re-extract the aqueous phase with the same solvent.
- Extract the remaining aqueous phase with 2 volumes of chloroform.
- Pool the organic extracts and dry under a stream of nitrogen.
- Separate carotenoids by HPLC on a reversed phase C_{18} column (Nova-pak HR 6μ C_{18}, 3.9 mm × 300 mm, Waters) using an isocratic mobile phase of acetonitrile/methanol/isopropyl alcohol (90:6:40, v/v). Record with an online photodiode array detector.

P. carotinifaciens cells were extracted with 0.5 ml dichloromethane/methanol (4:1, v/v)+0.5 ml hexane, and the extract was chromatographed by HPLC, using a Wako-sil 5SIL-120 column, eluted with *n*-hexane/dichloromethane/methanol (10:8:1, v/v) (Tsubokura *et al.*, 1999). Both methods led to the identification of the pigment as astaxanthin, based on its comparison with authentic astaxanthin (Figure 4B). Astaxanthin production had not been documented before in *Paracoccus* species.

For final confirmation of the structure of the pigment, mass spectrometry was carried out using positive-ion electron ionization on a VG 7070H double focusing magnetic sector mass spectrometer, using a Finnigan INCOS 2300 data system for data acquisition and processing. Full-scan mass spectra were recorded over the m/z range 40–700 at an accelerating voltage of 2 kV in a total time of 3.5 s. The probe temperature was raised gradually from ambient to $<300°C$ in about 5 min. The spectra were recorded at an ionization potential of 70 eV (Harker *et al.*, 1998).

References

Achenbach, H. (1987). The pigments of the flexirubin-type. A novel class of natural products. *Fortschr. Chem. Org. Naturst.* **52**, 73–111.

Achenbach, H., Kohl, W. and Reichenbach, H. (1978). The flexirubin-type pigments—a novel class of natural pigments from gliding bacteria. *Rev. Latinamer. Quim.* **9**, 111–124.

Achenbach, H., Böttger, A., Kohl, W., Fautz, E. and Reichenbach, H. (1979a). Untersuchungen zur Biogenese des Flexirubins – Herkunft des Benzolringes a und der aromatischen C-Methylgruppen. *Phytochemistry* **18**, 961–963.

Achenbach, H., Kohl, W., Alexanian, S. and Reichenbach, H. (1979b). Untersuchungen an Stoffwechselprodukten von Mikroorganismen, XVII. Neue Pigmente vom Flexirubin-Typ aus *Cytophaga* spec. Stamm Samoa. *Chem. Ber.* **112**, 196–208.

Achenbach, H., Kohl, W., Böttger-Vetter, A. and Reichenbach, H. (1981). Untersuchungen an Stoffwechselprodukten von Mikroorganismen, XXII: Untersuchung der Pigmente aus *Flavobacterium* spec. Stamm C 1/2. *Tetrahedron* **37**, 559–563.

Antón, J., Oren, A., Benlloch, S., Rodríguez-Valera, F., Amann, R. and Rosselló-Mora, R. (2002). *Salinibacter ruber* gen. nov., sp. nov., a novel extreme halophilic member of the Bacteria from saltern crystallizer ponds. *Int. J. Syst. Evol. Microbiol.* **52**, 485–491.

Balashov, S. P., Imasheva, E. S., Boichenko, V. A., Antón, J., Wang, J. M. and Lanyi, J. K. (2005). Xanthorhodopsin: a proton pump with a light-harvesting carotenoid antenna. *Science* **309**, 2061–2064.

Béjà, O., Aravind, L., Koonin, E. V., Suzuki, M. T., Hadd, A., Nguyen, L. P., Jovanovich, S. B., Gates, C. M., Feldman, R. A., Spudich, J. L., Spudich, E. N. and DeLong, E. F. (2000). Bacterial rhodopsin: evidence for a new type of phototrophy in the sea. *Science* **289**, 1902–1904.

Bernardet, J. F., Nakagawa, Y. and Holmes, B. (2002). Proposed minimal standards for describing new taxa of the family *Flavobacteriaceae* and emended description of the family. *Int. J. Syst. Evol. Microbiol.* **52**, 1049–1070.

Biebl, H. and Drews, G. (1969). Das in-vivo Absorptions spektrum als taxonomisches Merkmal bei Untersuchungen zur Verbreitung von Athiorhodaceae. *Zentbl. Bakteriol. Parasitenkd. Infektionskr. Hyg.* **123**, 425–452.

Bolliger, H. R. and König, A. (1969). Vitamins, including carotenoids, chlorophylls and biologically active quinones. In: *Thin Layer Chromatography* (E. Stahl, ed), 2nd ed. pp. 259–311. Springer-Verlag, Berlin.

Bricaud, A., Allali, K., Morel, A., Marie, D., Veldhuis, M. J. W., Partensky, F. and Vaulot, D. (1999). Divinyl chlorophyll *a*-specific absorption coefficients and absorption efficiency factors for *Prochlorococcus marinus*: kinetics of photoacclimation. *Mar. Ecol. Progr. Ser.* **188**, 21–32.
Castenholz, R. W. and Garcia-Pichel, F. (2000). Cyanobacterial responses to UV-radiation. In: *The Ecology of Cyanobacteria. Their Diversity in Time and Space* (B. A. Whitton and M. Potts, eds), pp. 591–611. Kluwer Academic Publishers, Dordrecht.
Eimhjellen, K. E. and Liaaen-Jensen, S. (1964). The biosynthesis of carotenoids in *Rhodopseudomonas gelatinosa*. *Biochim. Biophys. Acta* **82**, 21–40.
Fautz, E. and Reichenbach, H. (1980). A simple test for flexirubin-type pigments. *FEMS Microbiol. Lett.* **8**, 87–90.
Fraser, P. D., Miura, Y. and Misawa, N. (1997). *In vitro* characterization of astaxanthin biosynthetic enzymes. *J. Biol. Chem.* **272**, 6128–6135.
Goericke, L. and Repeta, D. J. (1992). The pigments of *Prochlorococcus marinus*: the presence of divinyl chlorophyll *a* and *b* in a marine prokaryote. *Limnol. Oceanogr.* **37**, 425–433.
Goericke, L. and Repeta, D. J. (1993). Chlorophylls *a* and *b* and divinyl chlorophylls *a* and *b* in the open subtropical North Atlantic Ocean. *Mar. Ecol. Progr. Ser.* **101**, 307–313.
Harker, M., Hirschberg, J. and Oren, A. (1998). *Paracoccus marcusii* sp. nov., an orange Gram-negative coccus. *Int. J. Syst. Bacteriol.* **48**, 543–548.
Imhoff, J. F. and Caumette, P. (2004). Recommended standards for the description of new species of anoxygenic phototrophic bacteria. *Int. J. Syst. Evol. Microbiol.* **54**, 1415–1421.
Jenkins, C. L. and Starr, M. P. (1982). The brominated aryl-polyene (xanthomonadin) pigments of *Xanthomonas juglandis* protect against photobiological damage. *Curr. Microbiol.* **7**, 195–198.
Kondratieva, E. N., Pfennig, N. and Trüper, H. G. (1992). The phototrophic prokaryotes. In: *The Prokaryotes. A Handbook on the Biology of Bacteria: Ecophysiology, Isolation, Identification, Applications* 2nd ed. (A. Balows, H. G. Trüper, M. Dworkin, W. Harder and K.-H. Schleifer, eds), Vol. I, pp. 312–330. Springer, New York.
Kühl, M., Chen, M., Ralph, P. J., Schreiber, U. and Larkum, A. W. D. (2005). Ecology: a niche for cyanobacteria containing chlorophyll *d*. *Nature* **433**, 820.
Lanyi, J. K. and Balashov, S. P. (2008). Xanthorhodopsin: a bacteriorhodopsin-like proton pump with a carotenoid antenna. *Biochim. Biophys. Acta* **1777**, 684–688.
Liaaen-Jensen, S. and Andrewes, A. G. (1985). Analysis of carotenoids and related polyene pigments. In: *Methods in Microbiology* (G. Gottschalk, ed), Vol. 18, pp. 235–255. Academic Press, London.
Lutnæs, B. F., Oren, A. and Liaaen-Jensen, S. (2002). New C_{40}-carotenoid acyl glycoside as principal carotenoid of *Salinibacter ruber*, an extremely halophilic eubacterium. *J. Nat. Prod.* **65**, 1340–1343.
Lutnæs, B. F., Strand, Å., Pétursdóttir, S. K. and Liaaen-Jensen, S. (2004). Carotenoids of thermophilic bacteria: *Rhodothermus marinus* from submarine Icelandic hot springs. *Biochem. Syst. Ecol.* **32**, 455–468.
Oelze, J. (1985). Analysis of bacteriochlorophyll. In: *Methods in Microbiology* (G. Gottschalk, ed), Vol. 18, pp. 257–284. Academic Press, London.
Oren, A. and Rodríguez-Valera, F. (2001). The contribution of *Salinibacter* species to the red coloration of saltern crystallizer ponds. *FEMS Microbiol. Ecol.* **36**, 123–130.
Overmann, J. and Garcia-Pichel, F. (2006). The phototrophic way of life. In: *The Prokaryotes. A Handbook on the Biology of Bacteria: Ecophysiology and Biochemistry 3rd ed.* (M. Dworkin, S. Falkow, E. Rosenberg, K.- H. Schleifer and E. Stackebrandt, eds), Vol. 2, pp. 32–85. Springer, New York.

Pfennig, N. (1978). General physiology and ecology of photosynthetic bacteria. In: *The photosynthetic bacteria* (R. K. Clayton and W. R. Sistrom, eds), pp. 3–18. Plenum Press, New York.

Reichenbach, H. (1989). Order I. *Cytophagales* Leadbetter 1974. In: *Bergey's Manual of Systematic Bacteriology* (J. T. Staley, M. P. Bryant and J. G. Holt, eds), Vol. 3, pp. 2011–2013. Williams & Wilkins, Baltimore.

Reichenbach, H., Kleinig, H. and Achenbach, H. (1974). The pigments of *Flexibacter elegans*: novel and chemosystematically useful compounds. *Arch. Microbiol.* **101**, 131–144.

Reichenbach, H., Kohl, W. and Achenbach, H. (1981). The flexirubin-type pigments, chemosystematically useful compounds. In: *The Flavobacterium-Cytophaga Group* (H. Reichenbach and O. B. Weeks, eds), pp. 101–108. Verlag Chemie, Weinheim.

Rettori, D. and Durán, N. (1998). Production, extraction and purification of violacein: an antibiotic pigment produced by *Chromobacterium violaceum*. *World J. Microbiol. Biotechnol.* **14**, 685–688.

Ronen, G., Cohen, M., Zamir, D. and Hirschberg, J. (1999). Regulation of carotenoid biosynthesis during tomato fruit development: expression of the gene for lycopene epsilon-cyclase is down-regulated during ripening and is elevated in the mutant. *Delta. Plant J.* **17**, 341–351.

Sabehi, G., Kirkup, B. C., Rosenberg, M., Stambler, N., Polz, M. F. and Béjà, O. (2007). Adaptation and spectral tuning in divergent marine proteorhodopsins from the eastern Mediterranean and the Sargasso Seas. *ISME J.* **1**, 48–55.

Sadowski, R. and Wójcik, W. (1983). Chromatography of chloroplast carotenoids on magnesium oxide thin layers. *J. Chromatogr.* **262**, 455–459.

Schmidt, K. (1978). Biosynthesis of carotenoids. In: *The Photosynthetic Bacteria* (R. K. Clayton and W. R. Sistrom, eds), pp. 729–750. Plenum Press, New York.

Shibata, K., Benson, A. A. and Calvin, M. (1954). The absorption spectra of suspensions of living micro-organisms. *Biochim. Biophys. Acta* **15**, 461–470.

Singh, H., John, J. and Cama, H. R. (1973). Separation of β-apocarotenals and related compounds by reversed-phase paper and thin-layer chromatography. *J. Chromatogr.* **75**, 146–150.

Starr, M. P., Jenkins, C. L., Bussey, L. B. and Andrewes, A. G. (1977). Chemotaxonomic significance of the xanthomonadins, novel brominated aryl-polyene pigments produced by bacteria of the genus *Xanthomonas*. *Arch. Microbiol.* **113**, 1–9.

Suzue, G., Tsukada, K. and Tanaka, S. (1967). A new triterpenoid from a mutant of *Staphylococcus aureus*. *Biochim. Biophys. Acta* **144**, 186–188.

Takaichi, S. (1999). Carotenoids and carotenogenesis in anoxygenic photosynthetic bacteria. In: *The Photochemistry of Carotenoids* (H. A. Frank, A. J. Young, G. Britton and R. J. Cogdell, eds), pp. 39–69. Kluwer Academic Publishers, Dordrecht.

Tandeau de Marsac, N. (1977). Occurrence and nature of chromatic adaptation in cyanobacteria. *J. Bacteriol.* **130**, 82–91.

Taylor, R. F. and Davies, B. H. (1974). Triterpenoid carotenoids and related lipids. The triterpenoid carotenes of *Streptococcus faecium* UNH 564P. *Biochem. J.* **139**, 751–760.

Tindall, B. J., Rosselló-Móra, R., Busse, H.-J., Ludwig, W. and Kämpfer, P. (2010). Notes on the characterization of prokaryote strains for taxonomic purposes. *Int. J. Syst. Evol. Microbiol.* **60**, 249–266.

Tsubokura, A., Yoneda, H. and Mizuta, H. (1999). *Paracoccus carotinifaciens* sp. nov., a new aerobic Gram-negative astaxanthin-producing bacterium. *Int. J. Syst. Bacteriol.* **49**, 277–282.

13 Characterization of Prokaryotes Using MALDI-TOF Mass Spectrometry

B.Bédis Dridi and Michel Drancourt
Unité de Recherche sur les Maladies Infectieuses et Tropicales Emergentes UMR CNRS 6236 IRD 198, IFR48, Institut Hospitalier Universitaire POLMIT, Université de la Méditerranée, Marseille, France

♦♦

CONTENTS

♦♦♦♦♦♦ I. INTRODUCTION

Matrix-assisted laser desorption/ionization time-of-flight (MALDI-TOF) mass spectrometry (MS) is a technique which gained importance over the last few years in prokaryote taxonomy. MALDI-TOF MS basically records the time of flight of proteins pulled up from the prokaryote cell in order to derive a protein profile to be compared with database (Figure 1). Pulling up prokaryote proteins is achieved by bombarding the prokaryote cells entrapped into a special matrix, with a laser beam, hence the designation matrix-assisted laser desorption/ ionization time of flight. MALDI-TOF MS mainly explores the polar, soluble proteins of the cell cytoplasm rather than apolar proteins anchored in the cell wall, and provides an m/z-based pattern of such proteins, where m designs the mass and z the electric charge of the proteins. Any modifications of the experimental conditions could potentially modify the pattern, but experience indicates

METHODS IN MICROBIOLOGY, VOLUME 38
0580-9517 DOI 10.1016/B978-0-12-387730-7.00013-9

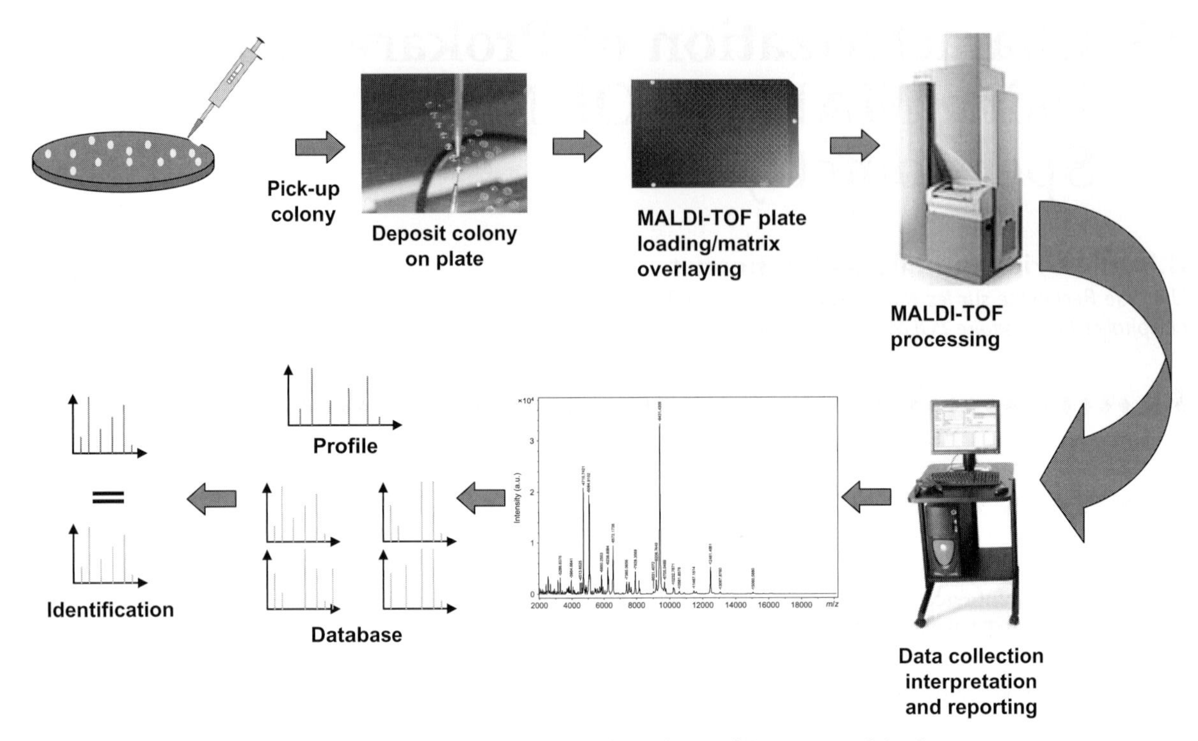

Figure 1. The principle of MALDI-TOF MS analysis of Bacteria and Archaea.

that not all of these parameters are equally important in determining the shape of the final protein pattern.

MALDI-TOF MS can contribute to the classification of members of the domains Bacteria and Archaea (La Scola *et al.*, 2010) and of viruses. It can also be used for the characterization of eukaryotic microorganisms such as the mold *Aspergillus* (Alanio *et al.*, 2010) and the yeast *Candida* (Ferroni *et al.*, 2010), for the identification of the source of cultured cell lines (Karger *et al.*, 2010), and for the classification of mammals based on the analysis of the dental pulp tissue (Tran *et al.*, 2011). The method is also useful for the characterization of human blood cells (Ouedraogo *et al.*, 2010).

Use of MALDI-TOF MS is particularly valuable to assess the nature of Bacteria and Archaea associated with inanimate and host-associated environments, including the human microbiota, where they can be regarded as either transient, commensal, or pathogen organisms (Dridi *et al.*, 2011). The low acidic pH of the matrix used in MALDI-TOF MS causes lysis of the cell envelope composed of peptidoglycans and glycerol-ester lipids in bacteria and of pseudomureins and glycerol-ether lipids in archaea (De Rosa *et al.*, 1986; Koga and Morii, 2007). There are few studies regarding the influence of the matrix on the performance of the MALDI-TOF MS analysis of bacteria (Liu *et al.*, 2007); in practice, a matrix formed by saturated α-cyano-4-hydroxycinnamic acid, 50% acetonitrile, and 2.5% trifluoroacetic acid is routinely used because it dries rapidly in 5 min without any further manipulation.

While MALDI-TOF MS could be a first step towards a fine, peptide-sequencing-based analysis of every protein based on the analysis of every peak, the MALDI-TOF MS profile could also be used as a barcode for the rapid identification of prokaryotes and to assess the phyletic relationships among organisms. These two applications rely on the setup of suitable MALDI-TOF MS profile databases and comparisons between a query profile and databases using appropriate software.

♦♦♦♦♦♦ II. INACTIVATING HARMFUL BACTERIA PRIOR TO MALDI-TOF ANALYSIS

Certain bacteria classified in the biosafety level (BSL) 3 are harmful for humans, exposing laboratory staff to unacceptable risk of contamination and infection (Baron and Miller, 2008). In addition, some of these organisms have been classified as group A or group B bioterrorism agents (Table 1). Inactivation prior to their manipulation is mandatory to protect laboratory staff and the environment. Although initial works on *Bacillus anthracis* (the anthrax agent), *Brucella melitensis* (the brucellosis agent), *Francisella tularensis* (the tularemia agent) and *Yersinia pestis* (the plague agent) used γ-irradiated organisms (Krishnamurthy *et al.*, 1996), further inactivation protocols have been developed for the routine practice (Table 1). Autoclaving *B. anthracis* spores has been used for the MALDI-TOF MS analysis of specific small acid-soluble proteins (Castanha *et al.*, 2006). A 30-min incubation in 80% TFA inactivates *B. anthracis* spores (Lasch *et al.*, 2008), *Y. pestis* (Lasch *et al.*, 2008, 2010) and *Burkholderia* spp. (Lasch *et al.*,

Table 1. List of pathogenic bacteria requiring appropriate inactivation prior to MALDI-TOF MS analysis

Organism	Inactivation treatment	Reference
Bacillus anthracis, spores	80% TFA, 30 min	Lasch *et al.*, 2008
	Autoclaving	Castanha *et al.*, 2006
	γ-Irradiation	Krishnamurthy *et al.*, 1996
Brucella melitensis	γ-Irradiation	Krishnamurthy *et al.*, 1996
Burkholderia spp.	80% TFA, 30 min	Lasch *et al.*, 2008
Francisella tularensis	γ-Irradiation	Krishnamurthy *et al.*, 1996
Mycobacterium tuberculosis	95°C, 30 min	Saleeb *et al.*, 2011
	95°C, 60 min	Djelouagji and Drancourt (2006); Drancourt M, unpublished
	70% Ethanol, 10 min	Lotz *et al.*, 2010
Yersinia pestis	γ-Irradiation	Krishnamurthy *et al.*, 1996
	70% Ethanol, 60 min	Ayyadurai *et al.*, 2010
	80% TFA, 30 min	Lasch *et al.*, 2008, 2010

2008). *Y. pestis* can also be inactivated by 60-min incubation in 70% ethanol (Ayyadurai *et al.*, 2010). *Mycobacterium tuberculosis* (the tuberculosis agent) can be inactivated after 10-min incubation in 70% ethanol (Lotz *et al.*, 2010) or by heating at 95°C for ≥30 min (Saleeb *et al.*, 2011; M. Drancourt, unpublished data). Our experience is that incubation in TFA significantly decreases the quality of MALDI-TOF MS spectra. We therefore propose it could rather be used for harmful bacteria in the absence of alternative inactivation protocol, i.e. sporulated organisms; heating bacteria at 95°C for ≥30 min could be used as a routine inactivation protocol for non-sporulating, harmful bacteria.

♦♦♦♦♦♦ III. PREPARING MALDI-TOF MS ANALYSIS OF PROKARYOTES

A. The Organisms

A great advantage of the MALDI-TOF MS approach is that it does not require any preparation of the organism to be analysed, and simply consists in picking-up a bacterial or archaeal colony from any solid medium, mixing the colony with the appropriate matrix on the MALDI plate, air-drying for 5 min, and performing the MALDI-TOF MS analysis (Figure 1) (see Standard Protocol, Section IV.A). Cultivation conditions on solid media appeared to have little effects for spectra generation and identification purposes, and several media including blood medium can be used as well as different growth states or growth temperatures. Accordingly, aerobic organisms grown under a microaerophilic atmosphere and under a 5% CO_2 atmosphere, as well as the anaerobic organisms can be used. Likewise, it was observed that the growth rate does not measurably influence the final profile. It is also possible to achieve MALDI-TOF identification of bacteria directly from a blood culture bottle (La Scola and Raoult, 2009). Preparation of bacteria recovered from blood-culture broths using various centrifugation and washing steps is beyond the scope of this chapter, and the reader can benefit from a recent review on that topic (Drancourt, 2010). In some cases of organisms with thick and resistant cell walls, have to be lysed using various combinations of mechanical, enzymatic and chemical breaking protocols. This holds true for mycobacteria (Mycobacteria Protocol, see Section IV.B) and for some archaea (Archaea Protocol, see Section IV.C).

B. The Matrix

The matrix contains acetonitrile (AN) and trifluoroacetic acid (TFA), and must therefore be manipulated in a chemical hood while wearing gloves. The matrix solution consists of a saturated solution of α-cyano-4-hydroxycinnamic acid (α-HCCA) in 50% AN, and 2.5% TFA. The composition of 50% AN/2.5% TFA is designated below as 'OS' (basic organic solvent). The pipetting scheme for 1 ml OS is 500 μl AN, 475 μl distilled water, 25 μl pure TFA; Add the α-HCCA crystals in a plastic Eppendorf tube and mix with 200–500 μl OS; Vortex for a few minutes at room temperature until the solution is saturated. Alternatively, use Bruker 'HCCA matrix portioned' and add 250 μl 'OS'. The prepared matrix can be stored at room temperature for up to 1 week.

C. The Plate

Colonies have to be deposited onto the MALDI-TOF plate, and the plate needs to be carefully washed prior to use in order to minimize the risk of false or uninterpretable protein spectra.

- Transfer the plate into a crystallizer (8 × 4 cm) and flood the plate with 70% of aqueous ethanol.
- Incubate for 5 min at room temperature.
- Remove the plate and rinse it carefully with warm tape water.
- Wipe the plate vigorously with a non-fluffy paper impregnated with 70% aqueous ethanol.
- Repeat the last two steps.
- Put the plate into the crystallizer and pour 150 μl 80% aqueous TFA using a pipette and vigorously wipe all the wells using a paper towel. (Note: To prepare 80% TFA, always add the acid to the water and not the opposite way.)
- Rinse the plate with demineralized water and wipe it with a paper towel.
- Let the plate completely dry for at least 15 min at room temperature before use.

D. The Controls

All plates must contain negative controls (a drop of non-inoculated matrix) in 1–4 wells. Positive controls may consist of a calibration solution of seven polypeptides of known molecular weight such as 2.00 μl bradykinin (1,060 Da), 2.60 μl angiotensin (1,296.5 Da), 3.14 μl Glu-fibrino peptide B (1570.6 Da), 3.52 μl renin substrate tetradecapeptide (1759 Da), 4.94 μl ACTH (18–39, 2465.7 Da), 22.93 μl insulin (bovine, 5733.5 Da) and 171.20 μl ubiquitin (bovine, 8564.9 Da) added to 789.67 μl 0.1% TFA solution. All chemicals can be purchased from Sigma/Aldrich Chemical (St. Louis, MO). A stock solution (1 mg/ml) of each standard is made up in HPLC-grade water, and stored at −20°C. Alternatively, a bacterial strain with a known MALDI-TOF MS profile can be used as a positive control, such as the Bacterial Test Standard (protein extract of *Escherichia coli* DH5alpha; Bruker Daltonics ref. 255343).

♦♦♦♦♦♦ IV. PROTOCOLS FOR MALDI-TOF MS ANALYSIS OF PROKARYOTES

A. Standard Protocol

The standard method consists of the direct transfer of a bacterial colony to the MALDI-TOF MS plate.

- The starting material is an isolated colony grown on solid medium. Initial manipulations involving potentially harmful organisms must be done under a safety hood in a laboratory at the appropriate safety level (2 or 3). Once the organism has been mixed with the matrix, it can be regarded as inactivated and may be manipulated outside the hood.
- Using the tip of a 10 μl pipette, collect one colony and deposit it as a thin layer in a well of the plate. The standard plate consists of polished steel and contains 96 wells.
- Add 1 μl of matrix.
- Dry in the chemical hood for 5 min at room temperature.
- Add negative control (a drop of non-inoculated matrix) into 1–4 wells, as well as positive controls mixed with the matrix to the plate. Positive control may either consist of a mixture of known proteins with a known MALDI-TOF profile (such mixtures are commercially available by the MALDI-TOF MS suppliers) or a bacterial strain with a known MALDI-TOF profile. Dry the matrix.
- Place the plate into the appropriate mass spectrometer compartment.
- Run the analysis.
- Interpret the data. Negative controls should be negative and the positive controls should yield the expected MALDI-TOF profiles, otherwise the analysis is invalidated. Experimental profiles must be compared with those of the database in order to get identification (Figure 2). Depending on the software, a cut-off value is provided for identification at the species level, identification at the genus level or absence of reliable identification. Experimental profiles can be incorporated into a hierarchic clustering analysis to get the position of the experimental profile among those of most closely related profiles (organisms) (Figure 3).

B. Mycobacteria Protocol

The use of MALDI-TOF MS for identifying mycobacteria has seldom been reported (Hettick *et al.*, 2004, 2006; Pignone *et al.*, 2006; Lotz *et al.*, 2010; Saleeb *et al.*, 2011). Mycobacteria have a different cell wall structure than other bacteria due to the presence of mycolic acids which render the mycobacterial cell wall more resistant to physical forces, chemicals and enzymes. In our experience, it is therefore necessary to increase the inoculum of intact mycobacterial cells or break the mycobacterial cell wall prior to MALDI-TOF MS analysis. A few papers have been published in which the inoculum used for analysis was not specified, or a large inoculum was mentioned. Here, we provide a protocol currently used in our laboratory that includes breaking of the mycobacterial cell, starting with either mycobacteria grown in liquid 7H9 Middlebrook medium or with mycobacteria grown in 7H10 Middlebrook or blood-agar-based solid medium (Drancourt and Raoult, 2007). The protocol starting with mycobacteria grown in 7H9 broth in MGIT tubes (Becton Dickinson) is as follows:

- Distribute 4 ml of 7H9 broth culture into two screw-capped tubes (2 ml each).
- Inactivate the mycobacteria by heating at 95°C for 1 h.

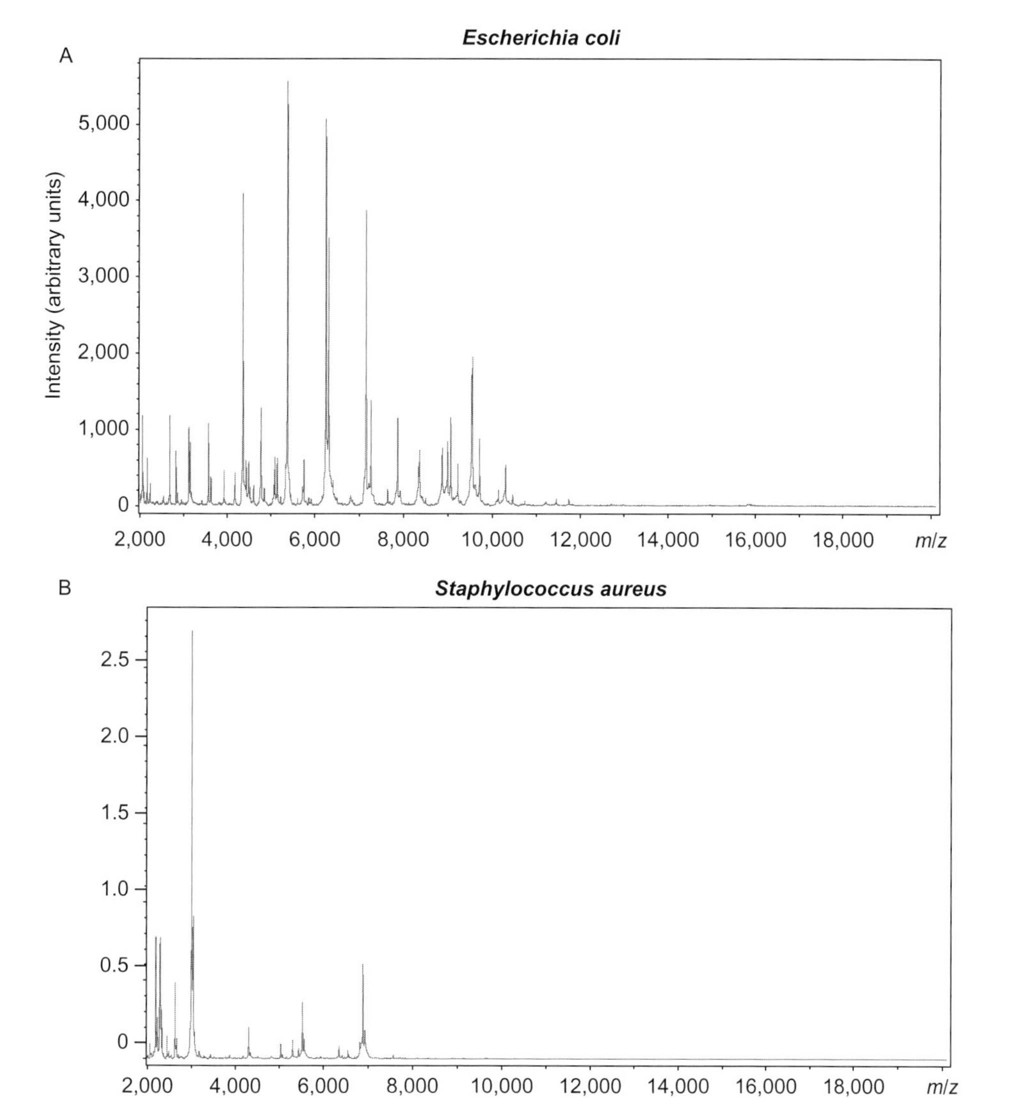

Figure 2. Typical MALDI-TOF MS spectra obtained for clinical isolates of *Escherichia coli* (A) and *Staphylococcus aureus* (B).

- Centrifuge the tubes for 10 min at 17,000 × *g*.
- Discard the supernatant and keep the pellet.
- Resuspend each pellet in 1 ml water (HPLC grade).
- Centrifuge for 10 min at 17,000 × *g*.
- Discard the supernatant and keep the pellet.
- Rinse each pellet twice with 1 ml water (HPLC grade).
- Mix the pellets with a pipette.
- Centrifuge for 10 min at 17,000 × *g*.
- Discard the supernatant and keep the pellet.

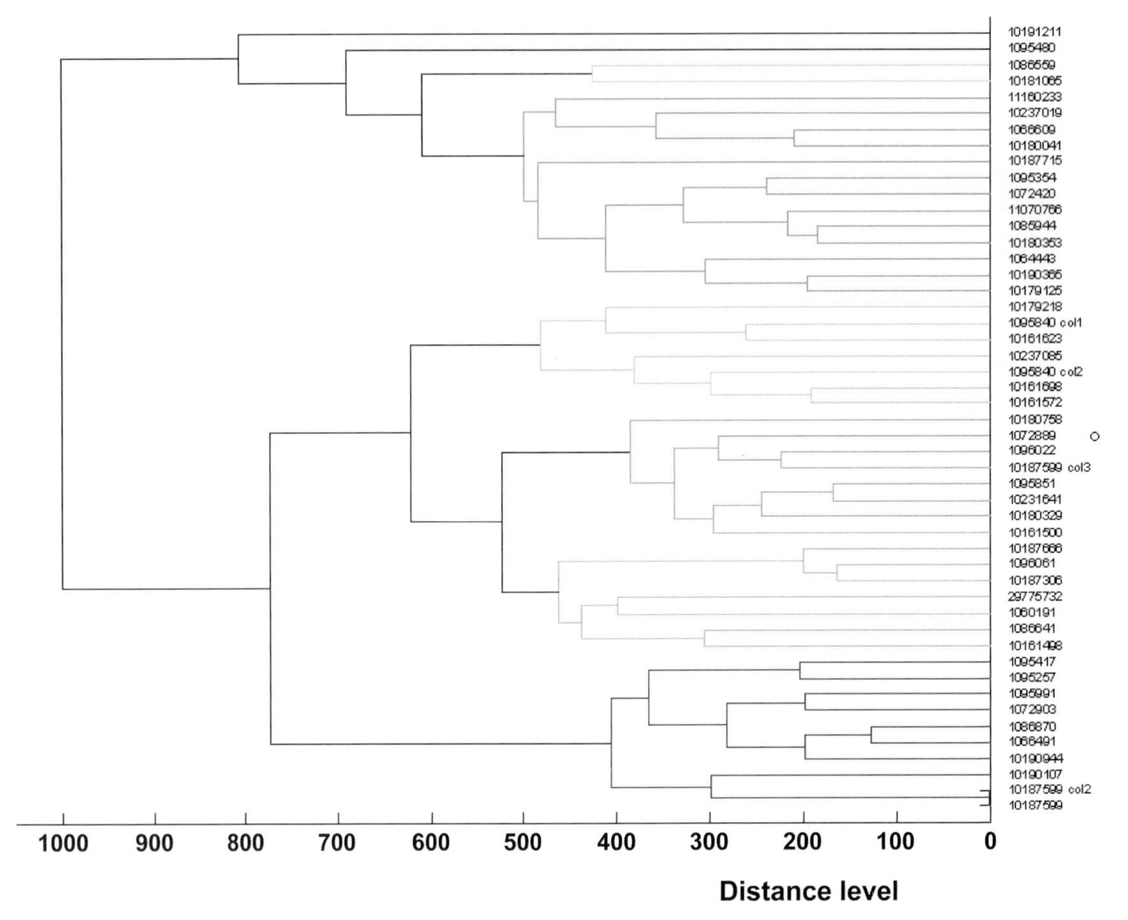

Figure 3. A dendrogram depicting MALDI-TOF MS protein profiles of *Escherichia coli* clinical isolates from Marseille.

- Resuspend the pellet in 500 μl water (HPLC grade) and add one spatula full of acid-washed glass beads (<106 μm; Sigma-Aldrich).
- Agitate using a FastPrep BIO 101 apparatus (Qbiogene, Strasbourg, France) (program 2), using four 45 s cycles.
- Transfer the supernatant to a new Eppendorf tube.
- Centrifuge 10 min at 17,000 × *g*.
- Withdraw the supernatant and resuspend the pellet in 5–10 μl of 70% HCOOH (adjust the volume according to the size of the bacterial pellet), and add between 5 and 10 μl of 100% AN (the same volume as the volume of formic acid used), and mix with a pipette.
- Centrifuge at 17,000 × *g* for 1 min.
- Deposit 1.5 μl of supernatant on a MALDI-TOF plate (four spots), let dry and cover with 1.5 μl of matrix as soon as it is dried.
- Analyse by MALDI-TOF mass spectrometer (Figure 4).

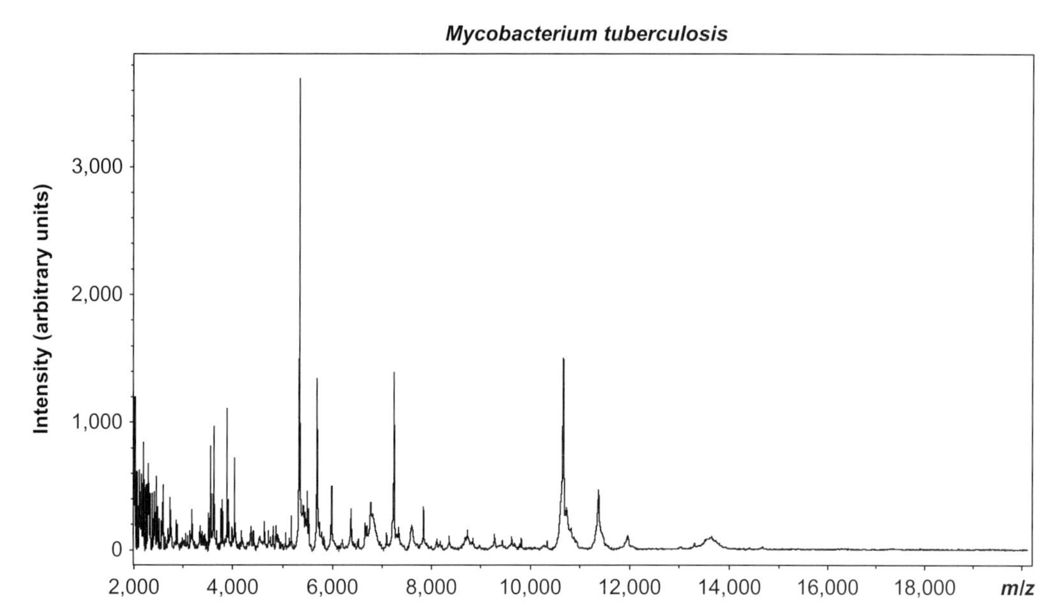

Figure 4. A typical MALDI-TOF MS spectrum obtained with a clinical isolate of *Mycobacterium tuberculosis* using the mycobacteria protocol described.

C. Archaea Protocol

Only one study has been published regarding the identification of archaea by MALDI-TOF MS (Krader and Emerson, 2004). This pioneering study was limited to distantly related, environmental organisms and explored a somewhat limited range of peaks <3,000 Da. We have developed a new method (B. Dridi, D. Raoult, M. Drancourt, unpublished data), which is presented below.

- Grow archaea in appropriate medium and under appropriate conditions.
- Transfer 1 ml of culture into a sterile screw-capped tube containing 0.3 g of acid-washed glass beads (diameter ≤106 μm, Sigma-Aldrich).
- Mechanically break the cells using a FastPrep BIO 101 apparatus (Qbiogene, Strasbourg, France) at level 6.5 (full speed) for 60 s.
- Centrifuge the supernatant at 13,000 × *g* for 5 min.
- Suspend the pellet in 500 μl of sterile water.
- Centrifuge at 13,000 × *g* for 5 min.
- Resuspend the pellet in 10 μl of sterile water.
- Deposit 1.5 μl of the suspension covered with 1.5 μl of matrix solution [saturated solution of α-cyano-4-hydroxycinnamic acid (α-HCCA) in 50% AN, 2.5% TFA] on a TP 384 target plate made of polished steel TF (Bruker Daltonics, Leipzig, Germany).
- Made four deposits for each isolate.
- Add negative and positive controls.
- Air-dry for 5 min.

- Record spectra, e.g. using the Microflex (Bruker Daltonics) in the positive linear mode for masses ranging from 2 to 20 kDa (parameter settings: ion source 1 (IS1), 20 kV; IS2, 18.5 kV; lens, 7 kV). A spectrum is obtained after 675 shots with a variable laser power. Automated data acquisition is achieved by using the software tool AutoXecute acquisition control. For spectra analysis and comparison, use FlexAnalysis 2.4 software (Bruker Daltonics).

D. Preparation for Freeze-Dried Bacteria

Freeze-dried bacteria are suitable for MALDI-TOF MS analysis (Krishnamurthy *et al.*, 1996), particularly with the aim of recalibrating a spectrometer after a breakdown. The method of freeze-drying consists in removing water from bacterial colonies by sublimation, i.e. by passing water from the solid state to the gaseous state in a lyophilizer. This technique makes it possible to preserve the properties of treated proteins. The freeze-drying cycle consists of:

- freezing, where the products are cooled at temperatures of −20°C to −80°C with water transformed into ice;
- the primary desiccation by vacuum sublimating free ice; and
- the secondary desiccation by desorption of water molecules trapped on the surface of the dried products.

At the end of the cycle, the product contains less than 1–5% water.

The method involves suspending microbial colonies to 200-μl HPLC grade water, homogenization, freeze to −80°C outside or inside the lyophilizer and freeze-drying.

For MALDI-TOF MS analysis of the freeze-dried stocks, either of the following two methods can be used:

- Suspend the freeze-dried bacteria in HPLC grade water and deposit 1 μl suspension and 1 μl of matrix HCCA on the MALDI-TOF plate.
- Poor the freeze-dried bacteria in matrix HCCA. Deposit 2 μl suspension on the MALDI-TOF plate.

♦♦♦♦♦♦ V. ANALYSING PROKARYOTES BY MALDI-TOF MS

A. Prokaryote Identification

Contrary to gene sequence databases, few prokaryote protein profile databases are commercially available; public databases, however, exist in which researchers may want to deposit their original prokaryotic protein profiles. The Proteome Commons database (http://proteomecommons.org/) is one of such databases. Depositing the protein profile of a newly described prokaryotic reference strain in such a database may become a recommendation as part as the initial description of a new prokaryote taxon. While several mass spectrometer suppliers exist, there are currently (March 2011) three companies providing databases and software directed to the identification and analysis of prokaryotes, i.e. Bruker Daltonics providing the Biotyper system, AnagnosTec providing the Saramis system and Andromas SAS providing the Andromas system(www.andomas.com). Using these systems, MALDI-TOF MS is currently the first line method for the rapid identification of bacteria and fungi in Europe (Seng *et al.*, 2009, 2010; Carbonnelle *et al.*, 2011) while being increasingly used in the United States (Dhiman *et al.*, 2011). It could also be used for the rapid identification of environmental Archaea (Krader and Emerson, 2004) as well as Archaea associated with host microbiota (Seng *et al.*, 2010).

The advantages of MALDI-TOF MS for identifying archaea and bacteria include the simplicity of the protocols which basically rely on the deposition of intact cells on an appropriate matrix on the MS plate and introduction of later plate in the mass spectrometer; the simplicity and rapidity of personnel training (2 h); the speed of the procedure which takes <5 min, being faster than the routine Gram staining (Seng *et al.*, 2009); and the low cost of the procedure which has been estimated as being of average <1 € per analysis. Current limitations of the approach include the need for a dense inoculum (average 10^8 colony-forming units) in order to achieve an interpretable spectrum; the necessity for cell preparation prior to mixing with matrix in about 10% of the bacterial (see Mycobacteria Protocol, Section IV.B) and archaeal isolates (see Archaea Protocol, Section IV.C); the requirement for at least 10 reference spectra in the database to achieve a correct species-level identification, thus temporarily limiting its usefulness for new and emerging species (Seng *et al.*, 2009) and the impossibility of distinguishing some closely related species such as *Streptococcus pneumoniae* from *Streptococcus mitis* (Seng *et al.*, 2009) or enterohemorrhagic *E. coli* and *Shigella* spp. isolates (He *et al.*, 2010). However, MALDI-TOF MS distinguishes between the subspecies of *Salmonella enterica* (Dieckmann *et al.*, 2008). None of these limitations is related to structural limitations of the method but rather to weak taxonomic relationships between species and on-going improvements in both the protocols, and software will continue to increase the power and field of applications of MALDI-TOF MS for the identification and characterization of Bacteria and Archaea.

B. Prokaryote Clustering

Few studies explored the clustering of MALDI-TOF MS protein profiles derived from prokaryotes. The interpretation of the MALDI-TOF MS profile-based dendrograms remains uncertain. While MALDI-TOF MS profile clustering revealed unsuspected clustering of *Corynebacterium pseudodiphtheriticum* strains among cystic fibrosis patients (Bittar *et al.*, 2010), other clustering analyses could be difficult to interpret: for example, the apparent clustering of Panton-Valentine leukocidin-positive *Staphylococcus aureus* isolates collected from a single microbial laboratory unravelled clustering of their genetic background rather than their secretion of toxin (Bittar *et al.*, 2009; Dauwalder *et al.*, 2010). As for Archaea, we observed that such clustering agreed with the one derived from the 16S rRNA gene sequencing of the same organisms (B. Dridi, D. Raoult, M. Drancourt, unpublished data).

♦♦♦♦♦♦ VI. QUALITY ASSURANCE IN MALDI-TOF MS ANALYSIS OF PROKARYOTES

Quality assurance in MALDI-TOF MS analysis of prokaryotes is based on four different levels. A first level is the regular maintenance of equipment and software according to the manufacturer's instructions. A second level is monitoring and recording of basic mass spectrometer parameters, i.e. laser power. A third level is based on the systematic introduction of controls in every plate. Negative controls consisting of non-inoculated matrix ensure that the plate is clean, otherwise leading to mixed, non-identifying spectra; negative controls are expected to give no spectrum or faint, non-identified spectra. Positive controls may consist of commercially available calibration suspensions with known m/z values, of commercially available *E. coli* protein extracts with known m/z values, and of reference strains such as reference *E. coli* and reference *S. aureus* with known profiles. As for reference strains, lyophilized stocks are suitable for such controls. These controls are expected to give a stable identification score and a reproducible spectrum in terms of number and intensity of peaks. A fourth level of quality assurance relies on determining key parameters of the analysis as we use in our laboratory. For each MALDI-TOF MS profile, we analyse peaks between 3 and 11 kDa by determining the following parameters: (1) the signal/noise ratio, (2) the resolution, (3) the relative intensity and (4) the area under the peak. We then calculate the total area under the peak for each organism. We next determine the percentage of peaks exhibiting a relative intensity of $\geq 10\%$. We then determine the highest values of signal/noise and resolution for each profile and retain the lowest value for the four profiles determined for each tested organism. The quality score includes four balanced parameters: (1) the

Table 2. Scoring the quality of the obtained MALDI-TOF MS protein profiles for 12 archaeal strains

	Signal/ noise	Resolution	Area	Percentage of peaks with relative intensity >10%	Score
Methanobrevibacter smithii DSM 861^{T}	57.5	1028	23,776	38.4	7
Methanobrevibacter smithii DSM 2374	55.3	1014	22,333	37	7
Methanobrevibacter smithii DSM 2375	58.2	1054	27,556	39.6	7
Methanobrevibacter smithii DSM 11975	61.3	1072	29,163	42.1	7
Methanobrevibacter oralis DSM 7256^{T}	61.5	1090	28,511	43	7
Methanosphaera stadtmanae DSM 3091^{T}	95.7	723	192,742	100	10
Staphylothermus hellenicus DSM 12710^{T}	62.2	3266	24,168	88.2	8
Methanospirillum hungatei DSM 864^{T}	98.3	475	139,1271	50	7
Natrinema pallidum DSM 15623^{T}	59.3	375	639,981	90	7
Sulfolobus solfataricus DSM 1616^{T}	22.9	622	224,403	100	6
Methanosarcina acetivorans DSM 2834^{T}	7.3	508	26,829	100	6
Natronomonas pharaonis DSM 2160^{T}	25.1	449	163,368	100	7
Average (standard deviation)					7.16 (1.03)

signal/noise ratio with a balance of 4 for values ≥25; (2) the resolution with a balance of 3 for values ≥500; (3) the area under the peak with a balance of 2 for values ≥50,000 and (4) the percentage of peaks with relative intensity of ≥10% with a balance of 1 for values ≥50. For example, excellent scores of 7.16 ± 1.03 (mean ± standard deviation) were obtained for archaea (Table 2).

References

Alanio, A., Beretti, J. L., Dauphin, B., Mellado, E., Quesne, G., Lacroix, C., Amara, A., Berche, P., Nassif, X. and Bougnoux, M. E. (2010). Matrix-assisted laser desorption ionization time-of-flight mass spectrometry for fast and accurate identification of clinically relevant *Aspergillus* species. *Clin. Microbiol. Infect.* **17**, 750–755.

Ayyadurai, S., Flaudrops, C., Raoult, D. and Drancourt, M. (2010). Rapid identification and typing of *Yersinia pestis* and other *Yersinia* species by matrix-assisted laser desorption/ionization time-of-flight (MALDI-TOF) mass spectrometry. *BMC Microbiol.* **10**, 285.

Baron, E. J. and Miller, J. M. (2008). Bacterial and fungal infections among diagnostic laboratory workers: evaluating the risks. *Diagn. Microbiol. Infect. Dis.* **60**, 241–246.

Bittar, F., Ouchenane, Z., Smati, F., Raoult, D. and Rolain, J. M. (2009). MALDI-TOF-MS for rapid detection of staphylococcal Panton-Valentine leukocidin. *Int. J. Antimicrob. Agents.* **34**, 467–470.

Bittar, F., Cassagne, C., Bosdure, E., Stremler, N., Dubus, J. C., Sarles, J., Reynaud-Gaubert, M., Raoult, D. and Rolain, J. M. (2010). Outbreak of *Corynebacterium pseudodiphtheriticum* infection in cystic fibrosis patients, France. *Emerg. Infect. Dis.* **16**, 1231–1236.

Carbonnelle, E., Mesquita, C., Bille, E., Day, N., Dauphin, B., Beretti, J. L., Ferroni, A., Gutmann, L. and Nassif, X. (2011). MALDI-TOF mass spectrometry tools for bacterial identification in clinical microbiology laboratory. *Clin. Biochem.* **44**, 104–109.

Castanha, E. R., Fox, A. and Fox, K. F. (2006). Rapid discrimination of *Bacillus anthracis* from other members of the *B. cereus* group by mass and sequence of "intact" small acid soluble proteins (SASPs) using mass spectrometry. *J. Microbiol. Meth.* **67**, 230–240.

Dauwalder, O., Carbonnelle, E., Benito, Y., Lina, G., Nassif, X., Vandenesch, F. and Laurent, F. (2010). Detection of Panton-Valentine toxin in *Staphylococcus aureus* by mass spectrometry directly from colony: time has not yet come. *Int. J. Antimicrob. Agents.* **36**, 193–194.

De Rosa, M., Gambacorta, A. and Gliozzi, A. (1986). Structure, biosynthesis, and physicochemical properties of archaebacterial lipids. *Microbiol. Rev.* **50**, 70–80.

Dhiman, N., Hall, L., Wohlfiel, S. L., Buckwalter, S. P. and Wengenack, N. L. (2011). Performance and cost analysis of matrix-assisted laser desorption ionization-time of flight mass spectrometry for routine identification of yeast. *J. Clin. Microbiol.* **49**, 1614–1616.

Dieckmann, R., Helmuth, R., Erhard, M. and Malorny, B. (2008). Rapid classification and identification of salmonellae at the species and subspecies levels by whole-cell matrix-assisted laser desorption ionization-time of flight mass spectrometry. *Appl. Environ. Microbiol.* **74**, 7767–7778.

Djelouagji, Z. and Drancourt, M. (2006). Inactivation of cultured *Mycobacterium tuberculosis* organisms prior to DNA extraction. *J. Clin. Microbiol.* **44**, 1594–1595.

Drancourt, M. (2010). Detection of microorganisms in blood specimens using matrix-assisted laser desorption ionization time-of-flight mass spectrometry: a review. *Clin. Microbiol. Infect.* **16**, 1620–1625.

Drancourt, M. and Raoult, D. (2007). Cost-effectiveness of blood agar for isolation of mycobacteria. *PLoS Negl. Trop. Dis.* **1**, e83.

Dridi, B., Raoult, D. and Drancourt, M. (2011). Archaea as emerging organisms in complex human microbiomes. *Anaerobe* **17**, 56–63.

Ferroni, A., Suarez, S., Beretti, J. L., Dauphin, B., Bille, E., Meyer, J., Bougnoux, M. E., Alanio, A., Berche, P. and Nassif, X. (2010). Real-time identification of bacteria and *Candida* species in positive blood culture broths by matrix-assisted laser desorption ionization-time of flight mass spectrometry. *J. Clin. Microbiol.* **48**, 1542–1548.

He, Y., Li, H., Lu, X., Stratton, C. W. and Tang, Y. W. (2010). Mass spectrometry biotyper system identifies enteric bacterial pathogens directly from colonies grown on selective stool culture media. *J. Clin. Microbiol.* **48**, 3888–3892.

Hettick, J. M., Kashon, M. L., Simpson, J. P., Siegel, P. D., Maruzek, G. H. and Weissman, D. N. (2004). Proteomic profiling of intact mycobacteria by matrix-assisted laser desorption/ionization time-of-flight mass spectrometry. *Anal. Chem.* **76**, 5769–5776.

Hettick, J. M., Kashon, M. L., Slaven, J. E., Ma, Y., Simpson, J. P., Siegel, P. D., Maruzek, G. N. and Weissman, D. N. (2006). Discrimination of intact mycobacteria at the strain level: a combined MALDI-TOF MS and biostatical analysis. *Proteomics* **6**, 6416–6425.

Karger, A., Bettin, B., Lenk, M. and Mettenleiter, T. C. (2010). Rapid characterisation of cell cultures by matrix-assisted laser desorption/ionisation mass spectrometric typing. *J. Virol. Meth.* **164**, 116–121.

Koga, Y. and Morii, H. (2007). Biosynthesis of ether-type polar lipids in archaea and evolutionary considerations. *Microbiol. Mol. Biol. Rev.* **71**, 97–120.

Krader, P. and Emerson, D. (2004). Identification of archaea and some extremophilic bacteria using matrix-assisted laser desorption/ionization time-of-flight (MALDI-TOF) mass spectrometry. *Extremophiles* **8**, 259–268.

Krishnamurthy, T., Ross, P. L. and Rajamani, U. (1996). Detection of pathogenic and non-pathogenic bacteria by matrix-assisted laser desorption/ionization time-of-flight mass spectrometry. *Rapid Commun. Mass. Spectrom.* **10**, 883–888.

Lasch, P., Nattermann, H., Erhard, M., Stämmler, M., Grunow, R., Bannert, N., Appel, B. and Naumann, D. (2008). MALDI-TOF mass spectrometry compatible inactivation method for highly pathogenic microbial cells and spores. *Anal. Chem.* **80**, 2026–2034.

Lasch, P., Drevinek, M., Nattermann, H., Grunow, R., Stämmler, M., Dieckmann, R., Schwecke, T. and Naumann, D. (2010). Characterization of *Yersinia* using MALDI-TOF mass spectrometry and chemometrics. *Anal. Chem.* **82**, 8464–8475.

La Scola, B. and Raoult, D. (2009). Direct identification of bacteria in positive blood culture bottles by matrix-assisted laser desorption ionisation time-of-flight mass spectrometry. *PLoS One* **4**, e8041.

La Scola, B., Campocasso, A., N'Dong, R., Fournous, G., Barrassi, L., Flaudrops, C. and Raoult, D. (2010). Tentative characterization of new environmental giant viruses by MALDI-TOF mass spectrometry. *Intervirology* **53**, 344–353.

Liu, H., Du, Z., Wang, J. and Yang, R. (2007). Universal sample preparation method for characterization of bacteria by matrix-assisted laser desorption ionization-time of flight mass spectrometry. *Appl. Environ. Microbiol.* **73**, 1899–1907.

Lotz, A., Ferroni, A., Beretti, J. L., Dauphin, B., Carbonnelle, E., Guet-Revillet, H., Veziris, N., Heym, B., Jarlier, V., Gaillard, J. L., Pierre-Audigier, C., Frapy, E., Berche, P., Nassif, X. and Bille, E. (2010). Rapid identification of mycobacterial whole cells in solid and liquid culture media by matrix-assisted laser desorption ionization-time of flight mass spectrometry. *J. Clin. Microbiol.* **48**, 4481–4486.

Ouedraogo, R., Flaudrops, C., Ben Amara, A., Capo, C., Raoult, D. and Mege, J. L. (2010). Global analysis of circulating immune cells by matrix-assisted laser desorption ionization time-of-flight mass spectrometry. *PLoS One.* **5**, e13691.

Pignone, M., Greth, K. M., Cooper, J., Emerson, D. and Tang, J. (2006). Identification of mycobacteria by matrix-assisted laser desorption ionization-time-of-flight mass spectrometry. *J. Clin. Microbiol.* **44**, 1963–1970.

Saleeb, P. G., Drake, S. K., Murray, P. R. and Zelazny, A. M. (2011). Identification of mycobacteria in solid-culture media by matrix-assisted laser desorption ionization-time of flight mass spectrometry. *J. Clin. Microbiol.* **49**, 1790–1794.

Seng, P., Drancourt, M., Gouriet, F., La Scola, B., Fournier, P. E., Rolain, J. M. and Raoult, D. (2009). Ongoing revolution in bacteriology: routine identification of bacteria by matrix-assisted laser desorption ionization time-of-flight mass spectrometry. *Clin. Infect. Dis.* **49**, 543–551.

Seng, P., Rolain, J. M., Fournier, P. E., La Scola, B., Drancourt, M. and Raoult, D. (2010). MALDI-TOF-mass spectrometry applications in clinical microbiology. *Future Microbiol.* **5**, 1733–1754.

Tran, T. N., Aboudharam, G., Gardeisen, A., Davoust, B., Bocquet-Appel, J. P., Flaudrops, C., Belghazi, M., Raoult, D. and Drancourt, M. (2011). Classification of ancient mammal individuals using dental pulp MALDI-TOF MS peptide profiling. *PLoS One.* **6**, e17319.

14 Determination of the G+C Content of Prokaryotes

Noha M. Mesbah[1], William B. Whitman[2], and Mostafa Mesbah[3]

[1] *Department of Biochemistry, Faculty of Pharmacy, Suez Canal University, Ismailia, Egypt;* [2] *Department of Microbiology, University of Georgia, Athens, GA 30602, USA;* [3] *Department of Pharmacognosy, Faculty of Pharmacy, Suez Canal University, Ismailia, Egypt*

♦♦♦

CONTENTS

♦♦♦♦♦♦ I. INTRODUCTION

Polyphasic taxonomy refers to the delineation of taxa at all levels using a combination of genotypic, phenotypic and phylogentic information (Vandamme *et al.*, 1996). Genotypic information is derived from the nucleic acids (DNA and RNA) present in the cell and, in general, genotypic methods, such as 16S rRNA gene or internal transcribed spacer sequence analysis, dominate modern taxonomic studies. Determination of the DNA base composition (mole% G+C) is one of the classical genotypic methods and is considered a part of a standard description of bacterial taxa (Vandamme *et al.*, 1996). DNA base composition is also an important property of cellular DNA, and the genomic G+C content has a major impact on codon usage (Sharp *et al.*, 2005).

Genomic G+C content varies greatly in the prokaryotic world, the gammaproteobacterial endosymbiont *Carsonella ruddii* has a G+C-content of 16.5% (Nakabachi *et al.*, 2006), whereas the myxobacterium *Anaeromyxobacter dehalogenans* has a G+C-content of 75% (Hildebrand *et al.*, 2010). Differences in genomic G+C

METHODS IN MICROBIOLOGY, VOLUME 38
0580-9517 DOI 10.1016/B978-0-12-387730-7.00014-0

content often follow major phylogenetic divisions, such as between the *Actinobacteria* (high-GC) and the *Firmicutes* (low-GC). However, large differences have also been observed within certain lineages such as the *Gammaproteobacteria, Spirochaetes* and *Euryarchaeota* (Haft *et al.*, 2005). In taxonomic studies however, genomic G+C content is commonly used as an exclusionary determinant in bacterial classification, where two strains that differ by more than 10 mol% are not considered as members of the same genus but similar DNA base composition does not necessarily imply that the two strains are closely related (Schleifer, 2009). The genomic G+C content is commonly used to differentiate between bacterial taxa at the genus and species levels. In general, the mol% G+C range is not larger than 3% within a species and not more than 10% within a genus (Vandamme *et al.*, 1996).

Genomic G+C content of DNA was initially determined by two methods, buoyant density centrifugation (Schildkraut *et al.*, 1962) and thermal denaturation (Marmur and Doty, 1962). Since then, new methods and techniques have been described to measure the genomic G+C content, including separation of nucleosides by high-performance liquid chromatography (HPLC) (Ko *et al.*, 1977; Mesbah *et al.*, 1989; Mischke and Wickstrom, 1980; Tamaoka and Komagata, 1984) and fluorometic determination of melting temperature (González and Saiz-Jimenez, 2002). In this chapter, detailed descriptions of the different methods employed are described and a brief comparison of G+C content values obtained by different methods is presented.

♦♦♦♦♦♦ II. EXTRACTION AND QUANTIFICATION OF DNA

Extraction of high-molecular weight genomic DNA can be a laborious and time-consuming process, but it is critical to ensure the accuracy and reproducibility of the genomic G+C content determination, regardless of the method used. All methods for determination work best with high-molecular weight DNA. In addition, methods based on chromatography and spectrophotometric analyses require DNA that is free from RNA contamination.

Presented below is a scaled up CTAB method that can be used to extract large quantities of high-molecular weight genomic DNA from prokaryotes. This method is based on that described in *Current Protocols in Molecular Biology* (Wilson, 1997). The procedure can be done using either 1.5 ml microfuge, or 30 ml centrifuge tubes. Quantities for both methods are indicated.

A. Materials Needed

TE buffer (10 mM Tris, 1 mM EDTA and pH 8.0)
Lysozyme
20 mg/ml proteinase K
5 M NaCl
CTAB/NaCl
24:1 chloroform:isoamyl alcohol

25:24:1 phenol:chloroform:isoamyl alcohol
Isopropanol
0% ethanol
RNase A (100 mg/ml)

B. Reagent Preparation

1. CTAB/NaCl (hexadecyltrimethyl ammonium bromide)

Dissolve 4.1 g of NaCl in 80 ml of water and slowly add 10 g of CTAB while heating (to 65°C) and stirring. Adjust the final volume to 100 ml and sterilize by autoclaving or filtration. It can take more than 3 h to dissolve CTAB.

2. Procedure

	1.5 ml	30 ml
1. Inoculate a liquid culture with the prokaryotic strain under study and incubate till the culture reaches late-exponential phase.		
2. Centrifuge the culture to form a compact pellet. For optimal results, DNA should be extracted from at least 500 mg (wet weight) of cells.		
3. Resuspend the cell pellet in TE buffer.	740 µl	14.8 ml
4. Adjust to optical density (at 600 nm) of 1.0 with TE buffer.		
5. Transfer the indicated amount of cells' suspension to a clean centrifuge tube.	20 µl	400 µl
6. Add lysozyme (100 mg/ml) and mix well.		
7. Incubate for 5 min at room temperature.		
8. Add 10% SDS. Mix well. Do not vortex.	40 µl	800 µl
9. Add proteinase K (10 mg/ml). Mix well.	8 µl	160 µl
10. Incubate at 37°C for 1 h		
11. Add 5-M NaCl. Mix well.	100 µl	2 ml
12. Add CTAB/NaCl (pre-heated to 65°C). Mix well by inverting several times. Do not vortex.	100 µl	2 ml
13. Incubate at 65°C for 10 min.		
14. Add chloroform:isoamyl alcohol (24:1). Mix by inverting several times.	500 µl	10 ml
15. Centrifuge at maximum speed for 10 min at room temperature.		
16. Transfer upper aqueous phase to clean centrifuge tube.		
17. Add phenol:chloroform:isoamyl alcohol (25:24:1). Mix well by inverting several times.	500 µl	10 ml
18. Centrifuge at maximum speed for 10 min at room temperature.		
19. Transfer aqueous phase to a clean centrifuge tube. Add 0.6 volume of isopropanol (at room temperature).		
20. Incubate at room temperature for 30 min.		
21. Centrifuge at maximum speed for 15 min.		
22. Wash the pellet with 70% ethanol and centrifuge at maximum speed for 5 min.		

23. Discard the supernatant and let pellet dry for 5–10 min at room temperature.		
24. Resuspend the pellet in TE buffer plus RNase (99 μl TE+1 μl RNase).	20 μl	400 μl
25. Incubate at 37°C for 30 min		

In all cases, mixing should be done by inverting the tube. Vortexing or over mixing will result in shearing of the DNA. If using 30 ml centrifuge tubes, it is necessary to use Oak Ridge tubes (Nalgene®) made from Teflon, to avoid interaction with phenol:cholorform. For determination of genomic G+C content, it is preferred to use DNA samples with concentrations of at least 10 μg (preferably 10–25 μg) to ensure reproducible results. Agarose gel electrophoresis (1–1.2% wt/vol) along with molecular weight standards is the method of choice for analysis of DNA quality and concentration. DNA samples that show excessive degradation, shearing or contaminating RNA should not be used as this can lead to inaccurate results.

In our experience, commercial kits for genomic DNA extraction yield small amounts (~50 μg) of DNA with a significant amount of shearing. Care should be taken with any extraction methods based on phenol:chloroform; residual organic solvents must be completely removed as they are inhibitory to enzymes used in later digestion and degradation procedures and interfere with spectrophotometric measurement. Likewise, ethanol used in precipitation of DNA must be completely removed to avoid interferences with downstream enzymatic procedures.

Excessive amounts of RNA present in the DNA sample result in incomplete DNA degradation and the presence of non-specific peaks on resultant chromatograms, which can lead to inaccurate base content calculation. Therefore, it is imperative to treat extracted DNA with RNase enzyme or equivalent to minimize contaminating RNA.

♦♦♦♦♦♦ III. DETERMINATION OF GENOMIC G+C CONTENT BY BUOYANT DENSITY CENTRIFUGATION

The density of DNA in CsCl increases linearly with the mol% G+C content. An equation correlating the density of DNA preparations with the mol% G+C values determined chemically was introduced by Schildkraut *et al.* (1962) and was subsequently applied and validated (De Ley, 1970).

The protocol described below is based on that described by Mandel *et al.* (1968).

A. Materials Needed

Caesium chloride (CsCl), optical grade, free of rubidium and divalent cations, particularly calcium and magnesium.

2 M Tris buffer, pH 8.5

CsCl stock solution: Prepared by dissolving 30 g of CsCl in 70 ml of the Tris buffer. The solution is filtered to remove insoluble material.

Standard DNA: A standard DNA that will band in a position outside the region of the density gradient in which most DNA samples band should be used. DNA from *Escherichia coli* strains B or K12 is typically used as the primary standard. Its density is 1.704 g/ml. DNA can also be extracted from virulent *Bacillus subtilis* phages (e.g. SP8). The DNA from these phages is dense, 1.742 ± 0.001 g/ml.

B. Preparation of DNA Sample

1. In a test tube, mix:
   a. 0.84 ml CsCl stock solution
   b. 0.18 ml H_2O
   c. 0.01 ml of standard DNA (50 μg/ml) with known density
   d. 0.04 ml of test DNA (50 μg/ml)

   The volume of water can be reduced if more dilute DNA solutions are used.

2. Remove 15 μl of the sample for determination of the refractive index of the final solution. The density of the CsCl solution containing the DNA should be approximately 1.71 g/ml. Based on the linear relation between refractive index and density (Eq. (1)),

$$\rho_{25.0^\circ} = 10.8601_{\eta D}{}^{25.0^\circ} - 13.4974 \quad (1)$$

   the refractive index should be approximately 1.4000. Adjustment to this value should be made with small amounts of water or solid CsCl.

3. Add 0.7 ml of the final CsCl solution to the centrifuge tube containing a plastic (Kel-F) centrepiece. This can be done by a 1 ml syringe.

C. Ultracentrifugation and Photography

Density gradient centrifugation is carried out at 44,770 rpm at 25°C. Equillibrium is reached in approximately 20 h. Lower centrifugation speeds can be used, but longer times must be allowed to attain equilibrium.

Ultraviolet absorption photographs are taken using a conventional camera fitted with a UV confocal for visible and UV wavelengths and commercial film. Light in the 260 μm range can be obtained using a broad-band xenon lamp and a monochromator.

Density tracings are measured with a recoding microdensitometer with a slit width 50 μm within the film dimension.

D. Calculation of Buoyant Density and G+C Content

Densities are calculated using the position of the standard DNA as a reference. To determine the buoyant density of the DNA at a distance *r* from the centre of rotation:

$$\rho = \rho_o + 4.2\omega^2(r^2 - r_o^2) \times 10^{-10}\ \text{g/cm}^3 \quad (2)$$

where

ρ_o = density of the standard DNA
ω = speed of rotation in radians/s
r_o = distance of the standard DNA from the centre of rotation
r = distance of the test DNA from the centre of rotation

Assuming that standard DNA from *B. subtilis* phage SP8 is used. The distance between the standard DNA band and the centre of rotation (r_o) is 6.818 cm (ρ_o=1.742 g/ml). The test DNA banded at 6.590 cm (r) from the centre of rotation.

At 44,770 rpm:

$$4.2\omega^2 \times 10^{-10} = 0.0092 \tag{3}$$

By substituting in Eq. (2):

$$\rho = 1.742 + 0.0092(6.590^2 - 6.818^2) = 1.714 \text{ g/ml} \tag{4}$$

The corresponding G+C content of the test DNA is obtained from the linear relation described by Schildkraut *et al.* (1962):

$$\text{G} + \text{C} = \frac{\rho - 1.660 \text{ g/ml}}{0.098} \tag{5}$$

where G+C is the mol fraction of guanine plus cytosine in the test DNA. In the example above, the G+C content is 0.55, or 55%.

♦♦♦♦♦♦ IV. DETERMINATION OF GENOMIC G+C CONTENT BY THERMAL DENATURATION

When DNA is heated, its strands separate, or denature. Denaturation can be measured with a spectrophotometer. As DNA goes from the native, double-stranded state to the denatured, single-stranded state, its absorbance at 260 nm will increase by approximately 40% (Johnson, 1985). A thermal melting profile is constructed, which correlates the temperature with change in absorbance of a DNA sample at 260 nm. The thermal melting profile is generated by gradually heating a DNA sample in the cuvette chamber of a spectrophotometer and continuously measuring the absorbance of the sample and temperature within the cuvette. The midpoint temperature of the thermal melting profile of a DNA sample increases with increasing mol% G+C content. The melting temperature (T_m) is the temperature at which half of the DNA sample is denatured.

DNA samples must be dialysed against the same buffer to control the ionic strength of the samples, as the ionic strength affects the melting temperature of DNA. A DNA standard must be incorporated into every spectrophotometer run to standardize the temperature.

The protocol below is based on that described by Johnson (1985).

A. Materials Needed

0.5X standard saline citrate (SSC): 0.15 M NaCl, 0.015 M sodium citrate and pH 7.0

Dialysis tubing (*1.25* cm × *125* cm): Tubing should be washed by placement in a 1–2% solution of sodium carbonate and boiling for 5 min. Tubing is then rinsed extensively with tap water and placed into a beaker of distilled water.

Standard DNA: DNA from *E. coli* strain B has a G+C content of 51 mol% and a T_m of 90.5% in SSC.

B. Procedure

- Prepare 2–4 l of 0.5X SSC.
- Dilute DNA samples to 50 μg/ml using 0.5X SSC. Samples can be 0.1–2 ml in volume, depending on the size of the cuvettes used. Prepare 5–10 ml of the standard DNA, since it will be used in each instrument run.
- Dialyse all DNA preparations overnight in the 0.5X SSC at 4°C. Change the buffer once during the dialysis time.
- Determine the melting profiles with an automatic recording spectrophotometer having a sample chamber heated by a circulating bath or an electrically heated cuvette holder. A thermometer should be located ideally within the cuvette, or within the cuvette chamber or cuvette holder. Samples should be analysed three to four times and be placed in different cuvette positions to control for temperature variations that may exist between different cuvette positions.
- Beginning at 60°C, determine the T_m values of the standard and test DNA samples as shown in Figure 1.
- Calculate the mol% G+C of each sample using Eq. (6):

$$\text{mol\%G} + \text{C} = \frac{[(A - B) + C] - 69.3}{0.41} \tag{6}$$

where

$A = T_m$ of DNA standard in 1X SSC (90.5 for *E. coli* B)
$B = T_m$ of DNA standard in 0.5X SSC
$C = T_m$ of test DNA in 0.5X SSC

The $A - B$ value is the difference between the T_m of the standard DNA in 1X SSC and that in 0.5X SSC. Adding this value to the T_m of the test DNA normalizes it to a T_m in SSC. The rest of the values are derived from the equation described by Marmur and Doty (1962). When plotting the T_m values of different microorganisms versus the mol% G+C, the line intersected the temperature axis at 69.3°C and its gradient was 0.41.

An advantage of this method is that it is not sensitive to RNA contamination. However, if there is secondary structure in RNA fragments, the resultant T_m might increase slightly. DNA fragment size affects the T_m. The T_m of high-molecular weight gDNA was 1–2°C higher than that of the same DNA

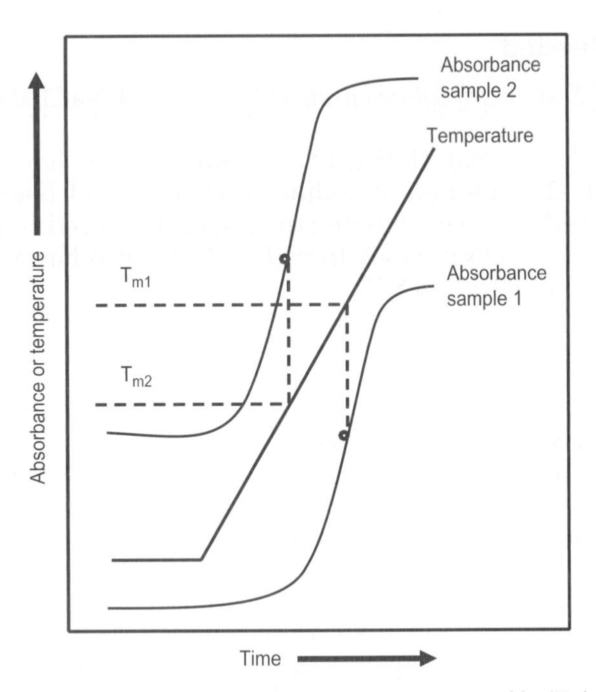

Figure 1. Thermal melting profile of DNA. The midpoint of the profile (T_m) is used for comparison with other T_m values. (Adapted from Johnson (1985)).

preparation after it was passed through a French pressure cell (Johnson, 1985). It is important that all DNA samples be prepared in the same way to avoid variation in measured T_m values.

♦♦♦♦♦♦ V. DETERMINATION OF GENOMIC G+C CONTENT BY FLUORIMETRIC ESTIMATION OF THERMAL DENATURATION TEMPERATURE

This method uses a fluorescent, double-strand specific dye and the software capabilities of a quantitative, real-time polymerase chain reaction (PCR) thermocycler (González and Saiz-Jimenez, 2002). Real-time PCR thermocyclers perform the melting curve experiments, where the genomic DNA is added to a thin-walled, optically clear tube along with a fluorescent dye that specifically binds to double-stranded DNA. The temperature is raised gradually, and the fluorescence of the sample is measured by the optical detection hardware of the real-time instrument. As the double-stranded DNA denatures, the fluorescent signal of the dye decreases. The systems optical detection software plots the melting curve of the DNA sample and can estimate the T_m.

SYBR Green I is the most commonly used fluorescent dye. It binds specifically to double-stranded DNA. Using this dye, double-stranded DNA molecules can be exclusively quantified in the presence of single-stranded DNA molecules during denaturation experiments. SYBR Green I shows maximum fluorescence at excitation and emission wavelengths of 497 and 520 nm, respectively. These peaks coincide with the fluoresence activated molecule (FAM) filter set available in the standard configuration of real-time PCR thermocyclers.

The protocol below can be used for determination of the denaturation temperature of DNA.

A. Materials Needed

Test DNA (~5 μg) per reaction
SYBR Green I (final dilution 1:1,000,000)
0.1X SSC:0.03 M NaCl, 0.03 M sodium citrate and pH 7.0
PCR tubes, plates, thin-walled and optically clear

B. Procedure

- Dilute genomic DNA in 0.1X SSC so that there is ~5 μg/tube,
- Add SYBR Green dye at a final dilution of 1:1,000,000 (final volume can range from 25 to 100 μl) depending on real-time instrument used).
- Set the real-time instrument to the following thermal conditions: 25–100°C at 1 °C/min, fluorescence measured at each step during the increase in temperature
- Analyse the melting curve of each sample, instrument optical detection software will calculate the T_m of each sample
- Calculate the mol% G+C content of the sample using the equation proposed by González and Saiz-Jimenez (2002):

$$\%G + C = 1.98T_m - 106.91 \quad (7)$$

Eq. (7) is derived from linear regression of calibration curves constructed from the T_m of total genomic DNA from different, reference microorganisms and their mol% G+C contents (González and Saiz-Jimenez, 2002).

♦♦♦♦♦♦ VI. DETERMINATION OF GENOMIC G+C CONTENT BY HIGH-PERFORMANCE LIQUID CHROMATOGRAPHY

The method involves digestion of DNA into individual nucleosides followed by separation of the individual nucleotides by reverse-phase HPLC. Genomic G+C content of DNA is then calculated from the ratio of deoxyguanosine (dGuo) to deoxythymidine (dThd). Although, in principle, the ratios of other nucleosides can be used, deoxycytidine (dCyt) and deoxyadenosine (dAde) are often modified in prokaryotes, and values based upon these bases are unreliable. The method described below is based on that described by Mesbah *et al.* (1989) with the modification described by Lee *et al.* (2005).

A. Digestion of DNA to Individual Nucleosides

Digestion involves first boiling DNA samples to separate the two strands, digestion of single-stranded DNA to nucleotides using S1 nuclease and digestion of nucleotides to nucleosides by bovine intestinal alkaline phosphatase. It is recommended that replicate digestions are performed to test for inhibitors of the degradation and ensure reproducibility.

1. Separation of DNA strands

It is important to boil DNA samples for a complete 2 min in order to ensure full strand separation. S1 nuclease cleaves double-stranded DNA poorly, and incomplete strand separation will result in irregularities in the calculated base ratios. Following boiling, DNA samples are immediately chilled on ice to prevent renaturation of the DNA strands and degradation of the DNA samples by contaminating nucleases.

Although G+C content determination can be carried out with as little as 0.5 μg of DNA, it is preferable that 10–25 μg of DNA be used to ensure accurate results. If the DNA contains high concentrations of salts or Tris–EDTA buffer, they can be removed by dialysis on a filter membrane. Removal of high concentrations of salts is essential because they interfere with digestion and result in errors in determining the base ratios.

2. Digestion of DNA

Single-stranded DNA is digested to nucleotides with S1 nuclease. Nucleotides are digested to nucleosides with bovine intestinal alkaline phosphatase.

S1 nuclease is a single-stranded endonuclease that hydrolyses single-stranded DNA and RNA into 5′-mononucleotides. The enzyme is added in excess to the digestion reaction to ensure complete degradation of double-stranded DNA. S1 nuclease requires zinc as a cofactor. Therefore, the concentration of zinc must exceed that of EDTA or other metal chelators that might be present in the DNA. The pH optimum for enzyme activity is 4.0–4.3, the pH of the digestion reaction is adjusted with 0.3-M sodium acetate.

The nucleotides are converted to nucleosides by incubating with bovine intestinal alkaline phosphatase. As with S1 nuclease, phosphatase activity is sensitive to pH and excessive salts and shows specificity among nucleotide monophosphates. For instance, deoxyguanosine monophosphate (dGMP) is hydrolysed more slowly than other deoxynucleotides, and the ratio of nucleosides released early in the degradation fluctuates greatly. Also, degradation with an insufficient amount of enzyme causes variable ratios of released nucleosides (Mesbah *et al.*, 1989).

Bovine intestinal alkaline phosphatase is usually supplied in 50% glycerol and is stored at −20°C. To prepare a working stock solution, the enzyme is diluted to 1 U/10 μl in 0.1-M glycine–HCl buffer (pH 10.4, adjusted with 5 M NaOH). The enzyme is stable for several days in this buffer, which is also used to adjust the pH of the acidic nuclease digestion reaction to the conditions optimal for the phosphatase. Ten microlitres of diluted enzyme is added to samples after completion of nuclease digestion, and the samples are incubated overnight.

A standard DNA must always be digested concomitantly with samples to serve as a control for calculation of the mol% and ensure the effectiveness of the digestion procedure. Salmon sperm or calf thymus DNA (2–5 μg/degradation) is routinely used as a standard in our laboratories.

Care must be taken in storing S1 nuclease enzyme as it is rapidly inactivated by even a few freeze-thaw cycles. S1 nuclease is commonly supplied as a concentrated solution in glycerol. Upon delivery to the laboratory, S1 nuclease should be diluted to 1 U/μl in 30-mM sodium acetate buffer (pH 5.1), dispensed to 25 μl aliquots and immediately stored at −20°C. Aliquots are thawed on ice and diluted to 1 U/5μl immediately before use. Aliquots must be discarded after use. If 1 U of enzyme hydrolyses 1 μg of DNA in 1 min, then 1–2 U should be sufficient to hydrolyse 25 μg of DNA in 30 min. To ensure complete degradation, the digestion reaction should be allowed to proceed for at least 1 h.

Incomplete nuclease degradation has a large effect on the ratios of nucleotide release. In general, dGMP is released much more slowly than other nucleotides, and the rate of nucleotide release varies greatly early in the degradation. Similarly, when small amounts of the enzyme are used in the digestion reaction, the ratios of nucleotides release fluctuate greatly (Mesbah *et al.*, 1989). In another source of error, residual deoxyadenosine monophosphate (dAMP) from incomplete degradation can co-elute with nucleosides critical for calculating the mol% and produce large errors. Therefore, enzymes with good activity are essential to obtaining accurate measurements.

The protocol below describes the procedure used in our laboratories for degradation of DNA.

(a) Materials needed

Sodium acetate, 0.3 M, pH 5.1 (adjusted with acetic acid)
Sodium acetate, 30 mM, pH 5.1 (adjusted with acetic acid)
Zinc sulfate, 20 mM
Glycine–HCl, 0.1 M, pH 10.4 (adjusted with 5 M NaOH)
S1 Nuclease
Bovine intestinal alkaline phosphatase

(b) Preparation of enzymes

- *S1 nuclease*: Dilute enzyme to 1-U/μl concentration with 30 mM sodium acetate. Distribute to 25 μl aliquots and store at −20°C. Prior to use, thaw aliquot on ice. Dilute to 1 U/5 μl with 0.3 M sodium acetate (pH 5.1). Keep on ice and discard after use.
- *Bovine intestinal alkaline phosphatase*: Dilute solution in 50% glycerol to 1 U/10 μl with 0.1 M glycine–HCl (pH 10.4). Diluted enzyme can be stored at 4°C for several days.

(c) DNA digestion procedure

- Prepare a boiling water bath.
- Dilute 10–25 μg of DNA to 70 μl with ultra-pure water in a 1.5 ml microfuge tube.
- Boil DNA for 2 min.
- Place DNA tubes immediately on ice.
- To each tube, add 5 μl sodium acetate (0.3 M, pH 5.1), 5 μl zinc sulphate (20 mM) and 5 μl S1 nuclease (1 U/5 μl).
- Incubate at 37°C for 2 h.
- Add 10 μl of bovine intestinal alkaline phosphatase (1 U/10 μl). pH of sample should be 8.0–8.5.
- Incubate at 37°C overnight.
- Centrifuge samples at maximum speed for 4 min directly before chromatographic analysis.

If multiple samples are being analysed, a master mix with sodium acetate, zinc sulphate and S1 nuclease can be prepared and 15 μl of this mix is added to each sample. Nuclease mix must be added immediately after boiling DNA to avoid renaturation of separated strands. Samples degraded with alkaline phosphatase can be stored at −20°C for up to 1 month prior to the chromatographic analysis.

B. Chromatographic Separation of Nucleosides

Nucleosides are separated by reverse-phase chromatography at 37°C using a C-18 column (Agilent Technologies). Eluted nucleosides are detected by ultraviolet (UV) spectrophotometry at 254 nm. Isocratic elution with trimethylamine phosphate mixed with methanol as running buffer is used. Under these conditions, the deoxynucleosides elute approximately with the following retention times: dCyt, 5 min; dGuo, 11 min; dThd, 13 min; and dAde, 26 min.

Temperature control is also critical. Without temperature control, the elution times and peak shapes fluctuate throughout the day, causing changes in the integration and systematic errors. Another concern is the elution of 5-methyl deoxycytidine, a common minor component of DNA that co-elutes with other bases under some conditions and produces errors in calculating the G+C content.

The presence of nucleoside monophosphates in the sample is an indicator of incomplete degradation. For instance, under these conditions, residual dAMP will elute shortly after dThd. If monophosphates are detected in the sample, it is strongly recommended that additional samples are purified further to remove inhibitors of the degradation and that the degradation be repeated.

The procedure below describes the protocol routinely used in our laboratory for chromatographic separation of nucleosides. Since the C-18 columns vary greatly, it is recommended that the buffers are optimized for a new column by systematic variation of the % methanol and temperature to ensure that the dGuo and dThd are well resolved from each other and other nucleosides that elute at similar times, especially Guo and 5-methyl deoxycytidine.

1. Buffers

- *Triethylamine phosphate (TEAP)*: a 1 M stock solution of TEAP is prepared by dilution of triethylamine in water. The pH is then adjusted to 5.1 with phosphoric acid. This solution should be wrapped in aluminium foil and stored in the dark. If the triethylamine has a yellow colour, it should be purified by vacuum distillation until it becomes colourless. If the 1 M TEAP stock solution turns yellow, it should be discarded. To prepare running buffer, 20 ml of 1-M TEAP is diluted in about 500 ml of ultra-pure water, 75 ml of methanol (HPLC grade) is added and the volume is adjusted to 1000 ml with ultra-pure water. The pH is adjusted to 5.1 with phosphoric acid. After passage through a 0.45-μm pore size cellulose triacetate membrane, the running buffer must be constantly stirred to avoid precipitation.
- Methanol (70%)
- Ultra-pure water
- All buffers are sparged with helium for at least 30 s. Alternatively, buffers can be degassed under vacuum.

2. Running conditions

- Reverse-phase chromatography column Agilent C18, 3-μm particle size, 2.1 mm × 150 mm
- Isocratic elution, flow rate 1 ml/min
- Column temperature 37°C
- Single-channel UV detection, absorbance 254 nm
- Injection volume 10–60 μl, run time 50 min
- Columns are usually stored in 70% methanol. Prior to the analyses, column must be conditioned with 70% methanol for 30 min, water for 15 min and then TEAP running buffer for at least 120 min.
- After all samples have been run, column is washed with water (15 min) and 70% methanol for 15 min.

Figure 2 shows the separation of nucleosides from 2 μg of digested salmon sperm DNA. When a C-18 reverse-phase column at 37°C and TEAP running buffer were used, dGuo and dThd are well separated. Degradation of this sample is sufficient, dAMP is not observed on the chromatogram. The small peaks eluting before dGuo represent small amounts of guanosine and 5-methyldeoxycytidine. The peaks eluting before dCyt represent unknown compounds

Figure 3 shows the separation of nucleosides from 10 μg of DNA from the extremophile *Natranaerobius thermophilus*. Deoxyguanosine and dThd are well separated. Small amounts of 5-methyldeoxycytidine (retention time 9.5 min) and adenosine (retention time 20.2 min) are seen. There are also cytosine and unidentified compounds eluting within 4–6 min. Thus, this sample had minor

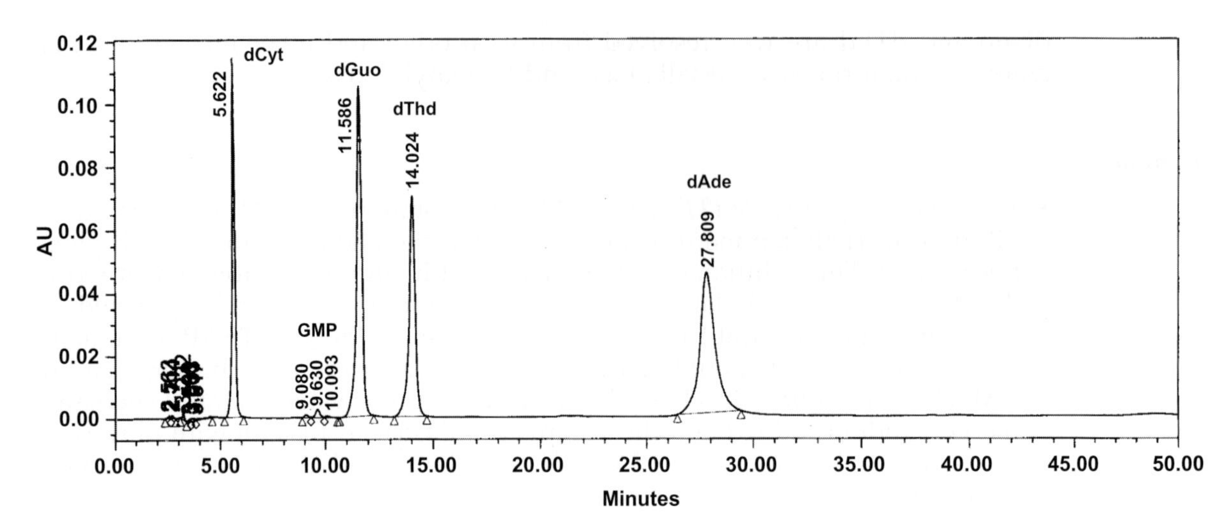

Figure 2. Separation of nucleosides produced by degradation of salmon sperm DNA (2 μg). Chromatography conditions are as described in the text. dGuo eluted with a retention time of 11.5 min and dThd eluted with retention time of 14.0 min. Peaks eluting at 9 min belong to GMP. Peaks eluting at 2 min belong to an unidentified compound. dCyt: deoxycytidine, GMP: guanosine monophosphate, dGuo: deoxyguanosine, dThd: deoxythymidine and dAde: deoxyadenosine.

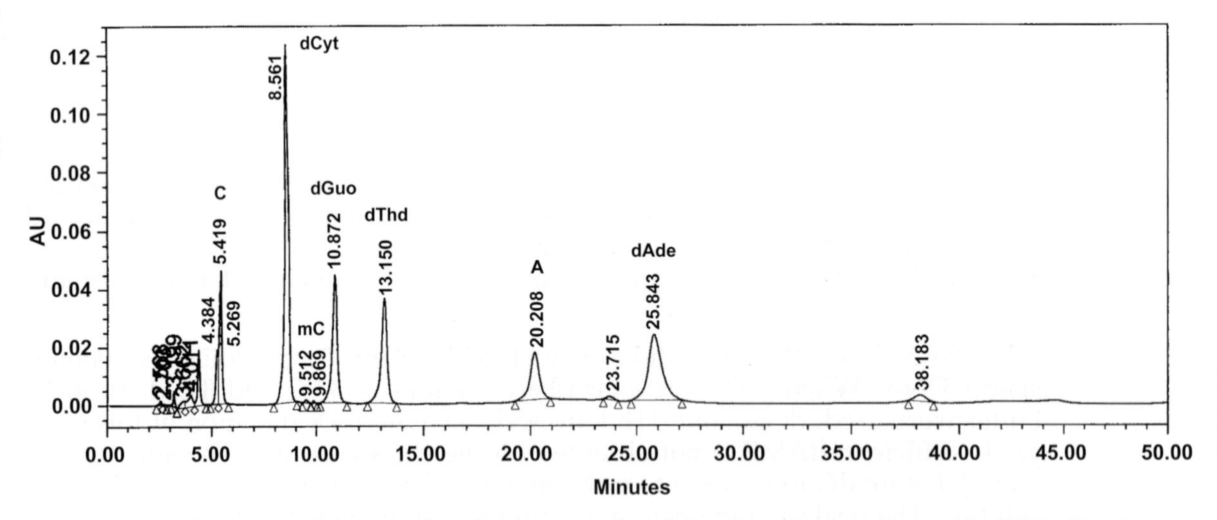

Figure 3. Separation of nucleosides produced by degradation of genomic DNA (10 μg) from *Natranaerobius thermophilus*. dGuo eluted with a retention time of 10.8 min and dThd eluted with a retention time of 13.1 min. Peaks eluting at 4–5 min belong to residual cytosine and unidentified compounds. Peaks eluting at 9 min belong to GMP. dAde eluted with a retention time of 25.8 min, and the peak at 20.2 min belongs to adenosine. mC: 5-methyldeoxycytidine, C: cytosine and A: adenosine.

RNA contamination. Estimation of the G+C content can be made only if dGuo and dThd are well separated from other bases and DNA components.

C. Calculation of G+C Content

The G+C content is estimated from the ratios of dThd and dGuo. To determine the G+C content of a sample, first the constant Y must be determined by chromatography of standard DNA. The constant Y is defined by Eq. (1):

$$Y = \left(\frac{\mathrm{wG}}{\mathrm{zT}}\right)\left(\frac{T}{G}\right) \tag{8}$$

where wG and zT are the apparent mole fractions for the DNA standards as determined experimentally by chromatography, and G and T are the true mole fractions of the nucleosides dGuo and dThd. The HPLC method reports values related to the apparent mole fraction, not the true mole fraction. The apparent mole fraction is a multiple of the true mole fraction and a constant which depends on the extinction coefficient of the bases and the chromatography conditions.

The (T/G) values for salmon sperm and calf thymus DNA are 1.3084 and 1.2712, respectively. Therefore, to calculate the Y factor for salmon sperm DNA, the ratio wG/zT, as determined by integration of the dG and dT peaks from the chromatography, is multiplied by 1.3084. The Y factor for salmon sperm DNA is usually about 2.60, but can change slightly with varying experimental conditions. It is strongly recommend that a standard is chromatographed after every three to four samples to determine if fluctuations in the chromatography are occurring during the run that might affect the determination.

The G+C content for an unknown sample is calculated from Eq. (9):

$$\mathrm{G} + \mathrm{C}\ \text{content} = \frac{1}{(1 + Y[dT/\mathrm{d}G])} \tag{9}$$

where dT and dG are the apparent mole fractions for the nucleosides dThd and dGuo, respectively, as determined by integration of the dThd and dGuo peaks. To report the G+C content as mol%, the value calculated by Eq. (9) is simply multiplied by 100. To ensure accuracy and reproducibility of digestion and separation processes, it is recommended that the mol% G+C content reported is the average of at least three separate determinations from at least two independent DNA digestions. Under optimal conditions, standard deviations of 0.01% can be achieved routinely. If large deviations are observed, they may reflect either inefficient degradation of DNA, impure DNA samples, or unstable chromatography conditions. For instance, old pumps with worn piston seals are common causes of high standard deviations because they cause small fluctuations in pressure that affect peak shapes and integration.

A useful control is to perform a degradation with water in place of DNA but with all the enzymes and buffers. Upon chromatography of the control, no UV absorbing peaks should elute near dGuo and dThd peaks. If they do, the degradation reagents should be purified to eliminate these peaks or the contribution of these peaks should be subtracted from the measured dGuo and dThd peaks.

♦♦♦♦♦♦ VII. COMPARISON OF GENOMIC G+C CONTENT DETERMINED EXPERIMENTALLY WITH THAT PREDICTED FROM WHOLE GENOME SEQUENCES

The accuracy of the buoyant density, thermal denaturation, qPCR and HPLC methods vary with sample purity, standard purity and individual laboratories (Marmur and Doty, 1962; Schildkraut *et al.*, 1962). The G+C contents determined by the HPLC method agree with values determined by the buoyant density and melting temperature methods only over a wide range (De Ley, 1970; Mesbah *et al.*, 1989). To facilitate comparison of G+C values between different strains, it is recommended that the method used for determination (buoyant density, thermal denaturation, qPCR and HPLC) be stated.

A comparison of experimentally determined mol% G+C values with those predicted from complete genome sequences of a number of microorganisms showed that, for the three main methods for mol% G+C determination (buoyant density, thermal denaturation and HPLC), the differences between experimentally determined values and predicted values ranged between 1.2% (thermal denaturation) and 2.0% (HPLC, buoyant density) (Table 1). Small differences are not unexpected given that some microorganisms contain single-stranded DNA, and cells harvested during the exponential growth phase may contain fractional copies of the genome. Thus, the predicted mol% G+C values may be, in fact, somewhat different from the actual values. There did not appear to be a difference in the variation between measured/predicted values for *Archaea* or *Bacteria*, nor was there a discernible difference in variation among different phyla of both domains. The small differences between experimentally determined and predicted mol% G+C values corroborate the accuracy of the methods used for determination, provided experimental conditions are standardized and replicated.

♦♦♦♦♦♦ VIII. CONCLUDING REMARKS

G+C content is a standard genotypic marker and is an essential part of the description of any novel prokaryote taxon. This chapter provides sufficient information to allow researchers to accurately determine G+C content of different organisms using a range of methods. Uniformity of the procedures used to determine the G+C content is essential to allow accurate comparison of the values among different taxa and to provide accurate descriptions of different phylogenetic divisions.

Table 1. Comparison of measured mol% G+C values and those predicted from whole genome sequences of selected bacteria and archaea. The method of measurement is indicated in parentheses. Predicted mol% G+C values were from the GenBank database, www.ncbi.nlm.nih.gov/genomes/lproks.cgi

Phylum	Species	Predicted G+C %	Measured G +C% (method)	Reference
Actinobacteria	*Acidimicrobium ferroxidans* DSM 10331^T	68.3	67–68.5 (T_m)	Clark and Norris (1996)
	Acidothermus cellulolyticus ATCC 43068^T	66.9	60.7 ± 0.6 (B.d.)	Mohagheghi *et al.* (1986)
	Archanobacterium haemolyticum DSM 20595^T	53	50–52 (T_m)	Collins *et al.* (1982)
	Arthrobacter chlorophenolicus DSM 12829^T	66	65.1 (T_m)	Westerberg *et al.* (2000)
	Brachybacterium faecium DSM 4810^T	72.0	69.4 (T_m)	Collins *et al.* (1988)
	Catenulispora acidiphila DSM 44928^T	69.8	71.5 (HPLC)	Busti *et al.* (2006)
	Coenixibacter woesei DSM 14684^T	72.7	71 (HPLC)	Monciardini *et al.* (2003)
	Corynebacterium aurimucosum ATCC 700975^T	60.6	58.0 (HPLC)	Yassin *et al.* (2002)
	Cryptobacterium curtum DSM 15641^T	50.9	50–51 (HPLC)	Nakazawa *et al.* (1999)
	Gardnerella vaginalis ATCC 14019^T	41.5	43.0 (B.d.)	Greenwood and Pickett (1980)
	Kocuria rhizophila DSM 11926^T	71.2	69.4 (HPLC)	Kovács *et al.* (1999)
	Kribbella flavida DSM 17836^T	70.6	70 (HPLC)	Park *et al.* (1999)
	Xylanimonas cellulosilytica DSM 15894^T	72.5	73 (T_m)	Rivas *et al.* (2003)
Aquificae	*Hydrogenobacter thermophilus* TK-6^T	44.0	43.5–43.9 (T_m)	Kawasumi *et al.* (1984)
	Persephonella marina DSM 14350^T	37.1	38.5 (HPLC, T_m)	Götz *et al.* (2002)
	Thermocrinis albus DSM 14484^T	46.9	49.6 (HPLC)	Eder and Huber (2002)
Bacteroidetes	*Bacteroides salanitronis* DSM 18170^T	46.5	46.9 (HPLC)	Thi Ngoc Lan *et al.* (2006)
	Chitinophaga pinensis DSM 2588^T	45.2	44.6 (T_m)	Sangkhobol and Skerman (1961)
	Croceibacter atlanticus ATCC BAA-628^T	33.9	34.8 ± 0.1 (HPLC)	Cho and Giovannoni (2003)
	Paludibacter propionicigenes DSM 17365^T	39.3	39.3 (HPLC)	Ueki *et al.* (2006)
	Parabacteroides distasonis ATCC 8503^T	45.1	44 (T_m)	Sakamoto and Benno (2006)

(*Continued*)

Table 1. (Continued)

Phylum	Species	Predicted G+C %	Measured G+C% (method)	Reference
	Pedobacter heparinus	42.0	42–43 (T_m)	Steyn *et al.* (1998)
	Porphorymonas gingivalis ATCC 55562[T]	48.4	41–42 (HPLC)	Hirasawa and Takada (1994)
	Rhodothermus marinus DSM 4252[T]	64.3	64 (T_m)	Alfredsson *et al.* (1988)
	Salinibacter ruber DSM 13855[T]	66.1	66.5 (HPLC)	Antón *et al.* (2002)
Firmicutes	*Acidaminococcus fermentans* DSM 20731[T]	55.8	56.6 (B.d.)	Rogosa (1969)
	Alicyclobacillus acidocaldarius DSM 446[T]	62	60.3 (T_m)	Da Costa *et al.* (2009)
			62.3 (B.d.)	
	Anoxybacillus flavithermus DSM 2641[T]	41.8	41.6 (HPLC)	Pikuta (2009)
	Bacillus amyloliquefaciens ATCC 233350[T]	46.5	44.4 ± 0.4 (T_m)	Priest *et al.* (1987)
	Bacillus atrophaeus 1942	43	44.2 ± 0.7 (B. d.)	Logan and De Vos (2009)
	Clostridium acetobutylicum ATCC 824[T]	30.9	28–29 (T_m)	Rainey *et al.* (2009)
	Clostridium perfringens ATCC 13124[T]	28.4	24–27 (T_m)	Rainey *et al.* (2009)
	Halothermothrix orenii H168[T]	37.9	39.6 (HPLC)	Cayol *et al.* (1994)
	Lactobacillus delbrueckii subsp. *bulgaricus* ATCC 11842[T]	49.7	49–51 (B.d., T_m)	Hammes and Hertel (2009)
	Listeria eslshimeri ATCC 35897[T]	36.4	36 (T_m)	McLauchlin and Rees (2009)
	Moorella thermoacetica ATCC 39073[T]	55.8	54 (T_m)	Wiegel (2009)
	Coprothermobacter proteolyticus DSM 5265[T]	45	43–45 (B.d.)	Ollivier and Garcia (2009)
Alphaproteobacteria	*Acidiphilum cryptum* JF-5	67.1	69.8 (T_m)	Harrison (1981)
	Azorhizobium caulinodans LMG 6465[T]	67.3	66 (T_m)	Dreyfus *et al.* (1988)
	Dinoroseobacter shibae DSM 16493[T]	65.5	64.8 (HPLC)	Biebl *et al.* (2005)
	Hisrchia baltica ATCC 49814[T]	45.2	45.6 (T_m)	Schlesner *et al.* (1990)
	Hyphomicrobium denitrificans ATCC 51888[T]	60.8	60.5 (HPLC)	Urakami *et al.* (1995)
	Ochrobactrum anthropi ATCC 49188[T]	56.1	57.2 (T_m)	Holmes *et al.* (1988)

(Continued)

Table 1. (Continued)

Phylum	Species	Predicted G+C %	Measured G+C% (method)	Reference
	Methylobacterium populi BJ001^T	69.4	70.4 ± 0.3 (HPLC)	Van Aken *et al.* (2004)
	Parvibaculum lavamentivorans DSM 13023^T	62.3	64 (HPLC)	Schleheck *et al.* (2004)
	Ruegeria pomeroyi DSS-3^T	64.1	68 ± 0.1 (HPLC)	González *et al.* (2003)
	Starkeya novella DSM 506^T	67.9	67.3–68.4 (T_m)	Kelley *et al.* (2000)
Gammaproteobacteria	*Acinetobacter baumannii* ATCC 17978^T	38.9	40–43 (T_m)	Bouvet and Grimont (1986)
	Alcanivorax borkumensis SK2	54.7	53.4 (HPLC)	Yakimov *et al.* (1998)
	Chromohalobacter salexigens DSM 3043^T	63.9	64.2–66.0 (T_m)	Arahal *et al.* (2001)
	Erwinia pyrifoliae DSM 12163^T	53.4	52 (T_m)	Kim *et al.* (1999)
	Kangiella koreensis DSM 16069^T	43.7	44 (HPLC)	Yoon *et al.* (2004)
	Pantoea vagans BCC 105 ^{T}	55	55.4 (HPLC)	Brady *et al.* (2009)
	Psychromonas ingrahamii CIP 37^T	40.1	40 (HPLC)	Auman *et al.* (2006)
	Shewanella pealeana ATCC 700345^T	44.7	45.0 (HPLC)	Leonardo *et al.* (1999)
Deltaproteobacteria	*Desulfatibacillum alkenivorans* DSM 16219 ^{T}	54.5	57.8 (HPLC)	Cravo-Laureau *et al.* (2004)
	Desulfohalobium retbaense DSM 5692^T	57.3	57.1 ± 0.2 (HPLC)	Ollivier *et al.* (1991)
	Geobacter lovleyi DSM 17278^T	54.7	56.7 ± 0.3 (HPLC)	Süng *et al.* (2006)
	Syntrophobacter fumaroxidans DSM 10017^T	59.9	60.6 ± 0.2 (HPLC)	Harmsen *et al.* (1998)
Betaproteobacteria	*Acidovorax avenae* subsp. *avenae* ATCC 19860^T	68.9	69.8 (T_m)	Willems *et al.* (1992)
	Burkholderia multivorans ATCC 17616^T	66.7	68–69 (T_m)	Vandamme *et al.* (1997)
	Polaromonas naphthalenivorans DSM 15660^T	61.7	61.5 (HPLC)	Jeon *et al.* (2004)
	Thiobacillus denitrificans ATCC 25259^T	66.1	63 (B.d.)	Kelley and Wood (2000)
Euryarchaeota	*Halogeometricum borinquense* DSM 11551^T	59.9	59.1 (HPLC)	Montalvo-Rodríguez *et al.* (1998)
	Haloquadratum walsbyi DSM 16790^T	47.9	46.9 (HPLC)	Burns *et al.* (2007)

(Continued)

Table 1. (Continued)

Phylum	Species	Predicted G+C %	Measured G+C% (method)	Reference
	Halorhabdus utahensis DSM 12940^T	62.9	64 (HPLC)	Wainø *et al.* (2000)
	Methanobrevibacter ruminantium DSM 1093^T	32.6	30.6 (B.d.)	Balch *et al.* (1979)
	Methanocella paludicola DSM 17711^T	54.9	56.6 (HPLC)	Sakai *et al.* (2008)
	Methanosphaerula palustris DSM 19958^T	55	58.9 ± 2 (qPCR)	Cadillo-Quiroz *et al.* (2009)
	Thermococcus gammatolerans DSM 15229^T	53.6	53.1 (T_m)	Joliver *et al.* (2003)
Crenarchaeota	*Acidilobus saccharovorans* DSM 16705^T	54.5	54.5 (T_m)	Prokofeva *et al.* (2009)
	Aeropyrum pernix K1^T	56.3	67 (HPLC)	Sako *et al.* (1996)
	Desulfurococcus kamchatkensis DSM 18924^T	45.3	44.4 (T_m)	Kublanov *et al.* (2009)
	Ignisphaera aggregans DSM 17230^T	52.9	52.9 (qPCR)	Niederberger *et al.* (2006)
	Pyrobaculum aerophilum DSM 7523^T	51.4	52 (T_m)	Völkl *et al.* (1993)
	Staphylothermus hellenicus DSM 12710^T	36.9	38 (HPLC)	Arab *et al.* (2000)

T_m: melting temperature, B.d.: buoyant density.

References

Alfredsson, G. A., Kristjansson, J. K., Hjörleifsdottir, S. and Stetter, K. O. (1988). *Rhodothermus marinus*, gen. nov., sp. nov., a thermophilic, halophilic bacterium from submarine hot springs in Iceland. *J. Gen. Microbiol.* **134**, 299–306.

Antón, J., Oren, A., Benlloch, S., Rodríguez-Valera, F., Amann, R. and Rosselló-Mora, R. (2002). *Salinibacter ruber* gen. nov., sp. nov., a novel, extremely halophilic member of the *Bacteria* from saltern crystallizer ponds. *Int. J. Syst. Evol. Microbiol.* **52**, 485–491.

Arab, H., Volker, H. and Thomm, M. (2000). *Thermococcus aegaeicus* sp. nov. and *Staphylothermus hellenicus* sp. nov., two novel hyperthermophilic archaea isolated from geothermally heated vents off Palaeochori Bay, Milos, Greece. *Int. J. Syst. Evol. Microbiol.* **50**, 2101–2108.

Arahal, D. R., Garcia, M. T., Vargas, C., Canovas, D., Nieto, J. J. and Ventosa, A. (2001). *Chromohalobacter salexigens* sp. nov., a moderately halophilic species that includes *Halomonas elongata* DSM 3043 and ATCC 33174. *Int. J. Syst. Evol. Microbiol.* **51**, 1457–1462.

Auman, A. J., Breezee, J. L., Gosink, J. J., Kämpfer, P. and Staley, J. T. (2006). *Psychromonas ingrahamii* sp. nov., a novel gas vacuolate, psychrophilic bacterium isolated from Arctic polar sea ice. *Int. J. Syst. Evol. Microbiol.* **56**, 1001–1007.

Balch, W. E., Fox, G. E., Magrum, L. J., Woese, C. R. and Wolfe, R. S. (1979). Methanogens: reevaluation of a unique biological group. *Microbiol. Rev.* **43**, 260–296.

Biebl, H., Allgaier, M., Tindall, B. J., Koblizek, M., Lünsdorf, H., Pukall, R. and Wagner-Döbler, I. (2005). *Dinoroseobacter shibae* gen. nov., sp. nov., a new aerobic phototrophic bacterium isolated from dinoflagellates. *Int. J. Syst. Evol. Microbiol.* **55**, 1089–1096.

Bouvet, P. J. M. and Grimont, P. A. D. (1986). Taxonomy of the genus *Acinetobacter* with the recognition of *Acinetobacter baumannii* sp. nov., *Acinetobacter haemolyticus* sp. nov., *Acinetobacter johnsonii* sp. nov. and *Acinetobacter junii* sp. nov. and emended descriptions of *Acinetobacter calcoaceticus* and *Acinetobacter lwoffii*. *Int. J. Syst. Bacteriol.* **36**, 228–240.

Brady, C. L., Venter, S. N., Cleenwerch, I., Engelbeen, K., Vancanneyt, M., Swings, J. and Coutinho, T. A. (2009). *Pantoea vagans* sp. nov., *Pantoea eucalypti* sp. nov., *Pantoea deleyi* sp. nov. and *Pantoea anthophila* sp. nov. *Int. J. Syst. Evol. Microbiol.* **59**, 2339–2345.

Burns, D. G., Janssen, P. H., Itoh, T., Kamekura, M., Li, Z., Jensen, G., Rodríguez-Valera, F., Bolhuis, H. and Dyall-Smith, M. L. (2007). *Haloquadratum walsbyi* gen. nov., sp. nov., the square haloarchaeon of Walsby, isolated from saltern crystallizers in Australia and Spain. *Int. J. Syst. Evol. Microbiol.* **57**, 387–392.

Busti, E., Cavaletti, L., Monciardini, P., Schumann, P., Rohde, M., Sosio, M. and Donadio, S. (2006). *Catenulispora acidiphila* gen. nov., sp. nov., a novel, mycelium-forming actinomycete, and proposal of *Catenulisporaceae* fam. nov. *Int. J. Syst. Evol. Microbiol.* **56**, 1741–1746.

Cadillo-Quiroz, H., Yavitt, J. B. and Zinder, S. H. (2009). *Methanosphaerula palustris* gen. nov., sp. nov., a hydrogenotrophic methanogen isolated from a minerotrophic fen peatland. *Int. J. Syst. Evol. Microbiol.* **59**, 928–935.

Cayol, J.-L., Ollivier, B., Patel, B. K. C., Prensier, G., Guezennec, J. and Garcia, J.-L. (1994). Isolation and characterization of *Halothermothrix orenii* gen. nov., sp. nov., a halophilic, thermophilic, fermentative, strictly anaerobic bacterium. *Int. J. Syst. Bacteriol.* **44**, 534–540.

Cho, J.-C. and Giovannoni, S. J. (2003). *Croceibacter atlanticus* gen. nov., sp. nov., a novel marine bacterium in the family *Flavobacteriaceae*. *Syst. Appl. Microbiol.* **26**, 76–83.

Clark, D. A. and Norris, P. R. (1996). *Acidimicrobium ferrooxidans* gen. nov., sp. nov.: mixed-culture ferrous iron oxidation with *Sulfobacillus* species. *Microbiology* **142**, 785–790.

Collins, M. D., Jones, D. and Schofield, G. M. (1982). Reclassification of '*Corynebacterium haemolyticum*' (MacLean, Liebow & Rosenberg) in the genus *Arcanobacterium* gen. nov. as *Arcanobacterium haemolyticum* nom. rev., comb. nov. *J. Gen. Microbiol.* **128**, 1279–1281.

Collins, M. D., Brown, J. and Jones, D. (1988). *Brachybacterium faecium* gen. nov., sp. nov. a coryneform bacterium from poultry deep litter. *Int. J. Syst. Bacteriol.* **38**, 45–48.

Cravo-Laureau, C., Matheron, R., Joulian, C., Cayol, J. -L. and Hirschler-Réa, A. (2004). *Desulfatibacillum alkenivorans* sp. nov., a novel n-alkene-degrading, sulfate-reducing bacterium, and emended description of the genus *Desulfatibacillum*. *Int. J. Syst. Evol. Microbiol.* **54**, 1639–1642.

Da Costa, M. S., Rainey, F. A. and Albuquerque, L. (2009). Genus I. *Alicyclobacillus*. In Bergey's Manual of Systematic Bacteriology (P. De Vos, G. M. Garrity, D. Jones, N. R. Krieg, W. Ludwig, F. A. Rainey, K. H. Schleifer and W. B. Whitman, eds), Vol. 3, pp. 229–243. Springer, New York.

De Ley, J. (1970). Reexamination of the association between melting point, buoyant density and chemical base composition of deoxyribonucleic acid. *J. Bacteriol.* **101**, 738–754.

Dreyfus, B., Garcia, J. -L. and Gillis, M. (1988). Characterization of *Azorhizobium caulinodans* gen. nov. sp. nov., a stem-nodulating nitrogen-fixing bacterium isolated from *Sesbania rostrata*. *Int. J. Syst. Bacteriol.* **38**, 89–98.

Eder, W. and Huber, R. (2002). New isolates and physiological properties of the *Aquificales* and description of *Thermocrinis albus* sp. nov. *Extremophiles* **6**, 309–318.

González, J. M. and Saiz-Jimenez, C. (2002). A fluorimetric method for the estimation of G+C mol% content in microorganisms by thermal denaturation temperature. *Environ. Microbiol.* **4**, 770–773.

González, J. M., Covert, J. S., Whitman, W. B., Henriksen, J. R., Mayer, F., Scharf, B., Schmitt, R., Buchan, A., Fuhrman, J. A., Kiene, R. P. and Moran, M. A. (2003). *Silicibacter pomeroyi* sp. nov. and *Roseovarius nubinhibens* sp. nov., dimethylsulfoniopropionate-demethylating bacteria from marine environments. *Int. J. Syst. Evol. Microbiol.* **53**, 1261–1269.

Götz, D. K., Banta, A., Beveridge, T. J., Rushdi, A. I., Simoneit, B. and Reysenbach, A. L. (2002). *Persephonella marina* gen. nov., sp. nov. and *Persephonella guaymasensis* sp. nov., two novel, thermophilic, hydrogen-oxidizing microaerophiles from deep-sea hydrothermal vents. *Int. J. Syst. Evol. Microbiol.* **52**, 1349–1359.

Greenwood, J. R. and Pickett, M. J. (1980). Transfer of *Haemophilus vaginalis* Gardner and Dukes to a new genus, *Gardnerella: G. vaginalis* (Gardner and Dukes) comb. nov. *Int. J. Syst. Bacteriol.* **30**, 170–178.

Haft, D. H., Selengut, J. D., Brinkack, L. M., Zafar, N. and White, O. (2005). Genome properties: a system for the investigation of prokaryotic genetic content for microbiology, genome annotation and comparative genomics. *Bioinformatics* **21**, 293–306.

Hammes, W. P. and Hertel, C. (2009). Genus I. *Lactobacillus*. In *Bergey's Manual of Systematic Bacteriology* (P. De Vos, G. M. Garrity, D. Jones, N. R. Krieg, W. Ludwig, F. A. Rainey, K. H. Schleifer and W. B. Whitman, eds), Vol. 3, pp. 465–511. Springer, New York..

Harmsen, H. J. M., van Kuijk, B. L. M., Plugge, C. M., Akkermans, A. D. L., de Vos, W. M. and Stams, A. J. M. (1998). *Syntrophobacter fumaroxidans* sp. nov., a syntrophic propionate-degrading sulfate-reducing bacterium. *Int. J. Syst. Bacteriol.* **48**, 1383–1387.

Harrison, A. P. (1981). *Acidiphilium cryptum* gen. nov., sp. nov., heterotrophic bacterium from acidic mineral environments. *Int. J. Syst. Bacteriol.* **31**, 327–332.

Hildebrand, F., Meyer, A. and Eyre-Walker, A. (2010). Evidence of selection upon genomic GC-content in bacteria. *PloS Genet.* **6**, e1001107.10.1371/journal.pgen.1001107

Hirasawa, M. and Takada, K. (1994). *Porphyromonas gingivicanis* sp. nov. and *Porphyromonas crevioricanis* sp. nov., isolated from beagles. *Int. J. Syst. Bacteriol.* **44**, 637–640.

Holmes, B., Popoff, M., Kiredjian, M. and Kersters, K. (1988). *Ochrobactrum anthropi* gen. nov., sp. nov. from human clinical specimens and previously known as group Vd. *Int. J. Syst. Bacteriol.* **38**, 406–416.

Jeon, C. O., Park, W., Ghiorse, W. C. and Madsen, E. L. (2004). *Polaromonas naphthalenivorans* sp. nov., a naphthalene-degrading bacterium from naphthalene-contaminated sediment. *Int. J. Syst. Evol. Bacteriol.* **54**, 93–97.

Johnson, J. L. (1985). Determination of DNA base composition. In *Methods in Microbiology* (G. Gottschalk, ed), Vol. 18, pp. 1–31. Academic Press, London.

Joliver, E., L'Haridon, S., Corre, E., Forterre, P. and Prieur, D. (2003). *Thermococcus gammatolerans* sp. nov., a hyperthermophilic archaeon from a deep-sea hydrothermal vent that resists ionizing radiation. *Int. J. Syst. Evol. Microbiol.* **53**, 847–851.

Kawasumi, T., Igarashi, Y., Kodama, T. and Minoda, Y. (1984). *Hydrogenobacter thermophilus* gen. nov., sp. nov., an extremely thermophilic, aerobic, hydrogen-oxidizing bacterium. *Int. J. Syst. Bacteriol.* **34**, 5–10.

Kelley, D. P. and Wood, A. P. (2000). Confirmation of *Thiobacillus denitrificans* as a species of the genus *Thiobacillus*, in the β-subclass of the *Proteobacteria*, with strain NCIMB 9548 as the type strain. *Int. J. Syst. Evol. Microbiol.* **50**, 547–550.

Kelley, D. P., McDonald, I. R. and Wood, A. P. (2000). Proposal for the reclassification of *Thiobacillus novellus* as *Starkeya novella* gen. nov., comb. nov., in the α-subclass of the *Proteobacteria*. *Int. J. Syst. Evol. Microbiol.* **50**, 1797–1802.

Kim, W.-S., Gardan, L., Rhim, S.-L. and Geider, K. (1999). *Erwinia pyrifoliae* sp. nov., a novel pathogen that affects Asian pear trees (*Pyrus pyrifolia* Nakai). *Int. J. Syst. Bacteriol.* **49**, 899–906.

Ko, C. Y., Johnson, J. L., Barnett, L. B., McNair, H. M. and Vercellotti, J. R. (1977). A sensitive estimation of the percentage of guanine plus cytosine in deoxyribonucleic acid by high performance liquid chromatography. *Anal. Biochem.* **80**, 183–192.

Kovács, G., Burghardt, J., Pradella, S., Schumann, P., Stackebrandt, E. and Màrialigeti, K. (1999). *Kocuria palustris* sp. nov. and *Kocuria rhizophila* sp. nov., isolated from the rhizoplane of the narrow-leaved cattail (*Typha angustifolia*). *Int. J. Syst. Bacteriol.* **49**, 167–173.

Kublanov, I. V., Bidijieva, S. K., Mardanov, A. V. and Bonch-Osmolovskaya, E. A. (2009). *Desulfurococcus kamchatkensis* sp. nov., a novel hyperthermophilic protein-degrading archaeon isolated from a Kamchatka hot spring. *Int. J. Syst. Evol. Microbiol.* **59**, 1743–1747.

Lee, Y. J., Wagner, I. D., Brice, M. E., Kevbrin, V. V., Mills, G. L., Romanek, C. S. and Wiegel, J. (2005). *Thermosediminibacter oceani* gen. nov. sp. nov and *Thermosediminibacter litoriperuensis* sp. nov., new anaerobic thermophilic bacteria isolated from Peru Margin. *Extremophiles* **9**, 375–383.

Leonardo, M. R., Moser, D. P., Barbieri, E., Brantner, C. A., MacGregor, B. J., Paster, B. J., Stackebrandt, E. and Nealson, K. H. (1999). *Shewanella pealeana* sp. nov., a member of the microbial community associated with the accessory nidamental gland of the squid *Loligo pealei*. *Int. J. Syst. Bacteriol.* **49**, 1341–1351.

Logan, N. A. and De Vos, P. (2009). Genus I. *Bacillus*. In *Bergey's Manual of Systematic Bacteriology* (P. De Vos, G. M. Garrity, D. Jones, N. R. Krieg, W. Ludwig, F. A. Rainey, K. H. Schleifer and W. B. Whitman, eds), Vol. 3, pp. 21–128. Springer, New York.

Mandel, M., Schildkraut, C. L. and Marmur, J. (1968). Use of CsCl density gradient analysis for determining the guanine plus cytosine content of DNA. In *Methods in Enzymology, (L. Grossman and K. Moldave, eds), Vol. 12, Part 2*, pp. 184–195. Academic Press, New York.

Marmur, J. and Doty, P. (1962). Determination of the base composition of deoxyribonucleic acid from its thermal denaturation temperature. *J. Mol. Biol.* **5**, 109–118.

McLauchlin, J. and Rees, C. E. D. (2009). Genus I. *Listeria*. In *Bergey's Manual of Systematic Bacteriology* (P. De Vos, G. M. Garrity, D. Jones, N. R. Krieg, W. Ludwig, F. A. Rainey, K. H. Schleifer and W. B. Whitman, eds), Vol. 3, pp. 244–257. Springer, New York.

Mesbah, M., Premachandran, U. and Whitman, W. B. (1989). Precise measurement of the G+C content of deoxyribonucleic acid by high-performance liquid chromatography. *Int. J. Syst. Bacteriol.* **39**, 159–167.

Mischke, C. F. and Wickstrom, E. (1980). Deoxynucleoside composition of DNAs and modified nucleoside composition of tRNAs determined at nanomole sensitivity by reversed-phase liquid chromatography. *Anal. Biochem.* **105**, 181–187.

Mohagheghi, A., Grohmann, K., Himmel, M., Leighton, L. and Updegraff, D. M. (1986). Isolation and characterization of *Acidothermus cellulolyticus* gen. nov., sp. nov., a new genus of thermophilic, acidophilic, cellulolytic bacteria. *Int. J. Syst. Bacteriol.* **36**, 435–443.

Monciardini, P., Cavaletti, L., Schumann, P., Rohde, M. and Donadio, S. (2003). *Conexibacter woesei* gen. nov., sp. nov., a novel representative of a deep evolutionary line of descent within the class *Actinobacteria*. *Int. J. Syst. Evol. Microbiol.* **53**, 569–576.

Montalvo-Rodríguez, R., Vreeland, R. H., Oren, A., Kessel, M., Betancourt, C. and López-Garriga, J. (1998). *Halogeometricum borinquense* gen. nov., sp. nov., a novel halophilic archaeon from Puerto Rico. *Int. J. Syst. Evol. Bacteriol.* **48**, 1305–1312.

Nakabachi, A., Yamashita, A., Toh, H., Ishakawa, H., Dunbar, H. E., Moran, N. A. and Hattori, M. (2006). The 160-kilobase genome of the bacterial endosymbiont. *Carsonella Sci.* **314**, 267.

Nakazawa, F., Poco, S. E., Ikeda, T., Sato, M., Kalfas, S., Sundqvist, G. and Hoshimo, E. (1999). *Cryptobacterium curtum* gen. nov., sp. nov., a new genus of Gram-positive anaerobic rod isolated from human oral cavities. *Int. J. Syst. Bacteriol.* **49**, 1193–1200.

Niederberger, T. D., Götz, D. K., McDonald, I. R., Ronimus, R. S. and Morgan, H. W. (2006). *Ignisphaera aggregans* gen. nov., sp. nov., a novel hyperthermophilic crenarchaeote isolated from hot springs in Rotorua and Tokaanu, New Zealand. *Int. J. Syst. Evol. Microbiol.* **56**, 965–971.

Ollivier, B. and Garcia, J.-L. (2009). Genus II. *Coprothermobacter*. In *Bergey's Manual of Systematic Bacteriology* (P. De Vos, G. M. Garrity, D. Jones, N. R. Krieg, W. Ludwig, F. A. Rainey, K. H. Schleifer and W. B. Whitman, eds), Vol. 3, pp. 1271–1274. Springer, New York.

Ollivier, B., Hatchikian, C. E., Prensier, G., Guezennec, J. and Garcia, J.-L. (1991). *Desulfohalobium retbaense* gen. nov. sp. nov. a halophilic sulfate-reducing bacterium from sediments of a hypersaline lake in Senegal. *Int. J. Syst. Bacteriol.* **41**, 74–81.

Park, Y.-H., Yoon, J.-H., Shin, Y. K., Suzuki, K.-I., Kuko, T., Seino, A., Kim, H.-J., Lee, J.-S. and Lee, S. T. (1999). Classification of '*Nocardioides fulvus*' IF0 14399 and *Nocardioides* sp. ATCC 39419 in *Kribbella* gen. nov., as *Kribbella flavida* sp. nov. and *Kribbella sandrarnycini* sp. nov. *Int. J. Syst. Bacteriol.* **49**, 743–752.

Pikuta, E. V. (2009). Genus IV. *Anoxybacillus*. In *Bergey's Manual of Systematic Bacteriology* (P. De Vos, G. M. Garrity, D. Jones, N. R. Krieg, W. Ludwig, F. A. Rainey, K. H. Schleifer and W. B. Whitman, eds), Vol. 3, pp. 134–141. Springer, New York.

Priest, F. G., Goodfellow, M., Shute, L. A. and Berkeley, R. C. W. (1987). *Bacillus amyloliquefaciens* sp. nov. norn. rev. *Int. J. Syst. Bacteriol.* **37**, 69–71.

Prokofeva, M. I., Kostrikina, N. A., Kolganova, T. V., Tourova, T. P., Lysenko, A. M., Lebedinsky, A. V. and Bonch-Osmolovskaya, E. A. (2009). Isolation of the anaerobic thermoacidophilic crenarchaeote *Acidilobus saccharovorans* sp. nov. and proposal of *Acidilobales* ord. nov., including *Acidilobaceae* fam. nov. and *Caldisphaeraceae* fam. nov. *Int. J. Syst. Evol. Microbiol.* **59**, 3116–3122.

Rainey, F. A., Jo Hollen, B. and Small, A. (2009). Genus I. *Clostridum*. In *Bergey's Manual of Systematic Bacteriology* (P. De Vos, G. M. Garrity, D. Jones, N. R. Krieg, W. Ludwig, F. A. Rainey, K. H. Schleifer and W. B. Whitman, eds), Vol. 3, pp. 738–828. Springer, New York.

Rivas, R., Sánchez, M., Trujillo, M. E., Zurdo-Piñeiro, J. L., Mateos, P. F., Martinez-Molina, E. and Velázquez, E. (2003). *Xylanimonas cellulosilytica* gen. nov., sp. nov., a xylanolytic bacterium isolated from a decayed tree (*Ulmus nigra*). *Int. J. Syst. Evol. Microbiol.* **53**, 99–103.

Rogosa, M. (1969). *Acidaminococcus* gen. n., *Acidaminococcus fermentans* sp. n., anaerobic Gram-negative diplococci using amino acids as the sole energy source for growth. *J. Bacteriol.* **98**, 756–766.

Sakai, S., Imachi, H., Hanada, S., Ohashi, A., Harada, H. and Kamagata, Y. (2008). *Methanocella paludicola* gen. nov., sp. nov., a methane-producing archaeon, the first

isolate of the lineage 'Rice Cluster I', and proposal of the new archaeal order *Methanocellales* ord. nov. *Int. J. Syst. Evol. Microbiol.* **58**, 929–936.
Sakamoto, M. and Benno, Y. (2006). Reclassification of *Bacteroides distasonis, Bacteroides goldsteinii* and *Bacteroides merdae* as *Parabacteroides distasonis* gen. nov., comb. nov., *Parabacteroides goldsteinii* comb. nov. and *Parabacteroides merdae* comb. nov. *Int. J. Syst. Evol. Microbiol.* **56**, 1599–1605.
Sako, Y., Nomura, N., Uchida, A., Ishida, Y., Morii, H., Koga, Y., Hoaki, T. and Maruyama, T. (1996). *Aeropyrum pernix* gen. nov., sp. nov., a novel, aerobic hyperthermophilic archaeon growing at temperatures up to 100°C. *Int. J. Syst. Bacteriol.* **46**, 1070–1077.
Sangkhobol, V. and Skerman, V. B. D. (1961). *Chitinophaga*, a new genus of chitinolytic myxobacteria. *Int. J. Syst. Bacteriol.* **31**, 285–293.
Schildkraut, C. L., Marmur, J. and Doty, P. (1962). Determination of the base composition of DNA from its buoyant density in cesium chloride. *J. Mol. Biol.* **4**, 430–443.
Schleheck, D., Tindall, B. J., Rosselló-Mora, R. and Cook, A. M. (2004). *Parvibaculum lavamentivorans* gen. nov., sp. nov., a novel heterotroph that initiates catabolism of linear alkylbenzenesulfonate. *Int. J. Syst. Evol. Microbiol.* **54**, 1489–1497.
Schleifer, K. H. (2009). Classification of Bacteria and Archaea: past, present and future. *Syst. Appl. Microbiol.* **32**, 533–542.
Schlesner, H., Bartels, C., Sittig, M., Dorsch, M. and Stackebrandt, E. (1990). Taxonomic and phylogenetic studies on a new taxon of budding, hyphal proteobacteria, *Hirschia baltica* gen. nov., sp. nov. *Int. J. Syst. Bacteriol.* **40**, 443–451.
Sharp, P. M., Bailes, E., Grocock, R. J., Peden, J. F. and Sockett, R. E. (2005). Variation in the strength of selected codon usage bias among bacteria. *Nucleic Acids Res.* **33**, 1141–1153.
Steyn, P. L., Segers, P., Vancanneyt, M., Sandra, P., Kersters, K. and Joubert, J. J. (1998). Classification of heparinolytic bacteria into a new genus, *Pedobacter*, comprising four species: *Pedobacter heparinus* comb. nov., *Pedobacter piscium* comb. nov., *Pedobacter africanus* sp. nov. and *Pedobacter saltans* sp. nov. proposal of the family *Sphingobacteriaceae* fam. nov. *Int. J. Syst. Bacteriol.* **48**, 165–177.
Süng, Y., Fletcher, K. E., Ritalahti, K. M., Apkarian, R. P., Ramos-Hernandez, N., Sanford, R. A., Mesbah, N. M. and Löffler, F. E. (2006). *Geobacter lovleyi* sp. nov. strain SZ, a novel metal-reducing and tetrachloroethene-dechlorinating bacterium. *Appl. Environ. Microbiol.* **72**, 2775–2782.
Tamaoka, J. and Komagata, K. (1984). Determination of DNA base composition by reversed-phase high performance liquid chromatography. *FEMS Micrbiol. Lett.* **25**, 125–128.
Thi Ngoc Lan, P., Sakamoto, M., Sakata, S. and Benno, Y. (2006). *Bacteroides barnesiae* sp. nov., *Bacteroides salanitronis* sp. nov. and *Bacteroides gallinarum* sp. nov., isolated from chicken caecum. *Int. J. Syst. Evol. Microbiol.* **56**, 2853–2859.
Ueki, A., Akasaka, H., Suzuki, D. and Ueki, K. (2006). *Paludibacter propionicigenes* gen. nov., sp. nov., a novel strictly anaerobic, Gram-negative, propionate-producing bacterium isolated from plant residue in irrigated rice-field soil in Japan. *Int. J. Syst. Evol. Microbiol.* **56**, 39–44.
Urakami, T., Sasaki, J., Suzuki, K.-I. and Komagata, K. (1995). Characterization and description of *Hyphomicrobium denitrifcans* sp. nov. *Int. J. Syst. Bacteriol.* **45**, 528–532.
Van Aken, B., Peres, C. M., Doty, S. L., Yoon, J. M. and Schnoor, J. L. (2004). *Methylobacterium populi* sp. nov., a novel aerobic, pink-pigmented, facultatively methylotrophic, methane-utilizing bacterium isolated from poplar trees (*Populus deltoides* × *nigra* DN34). *Int. J. Syst. Evol. Microbiol.* **54**, 1191–1196.

Vandamme, P., Pot, B., Gillis, M., De Vos, P., Kersters, K. and Swings, J. (1996). Polyphasic taxonomy, a consensus approach to bacterial systematics. *Microbiol. Rev.* **60**, 407–438.

Vandamme, P., Holmes, B., Vancanneyt, M., Coenye, T., Hoste, B., Coopman, R., Revets, H., Lauwers, S., Gillis, M., Kersters, K. and Govan, J. R. W. (1997). Occurrence of multiple genomovars of *Burkholderia cepacia* in cystic fibrosis patients and proposal of *Burkholderia multivorans* sp. nov. *Int. J. Syst. Bacteriol.* **47**, 1188–1200.

Völkl, R., Huber, R., Drobner, E., Rachel, R., Burggraf, S., Trincone, A. and Stetter, K. O. (1993). *Pyrobaculum aerophilum* sp. nov., a novel nitrate-reducing hyperthermophilic archaeum. *Appl. Environ. Microbiol.* **59**, 2918–2926.

Wainø, M., Tindall, B. J. and Ingvorsen, K. (2000). *Halorhabdus utahensis* gen. nov., sp. nov., an aerobic, extremely halophilic member of the *Archaea* from Great Salt Lake, Utah. *Int. J. Syst. Evol. Microbiol.* **50**, 183–190.

Westerberg, K., Elvang, A. M., Stackebrandt, E. and Jansson, J. K. (2000). *Arthrobacter chlorophenolicus* sp. nov., a new species capable of degrading high concentrations of 4-chlorophenol. *Int. J. Syst. Evol. Microbiol.* **50**, 2083–2092.

Wiegel, J. (2009). Genus VI. Moorella. In *Bergey's Manual of Systematic Bacteriology* (P. De Vos, G. M. Garrity, D. Jones, N. R. Krieg, W. Ludwig, F. A. Rainey, K. H. Schleifer and W. B. Whitman, eds), Vol. 3, pp. 1247–1253. Springer, New York.

Willems, A., Goor, M., Thielemans, S., Gillis, M., Kersters, K. and De Ley, J. (1992). Transfer of several phytopathogenic *Pseudomonas* species to *Acidovorax* as *Acidovorax avenae* subsp. *avenae* subsp. nov., comb. nov., *Acidovorax avenae* subsp. *citrulli*, *Acidovorax avenae* subsp. *cattleyae*, and *Acidovorax konjaci*. *Int. J. Syst. Bacteriol.* **42**, 107–119.

Wilson, K. (1997). Preparation of genomic DNA from bacteria. In *Current Protocols in Molecular Biology* (F. M. Ausubel, R. Brent, R. E. Kingston, D. D. Moore, J. G. Seidman, J. A. Smith and K. Struhl, eds), pp. 2.4.1–2.4.5. John Wiley & Sons, New York. Vol.

Yakimov, M. M., Golyshin, P. N., Lang, S., Moore, E. R. B., Abraham, W.-R., Lunsdorf, H. and Timmis, K. N. (1998). *Alcanivorax borkumensis* gen. nov., sp. nov., a new, hydrocarbon-degrading and surfactant-producing marine bacterium. *Int. J. Syst. Bacteriol.* **48**, 339–348.

Yassin, A. F., Steiner, U. and Ludwig, W. (2002). *Corynebacterium aurimucosum* sp. nov. and emended description of *Corynebacterium minutissimum* Collins and Jones (1983). *Int. J. Syst. Evol. Microbiol.* **52**, 1001–1005.

Yoon, J.-H., Oh, T.-K. and Park, Y.-H. (2004). *Kangiella koreensis* gen. nov., sp. nov. and *Kangiella aquimarina* sp. nov., isolated from a tidal flat of the Yellow Sea in Korea. *Int. J. Syst. Evol. Microbiol.* **54**, 1829–1835.

15 DNA–DNA Hybridization

Ramon Rosselló-Móra*, Mercedes Urdiain, and Arantxa López-López
Marine Microbiology Group, Department of Ecology and Marine Resources, IMEDEA (CSIC-UIB), C/Miquel Marqués 21, E-07190 Esporles, Illes Balears, Spain

*E-mail: rossello-mora@uib.es

♦♦♦

CONTENTS

DNA–DNA Hybridization

♦♦♦♦♦♦ I. INTRODUCTION

DNA–DNA hybridization (DDH) techniques, also known as DNA–DNA reassociation techniques, are based on an attempt to make raw comparisons of whole genomes between different organisms in order to calculate their overall genomic similarities. Just after the discovery of the intrinsic properties of DNA (i.e. information content and secondary structure resilience), a number of techniques were developed and applied to microbial taxonomy in order to circumscribe its basic unit, the species. At that time, it was believed that such genetic comparisons would render more stable classifications than those simply based on phenotypic similarities (Krieg, 1988). The first attempt to elucidate taxonomic relationships based on single-stranded DNA reassociation was conducted by Schildkraut *et al.* (1961). This was a breakthrough for microbial systematics and the construction of the current microbial classification system. They demonstrated that duplex formation between the denatured DNA of one organism and that of another organism would only occur if the overall DNA base compositions were similar and if the organisms from which the DNA was extracted were genetically related. Since then, and for about 50 years, DDH has been used as the gold standard for numerical circumscription of genospecies in the classification of prokaryotes.

METHODS IN MICROBIOLOGY, VOLUME 38
0580-9517 DOI 10.1016/B978-0-12-387730-7.00015-2

Despite the extensive use of DDH approaches and data, these have often been criticized due to their high experimental error and their failure to generate cumulative databases (e.g. Sneath, 1989; Stackebrandt, 2003). There is a need to replace currently used DDH methodologies with other more accurate and reproducible techniques (Stackebrandt *et al.*, 2002). Novel approaches such as multilocus sequence analysis (MLSA) based on housekeeping gene sequence analyses (Gevers *et al.*, 2005), and average nucleotide identity (ANI) measurements based on genome-to-genome comparisons of orthologous genes (Konstantinidis and Tiedje, 2005) have been suggested as alternatives to DDH. ANI especially seems to reproduce DDH results with more accuracy, with the added value that studies can be done *in silico* using public databases, and also the results are reliable even with pair-wise comparisons using sequences covering about 20% of each genome (Richter and Rosselló-Móra, 2009). However, for the next few years, until whole genome sequencing will be economical and routine, DDH approaches will remain as one of the main parameters in circumscription of species of prokaryotes.

During the past half century of using DDH studies for microbial taxonomy, quite a few techniques have been developed (Table 1). All these techniques have in common the measurement of the extent and/or stability of the hybrid double-stranded DNA resulting from a denatured mixture of DNAs incubated under stringent conditions that allow only renaturation of complementary sequences. The use of these different techniques and their comparisons have been extensively discussed in the literature (Goris *et al.*, 1998; Grimont, 1988; Grimont *et al.*, 1980; Johnson, 1985, 1989; Owen and Pitcher, 1985; Stackebrandt and Liesack, 1993; Tjernberg *et al.*, 1989; and for a review, see Rosselló-Móra, 2006). All methods rely on a few common properties with the difference between them basically being variations in the DNA label type and/or the measurement technique. There are two main strategies for performing reassociation experiments: those where the hybridization reaction is carried out in free solution and those that employ previous fixation of the test DNA onto a solid surface (Table 1). The method of choice depends on the equipment and expertise of a given laboratory.

There are two main parameters that can be determined: the relative binding ratio (RBR) and the increment in melting temperature (ΔT_m). Sometimes the same procedure can provide/cover both parameters, but most of the techniques provide just one or the other (Table 1). The RBR is the measurement of the amount of the double-stranded hybrid DNA for a given pair of genomes relative to that measured for the reference DNA performed under identical renaturation conditions. RBR is expressed as a percentage based on the fact that the reference genome hybridizes 100% with itself. For those methods that use labelled DNA, large amounts of the labelled DNA may still remain as single-stranded DNA after the hybridization experiment, in which case the binding ratio (BR) is calculated as the amount of double-stranded hybrid DNA in relation to the total labelled DNA added in each single experiment. RBR is then determined by comparing the reassociation percentage of each heterologous reaction to that of the homologous reaction, which is considered to be 100%.

Table 1. Methods and labels used for DNA–DNA hybridization experiments, and type of measurement that each procedure allows

Free-solution methods	Label	Measurement	Reference
Buoyant density	Heavy isotopes	RBR	Schildkraut *et al.*, 1961
Hydroxyapatite	Radioactive isotope	RBR; ΔT_m	Brenner *et al.*, 1969b Lindh and Ursing, 1986
Hydroxyapatite/ Microtiter plate	Digoxigenin–biotin	RBR	Urdiain *et al.*, 2008
Spectrophotometry	none	RBR; ΔT_m	De Ley *et al.*, 1970 Huß *et al.*, 1983
Fluorimetric	none	ΔT_m	González and Sáiz-Jiménez, 2004
Endonuclease	Radioactive isotopes	RBR	Crosa *et al.*, 1973 Popoff and Coynault, 1980
Bound DNA	**Label**	**Measurement**	**Reference**
Agar embedded	Radioactive isotopes	RBR	Bolton and McCarthy, 1962 Brenner *et al.*, 1969a
Membrane filters	Radioactive isotopes	RBR; ΔT_m	Johnson, 1981 Owen and Pitcher, 1985 Tjernberg *et al.*, 1989
Membrane filters	non-radioactive labels	RBR	Jahnke, 1994 Cardinali *et al.*, 2000 Gade *et al.*, 2004
Microtiter plate bound DNA	photobiotin	RBR	Ezaki *et al.*, 1989 Adnan *et al.*, 1993 Kaznowski, 1995 Christensen *et al.*, 2000
Microtiter plate bound DNA	digoxigenin	ΔT_m	Mehlen *et al.*, 2004

RBR = relative binding ratio.

Spectrophotometric methods calculate the amount of hybrid DNA by basically comparing the reassociation kinetics with those of homologous DNA. It is important to note that RBR values are highly dependent on the stringency of the method used. At a given ionic strength, hybridizations may be carried out under what are considered to be optimal conditions (25–30°C below the melting point of the reference native DNA, i.e. T_m); stringent or exacting conditions (10–15°C below T_m), or relaxed, non-exacting conditions (30–50°C below T_m), although most of the results correspond to optimal condition experiments (Schleifer and Stackebrandt, 1983). The RBR is the most used parameter in the circumscription of species.

A more reliable parameter to determine is the ΔT_{m} because it is independent of the quantity and quality of the DNAs used for the experiment (Tjernberg *et al.*, 1989). This parameter is a reflection of the thermal stability of the DNA duplexes. ΔT_{m} is actually the difference between the melting temperatures of a given homologous DNA and a hybrid DNA. At a given ionic strength, the melting temperature of a DNA (or thermal denaturation midpoint, T_m, where 50% of DNA strands appear denatured) is directly related to its GC content (Schildkraut and Lifson, 1965; Turner, 1996). Hybrid DNAs will tend to melt earlier. The less related a pair of DNAs, the higher the difference between their melting points in degrees Celsius in comparison to their corresponding homologues. The reason is that a lower base pairing will render a less thermally stable base complementation. When the measurements are carried out with a labelled reference DNA, the melting temperatures are solely related to the extent of base pairing, and remain independent of the quality and quantity of each of the DNAs used for the hybridization. Consequently, the results of analysing melting profiles are very reproducible and less subjected to experimental error than RBR. However, because of the technical difficulties, RBR is used much more when trying to calculate raw genome similarities. In principle, both parameters do not need to be related: RBR reflects the extent of double-stranded DNA with a base complementarity of less than 15% base mispairing (Ullman and McCarthy, 1973; Stackebrandt and Goebel, 1994), and ΔT_m reflects the extent of sequence identity. However, it has been demonstrated empirically that there is indeed a linear correlation between them (Rosselló-Móra and Amann, 2001), and generally values of RBR above 50% will correlate with a ΔT_m below 4–5°C.

In order to provide easily implementable methods in all laboratories, we will outline three main procedures that do not use radiolabels. The three methods we have chosen to describe in detail are among the most widely used. We will also describe the method of DNA preparation as the DNA sample quality strongly influences the data obtained. It is of extreme importance that the DNA used for DDH is of the highest quality and free of traces of RNA.

♦♦♦♦♦♦ II. DNA PREPARATION

The DNA quality of the samples to be studied is of utmost importance in obtaining accurate and reproducible results. There are many methods described in the literature to purify genomic DNA from microorganisms (e.g. using French-Press lysis, sonication and purification with hydroxyapatite, or phenol–chloroform extraction). In addition, there are numerous commercial kits that can be used to prepare high-quality DNA. However, in our experience, the traditional method of Marmur (1961), improved and modified at a small scale for use with microfuge tubes (Urdiain *et al.*, 2008), yields enough DNA with adequate quality for DDH. It is important to note that DDH methods generally require large amounts of DNA (>15 µg per assay), and that most of the isolation procedures that had been designed for PCR amplification procedures generally yield too low amounts to guarantee the success of the experiment.

A. Isolation of High-Quality DNA

- Grow microbial cells in either liquid or solid media. Biomass grown on agar plates can be collected with a sterile swab to yield approximately 200 mg of fresh weight. Alternatively, the same biomass can be collected from liquid media by centrifugation at 13,000 rpm. Cell biomass (Figure 1A) should be washed once with 1X PBS by cell resuspension and centrifugation at 13,000 rpm. Discard the supernatant, and resuspend the pellet in 1 ml saline EDTA.
- Cell disruption depends on the nature of each microorganism. Some microorganisms are easy to lyse just using the saline EDTA and followed by the enzymatic lysis step. In some cases, the cells may be very difficult to disrupt due to their strong cell envelope. In such cases, several freeze (−80°C or liquid nitrogen) and thaw (between 37°C and 65°C in a water bath) cycles may be undertaken. Alternatively, for organisms very recalcitrant to cell disruption, cell biomass embedded in liquid nitrogen can be grounded in a porcelain mortar and the disrupted material resuspended in 1 ml of XS buffer. Vortex the samples resuspended in XS buffer for 2 min and incubate for 2 h at 65°C

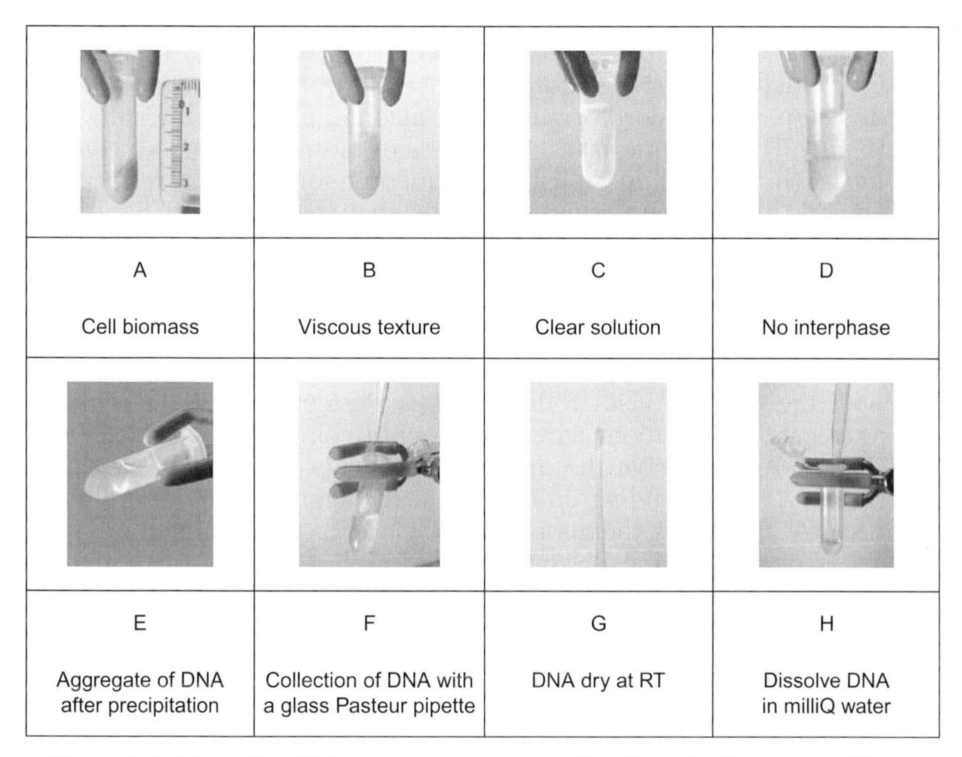

Figure 1. High-quality DNA preparation steps: (A) collected cell biomass; (B) viscous texture of the cellular extract after enzymatic lysis treatment; (C) clear solution after incubation with 1% SDS; (D) No interphase must be observed after the chloroform:isoamyl alcohol washing steps; (E) aggregate of DNA after precipitation; (F) recovery of the DNA with a glass Pasteur pipette; (G) DNA drying at room temperature; (H) dissolving the DNA in miliQ water.

(briefly mixing each ½ h). Following the incubation, vortex the cells for 10 s and transfer on ice for 10 min. Centrifuge the samples for 10 min at 13,000 rpm and collect the supernatant in a fresh tube to proceed with the enzymatic lysis.

- Perform enzymatic lysis by adding 3.3 μl of lysozyme (300 mg/ml) and 10 μl of proteinase K (10 mg/ml) to the 1 ml cell suspension on saline EDTA, and mixing vigorously. For some microorganisms, e.g. Gram-positives, better yields can be obtained by amending the mixture with 10 μl of mutanolysin (1000 U/ml). Incubate the mixture for 15–45 min at 37°C until the extract acquires a viscous texture (Figure 1B). Subsequently, add 80 μl of 25% SDS, and incubate the mixture for 10 min at 65°C, or until a clear lysate is observed. In general, lysis is considered optimal when the suspension turns from turbid to a clear solution (Figure 1C). Clear lysis cannot always be observed, but the viscosity indicates cell disruption. After lysis, add 250-μl 5 M NaCl and mix through inversion to precipitate proteins.
- To extract separate and purify the nucleic acids, add an equal volume of equilibrated phenol:chloroform:isoamyl alcohol (25:24:1) and mix by shacking until a homogeneous milky suspension is obtained. Alternatively, the extraction can be done avoiding the use of equilibrated phenol by adding just chloroform: isoamyl alcohol (24:1) and mixing until a milky suspension is obtained. Separate immiscible phases by centrifugation for 5 min at 13,000 rpm. Cautiously recover the aqueous phase. Repeat three times (at least) this washing step with chloroform:isoamyl alcohol until no interphase is observed (Figure 1D). In some cases, the excess of biomass can produce a viscosity that makes it difficult to collect the aqueous solution. To reduce the viscosity, amend the solution with saline EDTA (between 0.2 and 0.5 ml) and repeat the first extraction step.
- To precipitate the nucleic acids, measure the volume recovered of the aqueous supernatant. Add 1/9 volume of 3 M sodium acetate (pH 7) and mix. Add 0.6 volume of isopropanol. Gently mix by inversion until a white visible aggregate is formed (Figure 1E), store at −20°C for about 30 min. Recover the aggregate with the tip of a glass rod or a glass Pasteur pipette (Figure 1F). Leave the aggregate dry at room temperature until complete evaporation of the alcohol (Figure 1G). Dissolve the dried nucleic acids in 250 μl of sterile milliQ water (Figure 1H). Depending on the size of the aggregate, more or less water can be used. At this point, the dissolved nucleic acids can be stored at 4°C.
- To proceed with the RNA removal, add 5 μl RNAse A/T_1 mix and incubate for 1–2 h at 37°C.
- Purify DNA once or twice (or until no interphase is detected) with chloroform:isoamyl alcohol (24:1), as described above.
- Purify DNA by precipitation as outlined above. In the cases where the DNA aggregate is not visible, the mixture can be centrifuged for 30 min at 13,000 rpm and the supernatant discarded. The pellet should be washed once by adding 1 ml of 70% ethanol (in water) and centrifuged for 30 min at 13,000 rpm and the supernatant discarded. Leave the pellet dry at room temperature.
- Dissolve the final DNA precipitate in 100–200 μl of sterile milliQ water and keep at −20°C until use.

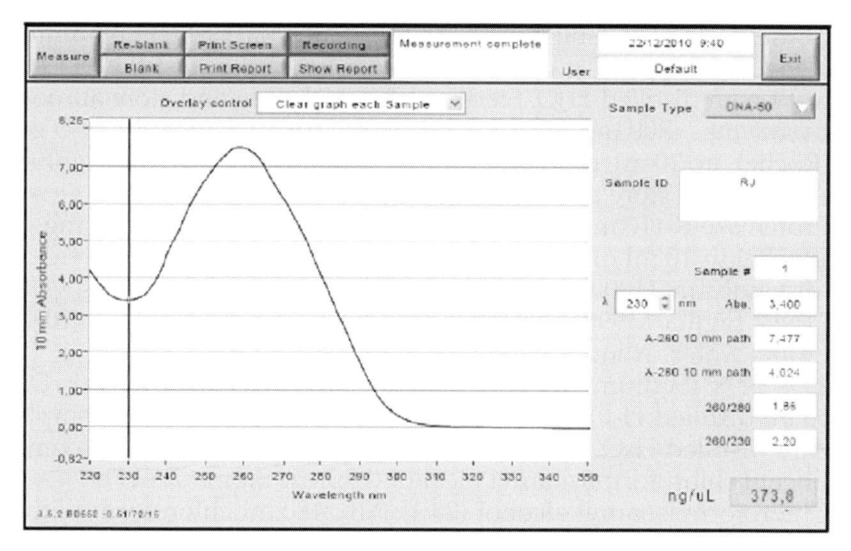

Figure 2. Nanodrop spectrum of a pure extracted DNA.

- The purity of the DNA should be determined spectrophotometrically by measuring OD_{280}, OD_{260} (maximal absorbance for double-stranded DNA) and OD_{230}. Most modern spectrophotometers (e.g. Nanodrop ND1000 (Nanodrop Technologies); BioPhotometer (Eppendorf); spectrophotometer Hitachi U-2900 UV-VIS) give already the purity ratios and the absolute DNA quantification. A correct purity of DNA is observed with the following ratios: $OD_{260}/OD_{280} \geq 1.8$ and $OD_{260}/OD_{230} \geq 2.0$ (Figure 2). DNA concentration can be calculated as 1 OD $\approx$ 50 µg/ml. Optimally, a DNA stock should be obtained of at least 250 µl of DNA with a concentration $\geq$0.3 µg/ml.

B. Reagents

10X PBS (Phosphate-buffered saline; 1.3 M NaCl, 0.027 M KCl, 0.1 M Na_2HPO_4, 0.02 M KH_2PO_4; pH 7.4): Dissolve in 700 ml of distilled H_2O: 80 g NaCl, 2 g KCl, 14.4 g Na_2HPO_4, 2.4 g KH_2PO_4; adjust the pH to 7.4 with 1 M of HCl. Add distilled H_2O to a final volume of 1 l, and sterilize by autoclaving. Store at room temperature.

NaCl 5 M: Add 292.2 g/l, autoclave and store at room temperature.

EDTA ($C_{10}H_{14}N_2Na_2O_8 \cdot 2H_2O$, Sigma, E5134) 0.25 M: Add 93 g/l to 800 ml of distilled H_2O, and add 5 g of NaOH to dissolve and adjust to a final pH 8 with 1 M NaOH, add distilled H_2O to a final volume of 1 l.

Saline EDTA (0.15 M NaCl, 0.1 M EDTA; pH 8): Add 30 ml of 5 M NaCl and 400 ml of EDTA 0.25 M, pH 8, bring to 1 l with distilled H_2O. Sterilize by autoclaving and store at room temperature.

XS buffer (1% potassium ethyl xanthogenate, 20 mM EDTA, 1% SDS, 800 mM ammonium acetate, 100 mM Tris-HCl; pH 7.4): In 600 ml of distilled H_2O, add 10 g of potassium ethyl xanthogenate (C_2H_5OCSSK), 5.8 g of EDTA, 10 g of SDS

(sodium dodecyl sulfate, $CH_3(CH_2)_{11}OSO_3Na$), 61.6 g of ammonium acetate ($CH_3CO_2NH_4$), 15.7 g of Tris-HCl; dissolve and adjust the pH at 7.4, and bring to 1 l with distilled H_2O. Sterilize by autoclaving and store at room temperature.
Lysozyme (300 mg/ml, Roche, 10837059001): Dissolve 5 g of lysozyme (Roche) in 10 ml of milliQ water and adjust to a final volume of 16.6 ml. Aliquot and store at −20°C.
Proteinase K (10 mg/ml, Roche, 03115879001): Dissolve 10 mg of proteinase K (Roche) in 10 ml of milliQ water. Aliquot and store at −20°C.
Mutanolysin (1000 U/ml, Mutanolysin from *Streptomyces globisporus* ATCC 21553; Sigma M9901-10KU): Reconstitute the lyophilisate (10KU) in 10 ml of milliQ water. Aliquot and store at −20°C.
25% SDS (Sodium dodecyl sulphate, $CH_3(CH_2)_{11}OSO_3Na$): Dissolve 25 g in 50 ml distilled H_2O, warm it to 65°C and stir gently until complete dissolution. Add distilled H_2O to a final volume of 100 ml and store at room temperature.
Phenol:chloroform:isoamyl alcohol (25:24:1; Sigma 77617).
Chloroform:isoamyl alcohol (24:1): Mix 480 ml chloroform with 20 ml isoamyl alcohol and store at room temperature.
Sodium acetate (3 M; pH 7): Dissolve 24.6 g of sodium acetate (CH_3COONa) in 60 ml of milliQ water and adjust the pH to 7. Add distilled H_2O to a final volume of 100 ml. Store at room temperature.
Isopropanol $(CH_3)_2CHOH$ for molecular biology.
RNAse A/T_1 mix (2 mg/ml of RNaseA, 5000 U/ml of RNase T_1; Fermentas EN0551).

♦♦♦♦♦♦ III. HYDROXYAPATITE METHOD USING DIGOXIGENIN–BIOTIN LABELLED DNA

This is a free-solution hybridization method that results as a modification of the original radioactive method of Brenner and colleagues (1969b) and Lindh and Ursing (1986). The procedure has been downscaled to facilitate manipulation and optimize the DNA quantities to be used. The most relevant points to pay attention to are (i) DNA of high purity is required with no RNA traces; (ii) the DNA concentrations of all of the different samples to be compared must be identical; (iii) the GC content of the 'homologous' DNA has to be known in order to use it for the stringency calculations (generally 30°C below the melting temperature); (iv) tube incubations should be done in a water bath or dry bath to ensure quick temperature equilibration of the mixtures. This method allows only the determination of the RBR, but not the increment in melting temperature (ΔT_m) (Urdiain *et al.*, 2008).

A. Labelling the Probe DNA

Reference DNA is double-labelled using DIG-11-dUTP and biotin-16-dUTP (75:25, by volume) by using the nick translation kit (Roche). It is recommended that the concentration of the reference DNA in solution is in the range 0.4 to 0.6 μg/μl.

- Prepare the labelling mix (on ice and in a 1.5 ml microfuge tube) with a final volume of 20 μl by adding 10 μl of dNTP mix (12 μl of dNTP mix contains 3 μl of each dATP, dGTP and dCTP, 2 μl of dTTP, 0.75 μl of DIG-11-dUTP and 0.25 μl of biotin-16-dUTP; Roche) + 2 μl of nick translation buffer + 2 μl of enzyme mixture and 2 μg of reference DNA. Increase volume to 20 μl with sterile milliQ water.
- Label the DNA for about 90 min at 15°C and store on ice.
- Precipitate DNA by adding 380 μl of sterile milliQ water + 45 μl of Na acetate (3 M, pH 7) + 890 μl of absolute ethanol. Gently mix after each addition step, and finally store at −20°C for about 30 min. Centrifuge at 13,000 rpm for 20 min and discard the supernatant. Dry the pellet at room temperature until completely dry. Resuspend the labelled DNA in 200 μl of sterile milliQ water.

B. Testing the Labelling Efficiency

Prior to any hybridization experiment, it is important to test the quality of the labelling reaction.

- Denature the 200 μl of resuspended and labelled DNA by boiling for 5 min and chill on ice. Mix well by vortexing and spin the microfuge tube briefly to collect the water drops.
- Prepare 10 ml of 1X PBS + 0.1% BSA. For each labelling mixture, prepare two tubes (A and B). Add 800 μl of 1X PBS + 0.1% BSA to tube A and 400 μl to tube B. Add 2 μl of denatured labelled DNA to tube A, mix well by vortexing and shortly spin the microfuge tube to collect the water drops. Transfer 400 μl from tube A to tube B, mix well by vortexing and spin briefly. Add 200 μl from each tube (A and B) to its respective well of the streptavidine-coated microtiter plate. Each microtiter plate well of tube A contains about 5 ng of labelled DNA, whereas those from tube B solution contain about 2.5 ng of labelled DNA. For each labelling evaluation, include one microtiter plate well with a control containing 200 μl of 1X PBS + 0.1% BSA.
- Cover the microtiter plate strips with adhesive film to avoid sample evaporation and incubate 30 min with vigorous shaking (we use 700 rpm on a microtiter plate shaker) at room temperature.
- Discard the solution from the microtiter wells and wash three times with 300 μl 1X PBS. Washing steps are done by filling the wells and discarding the solution. Washing does not need any incubation.
- Add 200 μl of the antibody solution, cover the microtiter plate strips with adhesive film and incubate 30 min with shaking (700 rpm) at room temperature.
- Discard the solution from the microtiter plate wells and wash three times with 300 μl of 1X PBS as mentioned above.
- Add 250 μl of developing solution (see below) and read the absorbance at time zero at 405 nm.

- Incubate the microtiter plate strips covered with adhesive film at 37°C for 20, 40 and 60 min. As a reference, an optimal labelling will render absorbances of about 0.6 OD units at 30 min; a good labelling will render about 0.3 OD U at 30 min and 0.6 OD units at 60 min. Low labelling efficiencies may cause difficulties in the hybridization experiments, and preparing a new labelling reaction is recommended if OD values are <0.3 OD units after 60 min incubation.

C. DDH Mixtures

For the hybridization step, it is of utmost importance that all unlabelled DNAs are prepared in the same concentrations. The hybridization solutions require 15 μg of unlabelled DNA. For this, we recommend that all DNAs are prepared at stock solutions of about 0.3 μg/μl. The best results are obtained if the hybridization stock solutions are prepared from more concentrated samples, and all dilutions are measured simultaneously with the same spectrophotometer. It is important that the solutions of DNA are homogeneous and the exact DNA concentrations determined. For each hybridization set (i.e. each labelled DNA with the corresponding set of unlabelled DNAs to be tested), it is important to add a homologous DNA mixture (i.e. labelled and unlabelled DNAs of the same organism). Prepare the hybridization mixtures as follows:

- Add 15 μg of unlabelled DNA to a microfuge tube (500 μl PCR tubes are recommended to perform the hybridization), and mix with 15 μl of labelled DNA (if the efficiency of the labelling reaction has been excellent (>0.6 U at 30 min incubation), the amount of unlabelled DNA can be reduced). Bring the DNA mixture total volume to 72 μl with sterile milliQ water.
- Boil the hybridization mixtures for 5 min to denature the DNAs and chill on ice for 5 min. Vortex and spin the microfuge tube briefly to collect the water drops.
- Add 28 μl of 1 M PB buffer to bring the solution to a final volume of 100 μl. The final concentration of the buffer is 0.28 M.
- Cover the hybridization solution with 50 μl of mineral oil to avoid evaporation. Mineral oil should cover the water solution completely with about 3–4mm overlay. To ensure that the phases are well separated, shortly spin the microfuge tubes prior to the hybridization incubation.

D. DDH Procedure

In the literature, several distinct hybridization stringency conditions may be assayed (read above). However, we recommend the use of 30°C below the melting temperature of the homologous DNA (the labelled DNA used) as it is the most widely used stringency condition. The DDH temperature used depends on the GC content of the homologous DNA and the ionic strength of the hybridization solution; here it is set at 0.28 M PB. The required T_m, DDH and washing temperatures using the hybridization conditions here are given in Table 2.

Incubate the DDH mixtures for 16 h in a water bath at $T_m - 30°C$, and store at 4°C until the hydroxyapatite treatment.

Table 2. GC mol% and their respective melting temperatures, and DDH and washing temperatures using the PB buffer or 2X SSC

GC mol% of the homologous DNA	Melting temperature (T_m)	Hybridization temperature (T_m −30)	Washing temperature (T_m −35)	Hybridization temperature using spectrophotometer (T_m −25)
30	87.0	57.0	52.0	62.0
31	87.4	57.4	52.4	62.4
32	87.8	57.8	52.8	62.8
33	88.2	58.2	53.2	63.2
34	88.7	58.7	53.7	63.7
35	89.1	59.1	54.1	64.1
36	89.5	59.5	54.5	64.5
37	89.9	59.9	54.9	64.9
38	90.3	60.3	55.3	65.3
39	90.7	60.7	50.7	65.7
40	91.1	61.1	56.1	66.1
41	91.5	61.5	56.5	66.5
42	91.9	61.9	57.0	66.9
43	92.3	62.3	57.4	67.3
44	92.8	62.8	57.8	67.8
45	93.2	63.2	58.2	68.2
46	93.6	63.6	58.7	68.7
47	94.0	64.0	59.1	69.1
48	94.4	64.4	59.5	69.5
49	94.8	64.8	59.9	66.3 (+5% formamide)
50	95.2	65.2	60.3	66.7 (+5% formamide)
51	95.6	65.6	60.7	67.1 (+5% formamide)
52	96.0	66.0	61.1	67.5 (+5% formamide)
53	96.4	66.4	61.5	67.9 (+5% formamide)
54	96.9	66.9	61.9	68.4 (+5% formamide)
55	97.3	67.3	62.3	68.8 (+5% formamide)
56	97.7	67.7	62.8	69.2 (+5% formamide)
57	98.1	68.1	63.2	69.6 (+5% formamide)
58	98.5	68.5	63.6	70.0 (+5% formamide)
59	98.9	68.9	64.0	70.4 (+5% formamide)
60	99.3	69.3	64.4	67.3 (+10% formamide)
61	99.7	69.7	64.8	67.7 (+10% formamide)
62	100.1	70.1	65.2	68.1 (+10% formamide)
63	100.5	70.5	65.6	68.5 (+10% formamide)
64	101.0	71.0	66.0	69.0 (+10% formamide)
65	101.4	71.4	66.4	69.4 (+10% formamide)
66	101.8	71.8	66.9	69.8 (+10% formamide)
67	102.2	72.2	67.3	70.2 (+10% formamide)
68	102.6	72.6	67.7	70.6 (+10% formamide)
69	103.0	73.0	68.1	71.0 (+10% formamide)
70	103.4	73.4	68.5	71.4 (+10% formamide)
71	103.8	73.8	68.9	71.8 (+10% formamide)

(Continued)

Table 2. (Continued)

GC mol% of the homologous DNA	Melting temperature (T_m)	Hybridization temperature (T_m −30)	Washing temperature (T_m −35)	Hybridization temperature using spectrophotometer (T_m −25)
72	104.2	74.2	69.3	72.2 (+10% formamide)
73	104.6	74.6	69.7	72.6 (+10% formamide)
74	105.1	75.1	70.1	73.1 (+10% formamide)
75	105.5	75.5	70.5	73.5 (+10% formamide)

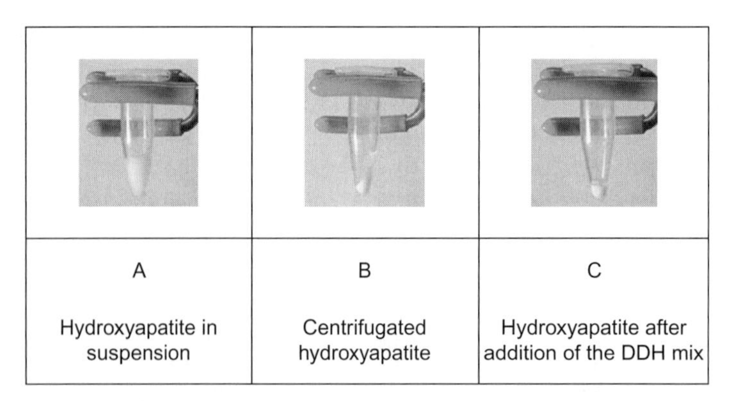

Figure 3. Equilibrated hydroxyapatite in suspension: (A) hydroxyapatite in suspension; (B) centrifuged hydroxyapatite; (C) Hydroxyapatite after the addition of the DDH mix.

E. Single-Stranded and Double-Stranded DNA Separation Using Hydroxyapatite (HA)

- For each DDH mixture, prepare two tubes (1.5 ml microfuge tubes) containing 200 μl of equilibrated HA (approximately 20 mg HA dry weight; Figure 3). Centrifuge and remove the clear supernatant. Store the wet HA at 4°C.
- Dilute each DDH mixture by adding 100 μl of sterile milliQ water to give a final volume of 200 μl bringing the solution to a final ion strength of 0.14 M PB. Mix well and spin briefly.
- Transfer 50 μl of the DDH mixture to each of the two HA tubes. Before pipetting into the HA tubes, the pipette tip should be carefully wiped with a clean tissue to remove adsorbed mineral oil.
- Mix the DDH mixture well with the HA and incubate for 15 min at 5°C below the reassociation temperature (i.e. T_m −35°C, Table 2).
- Add 450 μl of PB (0.14 M PB + 0.2% SDS), mix well and incubate with repeated mixing for 5 min at the same temperature. In this step, single-stranded DNA remains unbound in the supernatant whereas double-stranded DNA binds the HA.
- Centrifuge (2 min at 13,000 rpm) and store the supernatant in a clean tube.

- Add 500 μl of PB (0.14 M PB + 0.2% SDS) to the HA and incubate with repeated mixing for 5 min at 5°C below the reassociation temperature.
- Centrifuge (2 min at 13,000 rpm) and collect and mix the supernatant in the same tube with the first washing supernatant. The total amount of supernatant harbouring single-stranded DNA is 1 ml.
- Add 200 μl of 0.4 M PB to the HA pellet containing bound double-stranded DNA, mix well and keep at room temperature for 1–2 min.
- Centrifuge (2 min at 13,000 rpm) and store the supernatant in a clean tube.
- Add 200 μl of 0.4 M PB to the HA pellet, mix well and keep at room temperature for 1–2 min.
- Centrifuge (2 min at 13,000 rpm) and collect and mix the supernatant in the same tube with the washing supernatant. The total amount of supernatant harbouring double-stranded DNA is 0.4 ml.
- At this step, both kinds of samples can be stored at 4°C until the detection of the eluted DNA. For each DDH mixture, two sets of tubes have been produced, each set comprises 1 ml of single-stranded DNA and 0.4 ml of double-stranded DNA.

F. Detection of Eluted DNA

- Add 10 μl of 10% BSA to the tube containing the 1 ml of single-stranded DNA and 4 μl of 10% BSA to the tube containing 0.4 ml of double-stranded DNA. Mix well and spin briefly.
- Add 200 μl of each single eluate to a well of a streptavidin-coated microtiter plate. Cover the microtiter plate strips with adhesive film to avoid sample evaporation and incubate for 2 h at room temperature with vigorous shaking (700 rpm).
- Discard the solution from the microtiter plate wells and wash three times with 300 μl 1X PBS.
- Add 200 μl of the antibody solution, cover the microtiter plate strips with adhesive film and incubate for 1 h with vigorous shaking (700 rpm) at room temperature.
- Discard the solution from the microtiter wells and wash three times with 300 μl of 1X PBS.
- Add 250 μl of developing solution and read the absorbance at time zero at 405 nm.
- Incubate the microtiter plate strips covered with adhesive film at 37°C. Read the colour development each hour until the colour measures range between 0.3 and 2 U (Figure 4). In exceptional cases where the colour development is slow, the incubation can be done overnight. However, this is not recommended.

G. Treatment of Hybridization Data (Figure 4)

1. Calculate the binding ratio as the percentage of labelled DNA released with 0.4 M PB (DS) compared to the total labelled DNA (that of 0.14 M PB (SS) summed with that of 0.4 M PB (DS)). This is

$$\frac{(2 \times \text{absorbance}_{405}\text{DS} \times 100)}{(2 \times \text{absorbance}_{405}DS) + (5 \times \text{absorbance}_{405}\text{SS})}$$

2. Calculate the relative binding ratio of heterologous DNA as the percentage of the homologous binding calculated for the control DNA. This is

$$\frac{(BR_{heterologous} \times 100)}{BR_{homologous}}$$

3. Calculate the pooled standard deviation for each experiment. This is

$$PSD = \sqrt{\frac{\sum (M - \chi)^2}{n}}$$

where M is the mean of the duplicated BR values for DDH mixture, X each single value obtained for each DDH mixture, and n the number of all single values obtained in the experiment (Figure 4).

Final remarks: There are other methods using microtiter plates including that implemented by Ezaki and colleagues (Ezaki *et al.*, 1989). This method immobilizes the target DNA onto the microtiter plate wells and the test DNA is labelled with biotin. In the original description of the method, measurements were undertaken using fluorogenic substrates, but later these were substituted by a chemiluminescent substrate as well as by covalent binding onto the surface of microtiter plates (Adnan *et al.*, 1993). Similar methods have been developed that use colorimetric reactions for the detection (Kaznowski, 1995), which importantly reduces the cost of the equipment used. Lately, more sophisticated but reliable methods have been developed, which allow experimentation with fastidious organisms whose DNA is difficult to recover (Mehlen *et al.*, 2004) and, in this case, genomic DNA is previously amplified before being bound to the microtiter well. Then, digoxygenin labelled reference DNA is used to perform the hybridization, and the stringency is accomplished by washing with decreasing ion strength buffers that allows the determination of melting profiles for hybrid molecules. Detection is achieved colorimetrically.

H. Reagents

Nick translation kit (Roche, Cat. Nr. 10976776001).

DIG-11-dUTP (Digoxigenin-11-2′-deoxy-uridine-5′-triphosphate, alkali-stable, Roche, Cat. Nr. 11093088910).

Biotin-16-dUTP (Biotin-16–2′-deoxy-uridine-5′-triphosphate, Roche, Cat. Nr. 11093070910).

Sodium acetate (3 M; pH 7): See recipe above.

Absolute ethanol molecular biology grade.

10X PBS (1.3 M NaCl, 0.05 M Na_2HPO_4, 0.015 M KH_2PO_4; pH 7.4): See recipe above.

BSA 10% (Albumin from bovine serum; Sigma, Cat. Nr. A3059): Dissolve 1 g of BSA in 7 ml of milliQ water and adjust to a final volume of 10 ml. Aliquot and store at −20°C.

1X PBS + 0.1% BSA: Add 100 μl of 10% BSA to 9.9 ml of 1X PBS and mix well.

(A) Example of colour development and absorbances at 405 nm of three hybridization mixtures.

Labelled DNA*: Strain C3		Single Stranded (SS)	Double Stranded (SS)
1: Strain C3 (Homologous DNA)	a	1.4730 (A3)	1.1870 (A4)
	b	1.2540 (B3)	1.0540 (B4)
2: *Afifella marina* (A. m) (Hybrid DNA)	a	1.4050 (E3)	0.8520 (E4)
	b	1.4440 (F3)	1.0720 (F4)
3: *Afifella pfennigii* (A. p) (Hybrid DNA)	a	1.9800 (G1)	0.3250 (G2)
	b	1.7110 (H1)	0.3020 (H2)

(B) Calculate the total single-stranded (x5) and double stranded (x2) respective absorbances.
Calculate the binding ratio (BR) for each experiment and the mean of each pair of experiments.

	SS	DS	SS x 5	DS x 2	**BR %** (a,b)	Mean BR (M)
C3a	1.4730	1.1870	7.365	2.374	24.3762	24.77
C3b	1.2540	1.0540	6.27	2.108	25.1611	
A. m (a)	1.4050	0.8520	7.025	1.704	19.5211	21.21
A. m (b)	1.4440	1.0720	7.22	2.144	22.8961	
A. p (a)	1.9800	0.3250	9.9	0.65	6.1611	6.38
A. p (b)	1.7110	0.3020	8.555	0.604	6.5946	

(C) Calculate the relative binding ratio (RBR) of each hybrid in respect to the homologous DNA (100% BR).
Calculate the sum of deviations from each mean (M-each BR value)$_2$, and calculate the pooled standard deviation of the experiment with these values (see formulae).

RBR %	(M-BRa)^2	(M-BRb)^2
100.0	0.15	0.15
85.6	2.85	2.85
25.7	0.05	0.05
Σ =	6.10	
n =	6	
PSD =	1.01 %	

$$BR = \frac{DS \times 2}{(SS) \times 5 + DS \times 2} \times 100 \qquad RBR = \frac{BR_{hybrid}}{BR_{homologous}} \times 100$$

$$PSD = \sqrt{\frac{\Sigma (M - X)^2}{n}}$$

Figure 4. Example of calculation of RBR.

Streptavidine-coated microtiter plates (StreptaWell, High Bind (high binding capacity), 12 × 8-well strips and frame, transparent, C-bottom; Roche, Cat. Nr. 11645692001).

Anti-Digoxigenin-AP (Fab fragments from an anti-digoxigenin antibody from sheep, conjugated with alkaline phosphatase; Roche, Cat. Nr. 11093274910): The antibody is used as 1 part of antibody in 5000 parts of solution (1:5000, by volume).

Antibody solution: To 5 ml of 1X PBS + 0.1% BSA, add 1 μl of the anti-digoxigenin antibody, and mix well by vortexing.

Developing buffer (7.5 mM Na_2CO_3, 15.5 mM $NaHCO_3$, 1 mM $MgCl_2 \cdot 6H_2O$; pH 9.6): Dissolve in 80 ml of milliQ water 0.0794 g Na_2CO_3 + 0.13 g $NaHCO_3$ + 0.02 g $MgCl_2 \cdot 6H_2O$; adjust to pH 9.6 and fill with milliQ water to a final volume of 100 ml. Store at room temperature.

Developing solution: To 5 ml of developing buffer, add 1 tablet (5 mg) of 4-nitrophenyl phosphate disodium salt hexahydrate (Sigma, Cat. Nr. N9389), mix well until complete dissolution. This solution is transparent (slightly yellow) and should be freshly prepared and protect from light incidence.

Solution A PB (Na_2HPO_4; 1 M; pH 9.1): Dissolve 70.9 g of Na_2HPO_4 in 400 ml of milliQ water and bring it to a final volume of 500 ml.

Solution B PB (NaH_2PO_4; 1 M; pH 3.8): Dissolve 24 g of NaH_2PO_4 in 125 ml of milliQ water and bring it to a final volume of 200 ml.

1 M PB buffer (Phosphate buffer 1 M; pH 6.9): Mix 300 ml of solution A with 150 ml of solution B, and adjust the pH at 6.9 with solution B.

Mineral oil for molecular biology (Sigma, Cat. Nr. M5904).

Equilibrated HA: (Hydroxyapatite, DNA grade Bio-Gel HTP; Bio-Rad, Cat. Nr. 130-0520): Mix 1 g of HA with 10 ml of 0.14 M PB, shake vigorously and store at 4°C.

PB 0.14 M: Mix 1.4 ml of 1 M PB (see above) with 8.6 ml of milliQ water.

PB 0.14 M −0.2% SDS: Mix 200 μl of 10% SDS with 9.8 ml of PB 0.14 M.

PB 0.4 M: Mix 4 ml of 1 M PB (see above) with 6 ml of milliQ water.

♦♦♦♦♦♦ IV. DDH FROM MELTING CURVES USING THE REAL-TIME PCR

This is a free-solution hybridization method that uses quantitative real-time PCR thermocyclers. The protocol measures the thermal stability of double-stranded (homologous or heterologous) DNA by means of a decrease in the fluorescence (bound to double-stranded DNA) with increasing temperatures. This method involves estimating the melting temperatures (T_m), and thus the increment in melting temperatures between homologous and hybrid DNA structures (ΔT_m). It is an easy and quick method that can be used as an alternative DDH measurement (González and Sáiz-Jiménez, 2004).

- Prepare genomic DNA from the strains of interest and estimate the DNA concentration (e.g. spectrophotometrically). Dilute the stock solutions in 2XSCC to a concentration of about 1 μg/μl. For accurate results, it is recommendable that all the stock solutions are measured with the same spectrophotometer and brought to similar concentrations.
- Prepare triplicate tubes with the DNA of each homologous combination (i.e. single strains), and the combination (also in triplicates) of each of the DNA pairs from the strains that you want to compare. Use 5 μg of each DNA per

Table 3. Recommended temperatures (T_{or}) for the renaturation experiments using real-time PCR

GC mol%	30–45	45–60	60 to 75
T_{or}	60°C	68°C	75°C
T_{or} (+15% formamide)	40°C	48°C	55°C

tube or well. The final DNA concentration in the mixture is 10 µg. Alternatively, the DNA amounts can be reduced to 5 µg by adding 2.5 µg of each DNA in the mixture.

- The optimum temperature of renaturation (T_{or}) depends on the GC content of the organisms under investigation. The temperature to be used in the DDH experiments can be obtained from Table 3.
- Apply the following protocol for iCycler for the hybridization conditions: (i) Leave the mixture at room temperature for 10 min. (ii) Denature at 99.9°C for 10 min. (iii) Renature the DNAs at the calculated T_{or} for 8 h. (iv) Optimize renaturation at T_{or} −10°C for 30 min. (v) Optimize renaturation at T_{or} −20°C for 30 min. (vi) Optimize renaturation at T_{or} −30°C for 30 min. (vii) Leave at 4°C (or at room temperature) until the measurements of the melting curves are carried out.
- Add SYBR Green I (1/100,000 final dilution of the hybridization mixture of 50 µl).
- Generate the melting curves with the iCycler applying a melting ramp of 0.2°C/s starting at 20°C up to 99°C. This protocol implies 395 cycles. Finally, leave the mixture at 4°C (or at room temperature). Note that some thermocyclers automatically add some 95°C cycles before the programmed protocol. To avoid the denaturation of your DNA during these cycles, place a control plate (no DNA and diluted [around 1/100,000] fluorescein or ROX as recommended by the manufacturers). After the added cycles, pause the run, exchange the plates or tubes and continue.
- Estimate the temperature for each DNA mixture at which 50% of the DNA is still double-stranded (T_m). Compare each homologous denaturation with their respective hybrid and calculate the difference in melting temperature between them. This is increment in melting temperature (ΔT_m) = T_m (homologous) − T_m (hybrid) in degrees Celsius (Figure 5).

Final remarks: The results of analysing melting profiles are very reproducible and less subjected to experimental error than RBR. However, because of the technical difficulties (higher amount of denaturation and washing steps or the need of special equipment as iCyclers or special spectrophotometers), RBR is much more popular when trying to calculate raw genome similarities. In principle, both parameters do not need to be related: RBR reflects the extent of double-stranded DNA with a base complementarity of less than 15% base mispairing (Ullman and McCarthy, 1973; Stackebrandt and Goebel, 1994), and ΔT_m reflects the extent of sequence identity. However, it has been demonstrated empirically that there is indeed a linear correlation between them (e.g. Grimont, 1988; Johnson, 1989; Rosselló-Móra and Amann, 2001; Tjernberg and Ursing,

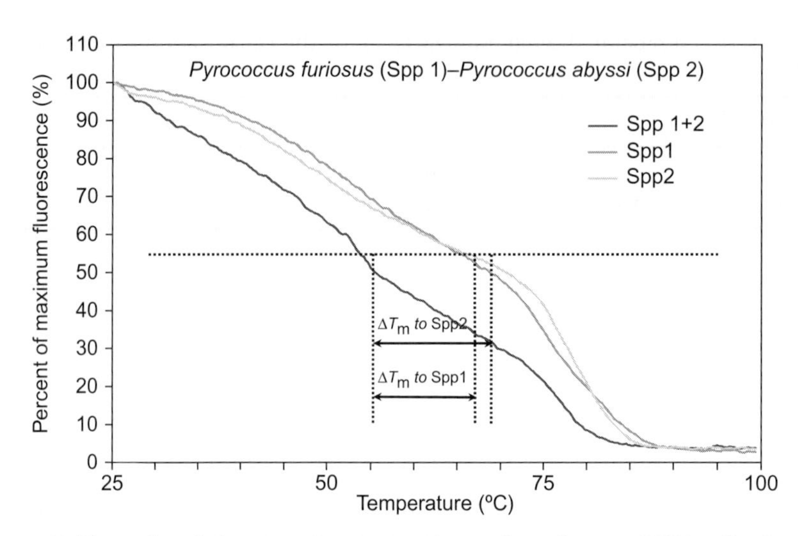

Figure 5. Example of denaturation rates of two homologous DNAs Spp1 and Spp2, and the resulting denaturation curve of the equimolar mixture of both DNAs. The increment in melting temperature (ΔT_m) is the difference between the melting temperatures of the homologous and hybrid DNAs. Reproduced with courtesy of Juan González Grau.

1989), and generally values of RBR above 50% will correlate with a ΔT_{m} below 4–5°C. Taking into account published regression lines (Rosselló-Móra and Amann, 2001), an approximate correlation will be given by this transformation: RBR % = $-5.0501\ \Delta T_m + 90.329$.

Reagents:

20X SSC (3 M NaCl, 0.3 M Na_3Citrate.$2H_2O$; pH 7.0): Dissolve 175 g NaCl + 88 g Na_3Citrate · $2H_2O$ to a final volume of 1 l; adjust pH 7.0 with 1 M HCl.
2X SSC: Dilute 1 ml of 20X SSC with 9 ml of sterile milliQ water.
SYBR Green I (Roche)

◆◆◆◆◆◆ V. SPECTROPHOTOMETRIC MEASUREMENT OF DDH

This is a free-solution hybridization method that is based on the analysis of the renaturation kinetics of a mixture of DNA originally described by De Ley and colleagues (1970) and modified by Huß and colleagues (1983). The method is based on recording the decrease in absorbance at 260 nm caused by the formation of the double helix of the DNA. In general, this method is not widely used due to the fact that it requires special spectrophotometers that can perform simultaneous measures of several DNA mixtures, and can incubate the cuvettes with controlled temperature including a denaturation step at 99°C. In addition, the method requires large amounts of pure DNA (De Ley *et al.*, 1970; Huß *et al.*, 1983).

- Prepare genomic DNA from the strains of interest and estimate the DNA concentration. To optimize the hybridization protocol, prepare a 10 ml stock DNA solution of about OD = 1 (±5%) at 260 nm in 2X SSC.

- As the measurements are sensitive to the ionic strength of the solution of DNA, it is necessary to adjust the salt content of all DNAs by means of dialysis. For dialysis: (i) Prepare the dialysis bags in EDTA-Na_2 as indicated below. The bags to be used are dried by pressing them with a clean paper, but be careful not to overdry them. (ii) Knot on one side of the bag. (iii) Transfer the 10 ml of stock DNA into the dialysis bags, remove the air by pressing the bag between fingers and knot to close. (iv) Submerge the bags in 2 l of 2X SSC for 2 days at 8°C stirring at a very slow rpm using a magnetic stirring bar. Alternatively, the buffer can be changed each 6 h/overnight. The quantity of buffer used depends on the dialysis time and the number of dialysis bags. (v) Transfer the dialyzed DNAs into clean 15 ml tubes and remeasure the absorbance$_{260}$. Adjust OD to =1 (±5%) at 260 nm with 2X SSC.
- Prepare the DNA mixtures for the hybridization protocol. For each pair of DNAs to be hybridized, each single homologous DNA is included and treated in the same way. Prepare three microfuge tubes (A, B, C) with A = 400 μl homologous DNA A; B = 400 μl homologous DNA B; and C = 200 μl DNA A + 200 μl DNA B. When required, 5% or 10% of formamide may be added.
- Calculate the formamide required for the hybridization. The hybridization temperature with this protocol is T_m −25°C. T_m can be calculated from the GC content (Table 2). In order to decrease the incubation temperature below 70°C (temperatures >70°C may result in DNA degradation), formamide may be added to the mixture in order to decrease the melting temperature of the DNA. The addition of 1% of formamide decreases the T_m by about 0.7°C. Alternatively, DMSO can be used to decrease the T_m, as 1% of DMSO reduces the T_m by 0.5°C. In order to simplify the calculations, you can use the following conditions: (i) For any DNA with GC mol% ≤ 48%, no formamide is added; (ii) for any DNA with GC mol% ranging from 49% to 59% (melting temperatures ranging from 94.8°C to 98.9°C), 5% formamide (final v/v concentration) may be added to the mixture; and (iii) for any DNA with GC mol% ≥60% (melting temperatures ≥99.3°C), 10% formamide (final v/v concentration) may be added to the mixture.
- Mix the solutions by vortexing for 1 min and transfer 400 μl of tubes A, B and C to three different cuvettes. Degas the cuvettes for 5 min in a vacuum container. Remove bubbles formed by shaking the cuvettes by hand or with the help of a plastic loop or tip. Cover the top of the DNA mixture with mineral oil and cap the cuvette with Teflon cap.
- Run the hybridization protocol always using a blank with 2X SSC containing the same formamide concentration. The protocol of hybridization is as follows: (i) Denature at 99°C for 15 min; (ii) decrease the temperature to the calculated T_m −25°C (Table 2) and allow to hybridize for 30 min. This procedure is performed twice.
- DDH similarity percentage is calculated with the following formula:

$$\frac{4xVMix - (VA + VB)}{2x\sqrt{VAxVB}}x100$$

where *VA* and *VB* are the reaction rates of the homologous reassociations of DNAs A and B, respectively, and *VMix* is the reaction rate of the heterologous reassociation of the equimolar mixture of DNAs A and B (De Ley *et al.*, 1970).

Reagents:

EDTA for the dialysis bags (5 mM EDTA): Dissolve 2.23 g of EDTA in 1.2 l of milliQ water.

Stock of dialysis bags: Boil about 40 pieces of dialysis tubing (20 cm) in 5% Na_2CO_3 for 30 min. Discard the liquid and rinse with boiling (simmering) water. Repeat this wash step three times. Boil 30 min in water and discard the liquid. Boil 30 min in 5 mM of EDTA, leave it cool and store at 4–8°C in this buffer for a maximum of 6 months.

20X SSC (3 M NaCl, 0.3 M $Na_3Citrate \cdot 2H_2O$; pH 7.0): Dissolve 175 g NaCl + 88 g $Na_3citrate \cdot 2H_2O$ to a final volume of 1 l; adjust pH 7.0 with 1 M HCl.

2X SSC: Dilute 1 ml of 20X SSC with 9 ml of sterile milliQ water.

DMSO: Dimethyl sulfoxide (Sigma, Cat. Nr. D8418).

Formamide (Sigma, Cat. Nr. D4551).

♦♦♦♦♦♦ VI. CONCLUDING REMARKS

DDH experiments have been used for almost half a century to circumscribe prokaryotic species. Much has been written about the benefits and pitfalls of the technique (Rosselló-Móra, 2006), but its use has been highly relevant for the construction of the current taxonomic schema of the prokaryotes. Here we have detailed the three most common and currently used techniques, and the selection of the technique to be used will depend on the equipment of each laboratory that wants to implement DDH. Contrary to the opinion of many scientists, these techniques are no more cumbersome than any other molecular biology methods commonly used, e.g. 'southern blotting' or 'pulsed field gel electrophoresis' (PFGE). It is true that one of the major drawbacks of DDH is that no cumulative databases can be constructed for interactive use. The current development of genome sequencing programs may result in the substitution of DDH by the calculation of ANI (Richter and Rosselló-Móra, 2009). ANI calculations can be performed with partial and random sequences of query genomes that have at least 20% sequence coverage. For the time being, and while we wait for the reduction in sequencing expenses, DDH is still the method of choice when trying to understand the genomic coherence of a putative prokaryotic species.

♦♦♦♦♦♦ ACKNOWLEDGEMENTS

The authors are indebted to Juan González of the Sevillian Institute of Natural Resources and Agrobiology (IRNAS), Spanish Research Council (CSIC) and Cathrin Spröer and Peter Schumann of the German Collection of Microorganisms and Cell Cultures (DSMZ) for sharing protocols and updating the information.

The authors also acknowledge their funding received from the Spanish Ministry of Science and Innovation with the projects Consolider Ingenio 2010 CE-CSD2007-0005 and CLG2009_12651-C02-01 and 02 (all co-financed with FEDER funding).

References

Adnan, S., Li, N., Miura, H., Hashimoto, Y., Yamamoto, H. and Ezaki, T. (1993). Covalently immobilized DNA plate for luminometric DNA-DNA hybridization to identify viridans streptococci in under 2 hours. *FEMS Microbiol. Lett.* **106**, 139–142.

Bolton, E. T. and McCarthy, B. J. (1962). A general method for the isolation of RNA complementary to DNA. *Proc. Natl. Acad. Sci USA* **48**, 1390–1397.

Brenner, D. J., Fanning, G. R., Johnson, K. E., Citarella, R. V. and Falkow, S. (1969a). Polynucleotide sequence relationships among members of *Enterobacteriaceae*. *J. Bacteriol.* **98**, 637–650.

Brenner, D. J., Fanning, G. R., Rake, A. V. and Johnson, K. E. (1969b). Batch procedure for thermal elution of DNA from hydroxyapatite. *Anal. Biochem.* **28**, 447–459.

Cardinali, G., Liti, G. and Martini, A. (2000). Non-radioactive dot-blot DNA reassociation for unequivocal yeast identification. *Int. J. Syst. Evol. Microbiol.* **50**, 931–936.

Christensen, H., Angen, Ø., Mutters, R., Olsen, J. E. and Bisgaard, M. (2000). DNA-DNA hybridization determined in micro-wells using covalent attachment of DNA. *Int. J. Syst. Evol. Microbiol.* **50**, 1095–1102.

Crosa, J. H., Brenner, D. J. and Falkow, S. (1973). Use of a single-strand specific nuclease for analysis of bacterial and plasmid deoxyribonucleic acid homo- and heteroduplexes. *J. Bacteriol* **115**, 904–911.

De Ley, J., Cattoir, H. and Reynaerts, A. (1970). The quantitative measurement of DNA hybridization from renaturation rates. *Eur. J. Biochem.* **12**, 133–142.

Ezaki, T., Hashimoto, Y. and Yabuuchi, E. (1989). Fluorometric deoxyribonucleic acid-deoxyribonucleic acid hybridization in microdilution wells as an alternative to membrane filter hybridization in which radioisotopes are used to determine genetic relatedness among bacterial strains. *Int. J. Syst. Bacteriol* **39**, 224–229.

Gade, D., Schlesner, H., Glöckner, F. O., Amann, R., Pfeiffer, S. and Thomm, M. (2004). Identification of Planctomycetes with order-, genus, and strain-specific 16S rRNA-targeted probes. Microb. Ecol. **47**, 243–251.

Gevers, D., Cohan, F. M., Lawrence, J. G., Spratt, B. G., Coenye, T., Feil, E. J., Stackebrandt, E., Van de Peer, Y., Vandamme, P., Thompson, F. L. and Swings, J. (2005). Re-evaluating prokaryotic species. *Nat. Rev. Microbiol* **3**, 733–739.

González, J. M. and Sáiz-Jiménez, C. (2004). A simple fluorimetric method for the estimation of DNA-DNA relatedness between closely related microorganisms by thermal denaturation temperatures. *Extremophiles* **9**, 75–79.

Goris, J., Suzuki, K.-I., De Vos, P., Nakase, T. and Kersters, K. (1998). Evaluation of a microplate DNA-DNA hybridization method compared with the initial renaturation method. *Can. J. Microbiol.* **44**, 1148–1153.

Grimont, P. A. D. (1988). Use of DNA reassociation in bacterial classification. *Can. J. Microbiol.* **34**, 541–546.

Grimont, P. A. D., Popoff, M. Y., Grimont, F., Coynault, C. and Lemelin, M. (1980). Reproducibility and correlation study of three deoxyribonucleic acid hybridization procedures. *Curr. Microbiol.* **4**, 325–330.

Huß, V. A. R., Festl, H. and Schleifer, K. H. (1983). Studies on the spectrometric determination of DNA hybridization from renaturation rates. *Syst. Appl. Microbiol.* **4**, 184–192.

Jahnke, K.-D. (1994). A modified method of quantitative colorimetric DNA-DNA hybridization on membrane filters for bacterial identification. *J. Microbiol. Meth* **20**, 237–288.

Johnson, J. L. (1981). Genetic characterization. In *Manual of Methods for General Microbiology* (P. Gerhardt, R. G. E. Murray, R. N. Costilow, E. W. Nester, W. A. Wood, N. R. Krieg and G. B. Philips, eds), pp. 450–472. ASM Press, Washington, DC.

Johnson, J. L. (1985). DNA reassociation and RNA hybridisation of bacterial nucleic acids. *Meth. Microbiol.* **18**, 33–74.

Johnson, J. L. (1989). Nucleic acids in bacterial classification. In *Bergey's Manual of Systematic Bacteriology* (S. T. Williams, M. E. Sharpe and J. G. Holt, eds), Vol. 4, pp. 2306–2309. Williams & Wilkins, Baltimore, MD.

Kaznowski, A. (1995). A method of colorimetric DNA-DNA hybridization in microplates with covalently immobilized DNA for identification of *Aeromonas* spp. *Med. Microbiol. Lett.* **4**, 362–369.

Konstantinidis, K. and Tiedje, J. M. (2005). Genomic insights that advance the species definition for prokaryotes. *Proc. Natl. Acad. Sci. USA* **102**, 2567–2592.

Krieg, N. (1988). Bacterial classification: an overview. *Can. J. Microbiol.* **34**, 536–540.

Lindh, E. and Ursing, J. (1986). Clinical strains of *Enterobacter agglomerans* (synonyms: *Erwinia herbicola, Erwinia mellitiae*) identified by DNA-DNA hybridization. *Acta Path. Microbiol. Immunol. Scand. Sect. B* **94** 250-231

Marmur, J. (1961). A procedure for the isolation of DNA from microorganisms. *J. Mol. Biol.* **3**, 208–218.

Mehlen, A., Goeldner, M., Ried, S., Stindl, S., Ludwig, W. and Schleifer, K.-H. (2004). Development of a fast DNA-DNA hybridization method based on melting profiles in microplates. *Syst. Appl. Microbiol.* **27**, 689–695.

Owen, R. J. and Pitcher, D. (1985). Current methods for estimating DNA base composition and levels of DNA-DNA hybridization. In *Chemical Methods in Bacterial Systematics* (M. Goodfellow and E. Minnikin, eds), pp. 67–93. Academic Press, London.

Popoff, M. and Coynault, C. (1980). Use of DEAE-cellulose filters in the S1 nuclease method for bacterial deoxyribonucleic acid hybridization. *Ann. Microbiologie* **131a**, 151–155.

Richter, M. and Rosselló-Móra, R. (2009). Shifting the genomic gold standard for the prokaryotic species definition. *Proc. Natl. Acad. Sci. USA* **106**, 1926–1931.

Rosselló-Móra, R. (2006). DNA-DNA reassociation methods applied to microbial taxonomy and their critical evaluation. In *Molecular Identification, Systematics, and Population Structure of Prokaryotes* (E. Stackebrandt, ed), pp. 23–50. Springer-Verlag, Heildelberg.

Rosselló-Móra, R. and Amann, R. (2001). The species concept for prokaryotes. *FEMS Microbiol. Rev.* **25**, 39–67.

Schildkraut, C. and Lifson, S. (1965). Dependence of the melting temperature of DNA on salt concentration. *Biopolymers* **3**, 195–208.

Schildkraut, C. L., Marmur, J. and Doty, P. (1961). The formation of hybrid DNA molecules and their use in studies of DNA homologies. J. Mol. Biol. **3**, 595–617.

Schleifer, K- H. and Stackebrandt, E. (1983). Molecular systematics of prokaryotes. *Ann. Rev. Microbiol.* **37**, 143–187.

Sneath, P. H. A. (1989). Analysis and interpretation of sequence data for bacterial systematics: the view of a numerical taxonomist. *Syst. Appl. Microbiol.* **12**, 15–31.

Stackebrandt, E. (2003). The richness of prokaryotic diversity: there must be a species somewhere. *Food Technol. Biotechnol.* **41**, 17–22.

Stackebrandt, E. and Goebel, B. M. (1994). Taxonomic note: a place for DNA-DNA reassociation and 16S rRNA sequence analysis in the present species definition in bacteriology. *Int. J. Syst. Bacteriol.* **44**, 846–849.

Stackebrandt, E. and Liesack, W. (1993). Nucleic acids and classification. In *Handbook of New Bacterial Systematics* (M. Goodfellow and A. G. O'Donnell, eds), pp. 151–194. Academic Press, London.

Stackebrandt, E., Frederiksen, W., Garrity, G., Grimont, P. A. D., Kämpfer, P., Maiden, M. C. J., Nesme, X., Rosselló-Móra, R., Swings, J., Trüper, H. G., Vauterin, L., Ward, A. C. and Whitman, W. B. (2002). Report of the *ad hoc* committee for the re-evaluation of the species definition in bacteriology. *Int. J. Syst. Evol. Microbiol.* **52**, 1043–1047.

Tjernberg, I. and Ursing, J. (1989). Clinical strains of *Acinetobacter* classified by DNA-DNA hybridization. *APMIS* **97**, 595–605.

Tjernberg, I., Lindh, E. and Ursing, J. (1989). A quantitative bacterial dot method for DNA-DNA hybridization and its correlation to the hydroxyapatite method. *Curr. Microbiol.* **18**, 77–81.

Turner, D. J. (1996). Thermodynamics of base pairing. *Curr. Opin. Struc. Biol.* **6**, 299–304.

Ullman, S. J. and McCarthy, B. J. (1973). The relationship between mismatched base pairs and the thermal stability of DNA duplexes. *Biochim. Biophys. Acta* **294**, 416–424.

Urdiain, M., López-López, A., Gonzalo, C., Busse, H.-J., Langer, S., Kämpfer, P. and Rosselló-Móra, R. (2008). Reclassification of *Rhodobium marinum* and *Rhodobium pfennigii* as *Afifella marina* gen. nov. comb. nov. and *Afifella pfennigii* comb. nov., a novel genus of photoheterotrophic Alphaproteobacteria and emended descriptions of *Rhodobium*, *Rhodobium orientis* and *Rhodobium gokarnense*. *Syst. Appl. Microbiol.* **31**, 339–351.

16 The Use of rRNA Gene Sequence Data in the Classification and Identification of Prokaryotes

Wolfgang Ludwig[a,*], **Frank Oliver Glöckner**[b,c], **and Pelin Yilmaz**[b,c]

[a]Lehrstuhl für Mikrobiologie, Technische Universität München, Am Hochanger 4, 85350 Freising, Germany; [b]Max Planck Institute for Marine Microbiology, Celsiusstrasse 1, 28359 Bremen, Germany; [c]Jacobs University Bremen GmbH, Campus Ring 1, 28759 Bremen, Germany

**E-mail: ludwig@mikro.biologie.tu-muenchen.de*

◆◆◆

CONTENTS

16S rRNA sequence

◆◆◆◆◆◆ I. HISTORY

The idea of molecular phylogeny was already formulated before the discovery of DNA, although first practical approaches were used by Zuckerkandl and Pauling (1965), and were intensified and combined with early bioinformatics by Dayhoff (Eck and Dayhoff, 1969). However, the protein markers used during that time did not allow comprehensive phylogenetic studies, as they were not present in all organisms. Carl Woese made the right choice by analysing the universally distributed 16S rRNA (18S rRNA for eukaryotes) (Fox *et al.*, 1977).

METHODS IN MICROBIOLOGY, VOLUME 38
0580-9517 DOI 10.1016/B978-0-12-387730-7.00016-4

The 16S rRNA was chosen because the smaller 5S rRNA did not provide sufficient information and the larger 23S rRNA could not be properly analysed at that time for technical reasons. Sequencing in that time was a laborious specialist method resulting in 'catalogs' of (RNase T1 generated) oligonucleotide sequences. No contiguous 16S rRNA primary structure could be determined. Dendrograms were derived by determining the fraction of identical oligonucleotides in the catalogs from different organisms. The method was slightly simplified and 'miniaturized' by Stackebrandt *et al.* (1981). The cataloging era came to an end by the introduction of primer directed reverse transcriptase sequencing of 16S rRNA (Lane *et al.*, 1985). Continuous primary structure stretches comprising up to 75% of the 16S rRNA full length could now be used for tree reconstructing. Random cloning of DNA restriction fragments containing (parts of) rRNA genes followed by sequence determination was also in use at that time (Brosius *et al.*, 1981; Toschka *et al.*, 1987). The invention of polymerase chain reaction technique (Saiki *et al.*, 1985) and its application for direct rRNA gene or clone sequencing (Böttger, 1989; Sogin, 1990) mark a breakthrough in the history of rRNA sequence analysis. Comparative rRNA gene sequence analysis currently represents the standard method for phylogenetic analyses, taxonomy and identification of prokaryotes.

♦♦♦♦♦♦ II. THE MOLECULE

As essential intrinsic components of the ribosome the rRNAs fulfill the requirements on phylogenetic markers for the most part. These requirements comprise ubiquitous distribution, functional constancy, evolutionary conservation, wide spectrum of sequence variation, large size, excluding paralogy and horizontal gene transfer. While the first four are almost ideally met, the latter three can only partly be taken as given.

A. Information Content

The phylogenetic information content of primary structures is limited by the size, the number of (evolutionary) allowed characters per position, and the conservation profile along the entire sequence. Although a huge 'space' of sequence variation (8.7×10^{15}, based upon the *Escherichia coli* standard comprising 1542 monomers and five positional character states: the four nucleotides plus the insertion - deletion event) would be possible in theory, this 'space' is drastically reduced in the case of real marker sequences (Ludwig and Klenk, 2001). Given the conserved nature of the rRNA markers, a remarkable fraction of primary structure positions is (almost) invariant and hence phylogenetically not informative. Furthermore, the degree of freedom with respect to character variability is reduced for many of the remaining informative individual sequence positions. Apparently, only part of the potential character changes during the course of evolution was 'allowed' for distinct sites as a consequence of functional pressure. Functional restrictions further constrict the information content by demanding coordinated character changes or preventing certain combinations of change at different sequence positions.

Given the lack of fossil records, only present day rRNA sequences can be compared for phylogenetic analyses. In principle, any individual character change that can be measured comparing contemporary primary structures reports on a single evolutionary event only. In the case of pairs of most diverged bacterial sequences only about 800 (number of variable sites) such events are documented. Against the background of three to four billion years of evolution of cellular life, only a minimum number of events can be directly measured. Besides these limitations, the conservation profile of rRNA sequences influences the amount and quality of phylogenetic conclusions, which can be deduced from comparative analyses. Such profiles help identifying more and less frequently changing primary structure positions which report on earlier and more recent evolutionary events, respectively. Obviously, large spans of evolutionary time might not be documented in present day sequence data. Further limitations that have to be taken into account interpreting comparative sequence data result from a high degree of noise in the data sets. Given the small number of possible character changes, undocumented and hence undetected multiple base changes during the course of evolution might have created 'false' identities (homoplasies) at certain positions in the present day sequences. The risk of misinterpreting identities correlates somehow with positional variability.

Despite all those limitations and pitfalls more phylogenetic information can be estimated than directly measured. The combination of models of evolution, conservation profiles and approaches estimating ancestor sequences extends the narrow spectrum of information that can directly be measured.

B. Sequence Determination

Nowadays, in the age of high throughput approaches rRNA sequence determination is the daily standard in microbiology. This is impressively documented by more than two million of entries in public sequence databases. Usually, the primary structure of the rRNA gene is determined and virtually transcribed into rRNA for database entries and comparative structure analyses.

The conservation profile of the rRNAs provide highly conserved islands lending targets for amplification and sequencing primers of broad specificity ranges with respect to phylogenetic diversity. None of the conserved islands, however, comprises a contiguous stretch of invariant positions for perfect matching of 18–20-mers commonly used as PCR or sequencing primers. Thus, the design of universal or domain specific primers is definitely impossible. To compensate such shortcomings, degenerated primer sequences are widely in use. Degeneration can be accomplished by using mixes of primer variations or oligonucleotides containing inosine substitutions (Ben-Dov *et al.*, 2011) at less conserved positions. Numerous primer sequences were proposed over the decades of comparative rRNA sequence analysis. A selection of primers recently used for almost complete rRNA gene amplification is given in Tables 1 and 2. The

Table 1. A selection of commonly used 16S rRNA primers for complete rRNA gene amplification. The numbering of the positions is according to that of the 16S rRNA of *Escherichia coli* (Brosius *et al.*, 1981)

Primer name	Sequence	Target group	*E. coli* position
GM3F[a]	AGAGTTTGATCMTGGC	*Bacteria*	8–24
E8F[b, c, d]	AGAGTTTGATCCTGGCTCAG	*Bacteria*	8–27
E334F[e]	CCAGACTCCTACGGGAGGCAGC	*Bacteria*	334–354
E341F[f]	CCTACGGGIGGCIGCA	*Bacteria*	341–356
E786F[g]	GATTAGATACCCTGGTAG	*Bacteria*	786–803
E939R[e]	CTTGTGCGGGCCCCCGTCAATTC	*Bacteria*	917–939
E1115R[b]	AGGGTTGCGCTCGTTG	*Bacteria*	1097–1115
GM4R[a]	TACCTTGTTACGACTT	*Bacteria*	1492–1507
E1541R[h]	AAGGAGGTGATCCANCCRCA	*Bacteria*	1522–1541
A2Fa[b, c]	TTCCGGTTGATCCYGCCGGA	*Archaea*	7–26
A2Fb[i]	TTCCGGTTGATCCTGCCGGA	*Archaea*	7–26
A3Fa[j]	TCCGGTTGATCCYGCCGG	*Archaea*	8–27
A109F[k]	ACKGCTCAGTAACACGT	*Archaea*	109–128
A1098F[b]	GGCAACGAGCGMGACCC	*Archaea*	1098–1114
Ab127R[l]	CCACGTGTTACTSAGC	*Archaea*	112–127
A348R[m]	CCCCGTAGGGCCYGG	*Archaea*	335–349
A934R[l]	GTGCTCCCCCGCCAATTCCT	*Archaea*	915–934
A1115R[b]	GGGTCTCGCTCGTTG	*Archaea*	1100–1114

[a]Muyzer *et al.*, 1995.
[b]Reysenbach and Pace, 1994.
[c]Reysenbach *et al.*, 1994.
[d]Martinez-Murcia *et al.*, 1995.
[e]Rudi *et al.*, 1997.
[f]Watanabe *et al.*, 2001.
[g]Coloqhoun, 1997.
[h]Suzuki and Giovannoni, 1996.
[i]López-García *et al.*, 2001.
[j]McInnery *et al.*, 1995.
[k]Whitehead and Cotta, 1999.
[l]Achenbach and Woese, 1995.
[m]Barns *et al.*, 1994. The naming of the primers from references [b] through [l] is according to Baker *et al.*, 2003.

parameters for standard PCR amplification procedures depend on the PCR kit or reagents applied as well as composition and sequence of the primer pairs used.

Two major variations of standard (Sanger) sequencing are commonly in use: sequencing of cloned rRNA genes or direct sequencing of rRNA gene amplicons. While analysing pure cultures the latter technique is preferentially applied, whereas, singularizing by cloning is needed in case of samples containing mixed populations. The cloning of amplified rRNA genes usually follows standard procedures applying commercial kits and host strains. The respective experimental parameters depend on the requirements specified by the manufacturer of the respective kits.

Commercial services are available offering both sequencing approaches. Alternatively, commercial kits and equipment are commonly used for performing

Table 2. A selection of commonly used 23S rRNA primers for complete rRNA gene amplification. The numbering of the positions is according to that of the 23S rRNA of *Escherichia coli* (Brosius *et al.*, 1981)

Primer name	Sequence	Target group	*E. coli* position
129f [a]	CYGAATGGGGVAACC	*Bacteria*	115–129
189f [a]	TACTDAGATGTTTCASTTC	*Bacteria*	189–207
457r [a]	CCTTTCCCTCACGGTACT	*Bacteria*	457–475
2490r [a]	CGACATCGAGGTGCCAAAC	*Bacteria*	2490–2508
2241r [b]	ACCGCCCCAGTHAAACT	*Bacteria*	2241–2257
23ar [c]	CGGTACTGGTTCACTATCGG	*Bacteria*	463–444
43a [c]	GGATGTTGGCTTAGAAGCAG	*Bacteria*	1055–1074
53a [c]	GGACAACAGGTTAATATTCC	*Bacteria*	1380–1399
62ar [c]	GGGGCCATTTTGCCGAGTTC	*Bacteria*	1668–1686
LSU190-F [d]	GAAYTGAARCATCTYAGTA	*Bacteria & Archaea*	189–207
LSU2445a-R [d]	CCCYGGGGTARCTTTTCTST	*Archaea*	2425–2447

[a]Hunt *et al.*, 2006.
[b]Lane, 1991.
[c]Van Camp *et al.*, 1993.
[d]DeLong *et al.*, 1999.

the standard sequencing procedures. The experimental parameters have to be adjusted according to the manufacturers recommendations in accordance with the kits and equipment used. Cloned rRNA gene sequencing is most appropriately performed in combination with vector specific primers; whereas primers targeting internal sites are preferably applied for direct sequencing PCR amplified rRNA genes. In the case of 16S rRNA genes usually the application of two sequencing primers is sufficient, while for 23S rRNA genes additional primers are needed. A selection of recently used sequencing primers is included in Tables 1 and 2.

C. Data Analysis

An optimized sequence alignment combined with prior and subsequent quality checking of the (raw) data are prerequisites for the majority of rRNA sequence based studies and conclusions.

1. Alignment

In a multiple alignment the individual positions of different sequences are arranged in common columns which are (vertically) examined in comparative analyses. In a meaningful alignment these columns join sequence positions most probably derived from homologous positions in the common ancestor structures. The most parsimonious approach which is used by almost all commonly applied alignment tools (Clustal series, Chenna *et al.*, 2003; MAFFT, Katoh and Toh, 2008; MUSCLE, Edgar, 2004) unifies sequence positions according to maximizing character identity at and around the respective sites. In the case of RNAs there is common agreement concerning the evolutionary importance of higher order structure preservation. This preservation is often not correlated with primary structure similarity. *In silico* as well as experimentally validated higher order structure models have been developed (Gutell *et al.*, 1992, 1993; Cannone *et al.*, 2002) and are commonly accepted. Although temporarily formed alternative higher order interactions in the working ribosome have been documented (Feng *et al.*, 2011), comparative primary structure based higher order structure prediction helps to substantially improve rRNA sequence alignments (Ludwig and Schleifer, 1994; Ludwig and Klenk, 2001; Chun and Hong, 2010). The underlying assumption postulates an evolutionary selection pressure upon higher order structure elements. This is documented by many examples of identical or similar potential intra-molecular folding despite differing primary structures from phylogenetically diverse sources. Furthermore, such findings have been interpreted as phylogenetic proofs of the respective structure elements. Agreement with or violation of the commonly accepted higher order structure models in multiple alignments can be checked using the respective visualization tools of software packages such as ARB (Ludwig *et al.*, 2004; http://www.arb-home.de), jPHYDIT (Jeon *et al.*, 2005; http://chunlab.snu.ac.kr/jphydit/), Bioedit (http://www.mbio.ncsu.edu/BioEdit/bioedit.html), MEGA (Kumar *et al.*, 2004; http://www.megasoftware.net/) (Table 3). Part of these visualization tools may highlight (phylogenetic) group consensus deviations in addition. Evaluating higher order structure and group consensus conformity not only helps optimizing alignments but also detecting sequencing errors and/or evolutionary peculiarities.

At the current stage of data availability it is not recommended to establish rRNA sequence alignments from scratch (Peplies *et al.*, 2008). Three special database initiatives - SILVA (Pruesse *et al.*, 2007; http://www.arb-silva.de); RDP (Cole *et al.*, 2009; http://rdp.cme.msu.edu), and Greengenes (DeSantis *et al.*, 2006; http://greeengenes.lbl.gov) - provide comprehensive sets of evaluated and processed rRNA sequence and associated metadata. The respective alignments have been established, improved and maintained by specialists over decades. Against this background, it certainly is reasonable to use the most similar entries in such databases as templates for adapting new sequence data to the reference alignments. Online services are provided by the three projects. The ARB software package (Ludwig *et al.*, 2004; http://www.arb-home.de) provides a standalone tool for doing so.

2. Quality checking

Unfortunately, the rapid and immense progress in sequencing technology which accompanied the history of rRNA based research did not only result in huge but valuable datasets but rather also in a lot of garbage data. This not only applies to the raw data in public databases such as EBI-ENA (Cochrane *et al.*, 2008; http://www.ebi.ac.uk), GenBank (Benson *et al.*, 2011; http://www.ncbi.nml.nih.gov/genbank), and DDBJ (Kaminuma *et al.*, 2011; http://www.ddbj.nig.ac.jp), but also the special rRNA databases SILVA (Pruesse *et al.*, 2007); RDP (Cole *et al.*, 2009), and Greengenes (DeSantis *et al.*, 2006) although to a much minor extent. Commonly found sequencing errors comprise terminal (with

Table 3. A selection of commonly used software for comparative rRNA based analyses

	OS	GUI	Ed	Dis	MP	ML	By	TD	OTU	PD	PE	URL	Reference
ARB	LM	+	+	+	+	+	−	+	+	+	+	http://www.arb-home.de	Ludwig *et al.* (2004)
Bioedit	W	+	+	+	+	+	−	−	+	−	−	http://www.mbio.ncsu.edu/BioEdit/	
jPHYDIT	WLM	+	+	+	+	+	+	−	−	−	−	http://chunlab.snu.ac.kr/jphydit/	Jeon *et al.* (2005)
MEGA	W	+	+	+	+	−	−	+	−	−	−	http://megasoftware.net	Kumar *et al.* (2004)
MrBayes	WLM	−	−	−	−	−	+	−	−	−	−	http://mrbayes.csit.fsu.edu/	Ronquist and Huelsenbeck (2003)
PHYLIP	WLM	−	+	+	+	+	−	+	−	−	−	http://evolution.genetics.washington.edu/	Felsenstein (1989, 1993)
PAUP*	WLM	−	−	+	+	+	+	+	−	−	−	http://paup.csit.fsu.edu/	Swofford (2003)
PhyML	L	−	−	−	−	+	−	−	−	−	−	http://atgc.lirmm.fr/phyml/	Guindon and Gascuel (2003)
RaxML	L	−	−	−	−	+	−	−	−	−	−	http://wwwkramer.in.tum.de/exelixis	Stamatakis (2006)
TREE-PUZZLE	WLM	−	−	+	−	+	−	−	−	−	−	www.treepuzzle.de	Schmidt *et al.* (2002)
Promrose	WLM	+	−	−	−	−	−	−	−	+	+	http://www.bioinformatics-toolkit.org	Ashelford *et al.* (2002)
mothur	WLM	−	−	−	−	−	−	−	+	−	−	http://www.mothur.org/wiki/Download_mothur	Schloss *et al.* (2009)

+, available; −, not provided.
OS, operating system: W, Windows; L, Linux; M, MacOS.
GUI, graphical user interface; Ed, sequence/structure editor; Ds, distance methods; MP, maximum parsimony; ML, maximum likelihood; By, Bayesian approach; TD, tree presentation/modification; OTU, operational taxonomic unit definition; PD, probe design; PE, probe evaluation.
Data partially from Chun and Hong (2010).

respect to the sequencing runs) and internal uncertainties as well as redundancies, remnants of cloning vector sequences, internal gaps due to missing data, erroneously assembled partial sequences (often resulting in inverted parts or large and small stretch duplications), and chimeric sequences. At least part of these errors can be recognized and eliminated by the data producers before submitting or inserting into public or private databases to avoid a burden of low quality data. Ambiguities, redundancies should be corrected, vector sequences removed and assembling performed carefully. Software tools for chimera checking (Bellerophon, Huber *et al.*, 2004, http://comp-bio.anu.edu.au/bellerophon/bellerophon.pl) and finding anomalies (Mallard, Ashelford *et al.*, 2006; Pintail, Ashelford *et al.*, 2005; http://www.cardiff.ac.uk/biosi/research/biosoft/) are available and could be applied during the sequence preprocessing phase.

Most advisably, a second round of quality checking should be performed after adding new sequence data to reference data sets. Assigning new sequences to established similarity (phylogenetic) groups not only helps to improve a preliminary alignment but also allows further quality checking by identifying potential deviations from group consensus. Furthermore, potential violations of (group consensus) higher order structure provide hints on primary structure dubieties. Overlaying well established (group specific) conservation profiles often helps distinguishing sequencing errors from evolutionary changes. Finally, chimera checking based upon aligned datasets can be done by virtually segmenting the alignment prior to statistical or phylogenetic analysis of the individual segments. Following the latter approach, a comparative analysis of tree topologies derived from the alignment segments may allow recognizing potential chimeric sequences as such. Statistical approaches estimate the probability of character (stretches) within the segments against the background of data (phylogenetic) environment and clustering.

Some software packages (e.g. ARB; Ludwig *et al.*, 2004; http://www.arb-home.de) provide tools for quality or chimera scoring. Knowledge databases like SILVA perform quality checks by systematically inspecting all public rRNA sequences for ambiguities, vector contaminations and anomalies. The results are intuitively visualized on the project website and the corresponding numerical values are available in the corresponding ARB files. A more detailed description is given by Pruesse *et al.* (2007) or at the SILVA web page (http://www.arb-silva.de).

♦♦♦♦♦♦ III. DATABASES

Nucleotide sequences are usually submitted to one of the three databases of INDSC (Cochrane *et al.*, 2011) EMBL (Cochrane *et al.*, 2008; http://www.ebi.ac.uk), Genbank (Benson *et al.*, 2011; http://www.ncbi.nml.nih.gov/genbank) or DDBJ (Kaminuma *et al.*, 2011; http://www.ddbj.nig.ac.jp). These databases, however, only host the primary sequence data along with further descriptive information. No quality checking is performed. Updating or correcting the descriptive information is the responsibility of the submitters and is often sporadically done. In the case of rRNA based methods and studies dedicated data resources providing processed sequences and up to date descriptive information – especially correct taxonomic assignment – are essential.

In the early nineties of the last century three initiatives offered first special rRNA databases to the community: the ribosomal database project in USA (RDP; Olsen *et al.*, 1992; http://rdp.cme.msu.edu) the ARB project in Germany (http://www.arb-home.de), and the European database of ribosomal RNAs (Wuyts *et al.*, 2004). The latter is no longer maintained. Later, the Greengenes database and workbench were made accessible to public (De Santis *et al.*, 2006; http://greeengenes.lbl.gov). In 2007 the ARB database initiative has been taken over by the SILVA project (Pruesse *et al.*, 2007; http://www.arb-silva.de).

A. The SILVA Project

The SILVA project (Pruesse *et al.*, 2007; http://www.arb-silva.de) maintains databases of processed small and large subunit rRNA sequences from all three domains: *Eucarya*, *Archaea* and *Bacteria*. Special emphasis is given on comprehensive and up to date contextual data assigned to the individual sequence entries. The various types of meta-information and the respective tags or database fields are listed on the website. Updates are continuously provided at the SILVA web page (http://www.arb-silva.de). The sequence alignments (section II.C.1) are automatically updated applying the SINA aligner tool based on seed alignments created and continuously evaluated by experts. This non-redundant seed selection comprises only (almost) complete (section II.A) high quality (section II.C.2) primary structures. Every individual primary structure has to pass an automatic pipeline for pre- and post-alignment quality scoring. Low quality sequences are automatically dropped and problematic sequences found in the manual curation process are added to a black list. Quality scores are assigned to database entries successfully passing the pipeline. Sequences derived from (pure) cultures or from amplified and/or cloned mixed sample DNA are designated as such. Valid (Euzéby's list of prokaryotic names with standing in nomenclature (LPSN), http://www.bacterio.cict.fr/, the Living Tree Project (LTP), Yarza *et al.*, 2008, 2010, http://www.arb-silva.de/projects/living-tree/; straininfo, Dawyndt *et al.*, 2005, http://www.straininfo.net/), expert proposed (Bergey's taxonomic outlines, http://www.bergeys.org/outlines.html), and other partially used taxonomic frameworks (EMBL: Cochrane *et al.*, 2008, http://www.ebi.ac.uk; Greengenes: De Santis *et al.*, 2006, http://greeengenes.lbl.gov) are added and continuously updated. To meet the varying requirements of the users (sections IV, IX, X) two variants of the LSU and SSU rRNA databases are maintained: the (Ref) reference data set containing only higher quality at least 60% (in reference to the *E. coli* standard) complete sequences and the Parc data sets comprising of all available primary structures that passed the initial quality checks and comprise at least 300 monomers. Versatile search functions, including an advanced taxonomic browser, allow the user composing custom data sets according to (combinations of) database field entries. The complete databases or customized subsets can be downloaded in ARB or multi-FASTA formats. Online tools can be used for template oriented alignment of user provided sequences using the SINA aligner and for probe and primer evaluation against the full datasets and custom generated subsets.

The SILVA web page also hosts special LSU and SSU rRNA databases comprising only high quality bacterial and archaeal type strain sequences maintained by the living tree project (LTP, Yarza *et al.*, 2008, 2010, http://www.arb-silva.de/projects/living-tree).

B. The RDP Project

The RDP project (Cole *et al.*, 2009; http://rdp.cme.msu.edu) provides the users with quality-controlled 16S rRNA sequence alignments. Data access and selection is possible via a taxonomy browser. The user selected data can be downloaded in various formats. A number of online facilities for analysing user provided sequences are accessible at the RDP web page. Individual sequences can be uploaded and positioned in a taxonomic framework based on the taxonomic outlines of the Bergey's Manual Trust (http://www.bergeys.org/outlines.html). A 'seqmatch' tool calculates sequence identity scores for user selected data sets. The 'probe match' tool identifies sequence entries containing the targets for user provided probe strings. Online tree reconstruction (tree builder) based upon customized alignments applying a weighted neighbor-joining algorithm is possible. 'Taxomatic' is another online tool which allows evaluating 'taxonomic' assignment versus sequence similarity. Similarity matrices are visualized as heat maps with superimposed 'taxonomy'. The term taxonomy in this context does not necessarily correspond to the current valid taxonomy but may also address other clustering schemes. In response to the rapid progress of modern high throughput sequencing techniques, RDP provides a pyrosequencing pipeline for easier processing large partial 16S rRNA gene sequence libraries.

C. The Greengenes Project

The Greengenes web application (De Santis *et al.*, 2006; http://greeengenes.lbl.gov) provides access to curated 16S rRNA sequence alignment. Besides the primary structure alignments, metadata – most notably taxonomic information – are maintained and updated. A taxonomy browser provides access to the data. User defined data selections can be exported in different formats. Online facilities are provided for updating the metadata in the user's data collections. Raw sequences can be submitted for aligning and/or quality checking. Furthermore, rRNA targeted probes can be evaluated online with respect to the occurrence of the respective target strings in the database sequence entries.

◆◆◆◆◆◆ IV. PHYLOGENETIC ANALYSIS

Currently, phylogenetic analyses of rRNA sequence alignments are routinely performed to generate a backbone for microbial classification and/or identification. In general, such analyses attempt to reconstruct the evolutionary history of the marker molecules. Given the limited information content of the rRNA markers (section II.A) and the noise in the huge contemporary datasets (II.A.), the ultimate goal of finding 'exact phylogenies' might never be attained. Nevertheless, the ongoing progress in software development and optimization as well as the rapid achievements concerning computer hardware power, increasingly allow multiple comparative analyses of comprehensive data sets. Consequently, phylogenetic conclusions can be inferred on a more solid basis than in the past.

A. Visualization

Phylogenetic relationships inferred from comparative rRNA analyses are commonly visualized as trees or dendrograms. In both cases the ancestry is represented by the topology of internal nodes. In the classical dendrogram vertical lines have no meaning except visualizing topology, whereas horizontal internal and terminal branches contribute to the overall tree size (Figure 1). Nowadays, the 'bended' or 'circular' dendrogram presentation is en vogue. (Radial and bent lines correspond to vertical and horizontal lines, respectively). In radial trees the length of all lines directly connecting nodes as well as the terminal branches add up the overall tree size. The meaning of branch lengths and size bars depends on the treeing method and software applied. In most cases somehow the estimated (!) number of character changes is indicated while often primarily the significance of the tree topology is visualized by branch lengths.

B. Treeing Methods

Phylogeny inference is based on incomplete information of present day markers missing direct information about the past. Therefore various evolutionary scenarios could be assumed which would have produced the observed sequence data. Inferring phylogeny means making a "best estimate" of an evolutionary history by seeking one or more preferred trees among the set of possible phylogenies (Swofford *et al.*, 1996). In general, there are two groups of treeing approaches: the cluster methods (algorithms) dealing with distance data and the discrete character methods using optimality criteria. Four categories of treeing approaches are commonly used analysing rRNA data: the distance procedures belong to the first group whereas maximum parsimony, maximum likelihood and Bayesian methods represent the second group.

1. Models of evolution

None of the phylogenetic inference methods is free of systemic error because the real evolutionary processes are not completely known. To compensate for this drawback hypothetical models of the evolutionary process are used in combinations with the treeing methods. The parameters of such models concern the general frequency (rate) of character change, the distribution (relative frequency) of the possible character states, and rate parameters for the different

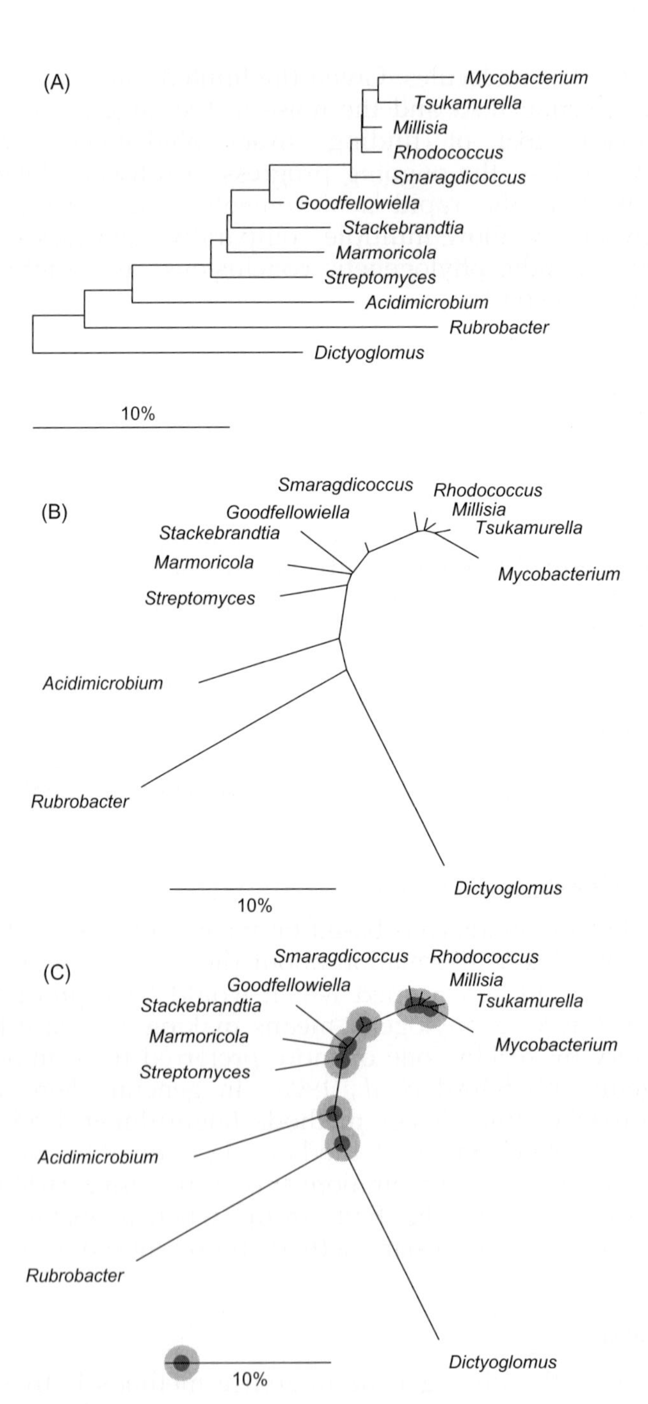

Figure 1. Tree visualization.
A: Dendrogram: The horizontal lines reflect significance and/or estimated changes. The sum of all horizontal lines connecting internal and terminal nodes indicates the distance of the respective sequences. The vertical lines reflect topology and do not contribute to distances.

types of character changes. These parameters are either intrinsic to the respective theoretical model or may be extracted from the data sets. Simple models assume equal frequency of the possible character states and same probability of any type of character change (Jukes and Cantor, 1969). Other commonly used models define different substitution rates for transition and transversion (Kimura, 1980) or allow individual substitution rates for all possible changes as well as unequal character distribution (Rodriguez *et al.*, 1990). Site specific differences of the general rate of substitution (reflected by conservation profiles) are a common characteristic of phylogenetic marker molecules (section II.A). There are models taking heterogeneous positional substitution rates into account (Yang *et al.*, 1994).

Software tools are available for finding established models which 'best' fit the respective data set. A comprehensive list along with bibliography and web sources is given on Felsenstein's web page (http://evolution.genetics.washington.edu/phylip/software.html#Modelselection).

2. Distance methods

Tree reconstruction applying distance methods follows a two step procedure (Swofford *et al.*, 1996; Chun and Hong, 2010). Pairwise distances are calculated for the respective data set and stored in a distance matrix. Clustering methods are used in the second step to properly represent the distance matrices as dendrograms or trees.

Hamming distances derived by simply counting positional differences of the sequence pairs of a given data selection certainly cannot properly reflect the actual number of evolutionary events. It has to be assumed that only a minor fraction of these events is documented in the present day sequences, given the limited number of potential character states (section II.A). Especially the number of earlier evolutionary changes will be drastically underestimated. Therefore, the measured distances are commonly transformed into phylogenetic distances. Several formulae for this transformation have been proposed. The Jukes Cantor correction (Jukes and Cantor, 1969) is the most popular among them. The effects of such corrections can be described as raising the measured distances in correlation to their values. The underlying assumption concerns an increasing underestimation of events with increasing distance values. Although such corrections make sense, one has to be aware that deeper branches in a tree

Figure 1. (*Continued*) B: Radial tree: All lines contribute to the tree size. The sum of all lines connecting internal and terminal nodes indicates the distance of the respective sequences.
C: Significance: areas of fuzziness are indicated by light (corresponding to 2% distance/significance) and dark (corresponding to 1% distance/significance) gray background around internal nodes.
Accession numbers: *Acidimicrobium ferrooxidans*, U75647; *Dictyoglomus thermophilum*, X69194; *Goodfellowiella coeruleoviolacea*, DQ093349; *Marmoricola aurantiacus*, Y18629; *Millisia brevis*, AY534742; *Mycobacterium tuberculosis*, X58890; *Rhodococcus rhodochrous*, X79288; *Rubrobacter radiotolerans*, U65647; *Smaragdicoccus niigatensis*, AB243007; *Stackebrandtia nassauensis*, AY650268; *Streptomyces lavendulae*, D85116; *Tsukamurella paurometabola*, AF283280.

are more concerned than peripheral ones. Thus those parts of trees may look more confident for which the weakest data are available. Modern versions of distance matrix based treeing methods account for this by giving lower weights to long branches (Bruno *et al.*, 2000). Distance matrices are calculated in combination with corrections, substitution tables and models. In addition, conservation profiles often are used for (alignment) column selection or positional weighting (section V.A.2).

The distance matrices provide the input for cluster analysis based or derived treeing methods. Most such methods produce so called additive trees. Length and positioning of branches in such trees are adapted to minimize the differences between the data in the matrix and their visualization in the tree. Ideal distance data would be ultrametric, meaning that in the respective trees the (summed) distances of terminal nodes to common internal nodes would always be equal, what is never the case with real data. In additive trees such problems are minimized by expressing unequal distances by adapting the respective branch lengths.

Most distance tree methods first seek for the pairs of lowest values followed by stepwise addition of the remaining data with increasing values. Obviously, the input order biases the resulting topology. Performing several runs while randomizing the input order may help recognizing such biases.

Variants of the neighbor-joining approach (Saitou and Nei, 1987; Gascuel, 1997; Bruno *et al.*, 2000) are the most popular distance tree methods used in comparative rRNA analyses. The major drawbacks of distance matrix methods concern the fact that the mode of character change is only indirectly taken into account and the tree space is not evaluated for the 'best' topologies.

3. Maximum parsimony

In contrast to distance methods the maximum parsimony approach (Swofford *et al.*, 1996; Chun and Hong, 2010) seeks the 'best' trees in the space of potential alternative topologies according to optimality criteria. The underlying model assumes that conservation is more likely than change. The optimality criterion is minimizing the overall tree length. The tree topologies that can be derived from the real alignment column data by assuming a minimum number of evolutionary changes win the competition among other candidate topologies. The parsimony values (number of changes to assume) can be estimated for a given tree topology by starting at the terminal nodes and reconstructing the ancestor sequences for the internal nodes while counting the minimum number of character changes. A simple example is shown for a single position in a quartette of sequences (Figure 2). Trees generated by the parsimony approach are generally dichotomous. A priori branches are of equal arbitrarily lengths with no meaning except visualizing the trees topology.

An exhaustive testing of all potential tree topologies is impossible for comprehensive rRNA data sets. (For only 20 sequences about 10^{20} topologies would be possible) Even modern supercomputers would not be able to analyse the huge number of possible tree topologies for comprehensive rRNA data sets. Therefore, heuristics are applied to reduce the number of topologies to be

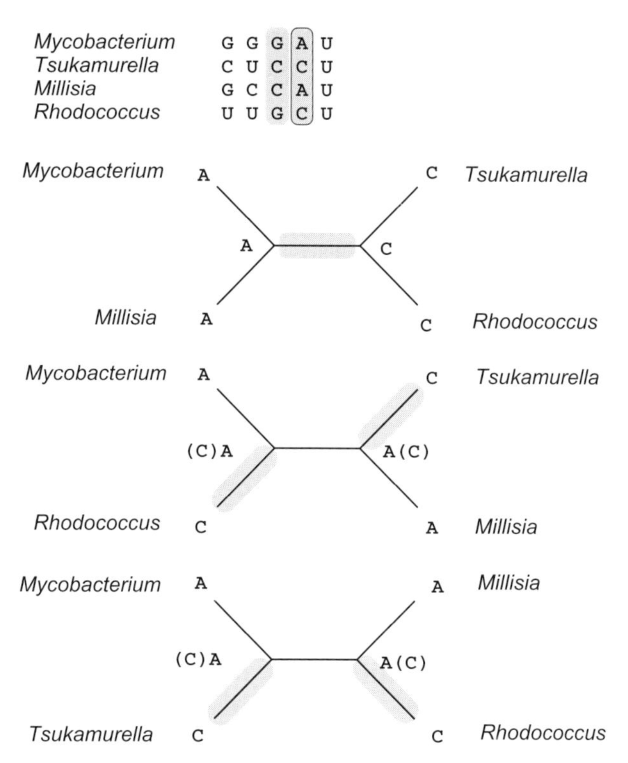

Figure 2. The maximum parsimony criterion.
A: Examples of informative (grey background) and not informative alignment columns (part of the respective 16S rRNA sequences).
B: The three possible tree topologies (shown for the boxed column). Nearest neighbor interchange (NNI) means testing the three topologies for minimum number of mutations. The changes to assume are indicated by gray background of the respective branches.

evaluated. Incremental addition is often used by popular maximum parsimony programs. New data are stepwise added to a starting tree and 'optimally' placed in the topology according to the parsimony criterion. Thus the number of trees to be tested is reduced to the number of potential positions of the sequence to add. Optimization tools for completed parsimony generated trees are in use. The nearest neighbor interchange procedure (NNI) tests the three possible topologies (Figure 2) for every internal branch and the four adjacent sub-trees or terminal branches. Branch swaps which reduce the overall parsimony value of the tree are maintained, others discarded. Other optimization methods allow 'moving' branches or sub-trees beyond the nearest neighbor nodes while adding a penalty to the parsimony value as a function of the number of leaped nodes. Parsimony based software commonly used for rRNA analyses provides these or similar optimization facilities (ARB, http://www.arb-home.de, Ludwig *et al.*, 2004; PAUP, http://paup.csi.fsu.edu, Swofford, 2003; PHYLIP, http://evolution.genetics.washington.edu/phylip.html).

The parsimony approach *per se* does only provide topologies. Meaningful branch lengths are calculated by associated tools in some programs (ARB, PAUP).

The ARB version expresses the difference between the 'best' and 'worse' parsimony values obtained by NNI as lengths indicating the significance of the respective branch.

The maximum parsimony approach allows analysing comprehensive data sets in reasonable time. However, some system inherited drawbacks have to be taken into consideration. In principle, all alignment columns are compared for all quartettes of sequences. However, part of the information carried by the sequence data cannot be utilized by parsimony methods. Comparing individual alignment columns for quartettes of sequences only those positions are informative which allow a bipartition into two clusters (Figure 2).

4. Maximum likelihood

Maximum likelihood methods (Cavalli-Sforza and Edwards, 1967; Felsenstein, 1981; Swofford *et al.*, 1996; Chun and Hong, 2010) like the maximum parsimony approaches, seek the 'best' tree by applying optimality criteria. This tree represents a hypothesis on the evolutionary history which according to the underlying model most likely would have given rise to the respective sequence data. Maximum likelihood procedures utilize much more of the sequence inherited information than maximum parsimony methods can do. Probabilities are calculated for the individual characters at every alignment position. The types of character changes are individually considered. Applying appropriate models and parameters positional variability will also be taken into account. The commonly applied treeing tools provide topologies as well as branch lengths defined according to the optimality criteria. These lengths indicate the significance of the respective topology and an estimate of change per site. As the technique is expensive in computing power and time, maximum likelihood methods have not been widely used for the analysis of comprehensive rRNA datasets until recently. The rapid progress in computer hardware development and the availability of powerful and fast implementations (RAxML, Stamatakis, 2006; PHYML, Guindon and Gascuel, 2003) rendered the maximum likelihood approach to a standard technique in rRNA based phylogeny reconstruction.

5. Bayesian analyses

Bayesian analysis was proposed in 1996 (Rannala and Yang, 1996) as a new approach and later reviewed by Huelsenbeck and Ronquist (2001), Huelsenbeck *et al.* (2002), and Lewis (2001). In principle, it works similar to maximum likelihood approaches in seeking the tree, which most likely reflects the evolutionary path resulting in the present day sequences. The peculiarity of Bayesian approaches is the inclusion of prior information on relationship (probability distribution of trees): The values seen on Bayesian phylogenies are the posterior probabilities for a particular clade, that is, the probability that the clade is "true" given the priors, model, and data (Archibald *et al.*, 2003).

♦♦♦♦♦♦ V. rRNA BASED TREEING

Given the different tree reconstruction methods, the various implementations, the number of alternative models and the many parameters, one could easily spend weeks to analyse a certain dataset reconstructing and evaluating trees. One should be aware that despite the availability of high quality comprehensive databases and improved treeing methods, a tree correctly reflecting the evolutionary history will presumably never be found. Even calling the best tree remains problematic, given the noise in the underlying data (II.A). Nevertheless, applying different treeing methods and parameters allow obtaining at least a 'feeling' on the significance or weakness of (local) tree topologies. In the following some recommendations for rRNA based tree reconstruction are given from the practitioner's point of view (Peplies *et al.*, 2008; Tindall *et al.*, 2010).

Basically there are two major subject areas of comparative rRNA gene sequence analyses: i) phylogenetic studies aiming in the reconstruction of a sound framework of relationships and ii) the assignment of data to clusters in this framework. That means in terms of organisms: i) reconstructing relationships and ii) identification.

A. Phylogeny Reconstruction

1. Data selection

Proper data selection is an essential prerequisite for reconstructing meaningful trees. Given the limited information content (section II.A) only complete sequences should be included. Obviously, only high quality data (section II.C.2) should be used. Data sets should be analysed to avoid misleading branch attraction effects (Ludwig *et al.*, 1998; Ludwig and Klenk, 2001), and balanced with respect to phylogenetic spectrum and the number of members of clusters. Single sequence branches from experience often cannot be stably positioned in trees as a result of branch attraction. Such effects often are compensated by including additional sequences in that phylogenetic neighborhood (if available). Practice has shown that inserting new sequences into trees may not only change neighboring but also remote local topologies. Again, comprehensive and well balanced data sets are less susceptible to such influences. In general, as many high quality and informative (non redundant) sequences should be included as the user's computing facilities can handle in reasonable time. The guide trees provided by the rRNA special databases (SILVA, Pruesse *et al.*, 2007; http://www.arb-silva.de; Greengenes, DeSantis *et al.*, 2006, http://greeengenes.lbl.gov; RDP, Cole *et al.*, 2009, http://rdp.cme.msu.edu) may help composing a balanced set of reference data. The current databases contain 'clouds' of almost identical or highly similar sequences. It makes sense, to reduce the burden of redundancy by defining OTUs (operational taxonomic units) and replacing them by the 'best' representative. The most popular software in use for defining OTUs currently is 'mothur' (Schloss *et al.*, 2009, http://www.mothur.org/wiki/Download_mothur). The ARB package (Ludwig *et al.*, 2004; Ludwig, 2005;

Kumar *et al.*, 2005, 2006; Westram *et al.*, 2011; http://www.arb-home.de) contains a tree based OTU tool which not only defines the OTUs for user selected (phylogenetic) levels but also proposes the 'best' representative. For convenience documented, non-redundant 16S/18S data sets based on clustering of SSURef are made available on the website (http://www.arb-silva.de/projects/nr) of the SILVA project.

2. Filters and masks

Attention should also be paid to the selection of alignment columns used in the analysis. Given the low number of possible character states, rRNA sequences from phylogenetic diverse sources may share 'false' identities (homoplasies) resulting from multiple base changes at homologous positions during the course of evolution. Such residues may cause branch attraction effects. Obviously, highly variable positions are more likely concerned than conserved ones. Discrete character treeing methods such as maximum parsimony or maximum likelihood approaches handle this problem to some extent. Nonetheless, it is recommended to test for potential branch attraction by performing multiple treeing runs applying positional filters or weighting masks. Filters virtually remove alignment columns, while weighting masks keep all positions, however, they assign different (according to the degrees of positional conservation) relative values. Commonly, conservation filters selecting positions invariant among 50% of all representatives related above the phylogenetic level (group) of interest are used. The comparison of tree topologies derived with and without including filters or masks helps to recognize a potential influence of homoplasies. Conservation profile based filters or weighting masks can be established by simply determining the fraction of identical residues for the individual alignment positions globally or confined to defined clusters. More sophisticated procedures apply parsimony or likelihood approaches in combination with global trees. The ARB package (Ludwig *et al.*, 2004; Ludwig, 2005; Kumar *et al.*, 2006; Westram *et al.*, 2011) provides such tools and allows combining different filters and/or masks for treeing. Besides conservation profiles higher order structure filters or weighting masks are sometimes used. The selection pressure on higher order structure enforces compensating changes at positions involved in canonical base pairing. Some investigators prefer to assign the two changes a lower weight. However, rRNA higher order structure at many positions is maintained involving non-canonical base pairing not necessarily requiring compensating changes. Higher order structure masks are provided with rRNA databases in ARB format available at SILVA (Pruesse *et al.*, 2007; http://www.arb-silva.de) and Greengenes (DeSantis *et al.*, 2003; http://greeengenes.lbl.gov) web pages. Customized filters can also be applied to virtually remove parts of the alignment which are present or absent in only part of the aligned sequences (insertions and deletions) or cannot be properly aligned. Again, performing comparative treeing runs (with or without including such filters) help to recognize the potential information content of such regions and potential influences on local or global topologies.

3. Software and models

Numerous implementations and software packages are available for phylogeny reconstruction. A comprehensive compilation along with short descriptions, web sources, and bibliography is hosted by J. Felsenstein on his web page (http://evolution.genetics.washington.edu/phylip/software.html). A selection of the most popular and currently used software (packages) is given in Table 3. Irrespective which treeing approach or software the users prefer, a careful evaluation of the resulting tree topologies is inevitable. None of the models are really correct for all data and positions. Even if the users apply software for finding the best fitting model, the results might be valid for parts of the dataset (tree) only. Similar to the application of filters and masks (section V.A.2) testing different models (if the preferred software provides such facilities) helps to get a 'feeling' on their potential influence on the resulting topologies.

Maximum likelihood based methods (section IV.B.4) are more and more in favor, as taking more sequence inherited and associated (profiles) information into account than parsimony and distance methods. However, heuristic based tree seeking according to optimality criteria may get stuck in suboptimal solutions. Therefore, topology evaluation by applying alternative approaches such as maximum parsimony and distance matrix methods should be performed. Although these methods are less sophisticated, the comparative treeing approach proved very efficient in estimating the significance and robustness of global and local topologies. The underlying assumption is if (local) topologies are in agreement while basically differing approaches were applied, these topologies represent the optimum solution at the current stage of methodology. Whereas topology differences highlight the problematic regions of a tree for which significant relative branching orders cannot significantly be determined.

Besides comparative tree evaluation, maximum parsimony and distance methods - as rapid working procedures - are also predestined methods for preliminary data sorting, inserting and analysing.

4. Resampling techniques

As pointed out in the preceding paragraph tree topology evaluation is mandatory, given the shortcomings of models, methods and data. Besides the comparative approach (favored by the authors) resampling procedures are widely in use. Bootstrapping and jackknifing mean randomly sampling or deleting alignment columns (or distance values), respectively (Lapointe *et al.*, 1994; Efron *et al.*, 1996; Swofford *et al.*, 1996). Thus, 100–1000 slightly differing data sets are generated and subjected to tree reconstruction. In a consensus tree, numbers assigned to internal nodes indicate how often the respective group appeared monophyletic in all trees generated. Simplified, if many sequence positions support the separation of internal nodes, most random data sets retain some of them resulting in a high bootstrap value. The other way round, nodes supported by a few positions only will achieve a low value, given that the majority of the randomly generated data sets do not contain supporting residues. These methods borrowed from statistics certainly help to recognize where noise in the data may obscure treeing analyses. However, the meaningfulness of the approach is under debate. It is criticized that in contrast to the comparative treeing approach, systemic error caused by the treeing method cannot be detected. The data set is changed, not the treeing method. Furthermore, resampling techniques usually are not appropriate for analysing comprehensive data sets. Along with the rapidly increasing databases and the many environmental studies that are finding new groups (the 'missing links'), comprehensive trees more and more exhibit bush-like topologies. In bush-like regions (short internodes), there is low support for separating neighbored nodes. Thus, a reasonable bootstrap value will never be achieved. The branch length of trees, at least if they reflect significance (sections IV.B.3, IV.B.4) provide this information graphically. A long branch means high support (many supporting positions), whereas short branches indicate low support by the sequence data. The insisting of reviewers on bootstrap values may induce authors to slim the data set to a selection sufficiently diverse to result in long branches and reasonable bootstraps. This is counterproductive to the above claim for using comprehensive data sets.

♦♦♦♦♦♦ VI. INTERPRETATION OF TREES

Despite the many tools and procedures which can be used for critically evaluating tree topologies one should be aware an intrinsic uncertainty of any branching order in trees. None of the models commonly used is really correct for the whole rRNA sequence space. None of the treeing procedures is able to provide a perfect solution for the data inherited problems of noise, homoplasies, (not homogenous) intra- and intermolecular (e.g. interaction with ribosomal proteins) biases, and block-wise changes (e.g. conserved tetra loops). Furthermore, there are biases in the data which cannot be virtually masked by filters because the respective sites are distributed and no hot spot columns can be defined for filtering. The high or low GC content bias is an example. Given all the shortcomings of the current methodologies a range of fuzziness has to be assumed concerning the position of the nodes in a tree. As a measure the range of prokaryotic paralogous rRNA sequence divergence could be applied. The genome projects provided valuable data on number and sequence variation of multiple rRNA genes of the organisms (Lee *et al.*, 2009; Pei *et al.*, 2010). Besides a few exceptions (*Desulfotomaculum kuznetsovii*, 8.3%; *Thermobispora bispora*, 6.4%; *Thermoanaerobacter tengcongensis*, 11.6%; *Thermoanaerobacter pseudoethanolicus*, 3.6%; *Thermomonopora chromogena*, 6%; *Haloarcula marismortui*, 5.6%; Acinas *et al.*, 2004; Mylvaganam and Dennis, 1992; Pei *et al.*, 2010, Wang *et al.*, 1997) the inter-operon heterogeneity does not exceed 2%. Consequently, any phylogenetic conclusions at this level of close relationship are questionable. Applying this criterion to bush-like tree regions none of the nodes would attain any significance. However, there is still phylogenetic information which can be deduced. Although any of the neighbored branches might be exchanged without hampering the tree's overall quality their separation from more distantly rooting edges will remain robust and hence can be regarded as significant.

Given the availability of modern rapid discrete character Methods (IV.B.3–5), nowadays the distance methods might become a bit outdated. Consequently, a tendency of preferring maximum parsimony and likelihood supports over distance based results becomes more often evident. Especially in cases of sequences which have preserved more evolutionary changes than others the discrete character methods based results might be closer to the evolutionary history. From experience, distance methods tend to place the respective branches where they don't 'interfere' the matrix based visualization; that means 'deeper' in the tree. The discrete character methods usually 'better' find the sister groups and express the higher amount of preserved changes by a longer branch.

♦♦♦♦♦♦ VII. TREE PRESENTATION

Following the comparative treeing concept (section V.A.3) usually a set of 'best' trees results, differing with respect to local topologies. (If discrepancies are more than local, the data should be carefully checked for inconsistencies.) Furthermore, if based on comprehensive data sets (section V.A.1) the resulting trees may contain many organisms (sequences) which are not within the scope of the current study. This leaves authors often with the dilemma how to present their results. If properly documented, it is certainly legitimate to perform the analyses upon a comprehensive data basis, and to remove those branches from the tree which are not of interest in the context of the study to present. Concerning the topology variations, one option is to present all tree variations relevant for the message that the authors want to communicate (Figure 3). However, this might rather confuse readers not familiar with rRNA based phylogeny inference. Whereas the insiders will redo the treeing analyses according to their own preferences. A second option (preferred by the authors) is to present a consensus tree showing only well resolved parts of the tree, which are supported by all or the majority of alternative trees, while conflicting local topologies are shown as collapsed multi-furcations (Figure 3). Thus the strengths and shortcomings of the tree collection are visible at a glance also for readers not so experienced with rRNA based treeing analyses. Viewing such trees, the reader should be aware, that scale bars might be slightly incorrect because of overall branch length loss due to collapsing.

♦♦♦♦♦♦ VIII. TREE-LESS VISUALIZATION OF RELATIONSHIPS

Comprehensive trees can only be visualized as dendrograms which cover many pages or screens. Although tree viewing software is available which allows collapsing branches sharing a common root to triangles or quadrangles (e.g. ARB; Ludwig *et al.*, 2004; http://www.arb-home.de) alternative methods for condensed visualization of relationships have been proposed.

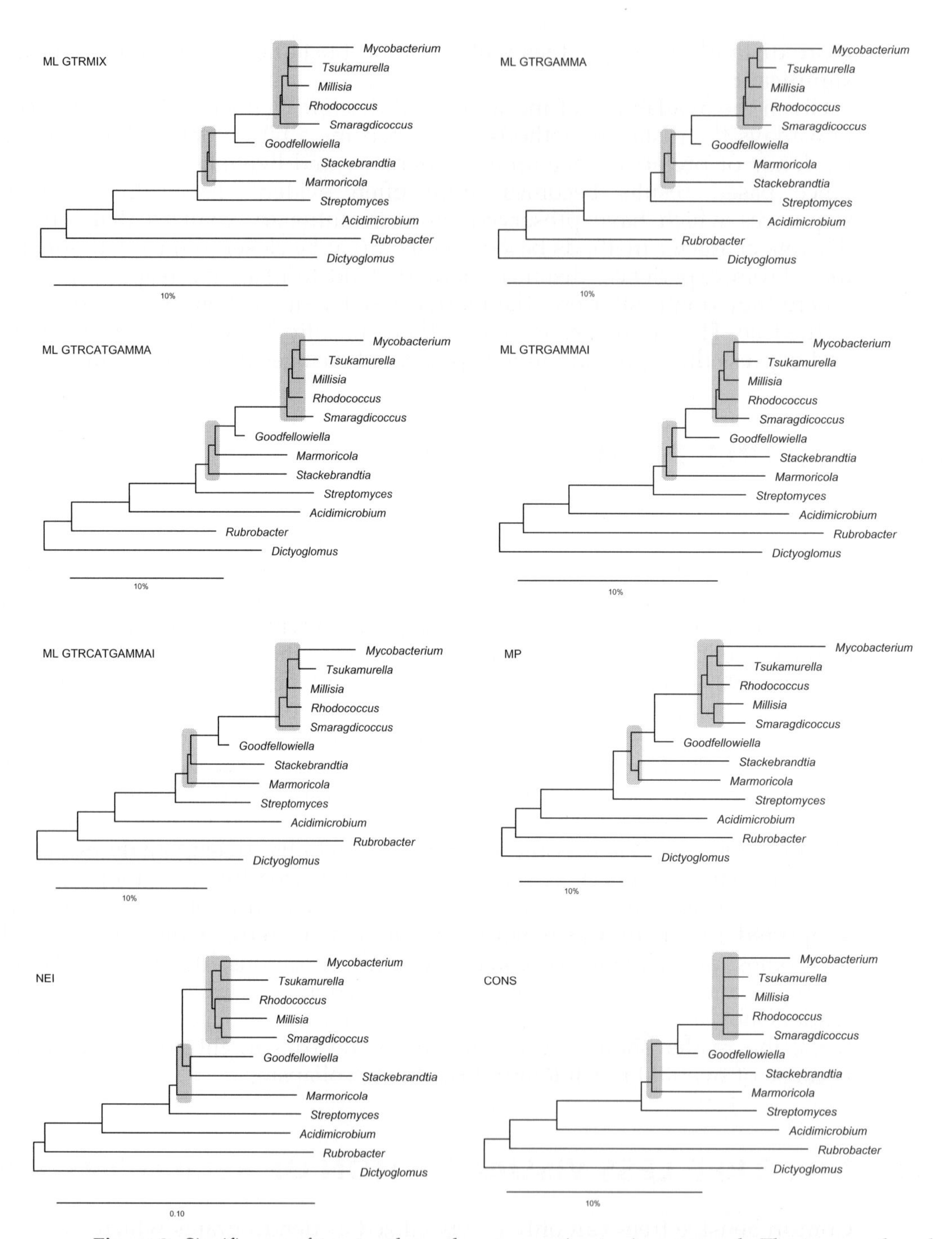

Figure 3. Significance of tree topology: the comparative treeing approach. The trees are based upon a dataset comprising 4000 high quality, and complete (not more than 10 missing terminal positions allowed) primary structures from *Actinobacteria* and representatives of other phyla. A column filter including only positions invariant in at least 50% of all *Actinobacteria* sequences was applied. RAxML (Stamatakis *et al.*, 2005, 2006), ARB-Parsimony and ARB-Dist (Ludwig *et al.*, 2004) were

A. Heat Maps

In heat maps phylogenetic distance or similarity matrices are visualized as colorized two dimensional plots. The colour code reflects the degrees of similarity or phylogenetic distance. In symmetric heat maps complete matrices are used (all values plotted against all) while asymmetric heat maps are based upon incomplete matrices containing the respective binary values for all sequences compared to a set of reference sequences. Lilburn and Garrity (2004) used such an approach for checking the congruency of the current taxonomy and 16S rRNA similarity. When arranging the data according to the current taxonomic framework 'misplaced' organisms (sequences) are indicated by colours not fitting in the environment in the heat map. Finding the appropriate colour environment may help assigning new data to existing taxa. This approach is used by the RDP (Cole *et al.*, 2009; http://rdp.cme.msu.edu) online 'Taxomatic' facility (section III).

B. Principal Component Analysis

Garritty *et al.* (2005) and Garrity and Lilburn (2002) used principal component analysis (PCA) for "mapping the taxonomic space". PCA allows "visualizing high-dimensional data in a lower-dimensional space by finding the uncorrelated single linear combinations of the original variables that explain most of the underlying variability in the data" (Mardia *et al.*, 1979; Venables and Ripley, 1994). PCA has been applied in taxonomic and ecological studies earlier (Sneath and Sokal, 1973; Dunn and Everitt, 1982). Garrity *et al.* (2005) adapted the methods for evaluating the current taxonomy with respect to phylogenetic distances derived from 16S rRNA data. The different existing or proposed higher taxa were located in PCA plots of the first and second or the second and third principal components. Highlighting the positions of members of the respective taxa in such plots helps recognizing the homogeneity of the taxa, finding 'misplaced' outliers, assigning new organisms (sequences) to taxa, and testing or proposing taxonomic concepts.

Both techniques, heat maps and PCA plots, are computational inexpensive and fast, however, are based upon distance data. Hence the advantages of discrete character based treeing procedures are missed.

♦♦♦♦♦♦ IX. TAXONOMY

With the introduction of comparative rRNA sequencing for the first time in the history of prokaryotic taxonomy the ultimate goal of a natural systematics became within reach. Nowadays, the determination and analysis of small

---

Figure 3. (*Continued*) used for tree reconstruction. ML GTRMIX, GTRGAMMA, GTRGAMMAI, GTRCATGAMMA, GTRCATGAMMAI maximum likelihood in combination with different models of evolution (RAxML manual: http://www.kramer.in.tum.de/exelixis); MP maximum parsimony; NEI neighbor joining; CONS consensus tree. Local topology differences are indicated by gray background. In the consensus tree these regions are collapsed to multi-furcations. Preference is given the discrete character methods derived topologies.

subunit rRNA sequences is an indispensable criterion when describing a new taxon. The first landmark of rRNA technology represented the cognition of 'archaebacteria' (now the domain *Archaea*) as fundamentally different from 'eubacteria' (now the domain *Bacteria*) and the eukaryotes (Woese and Fox, 1977). From the beginning the taxonomy of *Archaea* remained rather stable, given that the rRNA approach was already in use when most of the archaeal taxa were defined. The situation has been completely different in the case of *Bacteria*. Many of the taxa had been defined and described before the rRNA era. Consequently, the history of bacterial taxonomy is accompanied by many renamings, transfers, and emendations. Although any new descriptions include rRNA data (Stackebrandt *et al.*, 2002), the process of extending and restructuring taxonomy is still under work. The genus *Clostridium* is an example of a still polyphyletic taxon awaiting further clarity through the collection and analysis of phenotypic and chemotaxonomic data.

In the taxonomy of prokaryotes higher taxa were only scarcely defined in the pre- and early sequencing eras, given the lack of usable characteristics. Beginning with the second edition of Bergey's Manual of Systematic Bacteriology the road map (arrangement of content) is based upon the results of comparative 16S rRNA analyses (Garrity and Holt, 2001; Garrity *et al.*, 2005; Ludwig *et al.*, 2009, 2011). Along with preparing the descriptive material for the volumes of the 2nd edition of Bergey's Manual of Systematic Bacteriology a complete (with respect to higher taxa) prokaryote taxonomy was developed and proposed. Despite the epoch-making impact of rRNA techniques on prokaryotic taxonomy one has to be aware the shortcomings of the rRNA markers. Thus rRNA based phylogenetic networks can only provide a rough backbone for taxonomy, not more. In the context of systematics there are three major drawbacks: i) the resolution power of rRNAs, ii) the many continua in the rRNA based picture of evolution, and iii) 'organismal' phylogenies based on single genes.

It is well known (but often not taken into account) that the highly conserved rRNA sequences in many situations do not contain sufficient information for species delineation. The criterion for species definition is still based on DNA-DNA reassociation data (Rosselló-Mora and Amann, 2001; Stackebrandt *et al.*, 2002). At overall rRNA similarity values above 98.7 % it is recommended to base taxonomic conclusions upon genomic DNA hybridization (Stackebrandt and Ebers, 2006). This concerns the range of fuzziness in rRNA based trees postulated above (section VI). As an example the clearly monophyletic genus *Nesterenkonia* (*Micrococcaceae, Micrococcales, "Actinobacteria"*) currently comprises 11 validly described species. Only three of them (*N. alba*, *N. flava*, and *N. lacusechoensis*) can be defined according to the <98.7% rRNA similarity criterion. *N. aethiopica* and *N. xinjiangensis* represent a pair of 'borderline' species sharing 98.8% sequence identity. The pair *N. halobia – N. halophila* and the species group *N. halotolerans*, *N. lutea*, *N. jeotgali*, *N. sandarakina* share 99.8% and >99.4% rRNA sequence identity, respectively.

In the early days of the rRNA methodology it had been rather easy to establish higher taxa, given the clear cut clusters separated by relatively long 'naked' branches. The *Alpha-* to *Gammaproteobacteria* (now classes) is an example. As pointed out above (section VI), nowadays almost all former 'naked' branches have been filled up with bush-like structures making it difficult to recognize clear-cut thresholds for delineating taxa boundaries. Thus also for higher taxa the rRNA based picture may provide a road map, however, other criteria might have to be included for taxon definition (if available). Indeed, the currently valid or proposed (Garrity *et al.*, 2005; Ludwig *et al.*, 2009, 2011) higher taxa above the rank of family and with the exception of phyla and domains often are polyphyletic in rRNA based trees. In the Living Tree project (Yarza *et al.*, 2008, 2010) a data set of type strain sequences was analysed with respect to sequence similarity ranges correlating with the genus, family, and phylum levels.

One of the mostly mentioned points of critique against rRNA based phylogeny and taxonomy concerned the conveyance of gene phylogenies to the organisms. Certainly, the conserved core of prokaryote genomes only represents the minor part of genomes. However, the many accessory or life style genes do not fulfill the requirements for global phylogenetic markers. Thus only a limited number of core gene products such as ATPase subunits, RNA polymerases, translational initiation and elongation factors, DNA gyrases and recombinases, and heat shock proteins can be used for comprehensive phylogenetic analyses. Phylogentic analyses of these alternative markers roughly support the rRNA based picture (Ludwig and Schleifer, 1999, 2005; Ludwig and Klenk, 2001; Cicarelli *et al.*, 2005; Ludwig, 2010). Exact agreement of tree topologies derived from alternative markers cannot be expected (Ludwig, 2010). All the shortcomings of methods and data discussed above (section VI) also hold true for the alternative markers. Furthermore, alternative markers highly likely will not have preserved the information on exactly the same eras of evolutionary times.

In taxonomy, it is always the type taxon, the name of which has been validly published that dictates the assignment and naming of new isolates according to rRNA (and other) similarities. If the latter cannot be found at the appropriate (taxonomic) levels a new taxon might be proposed and described. There are important web resources which should be consulted when using rRNA sequences for taxon assignment. The rules and the currently valid taxa can be accessed at Euzeby's list of prokaryote names with standing in nomenclature' (LPSN, http://www.bacterio.cict.fr/). Not only links to the original publications, but also accession numbers and links for the respective rRNA sequences are provided. The taxonomic outlines of the volumes of the 2nd edition of Bergey's Manual of Systematic Bacteriology listing the taxa with validly published names and the proposed taxa are available at http://www.bergeys.org/outlines.html. The special rRNA databases (SILVA, Pruesse *et al.*, 2007; http://www.arb-silva.de; Greengenes, DeSantis *et al.*, 2006; RDP, Cole *et al.*, 2009) provide updated taxonomic information and type strain sequences labeled as such. Another important web service (http://www.straininfo.net/; Dawyndt *et al.*, 2005) provides information on all prokaryotic strains, strain numbers, history, and culture collections. Probably the most helpful support for rRNA based taxonomic studies comes from the 'Living Tree Project' (LTP; Yarza *et al.*, 2008, 2010) hosted at the SILVA web page (http://www.arb-silva.de/projects/

living-tree/). Regularly updated databases of aligned 16S and 23S rRNA sequences are maintained by experts for all validly published type strains for which such data are available. Besides evaluated alignments these databases contain correct names, strain numbers, taxonomy, and a lot of further relevant contextual data. Guide trees are provided with the databases. A manually curated database of type strain rRNA sequences (EZTAXON) is also provided by Chun et al. (2007; http://www.eztaxon.org/). Thus, the LTP databases in combination with the other mentioned web resources provide the scientist everything needed assigning a new isolate to a known taxon or recognizing it as new.

♦♦♦♦♦♦ X. IDENTIFICATION

Besides the phylogenetic aspects rRNA based identification at and above the species level currently plays a central role in microbiological studies. Basically, there are three types of rapid methods in use: i) comparative sequence analysis, ii) probe technology, and iii) diagnostic PCR. In the age of high throughput sequencing formerly applied pattern techniques such as amplified rRNA gene restriction analysis (ARDRA) are certainly outdated.

For rRNA based phylogenetic studies only full sequences should be used (section II.A), however, partial sequence data might be sufficient for identification. Identification is based upon high similarity of appropriate sequence (alignment) regions. The reference data set, however, has to consist of high quality full length sequences to ensure optimal reference cluster definition. Obviously, the variable rRNA regions are predestined candidates for diagnostic targets. However, the candidate regions have to be carefully checked for unifying or differentiating information for the group of interest. Different targets might have to be selected for different groups. Given the high degree of expert maintenance of the special rRNA databases (SILVA, Pruesse *et al.*, 2007; http://www.arb-silva.de; Greengenes, DeSantis *et al.*, 2006; RDP, Cole *et al.*, 2009; LTP, Yarza *et al.*, 2008, 2010, http://www.arb-silva.de/projects/living-tree/; EZTAXON, Chun *et al.*, 2007, www.eztaxon.org) blasting the general sequence databases (EMBL, http://www.ebi.ac.uk; Genbank, http://www.ncbi.nml.nih.gov/genbank; DDBJ, http://www.ddbj.nig.ac.jp) is not recommended. The rRNA databases provide online next relative searches for unaligned user provided data. The ARB software package (Ludwig *et al.*, 2004; http://www.arb-home.de) allows local scoring of unaligned query data. Similarly, online and local tools are provided for automated aligning the query data to the reference data bases. While partial sequences usually can readily be inserted into the alignments of reference databases, care has to be taken if identification by treeing is preferred. Full and partial sequences never should be combined for tree reconstruction nor should complete reference data be truncated to the region covered by the partials. Some tree reconstruction software such as the ARB parsimony tool (Ludwig *et al.*, 2004; http://www.arb-home.de) and RAxML (Stamatakis, 2006; http://wwwkramer.in.tum.de/exelixis/) allow adding (partial) sequence data to reference trees according to the parsimony criterion while keeping the reference tree's initial topology.

The diagnostic PCR and probe techniques might be regarded as special cases of partial sequence comparison. The targets of group (taxon) specific primer and probe oligonucleotides contain one to few signature characters for the target group of interest. Often combinations of probes and/or primers have to be used to 'cover' the respective target group. Specific probe and primer design as well as probe technologies are complex topics not within the scope of this chapter and will only superficially be addressed. Diagnostic conventional or real time PCR is widely used in medical and environmental microbiology. Commercial kits and services are available. Two different rRNA targeted probe techniques are currently in focus. The fluorescent in situ hybridization (FISH; Amann *et al.*, 1995, 2001) combines identification with enumeration and localization in fixed samples. Rapid and large scale (with respect to the number of different groups or taxa) identification can be achieved using micro arrays of diagnostic probes, so called phylochips (Loy *et al.*, 2002; Peplies *et al.*, 2003; Sagaram *et al.*, 2009). The ARB software package contains tools for probe design (Ludwig *et al.*, 2004) and evaluation (Kumar *et al.*, 2006). Further probe design software such as Primrose (Ashelford *et al.*, 2002) is available at http://www.bioinformatics-toolkit.org/Primrose/index.html. The RDP (Cole *et al.*, 2009; http://rdp.cme.msu.edu), Greengenes (DeSantis *et al.*, 2006; http://greeengenes.lbl.gov) and SILVA (http://www.arb-silva.de/search/probe-match/) rRNA web resources allow online probe target evaluation. A comprehensive and well-documented collection of proven and tested rRNA targeted probes is provided by probeBase (Loy *et al.*, 2007; www.microbial-ecology.net/probebase).

♦♦♦♦♦♦ XI. NEXT GENERATION SEQUENCING

The advent of next generation sequencing (NGS) technologies has brought the rRNA gene as the universal marker for the cultivation-independent analysis of the diversity of microorganisms even more into focus (Tringe and Hugenholtz, 2008). Pioneering studies using the hypervariable V6 region of the rRNA gene revealed an unprecedented biodiversity now addressed as the "rare biosphere" (Sogin *et al.*, 2006). Although the discussions about the validity of the approach are ongoing (Pedrós-Alió, 2006) there are no doubts that NGS has opened the door to investigate the geographical and temporal distribution and changes of diversity on an unprecedented scale. Standard NGS approaches range from amplicon based tag sequencing to total RNA sequencing and the extraction of rRNA gene fragments from metagenomes. Although the relatively short size of the NGS tags (100 to 400 bases) hampers in depth phylogenetic inference, classifying the sequences with respect to high quality reference databases, as described for identification (section X), has become a common procedure. Environmental biodiversity surveys with hundreds or even thousands of samples based on rRNA gene tags are already ongoing (see www.earthmicrobiome.org) and the democratization of sequencing capacities by bench top NGS machines will render tag based biodiversity analysis an everyday procedure. However, the amount of data produced with every sequencing run demands for new and computationally efficient bioinformatic tools for their analysis.

Initial de-replication, de-noising and clustering methods need to be applied to clean up and reduce the data stream. Common tools are Pyronoise (Quince *et al.*, 2011) for noise reduction, as well as mothur (Schloss *et al.*, 2009), Uclust (Edgar, 2010) or CD-hit (Li and Godzik, 2006) for clustering and OTU generation. The significantly reduced datasets can then be subject to e.g. Blast searches against classified reference datasets as e.g. provided in MultiFasta format by the SILVA or RDP II projects. RDP II and QIIME even offer pyrosequencing pipelines for the analysis of high-throughput community sequencing data (Cole *et al.*, 2009; Caporaso *et al.*, 2010). An important part when dealing with large scale data is the visualization of the results. Overviews of taxonomic breakdowns across many samples are not easy to archive. A recent tool that can help in this respect is the radiant package (http://sourceforge.net/apps/mediawiki/radiant/index.php?title=Main_Page) which provides a dynamic visualisation and zoom-in functions for any kind of hierarchical data.

A rich source of rRNA sequences and especially of 23S/28S fragments, which has so far not gained too much attention are the huge amount of metagenomic datasets. A recent study to find SSU and LSU sequences in the Global Ocean Survey (GOS) metagenomes has shown that in total around 12,000 and 22,000 high quality SSU and LSU sequences could be retrieved, respectively (Yilmaz *et al.*, 2011). These findings indicate that the metagenomes still harbour a wealth of cultivation-independent and PCR-unbiased sequences that just wait to be explored.

♦♦♦♦♦♦ XII. CURRENT STATUS AND FUTURE OF THE RRNA APPROACH

Currently, the prokaryotic taxonomy above the species level is mainly based upon rRNA based phylogenetic analyses. The structuring of the larger part of the current databases comprising validly published or unnamed potential organisms often only represented by sequence data is also based upon rRNA data. The criticism that rRNAs represent a minimal part of genomic information is as old as the rRNA technology itself. In the era of genomics it is enforced by the ongoing debate on the existence of any clonality, trees, rings, networks and root of life (Gogarten *et al.*, 2002; Bapteste *et al.*, 2005). Certainly, genome evolution is only partially clonal but driven largely by gene acquisition or loss (Ludwig, 2010). Major discrepancies have to be expected for potentially clonal and acquired gene phylogenies. In order to reconstruct organism phylogenies only orthologous markers can be used. Horizontal gene transfer and gene duplication events make a sound recognition of orthology difficult or even impossible for the major part of the genomes. Probably all genomes might be chimeric with respect to foreign DNA received and established during the course of evolution (Ludwig, 2010). However, stably established foreign DNA may provide useful – now clonal – markers for phylogeny reconstruction above a certain level of relationship. Finding sets of orthologous markers for the respective levels of relationship is a major task of genome based phylogeny inference.

Nevertheless, there exists a small set of conserved markers preserving a genealogical trace that goes back in time to the universal ancestor state. Although this trace is certainly obscured by a burden of noise and missing information, there is general support of the rRNA based picture by other core markers (Ludwig, 2010). At the current state of the art the rRNAs still seem to be among the most informative molecules for phylogenetic analyses. It is still justified to base the bacterial taxonomy at the higher ranks upon the results of comparative rRNA sequence analysis. However, taxonomists have to find concepts and (bioinformatic) tools for handling and interpreting the present and rapidly increasing data flood of non clonal genome parts for taxonomic purposes.

References

Achenbach, L. and Woese, C. (1995). 16S and 23S rRNA-like primers. In: *Archaea: A Laboratory Manual - Thermophiles* (F. T. Robb and A. R. Place, eds), pp. 269–272. Cold Spring Harbor Laboratory Press, New York.

Acinas, S. G., Marcelino, L. A., Klepac-Ceraj, V. and Polz, M. F. (2004). Divergence and redundancy of 16S rRNA sequences in genomes with multiple *rrn* operons. *J. Bacteriol.* **186**, 2629–2635.

Amann, R., Ludwig, W. and Schleifer, K. H. (1995). Phylogenetic identification and *in situ* detection of individual microbial cells without cultivation. *Microbiol. Rev.* **59**, 143–169.

Amann, R., Fuchs, B. M. and Behrens, S. (2001). The identification of microorganisms by fluorescence *in situ* hybridisation. *Curr. Opin. Biotechnol.* **12**, 231–236.

Archibald, J. K., Mort, M. E. and Crawford, D. J. (2003). Bayesian inference of phylogeny: a non-technical primer. *Taxon* **52**, 187–191.

Ashelford, K. E., Weightman, A. J. and Fry, J. C. (2002). PRIMROSE: a computer program for generating and estimating the phylogenetic range of 16S rRNA oligonucleotide probes and primers in conjunction with the RDP-II database. *Nucleic Acids Res.* **30**, 3481–3489.

Ashelford, K. E., Chuzhanova, N. A., Fry, J. C., Jones, A. J. and Weightman, A. J. (2005). At least one in twenty 16S rRNA sequence records currently held in public repositories estimated to contain substantial anomalies. *Appl. Environ. Microbiol.* **12**, 7724–7736.

Ashelford, K. E., Chuzhanova, N. A., Fry, J. C., Jones, A. J. and Weightman, A. J. (2006). New screening software shows that most recent large 16S rRNA gene clone libraries contain chimeras. *Appl. Environ. Microbiol.* **72**, 5734–5741.

Baker, G. C., Smith, J. J. and Cowan, D. A. (2003). Review and re-analysis of domain-specific 16S primers. *J. Microbiol. Meth.* **55**, 541–555.

Bapteste, E., Susko, E., Leigh, J., MacLeod, D., Charlebois, R. L. and Doolittle, W. F. (2005). Do orthologous gene phylogenies really support tree-thinking? *BMC Evol. Biol.* **5**, 33–43.

Barns, S. M., Fundyga, R. E., Jeffries, M. W. and Pace, N. R. (1994). Remarkable archaeal diversity detected in Yellowstone National Park hot spring environment. *Proc. Natl. Acad. Sci. USA* **91**, 1609–1613.

Ben-Dov, E., Siboni, N., Shapiro, O. H., Arotsker, L. and Kushmaro, A. (2011). Substitution by inosine at the 3′-ultimate and penultimate positions of 16S rRNA gene universal primers. *Microb. Ecol.* **61**, 1–6.

Benson, D. A., Karsch-Mizrachi, I., Lipman, D. J., Ostell, J. and Sayers, E. W. (2011). GenBank. *Nucleic Acids Res.* **39**, D32–D37.

Böttger, E. C. (1989). Rapid determination of bacterial ribosomal RNA sequences by direct sequencing of enzymatically amplified DNA. *FEMS Microbiol. Lett.* **65**, 171–176.

Brosius, J., Dull, T. J., Sleeter, D. D. and Noller, H. F. (1981). Gene organisation and primary structure of a ribosomal RNA operon from *Escherichia coli*. *J. Mol. Biol.* **148**, 107–127.

Bruno, W. J., Socci, N. D. and Halpern, A. L. (2000). Weighted neighbor joining: a likelihood-based approach to distance-based phylogeny reconstruction. *Mol. Biol. Evol.* **17**, 189–197.

Cannone, J. J., Supramanian, S., Schnare, M. N., Collett, J. R., D'Sousa, L. M., Du, Y., Feng, B., Lin, N., Madabusi, L. V., Müller, K. M., Pande, M., Shang, Z., Yu, N. and Gutell, R. R. (2002). The comparative RNA web (CRW) site: an online database of comparative sequence and structure information for ribosomal, intron and other RNAs. *BMC Bioinformatics* **3**, 2.

Caporaso, J. G., Kuczynski, J., Stombaugh, J., Bittinger, K., Bushman, F. D., Costello, E. K., Fierer, N., Pena, A. G., Goodrich, J. K., Gordon, J. I., Huttley, G. A., Kelley, S. T., Knights, D., Koenig, J. E., Ley, R. E., Lozupone, C. A., McDonald, D., Muegge, B. D., Pirrung, M., Reeder, J., Sevinsky, J. R., Turnbaugh, P. J., Walters, W. A., Widmann, J., Yatsunenko, T., Zaneveld, J. and Knight, R. (2010). QIIME allows analysis of high-throughput community sequencing data. *Nature Methods* **7**, 335–336.

Cavalli-Sforza, L. L. and Edwards, A. W. F. (1967). Phylogenetic analyses: Models and estimation. *Evolution* **32**, 550–570.

Chenna, R., Sugawara, H., Koike, T., Lopez, R., Gibson, T. J., Higgins, D. G. and Thompson, J. D. (2003). Multiple sequence alignment with the Clustal series of programs. *Nucleic Acids Res.* **31**, 3497–3500.

Chun, J. and Hong, S. G. (2010). Methods and programs for calculation of phylogenetic relationships from molecular sequences. In: *Molecular Phylogeny of Microorganisms* (A. Oren and R. T. Papke, eds), pp. 65–83. Caister Academic Press, Norfolk.

Chun, J., Lee, J. H., Jung, Y., Kim, M., Kim, S., Kim, B. K. and Lim, Y. W. (2007). EzTaxon: a web-based tool for the identification of prokaryotes based on 16S ribosomal RNA gene sequences. *Int. J. Syst. Evol. Microbiol.* **57**, 2259–2261.

Ciccarelli, F. D., Doerks, T., von Mering, C., Creevey, C. J., Snel, B. and Bork, P. (2005). Toward automatic reconstruction of a highly resolved tree of life. *Science* **311**, 1283–1286.

Cochrane, G., Akhtar, R., Bonfield, J., Bower, L., Demiralp, F., Faruque, N., Gibson, R., Hoad, G., Hubbard, T., Hunter, C., Jang, M., Juhos, S., Leinonen, L., Leonard, S., Lin, Q., Lopez, R., Lorenc, D., McWilliam, H., Mukherjee, G., Plaister, S., Radhakrishnan, R., Robinson, S., Sobhany, S., Hoopen, P. T., Vaughan, R., Zalunin, V. and Birney, E. (2008). Petabyte-scale innovations at the European Nucleotide Archive. *Nucleic Acids Res.* **37**, D19–D25.

Cochrane, G., Karsch-Mizrachi, I. and Nakamura, Y. (2011). The International Nucleotide Sequence Database Collaboration. *Nucleic Acids Res.* **39**, D15–D18.

Cole, J. R., Wang, Q., Cardenas, E., Fish, J., Chai, B., Farris, R. J., Kulam-Syed-Mohideen, A. S., McGarrell, D. M., Marsh, T., Garrity, G. M. and Tiedje, J. M. (2009). The Ribosomal Database Project: improved alignments and new tools for rRNA analysis. *Nucleic Acids Res.* **37**, D141–D145.

Coloqhoun, J. A. (1997). Discovery of Deep-Sea Actinomycetes. PhD Dissertation. Research School of Biosciences. University of Kent, Canterbury.

Dawyndt, P., Vancanneyt, M., De Meyer, H. and Swings, J. (2005). Knowledge accumulation and resolution of data inconsistencies during the integration of microbial information sources. *IEEE Transactions on Knowledge and Data Engineering* **17**, 1111–1126.

DeLong, E., Taylor, L., Marsh, T. and Preston, C. (1999). Visualization and enumeration of marine planktonic Archaea and Bacteria by using polyribonucleotide probes and fluorescent in situ hybridization. *Appl. Environ. Microbiol.* **65**, 5554–5563.

DeSantis, T. Z., Hugenholtz, P., Larsen, N., Rojas, M., Brodie, E. L., Keller, K., Huber, T., Dalevi, T., Hu, P. and Andersen, G. L. (2006). Greengenes, a chimera-checked 16S rRNA gene database and workbench compatible with ARB. *Appl. Environ. Microbiol.* **72**, 5069–5072.

Dunn, G. and Everitt, B. S. (1982). *An Introduction to Mathematical Taxonomy*. Cambridge University Press, Cambridge.

Eck, V. and Dayhoff, M. O. (1969). *Atlas of Protein Sequence and Structure*. Vol. 4, National Biomedical Research Foundation, Silver Springs, MD

Edgar, R. C. (2004). MUSCLE: a multiple sequence alignment method with reduced time and space complexity. *BMC Bioinformatics* **5**, 113.

Edgar, R. C. (2010). Search and clustering orders of magnitude faster than BLAST. *Bioinformatics* epub, 10.1093/bioinformatics/btq1461.

Efron, B., Halloran, E. and Holmes, S. (1996). Bootstrap confidence levels for phylogenetic trees. *Proc. Natl. Acad. Sci. USA* **93**, 13429–13434.

Felsenstein, J. (1981). Evolutionary trees from DNA sequences: A maximum likelihood approach. *J. Mol. Evol.* **17**, 368–376.

Felsenstein, J. (1989). PHYLIP – Phylogeny Inference Package (Version 3.2). *Cladistics* **5**, 164–166.

Felsenstein, J. (1993). PHYLIP – Phylogeny Inference Package (Version 3.5). University of Washington, Seattle.

Feng, S., Li, H., Zhao, J., Pervushin, K., Lowenhaupt, K., Schwartz, T. U. and Dröge, P. (2011). Alternate rRNA secondary structures as regulators of translation. *Nature Struct. Mol. Biol.* **18**, 169–176.

Fox, G., Pechman, K. and Woese, C. R. (1977). Comparative cataloging of 16S ribosomal ribonucleic acid: Molecular approach to prokaryotic systematic. *Int. J. Syst. Bacteriol.* **27**, 44–57.

Garrity, G. M. and Holt, J. G. (2001). The roadmap to the manual. In: *Bergey's Manual of Systematic Bacteriology* (G. M. Garrity, ed), Vol. 1, pp. 119–155. Springer, New York.

Garrity, G. M. and Lilburn, T. G. (2002). Mapping taxonomic space: an overview of the road map to the second edition of Bergey's Manual of Systematic Bacteriology *WFCC Newsl* **35**, 5–15.

Garrity, G. M., Bell, J. A. and Lilburn, T. (2005). The revised road map to the manual. In: *Bergey's Manual of Systematic Bacteriology* (G. M. Garrity, ed), Vol. 2, pp. 159–187. Springer, New York.

Gascuel, O. (1997). BIONJ: an improved version of the NJ algorithm based on a simple model of sequence data. *Mol. Biol. Evol.* **14**, 685–695.

Gogarten, J. P., Doolittle, W. F. and Lawrence, J. G. (2002). Prokaryotic evolution in light of gene transfer. *Mol. Biol. Evol.* **19**, 2226–2238.

Guindon, S. and Gascuel, O. (2003). A simple, fast, and accurate algorithm to estimate large phylogenies by maximum likelihood. *Syst. Biol.* **52**, 696–704.

Gutell, R. R., Power, A., Hertz, G., Putz, E. and Stormo, G. (1992). Identifying constraints on the higher-order structure of RNA: continued development and application of comparative sequence analysis methods. *Nucleic Acids Res.* **20**, 5785–5795.

Gutell, R. R., Gray, M. W. and Schnare, M. N. (1993). A compilation of large subunit (23S- and 23S-like) ribosomal RNA structures. *Nucleic Acids Res.* **21**, 3055–3074.

Huber, T., Faulkner, G. and Hugenholtz, P. (2004). Bellerophon: a program to detect chimeric sequences in multiple sequence alignments. *Bioinformatics* **20**, 2317.

Huelsenbeck, J. P. and Ronquist, F. (2001). MrBayes: Bayesian inference of phylogeny. *Bioinformatics* **17**, 754–755.

Huelsenbeck, J. P., Larget, B., Miller, R. E. and Ronquist, F. (2002). Potential application and pitfalls of Bayesian inference of phylogeny. *Syst. Biol.* **51**, 673–688.

Hunt, D. E., Klepac-Ceraj, V., Acinas, S. G., Gautier, C., Bertilsson, S. and Polz, M. F. (2006). Evaluation of 23S rRNA PCR primers for use in phylogenetic studies of bacterial diversity. *Appl. Environ. Microbiol.* **72**, 2221–2225.

Jeon, Y. S., Chung, H., Park, S., Hur, I., Lee, J. H. and Chun, J. (2005). jPHYDIT: a JAVA-based integrated environment for molecular phylogeny of ribosomal RNA sequences. *Bioinformatics* **21**, 3171–3173.

Jukes, T. H. and Cantor, C. R. (1969). Evolution of protein molecules. In: *Mammalian Protein Metabolism* (H. N. Munro, ed), pp. 21–132. Academic Press, New York.

Kaminuma, E., Kosuge, T., Kodama, Y., Aono, I., Mashima, J., Gojobori, T., Sugawara, H., Ogasawara, O., Takagi, T., Okubo, K. and Nakamura, Y. (2011). DDBJ progress report. *Nucleic Acids Res.* **39**, D22–D27.

Katoh, K. and Toh, H. (2008). Recent developments in the MAFFT multiple sequence alignment program. *Brief. Bioinform.* **9**, 286–298.

Kimura, M. (1980). A simple method for estimating evolutionary rate of base substitutions through comparative studies of nucleotide sequences. *J. Mol. Evol* **16**, 111–120.

Kumar, S., Tamura, K. and Nei, M. (2004). MEGA3: integrated software for Molecular Evolutionary Genetics and sequence Alignment. *Brief. Bioinform.* **5**, 150–163.

Kumar, Y., Westram, R., Behrens, S., Fuchs, B., Glöckner, F. O., Amann, R. and Ludwig, W. (2005). Graphical representation of ribosomal RNA probe accessibility data using ARB software package. *BMC Bioinformatics* **6**, 61–70.

Kumar, Y., Westram, R., Kipfer, P., Meier, H. and Ludwig, W. (2006). Evaluation of sequence alignments and oligonucleotide probes with respect to three-dimensional structure of ribosomal RNA using ARB software package. *BMC Bioinformatics* **7**, 240.

Lane, D. J. (1991). 16S/23S rRNA sequencing. In: *Nucleic Acid Techniques in Bacterial Systematics* (E. Stackebrandt and M. Goodfellow, eds), pp. 115–175. J. Wiley & Sons, Chichester.

Lane, D. J., Pace, B., Olsen, G. J., Stahl, D. A., Sogin, M. L. and Pace, N. R. (1985). Rapid determination of 16S ribosomal RNA sequences for phylogenetic analyses. *Proc. Natl. Acad. Sci. USA* **82**, 6955–6959.

Lapointe, F. J., Kirsch, J. A. W. and Bleiweiss, R. (1994). Jackknifing of weighted trees: Validation of phylogenies reconstructed from distance matrices. *Mol. Phyl. Evol.* **3**, 256–267.

Lee, Z. M. P., Bussema, C., III and Schmidt, T. M. (2009). rrnDB: Documenting the number of rRNA and tRNA genes in bacteria and archaea. *Nucleic Acids Res.* **37**, D489–D493.

Lewis, P. O. (2001). Phylogenetic systematics turns over a new leaf. *Trends Ecol. Evol.* **16**, 30–37.

Li, W. and Godzik, A. (2006). Cd-hit: a fast program for clustering and comparing large sets of protein or nucleotide sequences. *Bioinformatics* **22**, 1658–1659.

Lilburn, T. G. and Garrity, G. M. (2004). Exploring prokaryotic taxonomy. *Int. J. Syst. Evol. Microbiol.* **54**, 7–13.

López-García, P., Moreira, D., López-López, A. and Rodríguez-Valera, F. (2001). A novel haloarchaeal-related lineage is widely distributed in deep oceanic regions. *Environ. Microbiol.* **3**, 72–78.

Loy, A., Lehner, A., Lee, N., Adamczyk, J., Meier, H., Ernst, J., Schleifer, K. H. and Wagner, M. (2002). Oligonucleotide microarray for 16S rRNA gene-based detection of all recognized lineages of sulfate-reducing prokaryotes in the environment. *Appl. Environ. Microbiol.* **68**, 5064–5081.

Loy, A., Maixner, F., Wagner, M. and Horn, M. (2007). probeBase - an online resource for rRNA-targeted oligonucleotide probes: new features 2007. *Nucleic Acids Res.* **35**, D800–D804.

Ludwig, W. (2005). Bioinformatics and web resources for the microbial ecologist. In: *Molecular Microbial Ecology* (M. Osborn, ed), pp. 345–371. Taylor and Francis, London.

Ludwig, W. (2010). Molecular phylogeny of microorganisms: Is rRNA still a useful marker? In: *Molecular Phylogeny of Microorganisms* (A. Oren and R. T. Papke, eds), pp. 23–40. Caister Academic Press, Norfolk.

Ludwig, W. and Klenk, H. P. (2001). Overview: a phylogenetic backbone and taxonomic framework for prokaryotic systematics. In: *Bergey's Manual of Systematic Bacteriology* (G. M. Garrity, ed), Vol. 1, pp. 49–66. Springer, New York..

Ludwig, W. and Schleifer, K. H. (1994). Bacterial phylogeny based on 16S and 23S rRNA sequence analysis. *FEMS Microbiol. Rev.* **15**, 155–173.

Ludwig, W. and Schleifer, K. H. (1999). Phylogeny of *Bacteria* beyond the 16S rRNA standard. *ASM News* **65**, 752–757.

Ludwig, W. and Schleifer, K. H. (2005). Molecular phylogeny of bacteria based on comparative sequence analysis of conserved genes. In: *Microbial Phylogeny and Evolution, Concepts and Controversies* (J. Sapp, ed), pp. 70–98. Oxford University Press, Oxford.

Ludwig, W., Strunk, O., Klugbauer, S., Klugbauer, N., Weizenegger, M., Neumaier, J., Bachleitner, M. and Schleifer, K. H. (1998). Bacterial phylogeny based on comparative sequence analysis. *Electrophoresis* **19**, 554–568.

Ludwig, W., Strunk, O., Westram, R., Richter, L., Meier, H., Yadhukumar, Buchner, A., Lai, T., Steppi, S., Jobb, G., Förster, W., Brettske, I., Gerber, S., Ginhart, A. W., Gross, O., Grumann, S., Hermann, S., Jost, R., König, A., Liss, T., Lüßmann, R., May, M., Nonhoff, B., Reichel, B., Strehlow, R., Stamatakis, A., Stuckmann, N., Vilbig, A., Lenke, M., Ludwig, T., Bode, A. and Schleifer, K. H. (2004). ARB: a software environment for sequence data. *Nucleic Acids Res.* **32**, 1363–1371.

Ludwig, W., Schleifer, K. H. and Whitman, W. B. (2009). Revised roadmap to the phylum *Firmicutes*. In: *Bergey's Manual of Systematic Bacteriology* (W. B. Whitman, ed), Vol. 3, pp. 1–13. Springer, New York.

Ludwig, W., Euzéby, J. and Whitman, W. B. (2011). Road map to the phyla *Bacteroidetes, Spirochaetes, Tenericutes (Mollicutes), Acidobacteria, Fibrobacteres, Fusobacteria, Dictyoglomi, Gemmatimonadetes, Lentisphaerae, Verrucomicrobia, Chlamydiae,* and *Planctomycetes.* In: *Bergey's Manual of Systematic Bacteriology* (W. B. Whitman, ed), Volume 4, pp. 1–19. Springer, New York.

Mardia, K. V., Kent, J. T. and Bibby, J. M. (1979). *Multivariate Analysis.* Academic Press, London.

Martinez-Murcia, A. J., Acinas, S. G. and Rodriguez-Valera, F. (1995). Evaluation of prokaryotic diversity by restrictase digestion of 16S rDNA directly amplified from hypersaline environments. *FEMS Microbiol. Ecol.* **17**, 247–256.

McInnery, J. O., Wilkinson, M., Patching, J. W., Embley, T. M. and Powell, R. (1995). Recovery and phylogenetic analysis of novel archaeal rRNA sequences from deep-sea deposit feeder. *Appl. Environ. Microbiol.* **61**, 1646–1648.

Muyzer, G., Teske, A. and Wirsen, C. O. (1995). Phylogenetic relationships of *Thiomicrospira* species and their identification in deep-sea hydrothermalvent samples by denaturing gradient gel electrophoresis of 16S rDNA fragments. *Arch. Microbiol.* **164**, 165–172.

Mylvaganam, S. and Dennis, P. P. (1992). Sequence heterogeneity between the two genes encoding 16S rRNA from the halophilic archaebacterium *Haloarcula marismortui*. *Genetics* **130**, 399–410.

Olsen, G. J., Overbeek, R., Larsen, N., Marsh, T. L., McCaughey, M. J., Maciukenas, M. A., Kuan, W. M., Macke, T. J., Xing, Y. and Woese, C. R. (1992). The Ribosomal Database Project. *Nucleic Acids Res.* **20**, 2199–2200.

Pedrós-Alió, C. (2006). Marine microbial diversity: can it be determined? *Trends Microbiol.* **14**, 257–263.

Pei, A. Y., Oberdorf, W. E., Nossa, C. W., Agarwal, A., Chokshi, P., Gerz, E. A., Jin, Z. D., Lee, P., Yang, L. Y., Poles, M., Brown, S. M., Sotero, S., DeSantis, T., Brodie, E., Nelson, K. and Pei, Z. H. (2010). Diversity of 16S rRNA genes within individual prokaryotic genomes. *Appl. Environ. Microbiol.* **76**, 3886–3897.

Peplies, J., Glöckner, F. O. and Amann, R. (2003). Optimization strategies for DNA microarray-based detection of bacteria with 16S rRNA-targeting oligonucleotide probes. *Appl. Environ. Microbiol.* **69**, 1397–1407.

Peplies, J., Kottmann, R., Ludwig, W. and Glöckner, F. O. (2008). A standard operating procedure for phylogenetic inference (SOPPI) using (rRNA) marker genes. *Syst. Appl. Microbiol.* **31** 251-157

Pruesse, E., Quast, C., Knittel, K., Fuchs, B. M., Ludwig, W., Peplies, J. and Glöckner, F. O. (2007). SILVA: a comprehensive online resource for quality checked and aligned ribosomal RNA sequence data compatible with ARB. *Nucleic Acids Res.* **35**, 7188–7196.

Quince, C., Lanzen, A., Davenport, R. and Turnbaugh, P. (2011). Removing noise from pyrosequenced amplicons. *BMC Bioinformatics* **12**, 38.

Rannala, B. and Yang, Z. (1996). Probability distribution of molecular evolutionary trees: a new method of phylogenetic inference. *J. Mol. Evol.* **43**, 304–311.

Reysenbach, A.-L. and Pace, N. R. (1994). Reliable amplification of hyperthermophilic archaeal 16S rRNA genes by the polymerase chain reaction. In: *Archaea: A Laboratory Manual—Thermophiles* (F. T. Robb and A. R. Place, eds), pp. 101–107. Cold Spring Harbor Laboratory Press, New York.

Reysenbach, A.-L., Wickham, G. S. and Pace, N. R. (1994). Phylogenetic analysis of the hyperthermophilic pink filament community in Octopus Spring, Yellowstone National Park. *Appl. Environ. Microbiol.* **60**, 2113–2119.

Rodriguez, F., Oliver, J. L., Marin, A. and Medina, J. R. (1990). The general stochastic model of nucleotide substitution. *J. Theor. Biol.* **142**, 484–501.

Ronquist, F. and Huelsenbeck, J. P. (2003). MRBAYES 3: Bayesian phylogenetic inference under mixed models. *Bioinformatics* **19**, 1572–1574.

Rosselló-Móra, R. and Amann, R. (2001). The species concept for prokaryotes. *FEMS Microbiol. Rev.* **25**, 39–67.

Rudi, K., Skulberg, O. M., Larsen, F. and Jacobsen, K. S. (1997). Strain classification of oxyphotobacteria in clone cultures on the basis of 16S rRNA sequences from variable regions V6, V7 and V8. *Appl. Environ. Microbiol.* **63**, 2593–2599.

Sagaram, U. S., DeAngelis, K. M., Trivedi, P., Andersen, G. L., Lu, S. and Wang, N. (2009). Bacterial diversity analysis of Huanglongbing pathogen-infected citrus using phylochips and 16S rDNA clone library sequencing. *Appl. Environ. Microbiol.* **75**, 1566–1574.

Saiki, R. K., Scharf, S., Faloona, F., Mullis, K. B., Horn, G. T., Erlich, H. A. and Arnheim, N. (1985). Enzymatic amplification of β-globin genomic sequences and restriction site analysis for diagnosis of sickle cell anemia. *Science* **230**, 1350–1354.

Saitou, N. and Nei, M. (1987). The neighbor-joining method: a new method for reconstructing phylogenetic trees. *Mol. Biol. Evol.* **4**, 406–425.

Schloss, P. D., Westcott, S. L., Ryabin, T., Hall, J. R., Hartmann, M., Hollister, E. B., Lesniewski, R. A., Oakley, B. B., Parks, D. H., Robinson, C. J., Sahl, J. W., Stres, B., Thallinger, G. G., Van Horn, D. J. and Weber, C. F. (2009). Introducing mothur: open source, platform-independent, community-supported software for describing and comparing microbial communities. *Appl. Environ. Microbiol.* **75**, 7537–7541.

Schmidt, H. A., Strimmer, K., Vingron, M. and von Haeseler, A. (2002). TREE-PUZZLE: maximum likelihood phylogenetic analysis using quartets and parallel computing. *Bioinformatics* **18**, 502–504.

Sneath, P. H. A. and Sokal, R. R. (1973). *Numerical Taxonomy: The Principles and Practice of Numerical Classification*. W.H. Freeman and Co., San Francisco.

Sogin, M. L. (1990). Amplification of ribosomal RNA genes for molecular evolution studies. In: *In: PCR Protocols: A guide to Methods and Applications* (M. A. Innis , ed), pp. 307–314. Academic Press, San Diego.

Sogin, M. L., Morrison, G. H., Huber, J. A., Welch, D. M., Huse, S. M., Neal, P. R., Arrieta, J. M. and Herndl, G. J. (2006). Microbial diversity in the deep sea and the underexplored "rare biosphere". *Proc. Natl. Acad. Sci. USA* **103**, 12115–12120.

Stackebrandt, E. and Ebers, J. (2006). Taxonomic parameters revisited: tarnished gold standards. *Microbiol. Today* **33**, 152–155.

Stackebrandt, E., Ludwig, W., Schleifer, K. H. and Gross, H. J. (1981). Rapid cataloguing of ribonuclease T1 resistant oligonucleotides from ribosomal RNAs for phylogenetic studies. *J. Mol. Evol.* **17**, 227–236.

Stackebrandt, E., Frederiksen, W., Garrity, G. M., Grimont, P. A., Kämpfer, P., Maiden, M. C., Nesme, X., Rosselló-Móra, R., Swings, J., Trüper, H. G., Vauterin, L., Ward, A. C. and Whitman, W. B. (2002). Report of the ad hoc committee for the re-evaluation of the species definition in bacteriology. *Int. J. Syst. Evol. Microbiol.* **52**, 1043–1047.

Stamatakis, A. (2006). RAxML-VI-HPC: Maximum likelihood-based phylogenetic analyses with thousands of taxa and mixed models. *Bioinformatics* **22**, 2688–2690.

Suzuki, M. T. and Giovannoni, S. J. (1996). Bias caused by template annealing in the amplification of mixtures of 16S rRNA genes by PCR. *Appl. Environ. Microbiol.* **62**, 625–630.

Swofford, D. L. (2003). *PAUP*. Phylogenetic Analysis Using Parsimony (*and Other Methods). Version 4*. Sinauer Associates, Sunderland, Massachusetts.

Swofford, D. L., Olsen, G. J., Waddell, P. J. and Hillis, D. M. (1996). Phylogenetic inference. In: *Molecular Systematics* (D. M. Hillis, C. Moritz and B. K. Mable, eds), pp. 407–514. Sinauer Associates, Sunderland.

Tindall, B. J., Rosselló-Móra, R., Busse, H. J., Ludwig, W. and Kämpfer, P. (2010). Notes on the characterization of prokaryote strains for taxonomic purposes. *Int. J. Syst. Evol. Microbiol.* **60**, 249–266.

Toschka, H. Y., Höpfl, P., Ludwig, W., Schleifer, K. H., Ulbrich, N. and Erdmann, V. A. (1987). Complete nucleotid sequence of a 23S ribosomal RNA gene from *Pseudomonas aeruginosa*. *Nucleic Acids Res.* **15**, 7182.

Tringe, S. G. and Hugenholtz, P. (2008). A renaissance for the pioneering 16S rRNA gene. *Curr. Opin. Microbiol.* **11**, 442–446.

Van Camp, G., Chapelle, S. and De Wachter, R. (1993). Amplification and sequencing of variable regions in bacterial 23S ribosomal RNA genes with conserved primer sequences. *Curr. Microbiol.* **27**, 147–151.

Venables, W. N. and Ripley, B. D. (1994). *Modern Applied Statistics with S-plus*. Springer, New York.

Wang, Y. Z., Thang, Z. and Ramanan, N. (1997). The actinomycete *Thermobispora bispora* contains two distinct types of transcriptionally active 16S rRNA genes. *J. Bacteriol.* **179**, 3270–3276.

Watanabe, K., Kodama, Y. and Harayama, S. (2001). Design and evaluation of PCR primers to amplify 16S ribosomal DNA fragments used for community fingerprinting. *J. Microbiol. Meth.* **44**, 253–262.

Westram, R., Bader, K., Prüsse, E., Kumar, Y., Meier, H., Glöckner, F. O. and Ludwig, W. (2011). ARB; a software environment for sequence data. In: *Handbook of Molecular*

Microbial Ecology I: Metagenomics and Complementary Approaches (F. J. de Bruijn, ed), pp. 399–406, Wiley-Blackwell.
Whitehead, T. R. and Cotta, M. A. (1999). Phylogenetic diversity of methanogenic Archaea in swine waste storage pits. *FEMS Microbiol. Lett.* **179**, 223–226.
Woese, C. R. and Fox, G. E. (1977). Phylogenetic structure of the prokaryotic domain: the primary kingdoms. *Proc. Natl. Acad. Sci. USA* **74**, 5088–5090.
Wuyts, J., Perriere, G. and Van de Peer, Y. (2004). The European ribosomal RNA database. *Nucleic Acids Res.* **32**, D101–D103.
Yang, Z. N., Goldman, N. and Friday, A. E. (1994). Comparison of models for nucleotide substitution used in maximum likelihood phylogenetic estimation. *Mol. Biol. Evol.* **11**, 316–324.
Yarza, P., Richter, M., Peplies, J., Euzéby, J., Amann, R., Schleifer, K. -H., Ludwig, W., Glöckner, F. O. and Rosselló-Móra, R. (2008). The All-Species Living Tree Project: a 16S rRNA based phylogenetic tree of all sequenced type strains. *Syst. Appl. Microbiol.* **31**, 241–250.
Yarza, P., Ludwig, W., Euzéby, J., Amann, R., Schleifer, K.-H., Glöckner, F. O. and Rosselló-Móra, R. (2010). Update of the all-species living tree project based on 16S and 23S rRNA sequence analyses. *Syst. Appl. Microbiol.* **33**, 291–299.
Yilmaz, P., Kottmann, R., Pruesse, E., Quast, C., and Glöckner, F. O. (2011). 23S ribosomal RNA genes in metagenomes – a case study in the Global Ocean Sampling Expedition. *Syst. Appl. Microbiol.* (in press).
Zuckerkandl, E. and Pauling, L. (1965). Molecules as documents of evolutionary history. *J. Theor. Biol.* **8**, 357–366.

17 Multilocus Sequence Determination and Analysis

Paul De Vos
Laboratory of Microbiology, Ghent University, Gent, Belgium
E-mail: paul.devos@ugent.be

♦♦

CONTENTS

♦♦♦♦♦♦ I. INTRODUCTION

Multilocus Sequence Analysis

Since the late 1960s and early 1970s, when molecular (DNA-related) methods were introduced in microbiological research, bacterial taxonomy (also called bacterial systematics) has become more oriented towards genomic-based approaches. It is now generally accepted that a reliable taxonomic system for prokaryotes must be based on phylogeny. The phylogenetic backbone of the taxonomic system can be retrieved from 16S rRNA gene sequence comparisons (Woese, 1987). However, in many cases, these sequences do not allow species delineation, for which DNA:DNA hybridization (DDH) is still regarded as the gold standard (Wayne *et al.*, 1987; Stackebrandt *et al.*, 2002). The rRNA genes are indeed too conserved in evolution to allow species discrimination according to the present species concept. Other genes that evolve faster show a higher discriminatory power and hence reveal a finer taxonomic resolution. These genes, so-called housekeeping genes, are also universal to some extent and belong to the core of the genome. Furthermore, in the present species concept (and also in the higher taxonomic ranking), a phenotypic discrimination is also a prerequisite for taxonomic descriptions.

A prokaryote taxonomy needs a relatively strict nomenclature that complies with the Bacteriological Code (Lapage *et al.*, 1992). The allocation of a prokaryote strain or a set of prokaryote strains to one of the existing taxa or to a new taxon depends on the results of characterization at the strain level. This

METHODS IN MICROBIOLOGY, VOLUME 38
0580-9517 DOI 10.1016/B978-0-12-387730-7.00017-6

relatively complex process comprises both genotypic and phenotypic elements (Tindall *et al.*, 2010). However, at the same time, more profound phylogenetic insights are developing very fast due to (i) the increasing number of sequenced genomes of prokaryotes (more than 1500 at this moment, July 2011) and (ii) proliferating metagenomic studies on a larger scale aiming at understanding the composition of the microbial communities of ecosystems.

As a consequence, the discovery of new prokaryote diversity is speeding up as the number of prokaryote species is orders of magnitudes larger than previously envisaged. Furthermore, the present species concept does not reflect the enormous ecological variation of prokaryote populations for which a more ecological-oriented species concept, the so-called ecotype concept of Cohan would be better suited (Cohan, 2001, 2002, 2006). It is beyond the scope of this contribution to discuss whether detailed analysis and knowledge of whole genome sequences (WGS) will allow us to understand overall genome organization and regulation of gene expression that is at the basis of the phenotypic functionalities needed to fully understand the phenotype. Whether this is really congruent with the ecological behaviour of the overall organism in the real world and not just a reflection of the genomic potentialities needs also to be established. It is also beyond the scope of this contribution to discuss in depth the prokaryote species concept itself. Since decennia, the species concept has been the subject of a debate that has still not yet come to a final consensus among microbiologists.

When combining the present speed of discovery of new species (roughly between 600 and 1000 per year during the last 5 years) with the existing species concept and rules and guidelines of nomenclature of prokaryotes, it would take us hundreds of years to inventorize the whole prokaryote diversity, which is estimated to be most probably higher than 10^6 species (Curtis *et al.*, 2002). Then, a second corollary is that less than 1% of this diversity is presently described, a number that was already suggested by Hugenholtz et al. (1998) based on the ratio of cultivated versus uncultivated diversity. Microbiologists need to find a way to accurately handle the vast amounts of emerging molecular data within the existing taxonomic system to avoid the creation of a chaotic situation in which the general goals of prokaryote taxonomy are no longer achieved. Indeed, prokaryote taxonomy, although artificial to some extent, facilitates communication among microbiologists on the occurrence of and the characteristics of groups of prokaryotes (taxa). It also allows for the identification of unknowns in order to understand their potential roles in diseases (e.g. needed for quarantine, epidemiological aspects) as well as in industrial and bioremediation processes. Furthermore, we cannot underestimate the role of the evolution of prokaryote diversity and how it functions as a marker for larger scale ecological changes due to climate change or important pollution events.

The debate on the existence of 'prokaryote species' in the nature and the 'prokaryote species concept' that came along with it has been going on for many decades, and attempts to formulate bacterial (we should now refer to as prokaryote) species definitions or concepts were undertaken (Breed *et al.*, 1957; Staley and Krieg, 1984; Wayne *et al.*, 1987; Stackebrandt *et al.*, 2002). At least there is a general agreement that a prokaryote species comprises a group of strains that show a certain degree of genotypic and phenotypic similarities.

Based on the variation of community genes, a vertical zonation of taxonomic groups and functionalities has been shown for planktonic microbial communities (DeLong *et al.*, 2006). Whether these distinct groups also coincide with our present species concept is still to be proven, and the effect of these community studies on prokaryote taxonomy is still under discussion. It is clear that the discussions about the eventual taxonomic rank of these so-called ecotypes should also be taken into consideration (Cohan, 2001, 2002, 2006). Konstantinidis *et al.* (2006) have suggested that we wait for further data on gene content and ecological distinctiveness before a more stringent species definition can be proposed, although the data suggest that species delineation of pathogenic variants of *Burkholderia* and *Shewanella* that do form distinct clusters differ from the broader clusters containing the environmental representatives.

Microbiologists who are less familiar with the science of taxonomy of prokaryotes are certainly encouraged to read more on the evolution and implementation of molecular data and also on the so-called polyphasic taxonomy of prokaryotes that emerged in the seventies (Colwell, 1970), as it is still regarded as the firm basis for the characterization and description of new prokaryote strains (Vandamme *et al.*, 1996). This polyphasic approach to prokaryote taxonomy stems from three basic principles, namely (i) the classification of prokaryotes into groups, (ii) the naming of these groups and (iii) the identification of unknown isolates as members of one of the named groups. If the identification process results in the discovery of a new species, this new taxon can be described using a polyphasic approach which is a combination of genotypic and phenotypic data (see also Tindall *et al.*, 2010).

Taking into account that (i) DNA-based methodologies, in particular whole genome sequencing, can underpin prokaryote phylogenies more accurately than in the past; (ii) the gold standard for species delineation still relies on the laborious and time-consuming methodologies of DDHs of genomic DNA; and (iii) the speed with which new taxa of prokaryotes are presently being discovered, a faster and less laborious molecular species discriminatory tool than the DDH approach is needed. It has already been suggested by Stackebrandt *et al.* (2002) that sequencing of well-selected housekeeping genes (=multilocus sequences) could replace the DDH. The advantage of gene sequence analysis is that the data are easily exchangeable between laboratories and can be compiled in general databases.

This contribution aims at providing methodologies to implement multilocus sequences in such a way that they do not conflict with the present view on taxonomy of prokaryotes but rather integrate them into prokaryote systematics.

♦♦♦♦♦♦ II. MULTILOCUS SEQUENCING IN PRACTICE

A. The Choice of Strains and Genes for Study

1. Strain choice

Multilocus sequence typing (MLST) was originally developed for studying groups of prokaryotes at the infra-specific level and is therefore applied in epidemiological studies (Maiden *et al.*, 1998). The questions to be answered in these kinds of studies relate to the clonality of the strains. Hence, very closely related strains retrieved from a local (hospital) or larger scale disease outbreaks are linked with well-defined symptoms of diseased humans, animals or plants. The aim of such studies is that by typing the set of strains the relationship between the causative strains of symptoms on the one hand and their spreading or origin on the other can be traced. There are many examples of MLST schemes, e.g. on Melioidosis and Glanders (Godoy *et al.*, 2003) that have been introduced in such epidemiological studies. Often these MLST schemes are linked with a diagnostic website application. The websites (http://pubmlst.org/) – hosted by the University of Oxford – and more specifically for *Escherichia coli* (http://mlst.ucc.ie/mlst/dbs/Ecoli) – hosted by University College Cork – are of a high standard. The number of genes to be sequenced often turns out to be a compromise between practicalities and the discriminatory power that is required; typically seven loci appear to be the norm (Feil *et al.*, 2004; Gevers *et al.*, 2005; Maiden *et al.*, 1998). MLST has also turned out to be of importance for species discrimination between phylogenetically closely related species and to discriminate between entities that are more like clonal clusters that are phylogenetically not very diverse (Godoy *et al.*, 2003; Hanage *et al.*, 2005a, 2005b).

The use of multilocus sequencing to achieve species and supra species (generic, family) discrimination and to retrieve taxonomic relationships should be referred to as multilocus sequence analysis (MLSA) (Gevers *et al.*, 2005). The result of such studies should be an MLSA scheme that can be used for identification at the species level. A prerequisite for a successful result is the choice of the biological material to be included in the study. If one takes into account that a prokaryote species is represented by a group of strains, different strains with considerable phenotypic variation and, if possible, from different ecological niches must be included for each species. Furthermore, the strains must be well characterized taxonomically, which means that they must coincide with the prokaryote species concept that assigns them to a given species, according to the present species concept based on DDH or (if multiple WGS of closely related species are available) on average nucleotide identity (ANI), of which a 94–95% value corroborates the 70% DDH delineation (Konstantinidis and Tiedje, 2005). It is difficult to find sets of strains that fulfil these conditions. Even microbial resource centres (culture collections) generally keep only a very limited number of strains per species in their holdings, and if there are many strains per species, they are often insufficiently characterized even though they have been identified and named. Presently, most species are described on the basis of a single isolate (e.g. the February 2011 issue of IJSEM (http://ijs.sgmjournals.org/) contains 23 new species of which only eight were based on more than one strain). Such situations of single strain species do not allow the implementation of the strain variation in the phenotypic aspect of the description, and the species genetic variation remains unknown. As a consequence, species-related MLSA groups cannot be retrieved and their extent cannot be determined. So, the identification of new strains becomes impossible and the taxonomic system deteriorates. It is beyond the scope of this contribution to include an in-depth discussion of the problems associated with the evolution of the taxonomy of prokaryotes, but this should be taken into consideration in the future by the ICSP (International Committee on Systematics of Prokaryotes).

Fortunately, MLSA results in a clustering of the strains (see Section D) based on sequence trees of either a single gene or concatenated genes into units (also called geno groups) that need to be evaluated for their taxonomic status. Based on MLS trees alone, it is indeed difficult, if not impossible, to decide on the taxonomic status of the distinct clusters observed as the cut-off levels of these clusters must be verified using the 70% DDH rule (or in the future probably more and more the 94% ANI value).

This MLSA approach allows in-depth taxonomic studies of phylogenetically closely related species/genera that are at present composed of a large number of strains that have been misidentified and hence misnamed (i.e. do not conform to the present species definition). Before starting a taxonomic study, it is important to collect a set of strains that is allocated at the genus level (in most cases, this is relatively reliable based on partial 16S rRNA gene sequences) and that includes the type strains of the already described species of the genera under consideration as well as strains that originate from a variety of ecological niches preferentially with a geographical spread.

MLS also permits the grouping of strains after extensive isolation campaigns into 'MLSA clusters' together with type strains of closely related species (appointed after comparative 16S rRNA gene sequence analysis, which clearly remains the anchor between phylogeny and taxonomy).

Ultimately 'polyphasic characterization' is needed before descriptions of new taxa and/or reshuffling of existing classification become possible.

2. Candidate genes for MLSA approach

It is generally accepted that the driving forces behind evolution are various factors that can be summarized as effects of the environment. It would be far outside the scope of this chapter to discuss here the factors that affect environmental adaptations except to say that they involve temperature, pH, composition and ratio of organic and inorganic components, presence and action of eukaryote/prokaryote organisms etc. When it comes to prokaryotes, specific conditions for speciation must be seen on a micro-geographic scale. Molecular phenomena that result in adaptations of the prokaryote genome, although different from those of most eukaryote organisms, also show certain parallelisms. These phenomena result in the prokaryote genome shaping and concern chromosomal rearrangements, nucleotide substitutions, gene duplication, gene deletion, homologous recombination and horizontal gene transfer. For a more detailed overview and their effects on genetic evolution of prokaryotes, see the work of Coenye *et al.* (2005). The effect of lateral gene transfer (LGT) on the adaptive capability to the environment and its frequency of occurrence have led to the opinion among (at least certain) microbiologists that the genome of a prokaryote species is many a time more extended than the genomes of the constituting strains the so-called pan genome (Medini *et al.*, 2008). Although the debate on the effect of LGT on prokaryote phylogenies is far from finished, comparative genome analyses support the existence of genes that are (i) recalcitrant towards LGT; (ii) less resistant to LGT; and (iii) very sensitive to LGT of which the latter is linked to the flexible adaption of the prokaryote genotype to the environment. These genes that are sensitive to LGT are considered accessory or auxiliary genes and are at the origin of environmental adaptation, while the others are considered so-called 'core genes' and are involved in more primary functions of the prokaryotic cell like replication, transcription and translation. In a stable, more molecular-based taxonomic system that reflects the genetic evolution, core genes are better candidates than the accessory genes which reflect the 'ecotype' more and are therefore more appropriate candidates to help discriminate host–pathogen relationships in, e.g. epidemiological studies, i.e. at the infra-specific level.

So the first criterion for selecting valuable candidate genes to build MLSA databases for taxonomic purposes concerns those genes that are not or only at low frequencies subjective for LGT. Nowadays it is accepted that these genes are mostly key components of the basic metabolic functions. The best candidates among these genes should ideally be retrieved from WGS data on the conditions that we have sufficient insight in the metabolic functions of the genes and that we can consider a range of completely sequenced genomes that covers the phylogeny of the group of prokaryotes one is interested in. WGS of these so-called reference genomes are presently only rarely available to meet these criteria.

The candidate genes must further meet the following criteria: (i) 'single copy gene principle', which is a prerequisite to avoid sequencing of paralogous genes that are the result of gene duplications and which thus may have evolved to a different functionality; (ii) the principle of genetic independency versus other MLSA candidate genes in order to limit the chance that LGT has occurred among multiple genes; (iii) the selected genes should be present in all strains that belong to the taxa under study (i.e. belong to the core of the genome of that particular group of prokaryotes); and (iv) the sequence variability of the genes should coincide with the overall genome similarity. We refer the interested reader to the work of Zeigler (2003) for more details on this selection procedure. Zeigler developed a useful approach to yield high-scoring gene candidates by comparing WGS analysis with similarities of single orthologous genes (Zeigler, 2003). In the contribution of Debruyne *et al.* (2008), a possible selection procedure is described for selecting the best suited candidate genes for MLSA studies or identification schemes. In brief, after the determination of the core genes, a number of software tools allow for the selection of the best performing genes with tree topologies that come close to the whole genome-based tree. Selected genes can be used to eventually construct concatenated trees, if single-gene trees do not meet the required whole-genome-like resolution. The Cluster of Orthologous Genes database (Tatusov *et al.*, 1997) can be consulted for either checking the categories into which different genes can be assigned (e.g. metabolism, cellular processing and signalling etc.) or confirming the genes chosen are sufficiently independent. Recently, a selective flow chart to find *in silico* the most appropriate genes for an MLSA approach of subclass of the Actinobacteridae has been published (Adékambi *et al.*, 2011). The study is based on 50 different WGS of representatives of this subclass and takes into account the above-mentioned criteria. It is one of the first examples where such a large phylogenetic group was approached by MLSA.

The final two criteria for selection are (i) the number of genes to be included and (ii) the length of the sequences. Although at the start of the sequencing process, a relative broad approach towards a number of candidate genes to sequence is defendable, experience shows that often the primers developed (see Section B) are not as effective as expected on the basis of their design to allow amplification of a maximal number of templates; it is agreed that for MLST, seven genes should be sufficient to develop a good typing scheme (Maiden *et al.*, 1998). The study of Konstantinidis *et al.* (2006) showed that the three best performing marker genes do allow for the construction of a phylogenetic tree that is in good agreement with whole genome tree for the same organisms (*E. coli*). To obtain sufficient variation, preferably the length of the sequences to be used should be around 900 bp (or more) (Adékambi *et al.*, 2011). MLSA schemes that are based on sequences that are too short and/or on too few strains should be approached with the necessary caution concerning their reliability as identification tools. A flow chart for the selection approach described is given in Figure 1.

In practice, however, candidate genes for MLSA studies are selected on the basis of experience with other MLSA studies because of lack of sufficiently suited reference genomes. This, of course, may ignore the basic principles that have been discussed above, and there is no doubt that LGT will interfere with the taxonomic scheme, and hence the identification results if the loci that are sequenced do also meet the other selection criteria (independent location of the genes on the genome, sequence lengths definitely below 900 bp, very limited number of genes) at low stringency.

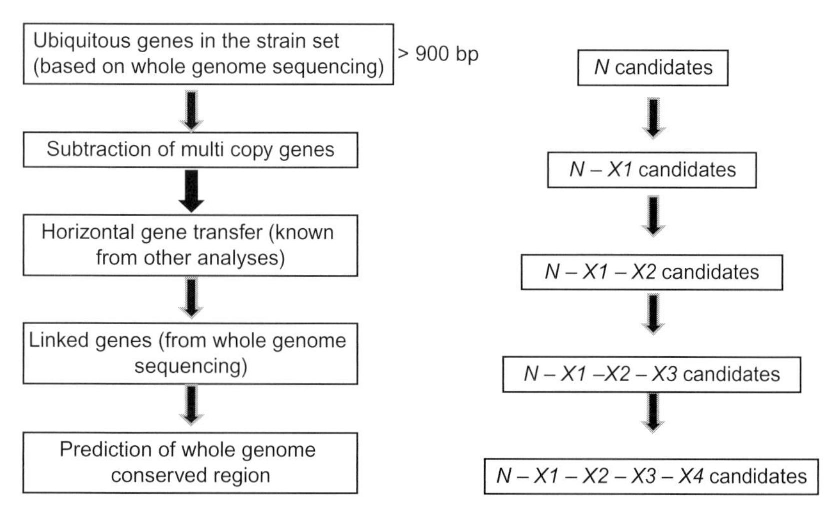

Figure 1. Different selection steps for candidate loci for multilocus sequencing based on Adékambi *et al.* (2011).

B. Development of Primers

The primer design can be performed *in silico* using various software packages (Geneious software [Drummond *et al.*, 2011], BioNumerics [Applied Maths NV, Belgium], primer 3 blast [from NCBI to find specific primers: http://www.ncbi.nlm.nih.gov/tools/primer-blast/] etc.).

Based on the *in silico* selection of genes (Section II.2) a further automated primer design can be implemented in order to obtain a set of primers with low degeneracy (Debruyne *et al.*, 2008). The primers must bind at conserved regions of the aligned selected genes under optimal conditions for PCR that generate an amplified fragment of around 900 bp long to give the best discriminative power. Of course, the primers must match with two conserved regions on the target. It is clear that the basis for a successful *in silico* study starts with the availability of sufficient WGS (reference genomes) of representatives of the taxon (species, genus and, if possible, family, depending on the phylogenetic diversity one wants to cover) under study.

Currently (mid-July 2011), 1566 prokaryote genomes are fully sequenced and publically available (http://www.genomesonline.org/) and another 6052 are being sequenced. Only about 50% of the sequenced genomes are underpinned by the biological material belonging to the holdings of public culture collections. From these, only 50% are type strains. This comes to about 400 type strains for which the WGS can be examined. This represents 5% of the presently classified prokaryote diversity and is therefore far less than the estimated overall prokaryote diversity (Section I) (data obtained from the StrainInfo team: http://www.StrainInfo.net).

So, in many cases, there are not sufficient WGS available and one has to work with relatively biased (i.e. degenerated) primers that allow for the amplification in a first round of as many as possible envisaged templates in order to cover the diversity of the taxa under study in a maximal way. One might then eventually obtain a sequence of amplicons that is much longer than the required sequences for the MLSA study and that must be narrowed down to specific amplicons in a second PCR round (i.e. the sequencing reaction), sometimes the result of a multiplex PCR with specific sequencing primers. It is common to have to work with more than one primer set to cover the complete strain set because of the insufficient coverage of the target templates.

The following general recommendations must be taken into account for obtaining this second primer set that will result in a more selective amplification of the target, i.e. the selected gene: (i) the melting temperature of the primer must be somewhere between 50 and 60°C (% G+C of the primer between 45 and 55% for a 20–25 mer), (ii) dimerization must be excluded (usually considered by primer design software), (iii) hairpin formation must be avoided, (iv) the binding should be specific, i.e. no secondary binding sites in the template, (v) there should be low specificity at the 3′ end to avoid mismatching and (vi) the melting temperature of the two primers should be within 5°C (the closer the better). In addition, the primers should be 'sticky' at their 5′ end and not at their 3′ end to avoid mispriming. Stickiness is improved by local high G+C content. Once the sequencing primer set(s) is designed, the PCR reaction must be optimized and this can still take some time and trial and error before the optimal combination of conditions for annealing, elongation and denaturation is achieved. These conditions may depend on the PCR apparatus used. PCR products obtained must be checked for their expected length via agarose gel electrophoresis.

A first screening of representatives of the various species in the taxon (e.g. the type strains), which are at the borders of the genus or family according to their 16S rRNA gene sequences, and thus cover the phylogenetic diversity, is a good indication for the suitability of the designed primer set(s). A number of primer pairs are given in Table 1 as examples; however, their application must be used with scrutiny because of the above-explained limitations.

C. Generating Sequences and their Quality Control

Generating a vast number of sequences, once the primer set(s) have been tested and approved, is more a routine technical issue that can be either outsourced or performed in-house if one has access to a sequencer. How consensus sequences are obtained is described in various papers (e.g. Naser *et al.*, 2005a, 2005b). Normally the sequences obtained are then stored in local databases and alignments for further analysis.

In most cases, the selected genes will code for proteins that have a key function in the overall metabolism of the organisms, i.e. coding genes that ultimately translate to a functional protein. The active molecules corresponding to the selected genes are mostly well known and described by primary, secondary and tertiary structures. It is of utmost importance that the DNA sequences that are obtained are checked for their coding performance. This can easily be performed via freely available software (Altschul *et al.*, 1990), and the quality of the amino acid sequence can be determined. Indeed, the match with the expected protein should reach a relatively high score. If this is not the case, the original sequence may be of insufficient quality because of deletions, stop codons etc. It is a prerequisite for further analysis that this quality control of the DNA sequences is underpinned by the expected gene product. If unexpected results are obtained, the sequencing procedure should best be repeated.

D. Analysis of the Sequences and Taxonomic Evaluation of the Results

In almost all cases, the sequences obtained are used to construct (phylogenetic) trees. Various trees can be constructed on the basis of different algorithms. It is beyond the scope of this contribution to discuss the impact of the algorithms used and the best suited algorithm to construct the most stable tree. I advise to use maximum likelihood or parsimony trees, although one has to realize that any tree is at some point biased by the quality of the sequences included, the penalties for gaps in the sequences etc. It is very important that all sequences have the same length before trees are calculated. Alignment of the sequences is another important issue. Although many software packages allow for automatic pairwise and multiple alignments, a visual inspection is in most cases required.

Table 1. Examples of primers used for gene amplification and/or sequencing of envisaged genes, their corresponding site on the gene (if known) and the references/or website if reference is unclear

Bacterial target	Gene	Primer name	Starting point/ stretch	Reference 1	Reference 2
Enterococcus, *Lactobacillus*, and other LAB	*pheS*	pheS-21-F	557	Naser et al. (2005) (a)	Naser et al. (2007)
	pheS	pheS-22-R	1031		
	pheS	pheS-23-R	968		
	rpoA	rpoA-21-F	1		
	rpoA	rpoA-23-R	802		
Enterococcus and other LAB	*atpA*	atpA-20-F	97	Naser et al. (2005) (b)	
	atpA	atpA-22-F	397		
	atpA	atpA-23-R	397		
	atpA	atpA-24-F	781		
	atpA	atpA-25-R	781		
	atpA	atpA-27-R	1219		
Vibrionaceae	*atpA*	atpA-01-F	37	Thompson et al. (2007)	
	atpA	atpA-02-F	760		
	atpA	atpA-03-R	760		
	atpA	atpA-04-R	1554		
	atpA	atpA-05-R	1204		
	atpA	atpA-06-F	55		
Leuconostoc, *Oenococcus*, and *Weissella*	*dnaA*,	dnaA445-F	445	Chelo et al. (2007)	
	dnaA,	dnaA445-Fs	445		
	dnaA,	dnaA1253-R	1253		
	dnaA,	dnaA1253-Rb	1253		
	dnaA,	dnaA1253-Rs	1253		
	gyrB,	UP-1-F	370		Yamamoto and Haryama (1995)
	gyrB,	UP-1s-F	370		Yamamoto and Haryama (1995)
	gyrB,	gyrBUP-1b-F	370		
	gyrB,	gyrB959-R	959		
	gyrB,	Gyr1412-R	1412		
	gyrB,	UP-2-R	1634		Yamamoto and Haryama (1995)
	gyrB,	UP-2S-R	1634		Yamamoto and Haryama (1995)
	rpoC	208-F	181		Morse et al. (1996)
	rpoC	rpoC233F	233		

	rpoC	1064F	1034		Morse et al. (1996)
	rpoC	rpoC1342F	1342		
	rpoC	rpoC1053R	1053		
	rpoC	1391R	1361		Morse et al. (1996)
	rpoC	rpoC1502F	1480		
	rpoC	rpoC2240F	2240		
	rpoC	rpoC2254R	2254		
	rpoC	rpoC2867R	2867		
	rpoC	3768R	3219		Morse et al. (1996)
	rpoC	3937R	3385		Morse et al. (1996)
	dnaK	DnaK1F	1		
	dnaK	DnaK7F	7		
	dnaK	DnaK337F	337		
	dnaK	DnaK592F	592		
	dnaK	DnaK611R	611		
	dnaK	DnaK1255R	1255		
	dnaK	DnaK1500F	1500		
	dnaK	DnaK1520R	1520		
	dnaK	DnaK1812R	1812		
	dnaK	DnaK1819R	1819		
Staphylococcus	*rpoB*	30F	30–50	Drancourt and Raoult (2002)	
	rpoB	192F	192–211		
	rpoB	806F	806–827		
	rpoB	920F	920–942		
	rpoB	1165F	1165–1186		
	rpoB	1302F	1302–1327		
	rpoB	1450F	1450–1471		
	rpoB	1741F	1741–1761		
	rpoB	1850F	1850–1870		
	rpoB	2245F	2245–2271		
	rpoB	2309F	2309–2329		
	rpoB	2334F	2334–2359		
	rpoB	2412F	2412–2432		
	rpoB	2534F	2534–2555		
	rpoB	2663F	2663–2681		
	rpoB	2995F	2995–3119		
	rpoB	2924F	2924–2944		
	rpoB	3200F	3200–3221		
	rpoB	3498F	3498–3520		

(*Continued*)

Table 1. (Continued)

Bacterial target	Gene	Primer name	Starting point/ stretch	Reference 1	Reference 2
	rpoB	3550F	3550–3570		
	rpoB	3843F	3843–3864		
	rpoB	4494F	4494–4514		
	rpoB	1759R	1779–1759		
	rpoB	1460R	1480–1460		
	rpoB	1910R	1930–1910		
	rpoB	2309R	2329–2309		
	rpoB	2334R	2354–2334		
	rpoB	2432R	2452–2432		
	rpoB	2573R	2593–2573		
	rpoB	2892R	2912–2892		
	rpoB	2915R	2935–2915		
	rpoB	2924R	2944–2924		
	rpoB	2995R	3015–2995		
	rpoB	Cm32b	3211–3191		
	rpoB	3321R	3341–3321		
	rpoB	3610R	3630–3610		
	rpoB	4139R	4159–4139		
	rpoB	4502R	4523–4502		
	rpoB	4508R	2428–4508		
	rpoB	4871R	4891–4871		
	rpoB	5000R	5021–5000		
	rpoB	5018R	5040–5018		
	rpoB	5030R	5051–5030		
	rpoB	5041R	5061–5041		
	rpoB	5085R	5105–5085		
	rpoB	5188R	5208–5188		
Escherichia coli	*aspC*	aspC-F4	57–76	http://www.shigatox.net/ecmlst/protocols/index.html	
	aspC	aspC-R7	631–650		
	clpX	clpX-F6	262–281		
	clpX	clpX-R1	914–933		
	fadD	fadD-F6	768–787		
	fadD	fadD-R3	1328–1347		
	icdA	icd-F2	352–371		
	icdA	icd-R2	1001–1020		
	lysP	lysP-F1	36–55		
	lysP	lysP-R8	644–663		

	mdh	mdh-F3	130–152	
	mdh	mdh-R4	760–779	
	uidA	uidA-277F	277–300	
	uidA	uidA-934R	911–934	
Campylobacter jejuni	*asp*	asp-A9-F	NS*	Dingle et al. (2001)
	asp	asp-A9-R	NS	
	asp	asp-S3-F	NS	
	asp	asp-S6-R	NS	
	gln	gln-A1-F	NS	
	gln	gln-A2-R	NS	
	gln	gln-S3-F	NS	
	gln	gln-S6-R	NS	
	glt	glt-A1-F	NS	
	glt	glt-A2-R	NS	
	glt	glt-S1-F	NS	
	glt	glt-S6-R	NS	
	gly	gly-A1-F	NS	
	gly	gly-A2-R	NS	
	gly	gly-S3-F	NS	
	gly	gly-S4-R	NS	
	pgm	pgm-A7-F	NS	
	pgm	pgm-A8-R	NS	
	pgm	pgm-S5-F	NS	
	pgm	pgm-S2-R	NS	
	tkt	tkt-A3-F	NS	
	tkt	tkt-A6-R	NS	
	tkt	tkt-S5-F	NS	
	tkt	tkt-S4-R	NS	
	unc	unc-A7-F	NS	
	unc	unc-A2-R	NS	
	unc	unc-S5-F	NS	
	unc	unc-S4-R	NS	
Gluconacetobacter and related taxa	*dnaK*	dnaK-01-F	NS	Cleenwerck et al. (2010)
	dnaK	dnaK-02-R	NS	
	dnaK	dnaK-03-F	NS	
	dnaK	dnaK-04-R	NS	
	groEL	groEL-10-F	NS	

(*Continued*)

Table 1. (Continued)

Bacterial target	Gene	Primer name	Starting point/ stretch	Reference 1	Reference 2
	groEL	groEL-11-R	NS		
			NS		
	groEL	groEL-12-F	NS		
	groEL	groEL-13-F	NS		
	groEL	groEL-14-R	NS		
	groEL	groEL-15-R	NS		
	rpoB	rpoB-01-F	NS		
	rpoB	rpoB-02-R	NS		
	rpoB	rpoB-03-F	NS		
	rpoB	rpoB-04-F	NS		
	rpoB	rpoB-05-R	NS		
	rpoB	rpoB-06-R	NS		
	rpoB	rpoB-07-R	NS		
	recA	recA_A_F	NS		
	recA	recA_A_R	NS		
	thrC	thrC_A_F	NS		
	thrC	thrC_A_R	NS		
	gltA	gltA_A_F	NS		
	gltA	gltA_A_R	NS		
Bradyrhizobium	atpD	atpD352Fb	352–372	Rivas et al. (2009)	
	atpD	atpD871Rb	890–871		
	recA	recA63Fc	63–85		
	recA	recA504Rc	504–523		
	gyrB	gyrB343Fc	343-364		
	gyrB	gyrB1043Rc	1061–1043		
	gyrB	gyrB846Fc	846–865		
	gyrB	gyrB846Rc	865–846		
	rpoB	rpoB83Fc	83–103		
	rpoB	rpoB1540R	1540–1557		
	rpoB	rpoB1061F	1061–1080		
	rpoB	rpoB1061Rc	1081–1061		
	rpoB	rpoB456Fc	456–475		
	rpoB	rpoB458Rc	478–458		
	dnaK	dnaK1466Fd	1466–1488		
Pantoea	*gyrB*	gyrB 01-F	NS	Brady et al. (2008)	
	gyrB	gyrB 02-R	NS		

	rpoB	rpoB CM7-F	NS	
	rpoB	rpoB CM31b-R	NS	
	atpD	atpD 01-F	NS	
	atpD	atpD 02-R	NS	
	infB	infB 01-F	NS	
	infB	infB 02-R	NS	
	gyrB	gyrB 07-F	NS	
	gyrB	gyrB 08-R	NS	
	rpoB	rpoB CM81-F	NS	
	rpoB	rpoB CM81b-F	NS	
	rpoB	rpoB CM32b-R	NS	
	atpD	atpD 03-F	NS	
	atpD	atpD 04-R	NS	
	infB	infB 03-F	NS	
	infB	infB 04-R	NS	
Ensifer	*gyrB*	gyrB343F	343–364	Martens et al. (2008)
	gyrB	gyrB1043R	1061–1043	
	gyrB	gyrB946F	846–865	
	gyrB	gyrB846R	865–846	
	rpoB	rpoB83F	83–103	
	rpoB	RpoB1061R	1081–1061	
	rpoB	RpoB456F	456–475	
	rpoB	RpoB458R	478–458	
	gap	gap109F	109–130	
	gap	gap940R	958–940	
	gap	gap528F	528–547	
	gap	gap528R	547–528	
	atpD	atpD352F	352–372	
	atpD	atpD871R	890–871	
	pup	Pup913F	913–930	
	pup	Pup1473R	1490–1473	
	rspA	rpsA350F	350–369	Martens et al. (2007)
	rspA	rpsA1582R	1582–1599	
	rspA	rpsA941F	941–960	
	rspA	rpsA941R	941–960	
	rspA	rpsA1127F	1127–1147	
	rspA	rpsA1127R	1127–1147	

*NS: not specified.

Researchers who are not familiar with phylogenetic tree construction and the use/bias of software packages are advised to read *Phylogenetic Trees Made Easy* (Hall, 2008).

The construction of the trees will only be relevant if sufficient representatives of each of the taxa (i.e. species) are available to cover their molecular variability. In many examples from the literature, this is not the case because (i) species descriptions are merely based on single isolates that are then *ipso facto* the type strains and researchers cannot obtain other strains and (ii) generating sequences of coding genes for different strains, even of the same species, is not always straightforward and the experimental efforts to obtain a sufficient number of sequences per gene may be laborious and time consuming. However, tree construction based on MLS of (type) strains only will not result in reliable species delineation. Such trees can only be indicative for further study; prokaryote taxonomists should realize that these kinds of data are of limited value, even if they support the existing taxonomy of the species at first sight. Indeed, in the long run a chaotic situation will be created for identification purposes because the boundaries of the species in such MLSA schemes are not determined. These species borders, whether they are interpreted according to the present species concept or a future concept, can only be delineated in the situation that the variation within each of the species of the genera or families under study is known, i.e. when a variety of strains (to put forward an absolute number is not possible, but at least 10 has been recommended) per species are included from different geographic/ecological origins. Only then an MLSA scheme that would allow identification at the species level can be the final outcome of such studies. Unfortunately, in many cases, the taxonomic approach as suggested above is not reached yet, even for those taxa for which a variety of strains is seemingly available via the public culture collections. Indeed, although named, many non-type strains in public culture collections are weakly characterized (in many examples only by partial 16S rRNA gene sequence or a few trivial phenotypic characteristics) and therefore misnamed (according to the present prokaryote species concept) at the time of deposition and thus erroneously assigned at the species level.

Including such strains in an MLSA scheme will provide a grouping that by no means corresponds to the given names. The situation described above generally does not hold for organisms of clinical importance if one works solely with clinical isolates. Indeed, if environmental isolates are included, the taxonomic situations again become unclear (Konstantinidis *et al.*, 2006).

The microbiologist who is confronted with a lot of misnamed and weakly characterized strains can make use of the MLSA approach to clarify the taxonomy of existing weakly characterized groups as well as of newly isolated strains. A pragmatic approach can be as shown in Figure 2.

Once the strains of the prokaryote group under study are selected according to criteria mentioned above, type strains of closely related taxa must be added and candidate MLS genes must be selected (using the above-described criteria), after which primers that have been designed can be tested. After the sequence protocols have been optimized, sequencing can start at high throughput. Of course, if one is working in a group of prokaryotes for which reasonably good

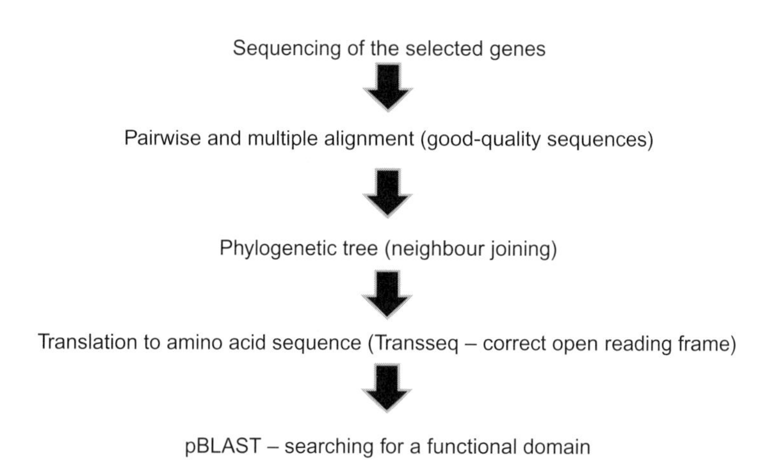

Figure 2. Analysis of multilocus sequences: a possible work flow.

MLSA schemes have been published, one can take advantage of the knowledge to build further on the existing data.

In the first phase, all sequences must be checked for quality according to the criteria discussed above and low-quality sequences must be removed from the database and repeated before further analysis. For each sequenced gene, a tree can be obtained showing a clustering of the strains that, at first sight, will be confusing because of incorrectly or preliminarily given names to the strains (this is at this point of no importance) in most cases. Normally, the sequences of the different loci will show a different resolution. This can be deduced from the branching level of the clusters when comparing the trees obtained for the different genes. The least complex situation occurs when each of the included type strains is located in different clusters in all of the gene trees. In most cases, however, this is not the case; it occurs that two or more type strains fall in the same cluster.

The next step is to choose the tree that has the highest visual resolution so that visually delineated clusters are appointed. On the resulting grouping, a TaxonGap analysis (Slabbinck *et al.*, 2008) can be performed on this first single-gene sequence tree. The composition of the clusters (groups) for the TaxonGap analysis is then as deduced from this first tree (with the highest resolution visually). When TaxonGap analysis is performed, the results are plotted in a graph for which all pairwise similarties of the sequences (pairwise sequence similarity matrix) from the first tree are used (e.g. one of the genes represented in Figure 3).

The graph can be interpreted as follows: if the clusters (chosen visually) are well delineated, the inner grey zone (i.e. the intra-group dissimilarity) will be smaller than the outer black zone (i.e. the inter-group dissimilarity). If this is not the case, the process must be repeated by lowering the cut-off level in the tree on a visual basis. This is somewhat trial and error, but in practice it is mostly straightforward. Once the graphic presentation of the clustering for that particular gene is as described, the following questions then arise: (i) is this

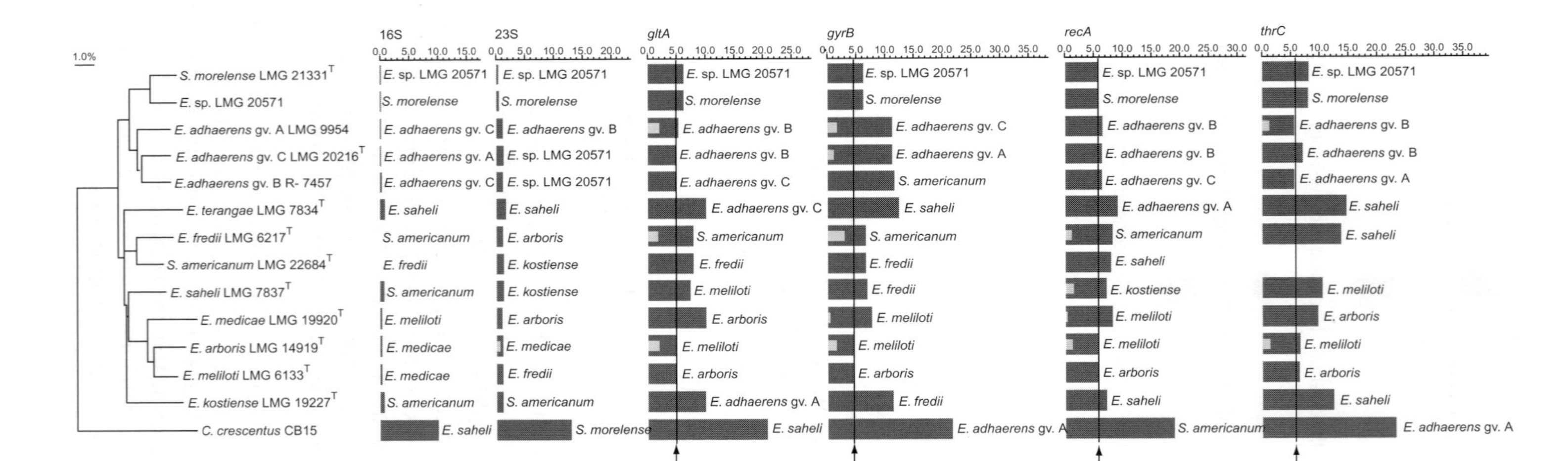

Figure 3. Matrix of heterogeneity (presented as light grey horizontal bar) and separability (presented by dark grey horizontal bar) values with different OTUs as matrix rows and different biomarker genes as matrix columns. The species/genomovars of the genus *Ensifer* were ordered according to their phylogenetic position in the maximum likelihood tree calculated from concatenated sequences of their type strains (when available) or another representative strain (see tree on the left). For each OTU and each biomarker, the closest neighbour (i.e. the taxon with the smallest separability with respect to this OTU and this biomarker) is listed at the right side of the dark grey bar For each biomarker, the vertical black line denotes the smallest separation. (from Slabbinck *et al.*, 2008, reproduced with permission).

clustering stable based on this single locus sequence? and (ii) what is the taxonomic meaning of the clustering?

To address the first question, a TaxonGap analysis based on the clusters of the already analysed gene is performed for the other loci. If this also results in a clustering for which the graphic presentation in TaxonGap is such that the inner grey zone is smaller than the outer black zone, we can conclude that there is a stable clustering of the strains in all gene loci that were sequenced (this is the ideal situation). However, it often occurs that some strains which cluster together for one gene are retrieved from different clusters for another gene. If it concerns a limited number of strains and if the phenomenon occurs for a limited number of genes (one or two), a concatenated tree construction will probably mask this aberrant behaviour of these strains. The concatenated clustering can then be considered the working base for further taxonomic analysis to address the second question.

Addressing the second question concerns the taxonomic meaning of the clustering. Hence, the clusters must be taxonomically evaluated taking into account (i) the rRNA gene sequence similarities, (ii) DDH data (according to the 70% rule) or (iii) WGS (according to the 94–95% ANI value) from a selection of the strains per cluster. Since type strains were included, they are the natural first choice as reference points for the experimental set-up. Various situations can occur. There are (i) none, (ii) one or (iii) more than one type strain in a given cluster.

For those clusters that do not contain type strains, the 16S rRNA similarity data are then indicative for the necessity of DDH studies.

When clusters contain only one type strain and the rRNA gene sequence analysis is not discriminative, DDHs (or WGS) need to be performed to meet the molecular taxonomic criteria for species delineation.

When the cluster contains more than one DNA relatedness group (DDH values definitely below 70%), the MLSA approach is not discriminative at the species level for this cluster. Further research (other genes, DNA fingerprinting) is needed. If the DDH values between the type strain and a number of strains within this cluster are 70% or higher, the strains of this cluster belong to the established species, polyphasic characterization can lead to an emended description.

When a cluster contains more than one type strain and the discriminatory power of 16S rRNA gene sequences, analysis reveals 97% similarity (increasingly 98% is regarded as the threshold) DDH or WGS are needed. Based on the data, it can be concluded that either both type strains are representatives of the same species and the species are objective synonyms (Lapage *et al.*, 1992) or both species do, indeed, merit a separate status and further analysis should then be performed to determine to which species the other strains of that cluster belong.

Finally, the taxonomic ranking of the different clusters that have been delineated by MLSA and confirmed by the TaxonGap analysis needs to be determined. The same procedure can be followed: 16S rRNA gene sequence similarities may delineate the species level. If this is not the case, DDH or WGS are needed.

In the ultimate taxonomic proposals, (i) a polyphasic characterization is a prerequisite, e.g. recommendations by Tindall *et al.* (2010) and (ii) the rules of the Bacteriological Code for naming of new taxa or recombination of taxa (Lapage *et al.*, 1992) should be respected. Subsequently, the taxonomy of the prokaryote group is improved in such a way that the molecular data are portable for further studies. Of course, in each clustering system, individual strains will be found that do not seem to cluster into one of the delineated groups. The lonely position of these strains may not be observed for all sequenced loci. The strains can be regarded as fuzzy strains or species (Hanage *et al.*, 2005a) that may be in a transient phase of speciation.

◆◆◆◆◆◆ III. SOME CONCLUSIONS AND FUTURE DEVELOPMENTS

There is no doubt that taxonomically evaluated MLSA studies can contribute to improve prokaryotic classification on the condition that the data are taxonomically evaluated. The MSLA schemes can then be used as identification tools not only in local applications (e.g. hospitals or special diagnostic labs), but also for more general identification purposes (e.g. Bishop *et al.*, 2009; Ramos *et al.*, 2011 etc.).

Although this evolution is positive in general, it also introduces certain risks if data can be added to the publically available MLS data schemes without curation. A curatorium should be instated that controls the authenticity of the biological material and the sequence data that accompany it. Quality control and preservation of the biological material (for later expansion of the data, e.g. additional sequences) is a prerequisite for a durable long-term taxonomic system that, although managed differently from the present system, should adhere to the same quality standards that are in place in the current taxonomic system.

References

Adékambi, T., Butler, R. W., Hanrahan, F., Delcher, A. L., Drancourt, M. and Shinnick, T. M. (2011). Core gene set as the basis of multilocus sequence analysis of the subclass of Actinobacteria. *Plos One* **6**, 10.

Altschul, S. F., Gish, W., Miller, W., Myers, E. W. and Lipman, D. J. (1990). Basic local alignment search tool. *J. Mol. Biol.* **215**, 403–410.

Bishop, C. J., Aanensen, D. M., Jordan, G. E., Kilian, M., Hanage, W. P. and Spratt, B. G. (2009). Assigning strains to bacterial species via the internet. *BMC Biol.* **7**, 3.

Brady, C., Cleenwerck, I., Venter, S., Vancanneyt, M., Swings, J. and Coutinho, T. (2008). Phylogeny and identification of *Pantoea* species associated with plants, humans and the natural environment based on multilocus sequence analysis (MLSA). *Syst. Appl. Microbiol.* **31**, 447–460.

Breed, R. S., Murray, E. G. D. and Smith, N. R. (eds.), (1957). *Bergey's Manual of Determinative Bacteriology*. Williams & Wilkins, Baltimore, MD.

Chelo, I. M., Zé-Zé, L. and Tenreiro, R. (2007). Congruence of evolutionary relationships inside the Leuconostoc-Oenococcus-Weissella clade assessed by phylogenetic analysis

of the 16S rRNA gene, dnaA, gyrB, rpoC and dnaK. *Int. J. Syst. Evol. Microbiol.* **57**, 276–286.
Cleenwerck, I., De Vos, P. and De Vuyst, L. (2010). Phylogeny and differentiation of species of the genus *Gluconacetobacter* and related taxa based on multilocus sequence analyses of housekeeping genes and reclassification of *Acetobacter xylinus* subsp. *sucrofermentans* as *Gluconacetobacter sucrofermentans* (Toyosaki *et al.*, 1996) sp. nov., comb. nov. *Int. J. Syst. Evol. Microbiol.* **60**, 2277–2283.
Coenye, T., Gevers, D., Van de Peer, Y., Vandamme, P. and Swings, J. (2005). Towards a prokaryotic genomic taxonomy. *FEMS Microbiol. Rev.* **29**, 147–167.
Cohan, F. M. (2001). Bacterial species and speciation. *Syst. Biol.* **50**, 513–524.
Cohan, F. M. (2002). What are bacterial species? *Ann. Rev. Microbiol.* **56**, 457–487.
Cohan, F. M. (2006). Towards a conceptual and operational union of bacterial systematics, ecology, and evolution. *Phil. Trans. R. Soc. B* **361**, 1985–1996.
Colwell, R. R. (1970). Polyphasic taxonomy of genus *Vibrio* – numerical taxonomy of *Vibrio cholerae*, *Vibrio parahaemolyticus*, and related *Vibrio* species. *J. Bacteriol.* **104**, 410–433.
Curtis, T. P., Sloan, W. T. and Scannell, J. W. (2002). Estimating prokaryotic diversity and its limits. *Proc. Natl. Acad. Sci. U.S.A.* **99**, 10494–10499.
Debruyne, L., Gevers, D. and Vandamme, P. (2008). Taxonomy of the family Campylobacteriaceae. In: *Campylobacter* (I. Nachamkin, C. M. Szymanski and M. J. Blaser, eds.), 3rd ed.. pp. 3–25. ASM Press, Washington, D.C.
DeLong, E. F., Preston, C. M., Mincer, T., Rich, V., Hallam, S. J., Frigaard., N.-U., Martinez, A., Sullivan, M. B., Edwards, R., Rodriguez Brito, B., Chisholm, S. W. and Karl, D. M. (2006). Community genomics among stratified microbial assemblages in the ocean's interior. *Science* **311**, 496–503.
Drancourt, M. and Raoult, D. (2002). rpoB gene sequence-based identification of *Staphylococcus* species. *J. Clin. Microbiol.* **40**, 1333–1338.
Drummond, A. J., Ashton, B., Buxton, S., Cheung, M., Cooper, A., Duran, C., Field, M., Heled, J., Kearse, M., Markowitz, S., Moir, R., Stones-Havas, S., Sturrock, S., Thierer, T., and Wilson, A. (2011). Geneious v5.4, http://www.geneious.com/.
Feil, E. J., Li, B. C., Aanensen, D. M., Hanage, W. P. and Spratt, B. G. (2004). eBURST: inferring patterns of evolutioanary descentamong clusters of related bacterial genotypes from multilocus sequence typing data. *J. Bacteriol.* **186**, 1518–1530.
Gevers, D., Cohan, F. M., Lawrence, J. G., Spratt, B. G., Coenye, T., Feil, E. J., Van de Peer, Y., Vandamme, P., Thompson, F. L. and Swings, J. (2005). Re-evaluating prokaryotic species. *Nat. Rev. Microbiol.* **3**, 733–739.
Godoy, D., Randle, G., Simpson, A. J., Aanensen, D. M., Pitt, T. L., Kinoshita, R. and Spratt, B. G. (2003). Multilocus sequence typing and evolutionary relationships among the causative agents of melioidosis and glanders, *Burkholderia pseudomallei* and *Burkholderia mallei*. *J. Clin. Microbiol.* **41**, 2068–2079.
Hall, B. G. (ed.), *Phylogenetic Trees Made Easy. A How-To Manual.* Sinauer Associates, Sunderland.
Hanage, W. P., Fraser, C. and Spratt, B. G. (2005a). Fuzzy species in recombinogenic bacteria. *BMC Biol.* **3**, 6.
Hanage, W. P., Kaijalainen, T., Herva, E., Saukkoriipi, A., Syrjanen, R. and Spratt, B. G. (2005b). Using multilocus sequence data to define the pneumococcus. *J. Bacteriol.* **187**, 6223–6230.
Hugenholtz, P., Goebel, B. M. and Pace, N. (1998). Impact of culture-independent studies on emerging phylogenetic view of bacterial diversity. *J. Bacteriol.* **180**, 4765–4774.
Konstantinidis, K. T. and Tiedje, J. M. (2005). Genomic insights that advance the species definition for prokaryotes. *Proc. Natl. Acad. Sci. U.S.A.* **102**, 2567–2572.

Konstantinidis, K. T., Ramette, A. and Tiedje, J. M. (2006). The bacterial species definition in the genomic era. *Phil. Trans. R. Soc. B* **361**, 1929–1940.

Lapage, S. P., Sneath, P. H. A., Lessel, E. F., Skerman, V. B. D., Seeliger, H. P. R. and Clark, W. A. (eds.), (1992). *International Code of Nomenclature of Bacteria (1990 Revision) Bacteriological Code.* American Society for Microbiology, Washington, D.C.

Maiden, M. C. J., Bygraves, J. A., Feil, E., Morelli, G., Russell, J. E., Urwin, R., Zhang, Q., Zhou, J., Zurth, K., Caugant, D. A., Feavers, I. M., Achtman, M. and Spratt, B. G. (1998). Multilocus sequence typing: a portable approach to the identification of clones within populations of pathogenic microorganisms. *Proc. Natl. Acad. Sci. U.S.A.* **95**, 3140–3145.

Martens, M., Dawyndt, P., Coopman, R., Gillis, M., De Vos, P. and Willems, A. (2008). Advantages of multilocus sequence analysis for taxonomic studies: a case study using 10 housekeeping genes in the genus *Ensifer* (including former Sinorhizobium). *Int. J. Syst. Evol. Microbiol.* **58**, 200–214.

Martens, M., Weidner, S., Linke, B., de Vos, P., Gillis, M. and Willems, A. (2007). A prototype taxonomic microarray targeting the rpsA housekeeping gene permits species identification within the rhizobial genus *Ensifer*. *Syst. Appl. Microbiol.* **30**, 390–400.

Medini, D., Serruto, D., Parkhill, J., Relman, D. A., Donati, C., Moxon, R., Falkow, S. and Rappuoli, R. (2008). Microbiology in the post-genomic era. *Nat. Rev. Microbiol.* **6**, 419–430.

Morse, R., Collins, M. D., O'Hanlon, K., Wallbanks, S. and Richardson, P. T. (1996). Analysis of the beta subunit of DNA-dependent RNA polymerase does not support the hypothesis inferred from 16S rRNA analysis that *Oenococcus oeni* (formerly *Leuconostoc oenos*) is tachytelic (fast-evolving) bacterium. *Int. J. Syst. Evol. Microbiol.* **46**, 1004–1009.

Naser, S. M., Thompson, F. L., Hoste, B., Gevers, D., Dawyndt, P., Vancanneyt, M. and Swings, J. (2005b). Application of multilocus sequence analysis (MLSA) for rapid identification of *Enterococcus* species based on rpoA and pheS genes. *Microbiology* **151**, 2141–2150.

Naser, S. M., Dawyndt, P., Hoste, B., Gevers, D., Vandemeulebroecke, K., Cleenwerck, I., Vancanneyt, M. and Swings, J. (2007). Identification of lactobacilli by pheS and rpoA gene sequence analyses. *Int. J. Syst. Evol. Microbiol.* **57**, 2777–2789.

Naser, S., Thompson, F. L., Hoste, B., Gevers, D., Vandemeulebroecke, K., Cleenwerck, I., Thompson, C. C., Vancanneyt, M. and Swings., J. (2005a). Phylogeny and identification of enterococci using atpA gene sequence analysis. *J. Clin. Microbiol.* **43**, 2224–2230.

Ramos, P. L., Moreira-Filho, C. A., Van Trappen, S., Swings, J., De Vos, P., Barbarosa, H. R., Thompson, C. C., Ribeiro Vasconcelos, A. T. and Thompson, F. L. (2011). An MLSA-based online scheme for the rapid identification of *Stenotrophomonas* isolates. *Mem. Inst. Oswaldo Cruz, Rio de Janeiro* **106**, 394–399.

Rivas, R., Martens, M., de Lajudie, P. and Willems, A. (2009). Multilocus sequence analysis of the genus *Bradyrhizobium*. *Syst. Appl. Microbiol.* **32**, 101–110.

Slabbinck, B., Dawyndt, P., Martens, M., De Vos, P. and De Baets, B. (2008). TaxonGap: a visualization tool for intra- and inter-species variation among individual biomarkers. *Bioinformatics* **24**, 866–867.

Stackebrandt, E., Frederiksen, W., Garrity, G. M., Grimont, P. A., Kämpfer, P., Maiden, M. C., Nesme, X., Rosselló-Mora, R., Swings, J., Trüper, H. G., Vauterin, L., Ward, A. C. and Whitman, W. B. (2002). Report of the ad hoc committee for the re-evaluation of the species definition in bacteriology. *Int. J. Syst. Evol. Microbiol.* **52**, 1043–1047.

Staley, J. T. and Krieg, N. (1984). Classification of prokaryotic organisms: an overview. In: *Bergey's Manual of Systematic Bacteriology* (N. R. Krieg and J. G. Holt, eds.), Vol. 1, pp. 1–3. Williams & Wilkins, Baltimore, MD.

Tatusov, R. L., Koonin, E. V. and Lipman, D. J. (1997). A genomic perspective on protein families. *Science* **278**, 631–637.

Thompson, C. C., Thompson, F. L., Vicente, A. C. and Swings, J. (2007). Phylogenetic analysis of vibrios and related species by means of *atpA* gene sequences. *Int. J. Syst. Evol. Microbiol.* **57**, 2480–2484.

Tindall, B. J., Rosselló-Mora, R., Busse, H. J., Ludwig, W. and Kämpfer, P. (2010). Notes on the characterization of prokaryote strains for taxonomic purposes. *Int. J. Syst. Evol. Microbiol.* **60**, 249–266.

Vandamme, P., Pot, B., Gillis, M., De Vos, P., Kersters, K. and Swings, J. (1996). Polyphasic taxonomy a consensus approach to bacterial systematics. *Microbiol. Rev.* **60**, 407–438.

Wayne, L. G., Brenner., D. J., Colwell, R. R., Grimont, P. A. D., Krichevsky, M. I., Moore, L. H., Moore, W. E. C., Murray, R. G. E., Starr, M. P. and Trüper, H. G. (1987). Report of the ad-hoc-committee on reconciliation of approaches to bacterial systematics. *Int. J. Syst. Bacteriol.* **37**, 463–464.

Woese, C. R. (1987). Bacterial evolution. *Microbiol. Rev.* **51**, 221–271.

Yamamoto, S. and Harayama, S. (1995). PCR amplification and direct sequencing of *gyrB* genes with universal primers and their application to the detection an taxonomic analysis of *Pseudomonas putida* strains. *Appl. Environ. Microbiol.* **61**, 1104–1109.

Zeigler, D. R. (2003). Gene sequences useful for predicting relatedness of whole genomes in bacteria. *Int. J. Syst. Evol. Microbiol.* **53**, 1893–1900.

18 Whole Genome Sequence Comparisons in Taxonomy

Rainer Borriss[1,*], Christian Rueckert[2], Jochen Blom[2], Oliver Bezuidt[3], Oleg Reva[3], and Hans-Peter Klenk[4]

[1]ABiTEP GmbH, Berlin, Germany; [2]Center for Biotechnology (CeBiTec), Bielefeld University, Bielefeld, Germany; [3]University of Pretoria, Department of Biochemistry, Bioinformatics and Computational Biology Unit, Pretoria, South Africa; [4]DSMZ - German Collection of Microorganisms and Cell Cultures, Braunschweig, Germany

**E-mail: rainer.borriss@rz.hu-berlin.de*

CONTENTS

I. INTRODUCTION

This chapter is devoted to the application of whole genome sequence comparisons in taxonomy. Driven by the rapid progress in sequencing technologies, 'low budget' bacterial genomes become increasingly available in a nearly unlimited number. At the time this chapter was finalized (April 2011), completed genomes representing 1604 bacterial and 85 archaeal species were present in the public data bank (http://www.ncbi.nlm.nih.gov/sutils/genom_table.cgi), reflecting the enormous progress made with the sequencing of microbial genomes in the recent years. With the advent of next generation sequencing, whole genome sequence comparisons will be more and more important for taxonomy, and they will become especially valuable in elucidating relationships of groups from closely related bacterial strains, which might form a single taxon, a subspecies or just an ecovar within a given species. The aim of this chapter is to hand out a tool set for taxonomists interested in applying genomics. These tools might prove useful especially in refining the classification of closely related strains, for which their taxonomic status has not been resolved by their 16S rRNA sequence. Here, we will exemplify this approach by selecting a specific group of plant-associated *Bacillus amyloliquefaciens* strains with plant growth promoting properties. In recent

METHODS IN MICROBIOLOGY, VOLUME 38
0580-9517 DOI 10.1016/B978-0-12-387730-7.00018-8

years these strains were increasingly applied as biological substitutes of agrochemicals, mainly to be used as biofertilizer and for biocontrol of phytopathogenic microorganisms and nematodes (Chen *et al.*, 2007).

Despite the enormous progress made in microbial whole genome sequencing in recent years, the recommended minimal standards for the description of new prokaryote taxa are mainly based on a set of microscopic and macroscopic features such as cell and colony morphology, physiological and biochemical characters, as well as profiles of fatty acid and cell wall constituents. In addition, 16S rRNA gene sequence analysis is required and, in the case of closely related strains/species, DNA–DNA hybridization is also recommended (Logan *et al.*, 2009). Unfortunately, those standards are not completely sufficient to discriminate closely related taxa in a satisfying manner, e.g. the members of the *Bacillus subtilis* species complex. For many years it has been recognized that these species cannot be discriminated on the basis of phenotypic characteristics and 16S rRNA gene nucleotide sequence alone. Besides fatty acid profiles that do not enable the discrimination of closely related bacterial groups, phylogenetic analysis of multiple protein-coding loci has been used as a complementary approach to detect and differentiate novel *Bacillus* taxa (Gatson *et al.*, 2006; Rooney *et al.*, 2009). We have successfully used the same approach to discriminate a group of plant-associated *Bacillus* strains related to the *B. amyloliquefaciens* DSM7^T and *B. subtilis* 168 (Borriss *et al.*, 2010). Two ecovars consisting of plant-associated and non-plant associated *B. amyloliquefaciens* strains were discriminated by variations in their partial *cheA* and *gyrA* sequences. Branching of the two clades was visible in the Neighbour-Joining (NJ) phylograms and was supported by bootstrap values of 76% and 100%, respectively. However, variations in selected marker gene sequences are not sufficient to discriminate taxonomic categories and establish novel subspecies. Therefore, we used several genomic methods, e.g. direct whole genome comparison, digital DNA–DNA hybridization and microarray-based comparative genomic hybridization (MCGH) as complementary approaches to justify that these ecovars represent two distinct subspecies. These methods will be described in more detail in the course of this chapter. The known genome sequence of the plant-associated strain FZB42 (Chen *et al.*, 2007) and the recently determined whole genome sequences obtained from *B. amyloliquefaciens* type strain DSM 7^T (Rueckert *et al.*, 2011) and of three Chinese plant-associated *B. amyloliquefaciens* strains known for their potential to promote plant growth, were included in our analysis. The differences detected in our genome comparisons, especially deviations in the core genomes, changes in the variable portion of the genomes, differences in values obtained in DDH and MCGH patterns were indicative for discriminating the members of the FZB42 subgroup (*B. amyloliquefaciens* subsp. *plantarum* subsp. nov.) from the strains related to the *B. amyloliquefaciens* DSM 7^T (Borriss *et al.*, 2010).

◆◆◆◆◆◆ II. SEQUENCING TECHNIQUES: NEXT GENERATION GENOME SEQUENCING

A. Sequencing Techniques

The key technology to enable taxonomic studies on the level of whole genomes respectively proteomes was the introduction of the so called next generation sequencing (NGS) technologies. Before the advent of these technologies, establishing a complete genome sequence using the classical Sanger sequencing approach required a huge amount of lab work to prepare the necessary clone libraries and a high amount of sequencing time, which prevented the use of whole genome sequencing for taxonomic purposes.

1. Next generation sequencing

With the commercial introduction of two platforms for high-throughput sequencing, the determination of the whole genome sequences of several strains of a certain species or several species of a genus for taxonomical purposes alone has become feasible. Both, the Genome Sequencer (GS) by 454 Life Sciences (Branford, CT, USA) and the Genome Analyzer (GA) by Solexa (San Diego, CA, USA) get rid of the need of clone libraries, as both rely on PCR-based library preparation techniques. Initially, both systems were not really suitable for *de novo* sequencing as only 20–30 bases could be reliably determined. As of 2011, the obtainable length has increased to about 450 bases with the GS-FLX platform (with Titanium reagents) and 2×150 bases with the GA-II*x* system, thus reaching a range reminiscent of the early automated Sanger sequencers. In contrast to the latest Sanger sequencers that could sequence only up to 384 samples in a single run, the NGS platforms provide millions (GS-FLX) to hundreds of millions (GA-II*x*) of sequences per run, driving the cost per assembled Mbase well below $1000.

While both systems allow *de novo* sequencing, several practical considerations have to be taken into account when deciding which one to use for a 'whole' genome project, due to the strengths and weaknesses of the two systems, as discussed below.

(a) Pyrosequencing (Genome Sequencer FLX, 454/Roche)

Being the first commercial NGS system in widespread use, the Genome Sequencer is based on the principle of pyrosequencing, first described by Ronaghi *et al.* (1996). Instead of using fluorescently labelled nucleotides or primers, the sequence read-out occurs via the conversion of pyrophosphate to ATP which is in turn converted to light by firefly luciferase. As with Sanger sequencing, it took almost a decade to create a viable commercial platform usable for whole genome sequencing. After a long series of optimizations and especially a high degree of miniaturization, the GS platform was introduced in 2005 (Margulies *et al.*, 2006).

Today, the GS-FLX platform with titanium reagents allows sequencing of approximately 500 Mbases in a single run taking about 5 h. This allows the *de novo* sequencing of a (hypothetical) bacterial genome of 20 million base pairs length with a coverage of 25-fold which is usually sufficient to correctly assemble 95–99% of the sequence with good quality. In standard practice, the picotiter plate used by the GS-FLX is segmented to allow two, four or eight bacterial genomes to be sequenced in parallel without tagging. While cost constraints normally prohibit the acquisition of a GS-FLX (or a GA-II*x*) by individual laboratories, the service can easily be bought from specialized companies and institutions. In addition, Roche has recently launched the GS Junior that is suitable for 'small scale' sequencing, i.e. a few dozen bacterial genomes per year.

When compared to the GA-II*x*, the main advantage of pyrosequencing with the GS-FLX system lies in the read length. While 450 bases compared to two times 150 bases does not appear to be much of a difference, one has to keep in mind that the subsequent sequence assembly will be interrupted by repetitive elements of approximately the size of the read length. Thus, the assembly of GS-FLX reads usually results in far fewer contigs than an assembly obtained from GA-II*x* reads. Another advantage, at least for researchers without a strong background in bioinformatics and without access to powerful computer clusters is the fact that the GS-FLX system comes with its own assembly software, including a graphical user interface (GUI). This usually allows the assembly of a draft genome consisting of a few to a few hundred contigs within a few hours, which is sufficient for many genomic and taxonomic studies, e.g. to produce an inventory of up to 99% of the protein coding genes, to calculate the core and pan genomes, and so on (see Section III for details). For establishing the complete genome sequence, the GS-FLX *de novo* assembler allows to automatically recognize and process reads from long paired-end libraries of up to 20 kbp. Thereby, usually all unique contigs of a genome can be assembled into one or a few scaffolds (arrays of contigs with known order) from a single library and single nucleotide polymorphisms in repetitive elements can often be resolved without the need for additional Sanger sequencing. In addition, the gsAssembler software can provide output in ACE format, which facilitates subsequent finishing approaches (see Section II.B).

On the downside, the per base pair cost for GS-FLX sequencing is (as of early 2011) approximately 100-fold that of GA-II*x* sequencing, which usually prohibits large scale sequencing of dozens or hundreds of genomes for purely taxonomic purposes. Another drawback of this technology is the so called homopolymer problem. Pyrosequencing of long stretches (>8 nucleotides) of identical bases often leads to an over- or underestimation of the correct number of the nucleotide in question. Therefore, GS-FLX sequencing genomes with either a high G + C or a high A + T content often results in a higher number of frame-shifts due to the increased probability of longer homopolymer stretches. A related problem is the presence of sequence gaps that result from PCR biases introduced during library preparation. Extremes in G + C content (in either direction) result in an increased probability of hairpin-loop formation that can inhibit or completely prevent PCR amplification. As the emPCR necessary during GS-FLX library preparation cannot be optimized as rigorously as the normal PCR needed for GA-II*x* libraries, genomes with an extremely high G + C content tend to have many poorly covered or even uncovered regions after GS-FLX sequencing.

(b) Sequencing by synthesis (Genome Analyzer IIx, Illumina/Solexa)

Based on the stepwise synthesis and subsequent detection of the incorporated nucleotide by fluorescence, the GA-II*x* platform allows sequencing of more than

200 million reads of up to 2×150 nucleotides in a single run. While considerably slower than the GS-FLX (the above example would take about 14 days to complete) the amount of data obtained far exceeds the output of a GS-FLX run, ranging between 60 and 90 Gbases, an amount more than doubled with the recent introduction of the HiSeq2000 platform. This drives the price per Mbase of assembled sequence well below $100, allowing for truly large scale studies. As mentioned in Section II.A.1.a, another advantage of GA-II*x* sequencing is the possibility to obtain less biased PCR libraries or completely forego PCR amplification (Kozarewa *et al.*, 2009), making it suitable for genomes with an extreme high G + C content.

The main drawback for researchers without access to bioinformatics support (people as well as compute capacity) is the lack of company-supplied assembly software for GA-II*x* data. A number of open source programs are available, e.g. ABySS (Simpson *et al.*, 2009), MIRA (Chevreux *et al.*, 1999) and velvet (Zerbino and Birney, 2008), but installation and utilization of these programs usually requires experienced users. This also complicates genome ordering and finishing using long paired end libraries, as many programs cannot utilize that kind of data. As mentioned above, another drawback is the still rather short read length which might cause problems when assembling genomes with many repetitive regions. As the length of GA-II*x* reads has steadily increased during the last years, this disadvantage might be resolved in the near future.

2. Sanger sequencing

Although the principal technique was discovered more than three decades ago (Sanger and Coulson, 1975), the sequencing of complete genomes was not started until 1990 when development and availability of automated sequencing machines had reached a critical mass. Today, if Sanger sequencing is used at all for new genome projects, it is applied during the finishing and polishing phases when the gaps and low quality regions left by assembling NGS data are resolved. For these applications it remains a necessary and valuable technique for the next future.

B. Assembly, Finishing and Annotation

1. Assembly and finishing

The choice of the NGS technique(s) to use for a project depends heavily on the information the researcher wants to obtain for the taxonomic comparison(s). If one is only interested in studying the phylogeny based on core and/or pan genomes (see Section III.A.2), scaffolding of contigs and finishing is usually not necessary, so the NGS technique might be selected primarily on the basis of costs and the availability of bioinformatics resources (see Section II.A).

On the other hand, if more detailed studies, like genomic rearrangements or the distribution and movement of mobile genetic elements is of interest, use of the GS-FLX platform, gsAssembler and a 10 kbp long paired end library is strongly recommended, at least for the near future. This will usually result in one or two scaffolds per replicon and provide a file in ACE format for finishing and polishing. For the latter two steps, consed (Gordon *et al.*, 1998) is perhaps still the best finishing package around. Developed as part of the Phred/Phrap package for the assembly and finishing of Sanger-based projects, consed has been updated to also handle GS-FLX and GA-II*x* data. Together with autofinish (Gordon *et al.*, 2001), the basic steps of finishing (primer selection, etc.) can easily performed based on the scaffolding data. The final task when attempting to completely close a genome is the creation/selection of suitable templates for Sanger sequencing. The straightforward approach relies on PCR to amplify the regions of interest; the indirect one consists of creating a large insert library and identifying of suitable templates by random sequencing or hybridization. Like selecting the best NGS technique for a project, choosing a technique for gap closure depends heavily on the circumstances like the organism to be finished, the number of gaps, and the genome size. PCR is most suitable when the number of gaps is small, the genome is large and the number of repetitive elements is low. As only the required templates are created, the amount of lab work is comparatively low. On the downside, all the usual problems of PCR like introduction of errors, formation of chimeric templates (in case of large repetitive regions), false priming and so on can severely hinder this approach, especially when a huge number of gaps have to be addressed. These pitfalls are avoided by utilizing a large insert library, but creating and screening it is expensive in time and money. Therefore, it is usually only useful if a large number of gaps have to be addressed, especially if they are caused by complex repetitive elements.

2. Gene prediction and annotation

Once the genome sequence has been established, either complete or at least on the level of a suitable draft (median contig size of at least 10 kbp), a crucial step for phylogenetic studies based on whole genome data is the correct prediction of genes and, to a lesser extent, their correct annotation (i.e. prediction of function). A number of gene prediction tools have been developed over the years, which can roughly be divided in two classes: *ab initio* predictors and comparison-based predictors. The former rely on intrinsic signals in the DNA that allow the differentiation of protein coding and non-protein coding regions, the latter search for protein sequences that are similar to those of other organisms. Examples for *ab initio* predictors include the widely used GLIMMER (Delcher *et al.*, 1999) and Prodigal (Hyatt *et al.*, 2010). The drawback of the *ab initio* approach is that these programs tend towards overprediction (i.e. many false-positives) and often miss the correct start codon. Comparison-based predictors like CRITICA (Badger and Olsen, 1999), on the other hand, can only find genes that are also present in other organisms and therefore tend to miss singletons. One way to combine the strengths of both approaches is by combining them in one tool, as demonstrated, e.g. for GISMO (Krause *et al.*, 2007), another way is to use several different tools and to weight and combine their output. The latter approach is, e.g. realized in the REGANOR web server (Linke *et al.*, 2006).

A good, freely available resource for gene prediction and subsequent annotation is the GenDB genome annotation system (Meyer *et al.*, 2003). The software is open source and can be obtained either as a standalone system or utilized as a web-based service. In the latter case, a user management system allows to handle confidential (i.e. unpublished) genome data and no additional software or databases have to be installed respectively maintained by the user (s). GenDB can handle complete microbial genomes as well as draft sequences and offers a number of useful pipelines and features. For gene prediction, the REGANOR pipeline is used, which in turn utilizes GLIMMER and CRITICA to do the actual gene prediction, but can be expanded to use other predictors of the user's choice. Once the coding regions in a genome have been identified, the Metanor pipeline can be used to predict and automatically annotate the functions of the encoded proteins. As the REGANOR pipeline, this pipeline uses the output of different tools like BLAST, hmmsearch and SignalP run against different databases (e.g. nr, SwissProt and PFAM) to create an automated gene annotation with as much depth as possible. Once again, a user can specify additional tools and/or databases to be used, thus tailoring the results to his specific needs. When the automated annotation is complete, the user can manually check and curate individual annotations, all of them are archived. This provides a reliable basis for further taxonomic analyses.

♦♦♦♦♦♦ III. COMPARATIVE GENOME ANALYSIS

This chapter describes tools and techniques for the comparison of microbial genomes.

A. Genome Comparison and Phylogeny

1. Global alignment of genomes

A first genome-based approach to gain insight into the evolutionary distance between two species is to inspect the synteny of the genome sequences. Two popular tools dedicated to whole genome comparisons are MUMmer (Kurtz *et al.*, 2004) and the Artemis Comparison Tool (ACT) (Carver *et al.*, 2005):

- **MUMMER** – MUMmer (MUM = Maximum Unique Match) is an open source software package for the rapid alignment of large genomic sequences on DNA and amino acid level. It provides a wide range of tools and utilities for alignments, filter steps and result visualization that can be combined to analyse pipeline. A typical pipeline for the comparison of two complete genomes would consist of:
  - **NUCmer** – Basic nucleotide alignment of the two sequences
  - **Show-coords and show-aligns** – Parsing of the alignment output of NUCmer
  - **Delta-filter** – Filtering of the alignments by length, identity, consistency etc.
  - **MUMmerplot** – Plotting and graphical representation of alignment results

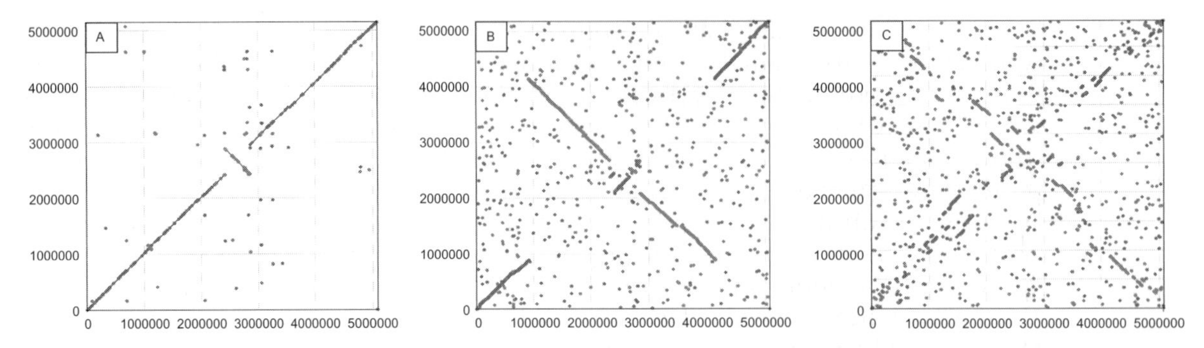

Figure 1. Synteny plots generated by MUMmer 3.07 comparing four *Xanthomonas campestris* pv. *campestris* B100 with three strains from the same genus, *Xanthomonas campestris* pv. *campestris* 8004 (A), *Xanthomonas axonopodis* pv. *citri* str. 306 (B), and *Xanthomonas oryzae* pv. *oryzae* KACC10331 (C). A dramatic increase in genomic rearrangements can be observed, whilst synteny is decreased corresponding to increasing phylogenetic distance.

Figure 1 shows three synteny plots generated by this pipeline.

- **ACT** – The Artemis Comparison Tool (ACT) is a visualization software for pair-wise genome alignments. It does not calculate alignments itself, but uses precalculated tabular output of the popular alignment software BLAST (Altschul *et al.*, 1997). Artemis allows the user to easily explore the synteny of the compared sequences and which genes are affected by genomic rearrangements.

2. Gene-based genome comparison

Availability of completely sequenced and well-annotated sequences for several members of a taxonomic group allows genotypic characterization of prokaryotes based on their gene content. Several terms have been defined in this field describing certain genomic subsets:

- **Core genome** – The term 'core genome' denotes the set of genes that is shared by a group of analysed organisms, usually all members of a certain taxonomic group (e.g. genus). This means that all genes of the core genome possess an orthologous gene in any other strain of the genome group.
- **Singleton genes** – The term 'singleton gene' describes a gene that is unique within a group of analysed organisms. This means that no orthologous genes can be identified in any other strain of the comparison set.
- **Pan genome** – The term 'pan genome' specifies the set of all independent genes within a group of analysed organisms. It comprises of the core genome, all singleton genes and all genes that can be identified in more than one, but not in all compared genomes. The pan genome describes the complete genetic potential of the analysed taxonomic group.

Several tools and databases have been developed to identify groups of orthologues within a set of genomes and to calculate the genomic subsets mentioned above, for example the Comprehensive Microbial Resource (CMR), the

Microbial Genome Database (MBGD) and EDGAR (Blom *et al.*, 2009; Peterson *et al.*, 2001; Uchiyama *et al.*, 2010).

The CMR provides comparative tools for a database of 723 microbial genomes (64 of them draft genomes). Thus, the multi-genome homology comparison tool allows the user to calculate the number of homologous genes between up to 15 selected comparison genomes. Special gene sets like the core genes or the singletons can be observed and exported in a tabular format. Another comparative tool included in the CMR is the genome homology graph, a dot plot showing the number of homologous genes between a selected reference genome and all genomes in the CMR database. The MBGD features comparative analyses for 1042 finished bacterial genomes. The genes of selected genomes can be clustered into homologous groups, resulting in a set of orthologue clusters. Additional analysis and visualization features are available for the clustered genes like multiple alignments or a comparison of the context of the genes on a genome map.

3. EDGAR

EDGAR, another resource for comparative genome analysis, is a dedicated approach for comparative and phylogenetic analysis of closely related genomes. EDGAR (Efficient Database framework for comparative Genome Analyses using BLAST score Ratios) provides several analysis and visualization features based on all-against-all BLAST comparisons of all genes of a set of analysed genomes. EDGAR uses a generic orthology criterion adjusted to the set of compared genomes based on BLAST score ratios (Lerat *et al.*, 2003), a technique where every BLAST hit is weighted in relation to the maximum score. Based on this generic threshold EDGAR creates project-specific databases storing the orthology information and serving as data source for subsequent analyses. EDGAR provides precalculated public databases for 95 bacterial genera with 846 genomes in total, but it is also possible to create private, access-controlled projects to analyse user-defined sets of genomes or unpublished data. Furthermore, it is possible to create EDGAR projects directly from GenDB projects.

EDGAR features the calculation of the core genome, the pan genome and the singleton genes of all or subset of genomes included in a project. It is also possible to calculate specific gene sets by defining Boolean operations on genomes. To visualize the distribution of shared and unique genes of compared genomes, Venn diagrams of up to five genomes can be created (see Figure 2). The comparative view provides a linear view of all orthologous genes in their genomic neighbourhood. To investigate large scale genomic events EDGAR provides an interface to create synteny plots.

Furthermore, EDGAR supports the differentiation between open and closed pan genomes (Medini *et al.*, 2005) by predicting the number of singletons introduced by each genome with increasing genome number. For this purpose, the number of singletons is calculated for each possible combination of genomes, subsequently a decay function is fitted to the averaged number of singletons for each quantity of genomes as described by Tettelin *et al.* (2005).

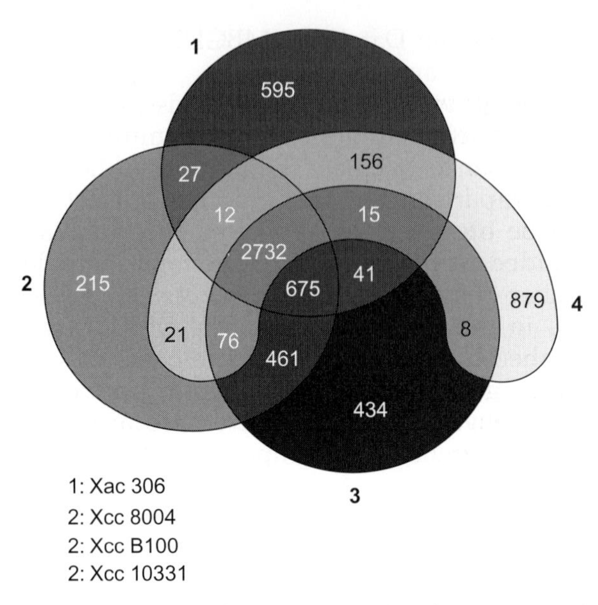

Figure 2. Venn diagram of the four *Xanthomonas* strains listed in Figure 1 showing the number of orthologous genes shared by the different strains.

4. Phylogenetic trees

Phylogenetic trees are branching diagrams illustrating the evolutionary relationships among species. Usually such trees are constructed based on sequence similarity between the highly conserved 16S rRNA genes or a set of housekeeping genes of several organisms. This limitation to a small set of input sequences can be problematic as the phylogeny of single genes does not necessarily reflect the phylogeny of the complete organisms. It is therefore highly desirable to use all genes of the core genome as input for the tree calculation, which dramatically increases its reliability (Gontcharov *et al.*, 2004). EDGAR creates multiple alignments of all orthologue-sets of the core genome by using MUSCLE (Edgar, 2004), removes unaligned parts with GBLOCKS (Talavera and Castresana, 2007), concatenates the multiple alignments of the single genes to one large alignment and finally creates a phylogenetic tree with the neighbour-joining implementation of the PHYLIP package (Felsenstein, 1995).

PHYLIP is a comprehensive collection of software tools that implement various algorithms for the creation of phylogenetic trees. Four of the most prominent algorithms are:

- **UPGMA:** Unweighted Pair Group Method with Arithmetic Mean: A simple clustering method that assumes a constant rate of evolution (molecular clock hypothesis). It needs a distance matrix of the analysed taxa that can be calculated from a multiple alignment.
- **Neighbour-joining (NJ):** Bottom-up clustering method that also needs a distance matrix. NJ is a heuristic approach that does not guarantee to find the perfect result, but under normal conditions has a very high probability to do

so. It has a very good computational efficiency, making it well suited for large datasets.

- **Maximum parsimony (MP):** This method tries to create a phylogeny that requires the least evolutionary change. It may suffer from long branch attraction, a problem that leads to incorrect trees in rapidly evolving lineages (Felsenstein, 1978).
- **Maximum-likelihood (ML):** ML uses a statistical approach to infer a phylogenetic tree. ML is well suited for the analysis of distantly related sequences, but is computationally expensive and thus not that well suited for larger input data.
  While phylogenetic trees calculated from large sets of orthologous genes are quite reliable, trees generated from smaller samples may need some further confirmation. In such cases the use of an outgroup and further bootstrapping support can be helpful:

In this context, two terms have to be defined:

Outgroups: When using distance matrix methods it is highly recommended to include at least one distantly related sequence for the analysis. This can be seen as a negative control. The outgroup should appear near the root of the tree and should have a longer branch length than any other sequence.
Bootstrapping: Bootstrapping is a resampling technique that is often used to increase the confidence that the inferred tree is correct. In a defined number of iterations (usually 100–1000) the multiple alignment that serves as input is permutated randomly and a phylogenetic tree is calculated. When the procedure is finished, a majority-rule consensus tree is constructed from the resulting trees of each bootstrap sample. The branches of the final tree are labelled with the number of times they were recovered during the procedure.

B. Electronic DNA–DNA Hybridization (DDH)

Nucleic acid hybridization methods were introduced into prokaryote systematics in the 1960s, long time before sequencing technologies were developed. Their purpose was to compare the degree of gene and genome similarities without knowing the actual sequences. With the current development towards a genome-sequence dominated microbiology/taxonomy the time of indirect measurements will rapidly come to an end. DNA–DNA hybridization (DDH) is generally applied when strains share more than 97% 16S rRNA gene sequence identity. DDH values not exceeding 70% are considered as an indication that the tested organisms belong to different species (Wayne *et al.*, 1987; Stackebrandt *et al.*, 2002; Tindall *et al.*, 2010).

In recent years, genome-based *'in silico'* alternatives to the experimental DDH estimate were developed. In the long run, the most important advantage of the novel genome sequenced-based *'in silico'* methods will be that each genome (sequence) can be compared with all then and thereafter available reference genome sequences, whereas the 'old and cumbersome' wet lab procedures required repeated extractions and hybridization of reference material whenever novel strains had to be analysed.

There are several indices that are obtained by comparing pairwise genomes that could be used in taxonomy. Noteworthy are the Average Nucleotide Identity (ANI; Konstantinidis *et al.*, 2006) and Maximal Unique Matches (MUM; Deloger *et al.*, 2009) indices as they have been posited to be able to substitute for DDH. ANI has been demonstrated to correlate with DDH, where the range of ~95–96% ANI may correspond to the current threshold of 70% DDH similarity (Goris *et al.*, 2007). A genome-to-genome distance comparison (GGDC) has recently been developed (Auch *et al.*, 2010a, 2010b). The method is based on whole genome data and allows also including unfinished draft genome sequences. We took advantage of this method to determine the genomic distances of strains FZB42 and DSM 7^T. The complete *B. subtilis* 168 genome and the draft genomes of three further plant-associated strains related to *B. amyloliquefaciens*, YAU Y2, CAU B946 and NAU B3, were also included in the analysis. The results demonstrated that *B. subtilis* and *B. amyloliquefaciens* can reliably be discriminated on the species level by their digital DDH values, which are much lower than 70%, whilst the DDH values between FZB42 and DSM 7^T were calculated as being around 77% (Table 1, Borriss *et al.*, 2010). Values, ranging between 70–80%, are considered as sufficient for discriminating subspecies (Goris *et al.*, 2007). GGDC analysis of the three draft genomes yielded DDH values of 86–88% with FZB42, but only 74–77% when

Table 1. '*In silico*' genome-to-genome distance comparison of genomic *Bacillus* DNAs using program BLAT. Regression-based DDH estimates (in %) are indicated (according to Borriss *et al.*, 2010)

Query/reference	Formula	DSM7	FZB42	*B. subtilis* 168
B. amyloliquefaciens DSM7	1[a]	97.71660	80.3271741	37.1054714
Length = 3980199 bp	2[b]	87.07480	64.4133336	14.9453016
FN597644	3[c]	≥100	77.6359098	30.2944702
B. amyloliquefaciens FZB42	1	80.3271741	97.71660	38.2065793
Length = 3918589 bp	2	64.4133336	87.07480	14.9819160
CP000560.1	3	77.6359098	≥100	31.1536318
B. amyloliquefaciens	1	75.9969741	84.561258	35.5791967
YAU-Y2	2	64.1295630	79.119362	14.2115092
Length = 4198660 bp	3	73.6625246	85.7222127	28.9901377
B. amyloliquefaciens	1	77.244344	86.316703	36.4790607
NAU-B3	2	64.0638979	79.1309892	14.3259432
Length = 4154898 bp	3	74.7654545	87.3710331	29.7046335
B. subtilis	1	80.761889	88.0674375	35.2256760
AU-B946	2	63.8177953	77.181397	13.9677772
Length = 3978635 bp	3	77.8597081	88.4282404	28.6752984
B. subtilis 168	1	37.1054714	38.2065793	97.71660
Length = 4215606 bp	2	14.9453016	14.9819160	87.07480
AL009126.3	3	30.2944702	31.1536318	≥100

[a]Formula: 1 (HSP length/total length).
[b]Formula: 2 (identities/HSP length).
[c]Formula: 3 (identities/total length).

compared with *B. amyloliquefaciens* DSM 7^T strain, supporting a closer taxonomic relatedness of the plant-associated *B. amyloliquefaciens* strains involved in our analysis (Borriss *et al.*, 2010).

1. Genome-to-genome distance comparison (GGDC)

The main steps are: (1) the determination of a set of HSPs (high-scoring segment pairs) or MUMs between two genomes; (2) the calculation of distances from these sets; and (3) the conversion of these distances in percent-wise similarities analogous to DDH (Figure 3).

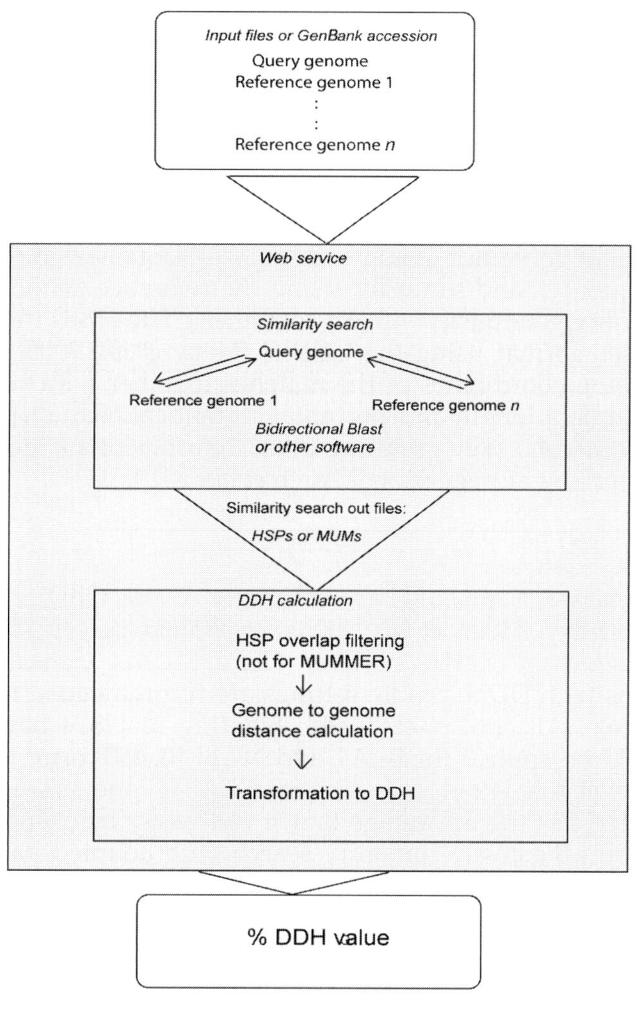

Figure 3. Flowchart outlining the steps required to calculate *in-silico* DDH values. Either GenBank accession numbers or FASTA files are uploaded on the server. The final values are received via e-mail (redrawn after Auch *et al.*, 2010a).

Whole Genome Sequence Comparisons in Taxonomy

(a) Requirements

The GGDC web server (http://ggdc.gbdp.org) allows multiple-FASTA files as input format. One file per genome is expected by default, containing each chromosome or plasmid as a single FASTA entry.

A single query genome can be compared to several reference genomes; organism names can be entered separately. The user can choose between several similarity search tools. Results are delivered via an e-mail to a user-specified address. The message routinely also contains a brief explanation of the results (Auch *et al.*, 2010a).

(b) Similarity search

Similarities between query and reference genomes are determined by using long-established tools for nucleotide-based sequence similarity searches. Currently, NCBI-BLAST, WU-BLAST (Altschul *et al.*, 1990), BLAT (Kent, 2002), BLASTZ (Schwartz *et al.*, 2003) and MUMmer (Kurtz *et al.*, 2004) are available on the web server. High-scoring segment pairs, HSPs, or maximally unique matches, MUMs, are determined by performing similarity searches for each combination of query genome and reference genome. Due to the asymmetric nature of heuristic similarity search strategies, the search is performed twice, first using the reference genome as 'subject sequence' and the query genome as 'query sequence', and secondly, using the reference genome as 'query sequence' and the query genome as 'subject sequence'. The HSPs (or MUMs) are stored in a condensed format using the CGVIZ (Henz *et al.*, 2005), which comprises the start and stop coordinates of the matches together with statistical data (e-value, score, alignment length and percentage identical characters for HSPs, alignment length for MUMs). The resulting data is sufficient for the distance calculation, while preserving storage space (Auch *et al.*, 2010a).

(c) Distance calculation

Distances between genomes are calculated using GBDP. The greedy-with-trimming algorithm (Henz *et al.*, 2005) is applied before calculating distances to remove regions of overlap between HSPs/MUMs. Considering error ratios and correlations with DDH, certain settings are recommended for any local alignment software (e.g. Altschul *et al.*, 1990). Filtering of HSPs having an e-value above 10^{-2} should be applied for BLAT and NCBI-BLAST prior to the distance calculation, however this is not necessary for BLASTZ and WU-BLAST. A downstream filtering step has the advantage that it can easily be changed without the necessity to re-run the costly similarity search with adapted parameters. This enables the user to reuse the data for further processing when required (Auch *et al.*, 2010a). Three types of distance functions can be applied to a set of HSPs: (i) relating the sum of the lengths of all HSPs to the length of the genomes; (ii) relating the number of identical nucleotides within the HSPs to the sum of the lengths of all HSPs; and (iii) relating the number of identical nucleotides within the HSPs to the length of the genomes. In the case of MUMs, within which all nucleotides are identical by construction, it makes only sense to apply function (i).

(d) *Conversion to percent-wise similarities*

The obtained distance values are converted into percent-wise similarities by using the corresponding values for intercept and slope. The percent-wise similarity can be used analogously to a DDH value. Values for intercept and slope are determined by applying the robust line fitting procedure as implemented in the R package (Version 2.6.2) to the dataset described in Auch *et al.* (2010b).

$$s(d) = md + c$$

Additionally, the corresponding distance threshold as determined in Auch *et al.* (2010b) can be used for species delimitation. Any distance value above the threshold can be regarded as indication that the two genomes analysed represent two distinct species (Auch *et al.*, 2010a).

C. Identification of Horizontally Transferred Genomic Islands

Horizontal gene transfer (HGT) is an important mechanism for the evolution of microbial genomes. Exchange of genomic islands (GEI) and islets was found to have occurred in different domains of life: Archaea, Bacteria and Eukarya (Choi and Kim, 2007). Mobile genetic elements possess genes that contribute to bacterial speciation and adaptation to different niches (Dobrindt *et al.*, 2004). Collectively, the latter factors form part of a gene organization known as the *flexible gene pool*. The flexible gene pools are named according to the types of functions they encode, and are as follows:

- **Pathogenicity islands (PAIs)** correlate with virulence and were first identified in uropathogenic *Escherichia coli* strains as distinct chromosomal regions in possession of genes encoding virulence factors (Oelschlaeger *et al.*, 2002). These factors enable bacteria to undergo several host–cell infection cycles, particularly, to adhere to host surfaces, attain protection against immune cells and produce toxins. Virulence factors are disseminated by plasmids and bacteriophages, for they play the most crucial role in mobilizing virulent cassettes across species boundaries (Betley and Mekalanos, 1985; Leplae *et al.*, 2006; Lima-Mendez *et al.*, 2008a).
- **Symbiosis islands** share similar structural properties with pathogenicity islands. They both use similar mechanisms that influence the integration and host–bacterial interaction. Unlike pathogenicity islands, symbiosis islands are not associated with bacterial virulence. They encode new proteins and functions that establish mutual relationships between bacteria and multicellular organisms. For example, *Mesorhizobium* carry chromosomally integrated nitrogen fixation islands that benefit their plant hosts (Uchiumi *et al.*, 2004).
- **Antibiotic resistance islands** endow bacteria with multiple drug resistance. Bacteria can develop resistance by either random mutation, transformation or transduction, but the most common way through which bacteria acquire drug resistance gene cassettes is conjugation. For example, most of the tetracycline resistance genes are identified in resistance plasmids, making

horizontal transfer the likely method of their transfer (Hartman *et al.*, 2003; Pezzella *et al.*, 2004).

- **Catabolic genomic islands** possess genes that enable bacteria to degrade xenobiotic chemicals that are difficult to consume or even harmful to living organisms. Genes encoding relevant enzymes frequently have been found to be located within these islands (Butler *et al.*, 2007).

GEIs are large DNA sequences which are distinguished from the core genome by their different G + C content. They are frequently associated with tRNA genes, and are often flanked by repeat sequences. Mobility genes that encode integrases or transposases are often detectable. It has been speculated that they evolved from former lysogenic bacteriophages and plasmids that have lost their ability for replication and self-transfer (Dobrindt *et al.*, 2004). The identification of genomic islands falls mainly on the basis of compositional features that distinguish them from native genes in the genome or they may be predicted by sequence similarity with previously identified genomic islands stored in databases.

1. Horizontal gene transfer database (HGT-DB)

Horizontal gene transfer database (HGT-DB) (http://genomes.urv.cat/HGT-DB/) is a composition-based web resource that provides pre-calculated averages and standard deviations for GC content, codon usage, relative synonymous codon usage and amino acid content of bacterial and archaeal complete genomes. It also provides lists of putative genomic islands, correspondence analyses of the codon usage and lists of extraneous genes in terms of their GC contents (Garcia-Vallve *et al.*, 2003). It uses a set of statistical approaches to determine the genes that deviate from the mean GC and/or average codon usage of the genome. HGT-DB provides no tools for analysis of genomes submitted by users.

2. Pathogenicity island database (PAI-DB)

Pathogenicity island database (PAI-DB) (http://www.gem.re.kr/paidb/) contains the comprehensive information of all reported and potential PAI regions in prokaryotic genomes. In total 1040 PAI-like regions were identified in 237 bacterial genomes by the PAI Finder tool. PAI Finder accepts input sequences of predicted ORFs in multi FASTA format. The query is limited to 400 ORFs per run (approximately 350 kb). The PAI-DB resource may be used as follows:

- Predicted ORFs must be saved in a FASTA file and each sequence in the file has to be named strictly according to the PAI Finder format: ORF id, name, coordinates in the genome (left..right) and the strand (+/−) separated by vertical lines (|). For example:
  >3|name3617|3406225..3406300| +
  ATGCGGATAGCTCAGTCGGTAGAGCAGGGGATTGAAAATCCCCGTGT
  CCTTGGTT

- Query sequences of total length below 30 kbp may be pasted in text box on the web page; otherwise, the sequences have to be stored locally and uploaded to the server.
- After clicking the button 'Analyze', the service returns lists of PAIs homologous to ones found in PAIDB for each ORF in the input.

As any other homology-based prediction tool, PAI Finder has limitations: it may only identify PAI if at least one similar genomic island is already present in the PAIDB. However, PAIDB is regularly updated, improving its reliability. Another limitation is that the genome of interest has to be pre-annotated and a PAI may be overlooked if the annotation is not appropriate. Also preparation of the input file may be time consuming.

3. A classification of mobile genetic elements (ACLAME) project

Aclame (http://aclame.ulb.ac.be/) is a comprehensive web resource that aids with the classification and annotation of proteins encoded by mobile genomic elements (MGEs) (Leplae *et al.*, 2004). It has a collection of protein families obtained from bacteriophages and plasmids. The proteins were clustered into families according to functional parameters they have in common by using TRIBE-MCL, a graph theory based Markov clustering algorithm. Genomic islands were identified by using Prophinder tool (Lima-Mendez *et al.*, 2008b). Prophinder was designed to detect prophages in bacterial genome sequences stored in GenBank formatted files (these files usually have extensions GB or GBK). Prophinder is available as an on-line tool and may be utilized as follows:

- Prepare a GenBank file of a bacterial genome of interest. GenBank files of many sequenced genomes are available for download from ftp://ftp.ncbi.nih.gov/genomes/Bacteria/.
- Go to the Prophinder home page http://aclame.ulb.ac.be/perl/Aclame/Prophages/prophinder.cgi.
- Use the button 'Browse' to upload the genome file.
- Accuracy of the analysis may be adjusted by setting the scanning window size; minimum number of phage related CDS in prophages; minimal number of ACLAME hits per scanning window; and Blast e-value threshold. Prophinder is homology based approach. It predicts prophages by blasting the predicted proteins encompassed with the sliding window against ACLAME database of phage associated proteins.
- The prediction may be refined by secondary search after masking all obvious hits (set by default) and by looking for flanking repeats.
- Provide your e-mail for the server feedback and click the button 'Submit genome'.

The extreme mutability of phage related genes in prophages may make them undetectable by Blast search. Fragmentation of prophages due to genome rearrangements complicates the detection by Prophinder even greater.

4. IslandViewer

IslandViewer (www.pathogenomics.sfu.ca/islandviewer/) is a web-resource that incorporates precomputed genomic islands that were identified by the three prediction methods: IslandPick (Langille and Brinkman, 2009), IslandPath (Hsiao *et al.*, 2003) and SIGI-HMM (Waack *et al.*, 2006). It provides a simple view of all genomic islands predictions for the latter methods through a single integrated interface.

To analyse the sequence of a newly sequenced bacterial chromosome, first write it to the file in GenBank or EMBL format. Then follow these steps:

- On the project web-site click 'Genome upload'.
- Choose the corresponding sequence file format and click the button 'Browse' to locate the file on the computer. Optionally, the genome name may be entered to facilitate the navigation through the resulted graphs if multiple genomes are going to be analysed.
- Click the button 'Upload'. When sequence upload is complete, enter the e-mail address and click 'Submit'. In a while you will be notified when the analysis is finished.
- Inspect your mail box. Eventually, you will get a message with a hyperlink that will bring you to the result of the analysis. The locations of identified genomic islands predicted by IslandPick, SIGI-HMM and IslandPath will be depicted by green, orange and blue boxes, respectively. A high-resolution graphical file and exact coordinates in an Excel file are available for download.

IslandViewer is superb in genomic island prediction by combining three alternative approaches based on genome comparison (IslandPick), codon usage comparison using a Hidden Markov Model algorithm (SIGI-HMM) and DNA composition comparison algorithm (IslandPath).

5. SeqWord genome browser and gene island sniffer

SeqWord Genome Browser (SWGB) was developed to visualize the natural compositional polymorphism of DNA sequences and to identify divergent genomic regions including horizontally transferred genomic islands (Ganesan *et al.*, 2008). The approach is based on the analysis of biased distribution of tetranucleotides in bacterial genomes. Several statistical parameters, – distances between local oligonucleotide usage (OU) patterns calculated for sliding windows and the global pattern of the whole genome, OU variance and pattern skew defined by Reva and Tümmler (2004, 2005) are superimposed by the program to distinguish between mobile genomic islands and other elements characterized by an alternative OU (clusters of genes encoding ribosomal RNA and proteins, tandem multiple repeats and so on). SWGB allows visual identification of genomic islands by browsing bacterial chromosomes, grouping genomic fragments by their compositional properties and a simultaneous referring to the genetic context. The SWGB resource may be used as follows:

- On the SWGB web-page (http://www.bi.up.ac.za/SeqWord/mhhapplet.php) select one bacterial chromosome, plasmid or phage in the list and click 'Display in the Applet'.
- Click on the tab 'Diagram' and choose the parameters n1_4mer:RV, ni_4mer:GRV and n0_mer:PS for the axes X, Y and Z, respectively (more about these parameters and abbreviations see in Ganesan *et al.*, 2008). Click 'Enter'. A diagram of distribution of 8 kbp long genomic fragments will appear on the plot. Using the mouse, draw a box around the group of dots on the plot (Figure 4) and click 'Get'. The program will return a list of genomic loci and annotations represented by the outlined dots on the plot, which correspond to the horizontally transferred genetic elements.
- Double-click a dot on the plot to refer to genetic content of the corresponding region on the 'Gene Map' diagram.

Sequences stored in FASTA or GenBank files may be analysed by SWGB locally.

- The command line Python program OligoWords is available for download from http://www.bi.up.ac.za/SeqWord/downloads.htm in several packages containing precompiled executable files. Input sequence files have to be copied to the 'OligoWords1.2.1/input/' folder.
- The command prompt window provides several parameters set by default that may be changed by the user (refer to the readme file). Type <Y> + <Enter> to perform the analysis.

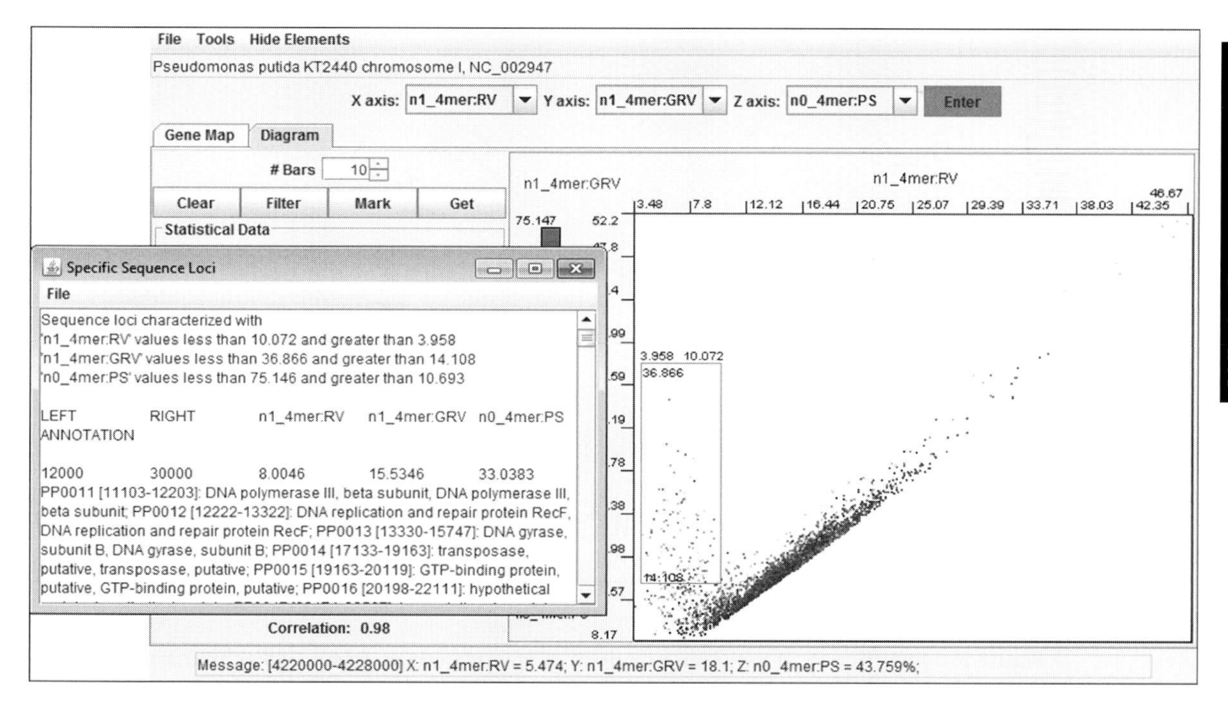

Figure 4. Identification of genomic islands in *Pseudomonas putida* KT2440 using SWGB.

- The program will analyse recursively all input files with the extensions FST, FAS, FNA and GBK, and store the results to the folder 'output' in text files with the extension OUT.
- Use the Java applet on the SWGB web-page to view these files using File- > Open menu command.

SWGB allows composition-based identification of genomic islands in annotated genomic sequences stored in GenBank files and in raw DNA sequences in FASTA format. The lengths of the sequences to be analysed have to be above 20 kbp. SWGB is not able to identify inserts of foreign DNA shorter that the half of the sliding window size. One common problem for all sliding window based approaches is that the resulted prediction may depend on the starting point of the analysis. To improve and automate the prediction of genomic islands a SeqWord Gene Island Sniffer (SWGIS) utility was developed (http://www.bi.up.ac.za/SeqWord/sniffer/index.html). This command prompt program uses a shorter sliding window in the areas where an insertion is suspected. A collection of genomic islands identified in bacterial genomes by SW Sniffer is present in GEI-DB (http://anjie.bi.up.ac.za/geidb/geidb-home.php).

SWGB and Sniffer cannot identify genomic islands if they share similar OU distribution with the host chromosome. For example, symbiotic islands of *Rhizobium* are not detectable by SWGB. Clusters of genes for 16S rRNA are

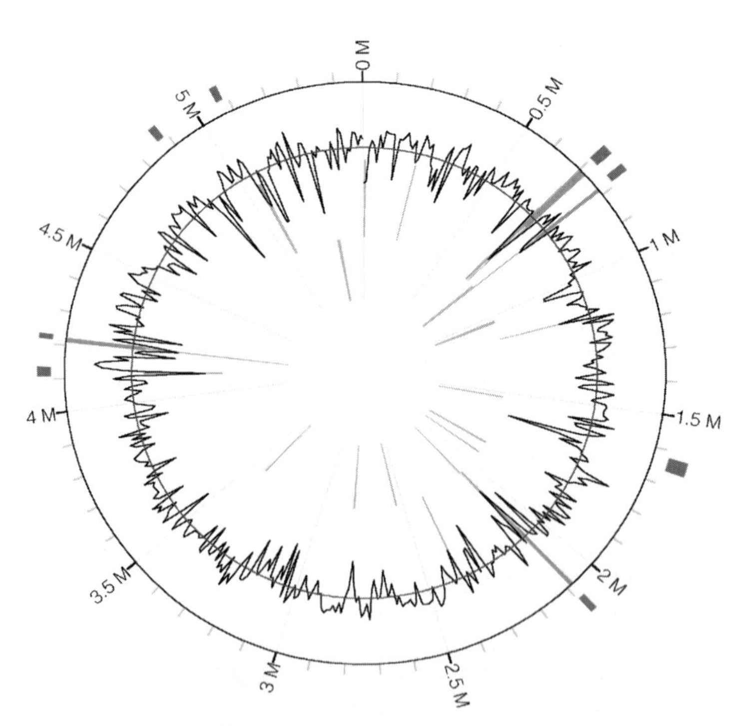

Figure 5. Identification of genomic islands in *Acidovorax avenae* ATCC 19860 by different prediction tools: SWGIS (red), IslandPath (blue), SIGI-HMM (orange) and IslandPick (green boxes). Black line histogram shows GC% variations (see colour insert).

often falsely predicted as genomic islands. To identify all genomic islands in a chromosome the best way is to combine the results of different prediction methods, as suggested by Langille and Brinkman (2009). In Figure 5 the results of predictions of horizontally transferred genetic elements in a newly sequenced genome of *Acidovorax avenae* ATCC 19860 by IslandPick, SIGI-HMM, IslandPath and SWGIS are superimposed on the chromosomal map.

♦♦♦♦♦♦ IV. MICROARRAY-BASED COMPARATIVE GENOMIC HYBRIDIZATION (MCGH)

MCGH is a powerful method for rapidly identifying regions of genomic diversity among closely related organisms in absence of complete genome datasets. The method can be applied in investigating a group of closely related strains given that a microarray prepared from a complete reference genome is available. The advantage is that no further whole genome sequencing of every strain under investigation is necessary. The disadvantage is that no genes unique in a single strain can be identified. The technique allows one to predict gene absence (or divergence) versus gene presence by measuring the relative hybridization efficiencies of two differentially Cy-labelled pools of genomic DNA taken from two strains. The method has been previously applied to discriminate different members of the *B. subtilis* clade (Earl *et al.*, 2007). We used a *B. amyloliquefaciens* FZB42 specific oligonucleotide microarray as reference to identify genes that are absent or divergent in the strains under investigation. As controls, FZB42–FZB42, and FZB42–CH40 (FZB42Δ*nrs::cm* and Δd*hb::em*) hybridizations were performed. As expected, the self–self control experiments yielded no genes with a $\log_2$ fluorescence ratio greater than 1. The results of the FZB42-CH40 hybridization did, however, reveal a potential limitation of the array; the values for only 8 of the 11 genes known to be deleted in this strain were above the cut-off ratio for gene absence or divergence. This may have been a consequence of cross-hybridization between gene spots, because the missing genes *nrsA*, *nrsC* and *nrsF* contain peptide synthetase modules that exist quite frequently in the genome of FZB42. Strains FZB42, its derivative CH40 and FZB24 were nearly identical according to the heat map composite image (Borriss *et al.*, 2010). The other plant-associated *B. amyloliquefaciens* strains formed a cluster with only limited diversity in comparison to FZB42. Strain DSM 7^T, together with the two other non-plant-associated *B. amyloliquefaciens* strains, ATCC15841 and S23, were more diverse and formed a separate cluster. Interestingly, the phylogeny obtained when the degree and pattern of gene variation measured by the arrays was used as a marker of relatedness was in almost perfect concordance with the phylogeny obtained when *gyrA* and *cheA* was used as markers of relatedness. In both cases the cluster of *B. amyloliquefaciens* FZB42 related strains were discriminated from the cluster of $DSM7^T$ related strains. As shown for *B. subtilis* (Earl *et al.*, 2007), MCGH may also prove to be a reliable phylogenetic tool for subtyping strains of *B. amyloliquefaciens* (Borriss *et al.*, 2010).

A. Microarray-based Comparative Genomic Hybridization—Procedure

1. Hybridization and microarray scanning

- The Bam4kOLI microarray is designed based on the sequenced complete genome of the 'master' strain *B. amyloliquefaciens* FZB42 (Chen *et al.*, 2007). The array contains 3931 50–70mer oligonucleotides representing predicted protein-encoding genes (3816 spots) and small non-coding RNA genes (238 spots) of FZB42. In addition, the array includes stringency controls with 71%, 80%, and 89% identity to the native sequences of five genes, *dna*A, *rps*L, *rps*O, *rps*P and *rpm*I, to monitor the extent of cross hybridization. The array also contains alien DNA oligonucleotides for four antibiotic resistance genes (em^R, cm^R, nm^R and spc^R) and eight spiking controls as well as one empty control. All oligonucleotides are printed in four replicates. Microarrays are produced and processed as described (Brune *et al.*, 2006).
- The oligonucleotides are designed using the Oligo Designer software from the Bioinformatics Resource Facility, CeBiTec, Bielefeld University. Melting temperatures of the oligonucleotides are calculated according to their %GC and oligo lengths, ranging from 73 to 83°C (optimal 78°C). Salt concentration is set to be 0.1 M. QGramMatch is used to analyse uniqueness of the designed oligos.
- The comparative hybridizations are repeated for each test strain two or three times and include at least one hybridization where the labelling regime is switched to rule out potential bias introduced by inherent differences in Cye dye incorporation. Five micrograms of purified, *Msp*I- and *Taq*I-digested genomic DNA is labelled with either Cy3- or Cy5-NHS ester as described by Giuntini *et al.* (2005). Unincorporated fluorescent nucleotides are removed by using Microcon 30 filter columns (Millipore, Milano, Italy). The appropriate Cy5 and Cy3 labelled probes are combined and mixed with 30 μl Cot-1 DNA (1 mg/ml), 20 ml yeast tRNA (5 mg/ml), 450 μl TE buffer to concentrate the samples until 40 μl using Microcon 30 filter columns. To each combined sample 8.5 μl of 20 × SSC and 0.74 μl of 10% SDS is added. The sample is denaturated at 100°C for 1.5 min, and then incubated for 37°C for 30 min. The hybridization probe is added to the microarray under a coverslip, and hybridization is then performed at 65°C for 16 h. Slides are washed at 60°C with 2 × SSC for 5 min. and then at 60°C with 0.2 × SSC containing 0.1% SDS for 5 min., and finally at room temperature with 0.2 × SSC for 2 min. The last step is conducted twice. The slides are immediately dried and scanned for fluorescence intensity by using a GenePix 4000B microarray scanner (Axon Instruments, Union City, CA), and the results are recorded in 16-bit multi-image TIFF files.

For each sample a total of four slides are hybridized (after dye swapping of the two different restriction enzyme DNA digests); considering that one slide carries three replicas of each gene, any sample is hybridized 12 times at each gene.

2. Normalization and significant hybridization differences

- Following hybridization and scanning, data analysis is done by applying the ImaGene 6.0 software (Biodiscovery Inc., Los Angeles, CA) for acquisition of the mean signal and mean local background intensity for each spot of the microarray and the EMMA 2.2 software for normalization and t-statistics (Dondrup *et al.*, 2003, 2009). A gene is considered to have a statistically significant difference in hybridization if the $\log_2$-ratio of the intensities (M value) is ≥ 1 or ≤ -1 and the mean intensity (A value; $A_i = \log_2(R_i G_i)^{0.5}$) is ≥ 7 and in two of the three repeats the $P_{adjusted}$ value was ≤ 0.1. A positive $\log_2$-ratio of the intensities (M value) indicates that the respective gene is missing in the genome of the tested strain.

3. EMMA and ArrayLIMS, useful platforms for microarray data processing

As a high throughput technique, microarray experiments produce large datasets, consisting of measured data, laboratory protocols and experimental settings. CeBiTEC from University Bielefeld has implemented the open source platform EMMA http://www.cebitec.uni-bielefeld.de/groups/brf/software/emma_info/ to store and analyse these data. EMMA gives access to all the transcriptomics datasets stored in the ArrayLIMS and provides automated pipelines for data processing, allowing an automated or manual analysis of expression profiles. In addition to routine data analysis algorithms, the system can be integrated with other components that contain additional data sources (e.g. genome annotation systems). In the design of the microarray experiments, special care must be taken in projects within the same network, to ensure comparability of these data and compliance to new and arising international standards. This system also provides automated tools to perform data normalizations, tests for the identification of statistically significant up or down-regulated genes, clustering algorithms and, in the long run, support for time-course analyses.

ArrayLIMS is a Microarray Laboratory Information Management system has been designed in order to streamline data acquisition and reporting processes. It provides a permanent and consistent storage of the microarray experiment data as well as a fast information retrieval, making the data rapidly available. The stored data is standardized, consisting of the hybridization steps (e.g. RNA production), production of the hybridization targets or the hybridization itself. It is also possible to store images of the hybridized and scanned slides as well as the corresponding data files.

◆◆◆◆◆◆ V. CONCLUDING REMARKS

The continued decrease in the price of sequencing whole genomes, together with the technical advances that have been made suggest that routine sequencing of prokaryote genomes is realistic from now on (Tindall *et al.*, 2010). A key issue that remains is the reliable annotation of all genes in a genome since identifying gene homologies (preferably orthologues) is of central importance in taxonomy. In principle there are three basic approaches: (1) genome indexes, increasingly used as an *'in silico'* alternative to the experimental DDH, (2) gene content, its successful application depends on the number of genome sequences available for this analysis, and (3) multiple aligned (gene) sequence datasets (3). In this chapter we have presented several methods, we have found practicable for the non-experienced scientist with background in microbial taxonomy, for estimating those genomic parameters. We are sure that further development in the field will facilitate use of genomics as an essential part of prokaryote taxonomy.

♦♦♦♦♦♦ ACKNOWLEDGEMENTS

Long-term financial support for R.B. in frame of the competence network Genome Research on Bacteria (GenoMik, GenoMikPlus and GenMikTransfer) and the Chinese-German collaboration program by the German Ministry for Education and Research (BMBF) is gratefully acknowledged.

References

Altschul, S. F., Gish, W., Miller, W., Myers, E. W. and Lipman, D. J. (1990). Basic local alignment search tool. *J. Mol. Biol.* **215**, 403–410.

Altschul, S. F., Madden, T. L., Schäffer, A. A., Zhang, J., Zhang, Z., Miller, W. and Lipman, D. J. (1997). Gapped BLAST and PSI-BLAST: a new generation of protein database search programs. *Nucleic Acids Res.* **25**, 3389–33402.

Auch, A. F., von Jan, M., Klenk, H.-P. and Göker, M. (2010a). Digital DNA-DNA hybridization for microbial species delineation by means of genome-to-genome sequence comparison. *Stand. Genomic Sci.* **2**, 142–148.

Auch, A. F., Klenk, H.-P. and Göker, M. (2010b). Standard operation procedure for calculating genome-to-genome distances based on high scoring segment pairs. *Stand. Genomic Sci.* **2**, 142–148.

Badger, J. H. and Olsen, G. J. (1999). CRITICA: Coding Region Identification Tool Invoking Comparative Analysis. *Mol. Biol. Evol.* **16**, 512–524.

Betley, M. J. and Mekalanos, J. J. (1985). Staphylococcal enterotoxin A is encoded by phage. *Science.* **229**, 185–187.

Blom, J., Albaum, S. P., Doppmeier, D., Pühler, A., Vorhölter, F. J., Zakrzewski, M. and Goesmann, A. (2009). EDGAR: asoftware framework for the comparative analysis of prokaryotic genomes. *BMC Bioinformatics.* **10**, 154.

Borriss, R., Chen, X. H., Rueckert, C., Blom, J., Becker, A., Baumgarth, B., Fan, B., Pukall, R., Schumann, P., Spröer, C., Junge, H., Vater, J., Pühler, A. and Klenk, H.-P. (2010). Relationship of *Bacillus amyloliquefaciens* clades associated with strains DSM 7^T and *Bacillus amyloliquefaciens* subsp. *plantarum* subsp. nov. based on their discriminating complete genome sequences. *Int. J. Syst. Evol. Microbiol.* **2010** [September 3; epub ahead of print].

Brune, I., Becker, A., Paarmann, D., Albersmeier, A., Kalinowski, J., Pühler, A. and Tauch, A. (2006). Under the influence of the active deodorant ingredient 4-hydroxy-3-methoxybenzyl alcohol, the skin bacterium *Corynebacterium jeikeium* moderately responds with differential gene expression. *J. Biotechnol.* **127**, 21–33.

Butler, J. E., He, Q., Nevin, K. P., He, Z., Zhou, J. and Lovley, D. R. (2007). Genomic and microarray analysis of aromatics degradation in *Geobacter metallireducens* and comparison to a *Geobacter* isolate from a contaminated field site. *BMC Genom.* **8**, 180.

Carver, T. J., Rutherford, K. M., Berriman, M., Rajandream, M. A., Barrell, B. G. and Parkhill, J. (2005). ACT: the Artemis comparison tool. *Bioinformatics.* **e21**, 3422–3423.

Chen, X. H., Koumoutsi, A., Scholz, R., Eisenreich, A., Schneider, K., Heinemeyer, I., Morgenstern, B., Voss, B., Hess, W. R., Reva, O., Junge, H., Voigt, B., Jungblut, P. R., Vater, J., Süssmuth, R., Liesegang, H., Strittmatter, A., Gottschalk, G. and Borriss, R. (2007). Comparative analysis of the complete genome sequence of the plant growth-promoting bacterium *Bacillus amyloliquefaciens* FZB42. *Nat. Biotechnol.* **25**, 1007–1014.

Chevreux, B., Wetter, T. and Suhai, S. (1999). Genome sequence assembly using trace signals and additional sequence information. *Comput. Sci. Biol., Proc. German Conference on Bioinformatics GCB'99 GCB*, 45–56.

Choi, I.-G. and Kim, S.-H. (2007). Global extent of horizontal gene transfer. *Proc. Natl. Acad. Sci. U.S.A.* **104**, 4489–4494.

Delcher, A. L., Harmon, D., Kasif, S., White, O. and Salzberg, S. L. (1999). Improved microbial gene identification with GLIMMER. *Nucleic Acids Res.* **27**, 4636–4641.

Deloger, M., El Karoui, M. and Petit, M.-A. (2009). A genomic distance based on MUM indicates discontinuity between most bacterial species and genera. *J. Bacteriol.* **191**, 91–99.

Dobrindt, U., Hochhut, B., Hentschel, U. and Hacker, J. (2004). Genomic islands in pathogenic and environmental microorganisms. *Nat. Rev. Microbiol.* **2**, 414–424.

Dondrup, M., Goesmann, A., Bartels, D., Kalinowski, J., Krause, L., Linke, B., Rupp, O., Sczyrba, A., Pühler, A. and Meyer, F. (2003). EMMA: a platform for consistent storage and efficient analysis of microarray data. *J. Biotechnol.* **106**, 135–146.

Dondrup, M., Hüser, A. T., Mertens, D. and Goesmann, A. (2009). An evaluation frame work for statistical tests on microarray data. *J. Biotechnol.* **140**, 18–26.

Earl, A. M., Losick, R. and Kolter, R. (2007). *Bacillus subtilis* genome diversity. *J. Bacteriol.* **189**, 1163–1170.

Edgar, R. C. (2004). MUSCLE: multiple sequence alignment with high accuracy and high throughput. *Nucleic Acids Res.* **32**, 1972.

Felsenstein, J. (1978). Cases in which parsimony or compatibility methods will be positively misleading. *Syst. Zool.* **27**, 401–410.

Felsenstein, J. (1995). *PHYLIP (Phylogeny Inference Package), Version 3.57 c.* University of Washington, Seattle.

Ganesan, H., Rakitianskaia, A. S., Davenport, C. F., Tummler, B. and Reva, O. N. (2008). The SeqWord Genome Browser: an online tool for the identification and visualization of atypical regions of bacterial genomes through oligonucleotide usage. *BMC Bioinformatics.* **9**, 333.

Garcia-Vallve, S., Guzman, E., Montero, M. A. and Romeu, A. (2003). HGT-DB: a database of putative horizontally transferred genes in prokaryotic complete genomes. *Nucleic Acids Res.* **31**, 187–189.

Gatson, J. W., Benz, B. F., Chandrasekaran, C., Satomi, M., Venkateswaran, K. and Hart, M. E. (2006). *Bacillus tequilensis* sp. nov., isolated from 2000-year-old Mexican shaft-tomb, is closely related to *Bacillus subtilis*. *Int. J. Syst. Evol. Microbiol.* **56**, 1475–1484.

Giuntini, E., Mengoni, A., De Filippo, C., Cavalieri, D., Aubin-Horth, N., Landry, C. R., Becker, A. and Bazzicalupo, M. (2005). Large-scale genetic variation of the symbiosis-required megaplasmid pSymA revealed by comparative genomic analysis of *Sinorhizobium meliloti* natural strains. *BMC Genom.* **6**, 158.

Gontcharov, A. A., Marin, B. and Melkonian, M. (2004). Are combined analyses better than single gene phylogenies? A case study using SSU rDNA and rbcL sequence comparisons in the Zygnematophyceae (Streptophyta). *Mol. Biol. Evol.* **21**, 612–624.

Gordon, D., Abajian, C. and Green, P. (1998). Consed: a graphical tool for sequence finishing. *Genome Res.* **8**, 195–202.

Gordon, D., Desmarais, C. and Green, P. (2001). Automated finishing with Autofinish. *Genome Res.* **11**, 614–625.

Goris, J., Konstantinidis, K. T., Klappenbach, J. A., Coenye, T., Vandamme, P. and Tiedje., J. M. (2007). DNA-DNA hybridization values and their relationship to whole-genome sequence similarities. *Int. J. Syst. Evol. Microbiol.* **57**, 81–91.

Hartman, A. B., Essiet, I. I., Isenbarger, D. W. and Lindler, L. E. (2003). Epidemiology of tetracycline resistance determinants in *Shigella* spp. and enteroinvasive *Escherichia coli*: characterization and dissemination of *tet(A)*-1. *J. Clin. Microbiol.* **41**, 1023–1032.

Henz, S. R., Huson, D. H., Auch, A. F., Nieselt-Struwe, K. and Schuster, S. C. (2005). Whole-genome prokaryotic phylogeny. *Bioinformatics.* **21**, 2329–2335.

Hsiao, W., Wan, I., Jones, S. J. and Brinkman, F. S. L. (2003). IslandPath: aiding detection of genomic islands in prokaryotes. *Bioinformatics.* **19**, 418–420.

Hyatt, D., Chen, G. L., Locascio, P. F., Land, M. L., Larimer, F. W. and Hauser, L. J. (2010). Prodigal: prokaryotic gene recognition and translation initiation site identification. *BMC Bioinformatics* **11**, 119.

Kent, W. J. (2002). BLAT – the BLAST-like alignment tool. *Genome Res.* **12**, 656–664.

Konstantinidis, K. T., Ramette, A. and Tiedje, J. M. (2006). The bacterial species definition in the genomic era. *Philos. Trans. R. Soc. Lond. B Biol. Sci.* **361**, 1929–1940.

Kozarewa, I., Ning, Z., Quail, M. A., Sanders, M. J., Berriman, M. and Turner, D. J. (2009). Amplification-free illumina sequencing-library preparation facilitates improved mapping and assembly of (G + C)-biased genomes. *Nat. Methods* **6**, 291–295.

Krause, L., McHardy, A. C., Nattkemper, T. W., Pühler, A., Stoye, J. and Meyer, F. (2007). GISMO – gene identification using a support vector machine for ORF classification. *Nucleic Acids Res.* **35**, 540–549.

Kurtz, S., Phillippy, A., Delcher, A. L., Smoot, M., Shumway, M., Antonescu, C. and Salzberg, S. L. (2004). Versatile and open software for comparing large genomes. *Genome Biol.* **5**, R12.

Langille, M. G. I. and Brinkman, F. S. L. (2009). IslandViewer: an integrated interface for computational identification and visualization of genomic islands. *Bioinformatics* **25**, 664–665.

Leplae, R., Hebrant, A., Wodak, S. J. and Toussaint, A. (2004). ACLAME: a CLAssification of Mobile genetic Elements. *Nucleic Acids Res.* **32**, D45–D49.

Leplae, R., Lima-Mendez, G. and Toussaint, A. (2006). A first global analysis of plasmid encoded proteins in the ACLAME database. *FEMS Microbiol. Rev.* **30**, 980–994.

Lerat, E., Daubin, V. and Moran, N. A. (2003). From gene trees to organismal phylogeny in prokaryotes: the case of the gamma-Proteobacteria. *PLoS Biol.* **1**, E19.

Lima-Mendez, G., Helden, J. V., Toussaint, A. and Leplae, R. (2008a). Reticulate representation of evolutionary and functional relationships between phage genomes. *Mol. Biol. Evol.* **25**, 762–777.

Lima-Mendez, G., Helden, J. V., Toussaint, A. and Leplae, R. (2008b). Prophinder: a computational tool for prophage prediction in prokaryotic genomes. *Bioinformatics* **24**, 863–865.

Linke, B., McHardy, A. C., Neuweger, H., Krause, L. and Meyer, F. (2006). REGANOR: a gene prediction server for prokaryotic genomes and a database of high quality gene predictions for prokaryotes. *Appl. Bioinformatics* **5**, 193–198.

Logan, N. A., Berge, O., Bishop, A. H., Busse, H.-J., de Vos, P., Fritze, D., Henydrickx, M., Kämpfer, P., Rabinovitch, L., Salkinoja-Salonen, M. S., Seldin, L. and Ventosa, A. (2009). Proposed minimal standards for describing new taxa of aerobic, endospore-forming bacteria. *Int. J. Syst. Evol. Microbiol.* **59**, 2114–2121.

Margulies, M., Egholm, M., Altman, W. E., Attiya, S., Bader, J. S., Bemben, L. A., Berka, J., Braverman, M. S., Chen, Y.-J., Chen, Z., Dewell, S. B., Du, L., Fierro, J. M., Gomes, X. V., Goodwin, B. C., He, W., Helgesen, S., He, Ho. C., Irzyk, G. P., Jando, S. C., Alenquer, M. L. I., Jarvie, T. P., Jirage, K. B., Kim, J.-B., Knight, J. R., Lanza, J. R.,

Leamon, J. H., Lefkowitz, S. M., Lei, M., Li, J., Lohman, K. L., Lu, H., Makhijani, V. B., McDade, K. E., McKenna, M. P., Myers, E. W., Nickerson, E., Nobile, J. R., Plant, R., Puc, B. P., Ronan, M. T., Roth, G. T., Sarkis, G. J., Simons, J. F., Simpson, J. W., Srinivasan, M., Tartaro, K. R., Tomasz, A., Vogt, K. A., Volkmer, G. A., Wang, S. H., Wang, Y., Weiner, M. P., Yu, P., Begley, R. F. and Rothberg, J. M. (2006). Genome sequencing in open microfabricated high density picoliter reactors. *Nature* **437**, 376–380.

Medini, D., Donati, C., Tettelin, H., Masignani, V. and Rappuoli, R. (2005). The microbial pan-genome. *Curr. Opin. Genet. Dev.* **15**, 589–594.

Meyer, F., Goesmann, A., McHardy, A. C., Bartels, D., Bekel, T., Clausen, J., Kalinowski, J., Linke, B., Rupp, O., Giegerich, R. and Pühler, A. (2003). GenDB–an open source genome annotation system for prokaryote genomes. *Nucleic Acids Res.* **31**, 2187–2195.

Oelschlaeger, T. A., Dobrindt, U. and Hacker, J. (2002). Pathogenicity islands of uropathogenic *E. coli* and the evolution of virulence. *Int. J. Antimicrob. Agents* **19**, 517–521.

Peterson, J. D., Umayam, L. A., Dickinson, T., Hickey, E. K. and White, O. (2001). The comprehensive microbial resource. *Nucleic Acids Res.* **29**, 123–125.

Pezzella, C., Ricci, A., DiGiannatale, E., Luzzi, I. and Carattoli, A. (2004). Tetracycline and streptomycin resistance genes, transposons, and plasmids in *Salmonella enterica* isolates from animals in Italy. *Antimicrob. Agents Chemother.* **48**, 903–908.

Reva, O. N. and Tummler, B. (2004). Global features of sequences of bacterial chromosomes, plasmids and phages revealed by analysis of oligonucleotide usage patterns. *BMC Bioinformatics* **5**, 90.

Reva, O. N. and Tummler, B. (2005). Differentiation of regions with atypical oligonucleotide composition in bacterial genomes. *BMC Bioinformatics* **6**, 251.

Ronaghi, M., Karamohamed, S., Pettersson, B., Uhlén, M. and Nyrén, P. (1996). Real-time DNA sequencing using detection of pyrophosphate release. *Anal. Biochem.* **242**, 84–89.

Rooney, A. P., Price, N. P., Ehrhardt, C., Swezey, J. L. and Bannan, J. D. (2009). Phylogeny and molecular taxonomy of the *Bacillus subtilis* species complex and description of *Bacillus subtilis* subsp. *inaquosorum* subsp. nov. *Int. J. Syst. Evol. Microbiol.* **59**, 2420–2436.

Rueckert, C., Blom, J., Chen, X. H., Reva, O. and Borriss, R. (2011). Genome sequence of *B. amyloliquefaciens* type strain DSM7^T reveals differences to plant-associated *B. amyloliquefaciens* FZB42. *J. Biotechnol.* **2011** [January 22; epub ahead of print].

Sanger, F. and Coulson, A. R. (1975). A rapid method for determining sequences in DNA by primed synthesis with DNA polymerase. *J. Mol. Biol.* **94**, 441–448.

Schwartz, S., Kent, W. J., Smit, A., Zhang, Z., Baertsch, R., Hardison, R. C., Haussler, D. and Miller, W. (2003). Human-mouse alignments with BLASTZ. *Genome Res.* **13**, 103–107.

Simpson, J. T., Wong, K., Jackman, S. D., Schein, J. E., Jones, S. J. and Birol, I. (2009). ABySS: A parallel assembler for short read sequence data. *Genome Res.* **19**, 1117–1123.

Stackebrandt, E., Frederiksen, W., George, M., Garrity, G. M., Grimont, P. A. D., Kaempfer, P., Maiden, M. C. J., Nesme, X., Rossello-Mora, R., Jean Swings, J., Trueper, H. G., Vauterin, L., Ward, A. C. and Whitman, W. B. (2002). Report of the ad hoc committee for the re-evaluation of the species definition in bacteriology. *Int. J. Syst. Evol. Microbiol.* **52**, 1043–1047.

Talavera, G. and Castresana, J. (2007). Improvement of phylogenies after removing divergent and ambiguously aligned blocks from protein sequence alignments. *Syst. Biol.* **56**, 564–577.

Tettelin, H., Masignani, V., Cieslewicz, M. J., Donati, C., Medini, D., Ward, N. L., Angiuoli, S. V., Crabtree, J., Jones, A. L., Durkin, A. S., Deboy, R. T., Davidsen, T. M.,

Mora, M., Scarselli, M., Margarit, Y., Ros, I., Peterson, J. D., Hauser, C. R., Sundaram, J. D., Nelson, W. C., Madupu, R., Brinkac, L. M., Dodson, R. J., Rosovitz, M. J., Sullivan, S. A., Daugherty, S. C., Haft, D. H., Selengut, J., Gwinn, M. L., Zhou, L., Zafar, N., Khouri, H., Radune, D., Dimitrov, G., Watkins, K., O'Connor, K. J. B., Smith, S., Utterback, T. R., White, O., Rubens, C. E., Grandi, G., Madoff, L. C., Kasper, D. L., Telford, J. L., Wessels, M. R., Rappuoli, R. and Fraser, C. M. (2005). Genome analysis of multiple pathogenic isolates of *Streptococcus agalactiae*: implications for the microbial "pan-genome". *Proc. Natl. Acad. Sci. U.S.A.* **102**, 13950–13955.

Tindall, B. J., Rossello-Mora, R., Busse, H.-J., Ludwig, W. and Kämpfer, P. (2010). Notes on the characterization of prokaryote strains for taxonomic purposes. *Int. J. Syst. Evol. Microbiol.* **60**, 249–266.

Uchiumi, T., Ohwada, T., Itakura, M., Mitsui, H., Nukui, N., Dawadi, P., Kaneko, T., Tabata, S., Yokoyama, T., Tejima, K., Saeki, K., Omori, H., Hayashi, M., Maekawa, T., Sriprang, R., Murooka, Y., Tajima, S., Simomura, K., Nomura, M., Suzuki, A., Shimoda, Y., Sioya, K., Abe, M. and Minamisawa, K. (2004). Expression islands clustered on the symbiosis island of the *Mesorhizobium loti* genome. *J. Bacteriol.* **186**, 2439–2448.

Uchiyama, I., Higuchi, T. and Kawai, M. (2010). MBGD update 2010: toward a comprehensive resource for exploring microbial genome diversity. *Nucleic Acids Res.* **38** (Suppl. 1):D361–D365.

Waack, S., Keller, O., Asper, R., Brodag, T., Damm, C., Fricke, W. F., Surovcik, K., Meinicke, P. and Merkl, R. (2006). Score-based prediction of genomic islands in prokaryotic genomes using hidden Markov models. *BMC Bioinformatics* **16**, 142.

Wayne, L. G., Brenner, D. J., Colwell, R. R., Grimont, P. A. D., Kandler, O., Krichevsky, M. I., Moore, L. H., Moore, W. E. C., Murray, R. G. E., Stackebrandt, E., Starr, M. P. and Trüper, H. G. (1987). Report of the Ad Hoc Committee on Reconciliation of Approaches to Bacterial Systematics. *Int. J. Syst. Bacteriol.* **37**, 463–464.

Zerbino, D. R. and Birney, E. (2008). Velvet: algorithms for de novo short read assembly using de Bruijn graphs. *Genome Res.* **18**, 821–829.

19 How to Name New Genera and Species of Prokaryotes?

Aharon Oren
Department of Plant and Environmental Sciences, Institute of Life Sciences, and the Moshe Shilo Minerva Center for Marine Biogeochemistry, The Hebrew University of Jerusalem, Jerusalem, Israel

♦♦♦

CONTENTS

♦♦♦♦♦♦ I. THE IMPORTANCE OF NAMES

> I do not deny but nature, in the constant production of particular beings, makes them not always new and various, but very much alike and of kin one to another: but I think it nevertheless true, that the boundaries of the species, whereby men sort them, are made by men; since the essences of the species, distinguished by different names, are, as has been proved, of man's making, and seldom adequate to the internal nature of the things they are taken from. So that we may truly say, such a manner of sorting of things is the workmanship of men.
> **(Locke (1689), An essay concerning human understanding, Book III, Chapter VI)**

> Microbiologists who have occasion to use the scientific names of the microorganisms with which they deal generally prefer to use *correct* names and to use them *correctly*.
> **(Buchanan *et al.* (1948), Foreword to the first edition of the International Bacteriological Code of Nomenclature)**

When Antonie van Leeuwenhoek first viewed bacteria through the lens of his primitive microscopes in the last decades of the seventeenth century, he did not name the tiny organisms he had discovered. The first attempts to classify and

METHODS IN MICROBIOLOGY, VOLUME 38
0580-9517 DOI 10.1016/B978-0-12-387730-7.00019-X

name bacteria were made in the nineteenth century by pioneers such as Christian Ehrenberg (1795–1877) and Ferdinand Cohn (1828–1898). Some of the genus names that feature in their early classification schemes are still in use today (e.g. *Bacillus, Micrococcus, Spirillum, Spirochaeta, Vibrio* etc.).

The basic unit of classification of all living organisms is the species. Today the list of different species of prokaryotes (Bacteria and Archaea combined, but not including most cyanobacteria) whose names have been validly published based on the internationally accepted rules (see Section II) encompasses little more than 9000 species (May 2011). However, in spite of the achievements of over 200 years of research, there still is no clear concept of what a prokaryote species actually is. Definitions can be found in the literature that circumscribe the species as 'a distinct group of strains that have certain distinguishing features and that generally bear a close resemblance to one another in the more essential features of organization' or 'an assemblage of clonal populations that share a high degree of phenotypic similarity, coupled with an appreciable dissimilarity from other assemblages of the same general kind' (Colwell *et al.*, 1995). The most accepted concept (the so-called phylo-phenetic species concept) defines a species as a monophyletic and genomically coherent cluster of individual organisms (strains) that show a high degree of overall similarity in many independent characteristics and is diagnosable by one or more discriminative phenotypic properties (Rosselló-Mora and Amann, 2001). Such definitions provide no practical information on how close that resemblance and similarity should be for two strains to be classified in the same species, what features of organization should be considered essential and what degree of dissimilarity is required for two strains to warrant classification in different species. The delineation of species according to such definitions remains therefore highly subjective.

A pragmatic definition of the prokaryote species has emerged in the past decades, based on a 'polyphasic' approach, which includes description of diagnostic phenotypic features combined with genomic properties (Wayne *et al.*, 1987; Stackebrandt *et al.*, 2002). A widely accepted criterion defines a prokaryotic species as a group of strains, including the type strain, that share at least 70% total genome DNA:DNA hybridization and have less than 5°C ΔT_m (=the difference in the melting temperature between the homologous and the heterologous hybrids formed under standard conditions). The delineation value of 70%, as introduced around 1987, is artificial, but has proven satisfactory in most cases. The previous chapters in this book provide an in-depth description of the different properties relevant in polyphasic taxonomy of prokaryotes and of the tests to be performed to determine whether a newly discovered prokaryote may belong to an already described and named species or whether a new species (and if applicable, a new genus, family, order etc.) should be established to accommodate the new isolate(s).

Based on our current pragmatic species concept for the prokaryotes, new species are being described at an ever-growing rate. As of 10 May 2011, the numbers of prokaryotic names with standing in the nomenclature included rank of class: 77, rank of order: 128 (of which 1 is illegitimate), rank of family:

291 (of which 6 are illegitimate), rank of genus: 2010 (of which about 105 are considered as synonyms, and 20 are illegitimate) and rank of species: 10,706 (of which 31 are later homotypic synonyms cited in the *Approved Lists of Bacterial Names* (Skerman *et al.*, 1980, 1989), 1237 are new combinations, 13 are *nomina nova*, about 288 are considered as later heterotypic synonyms, and 67 are illegitimate). Although in the 1980s the numbers of genera and species annually added to the lists were around 23 and 130, respectively, in the period 2006–2010 about 120 new genera and about 620 new species were described each year (Euzéby, 1997; http://www.bacterio.net).

The nomenclature of the prokaryotes is governed by the general considerations, principles, rules and recommendations of the International Code of Nomenclature of Bacteria (The Bacteriological Code), a document approved by the International Committee on Systematic Bacteriology in 1990 (Lapage *et al.*, 1992). A new version of the Code will soon be published, to be named the International Code of Nomenclature of Prokaryotes. Nomenclature of prokaryotes made a new start on 1 January 1980. By that time, the nomenclature had become extremely confusing, as different names had often been given to identical bacteria so that synonyms abounded. The publication of the 'Index Bergeyana – An annotated alphabetic listing of names of the taxa of the bacteria' (Buchanan *et al.*, 1966), with about 28,900 entries, showed the urgent need to establish order in the increasing chaos. The situation was rectified in 1980 with the publication of the 'Approved list of bacterial names' (Skerman *et al.*, 1980, 1989). Rule 24a of the Bacteriological Code states: 'Priority of publication dates from 1 January 1980. On that date all names published prior to 1 January 1980 and included in the Approved Lists of Bacterial Names of the ICSB are treated for all nomenclatural purposes as though they had been validly published for the first time on that date'. With the publication of the 'Approved Lists', which encompassed 2335 names (7 classes, 1 subclass, 21 orders, 3 suborders, 66 families, 24 tribes, 290 genera, 1792 species and 131 subspecies), all other earlier published names lost their status in the nomenclature.

Central listing of new names validly published after 1 January 1980, according to the rules of the International Code of Nomenclature of Bacteria, ensured that the number of validly published names of species, genera, families and orders of prokaryotes is known at any time. All names with standing in the nomenclature are listed in the *International Journal of Systematic and Evolutionary Microbiology* (until 1999: the *International Journal of Systematic Bacteriology*). The information is also available online at http://www.bacterio.net, a website established and faithfully maintained by Prof. Jean Euzéby of the National Veterinary School of Toulouse, France (Euzéby, 1997).

New names can be added to the list of validly published names by publication of the description of the new taxa in the *International Journal of Systematic and Evolutionary Microbiology* while meeting different criteria as determined by the Code ('valid publication' – publication of a name in the *International Journal of Systematic and Evolutionary Microbiology/International Journal of Systematic Bacteriology*, accompanied by a description of the taxon or by a reference to a previous effectively published description of the taxon, and with the designation of a nomenclatural type in accordance with the rules of the International Code of Nomenclature of Prokaryotes, as stated in Rules 27 and 30 of the Code). Descriptions of new taxa and publication of new names are also possible in other journals. This is known as 'effective publication' of names, meaning publication of a name in printed matter and/or electronic material generally made available to the scientific community (Rule 25 of the Code). Names thus published obtain standing in the nomenclature only when they are also included in the 'Validation Lists' ('lists of new names and new combinations previously effectively, but not validly, published') that appear every two months in the *International Journal of Systematic and Evolutionary Microbiology* (Tindall *et al.*, 2006).

♦♦♦♦♦♦ II. THE CODES OF NOMENCLATURE AND THE INTERNATIONAL COMMITTEE ON SYSTEMATICS OF PROKARYOTES

The rules of prokaryote nomenclature, as published in the International Code of Nomenclature of Bacteria ('The Bacteriological Code', published first in 1948 as the International Bacteriological Code of Nomenclature) (Buchanan *et al.*, 1948), are set by the International Committee on Systematics of Prokaryotes (ICSP, before 2000: the International Committee of Systematic Bacteriology; http://www.the-icsp.org). This committee, a constituent part of the International Union of Microbiological Societies, discusses nomenclatural problems that have arisen in different groups of prokaryotes and proposes changes and amendments to the rules of the Code. The Judicial Commission of the ICSP deals with problematic cases in bacterial nomenclature and renders judicial decisions in instances of controversy about the valid publication of a name, identity of type strains and cases of emerging problems with the interpretation of the rules of the Code. It also may propose amendments to the Code and consider exceptions that may be needed to certain rules. The ICSP has also established taxonomic subcommittees (currently 28) that discuss nomenclatural problems of specific groups of prokaryotes. The minutes of the meetings of the ICSP, its Judicial Commission and its taxonomic subcommittees are published in the *International Journal of Systematic and Evolutionary Microbiology*.

The last published version of the 'Bacteriological Code', as approved at the Ninth International Congress of Microbiology, Moscow, 1966 and revised in 1990 (Lapage *et al.*, 1992), has since been amended at subsequent meetings of the ICSP. A new revised version of the Code is currently in preparation, and will be named the International Code of Nomenclature of Prokaryotes.

The Code contains the rules and recommendations to guide authors of papers describing new species, genera, families etc. of prokaryotes how to properly name the new taxa. It covers the rules for the naming of species (and subspecies), genera (and subgenera), tribes (and subtribes), families (and subfamilies), orders (and suborders) and classes (and subclasses) of prokaryotes. No provisions are made by the Code for the naming of the higher taxa: phylum and kingdom, and infrasubspecific subdivisions (biovars, serovars etc.).

Here follows an annotated selection of some of the most relevant passages from the Code:

> **General Consideration 4**
> Rules of nomenclature do not govern the delimitation of taxa nor determine their relations. The rules are primarily for assessing the correctness of the names applied to defined taxa; they also prescribe the procedures for creating and proposing new names.

Nomenclature of prokaryotes is thus governed by internationally accepted rules, but there is no official taxonomy and classification of prokaryotes. Thus classification schemes such as given, e.g. in *Bergey's Manual of Systematic Bacteriology* (Garrity and Holt, 2001), and widely used among microbiologists do not have any official status.

> **Principle 1** (as modified at the meetings of the ICSB/ICSP in Sydney in 1999) (De Vos and Trüper, 2000)
>
> The essential points in nomenclature are as follows.

1. Aim at stability of names.
2. Avoid or reject the use of names which may cause error or confusion.
3. Avoid the useless creation of names.
4. Nothing in this Code may be construed to restrict the freedom of taxonomic thought or action.

Stability of names is a major aim of all Codes of Nomenclature. It may be necessary from time to time to change existing names on the basis of improved insights in the taxonomy of certain groups of prokaryotes, but confusion is likely if renaming of organisms occurs frequently. All Codes of Nomenclature therefore require conservation of older names wherever possible. It is essential that names should be stable, unambiguous and necessary (Sneath, 2001).

> **Principle 2** (as modified at the meetings of the ICSB/ICSP in Sydney in 1999) (De Vos and Trüper, 2000)
>
> The nomenclature of bacteria is not independent of botanical and zoological nomenclature. When naming new taxa in the rank of genus or higher, due consideration is to be given to avoiding names which are regulated by the Zoological Code and the International Code of Botanical Nomenclature. *Note*: This principle takes effect with publication of acceptance of this change by the ICSB (now ICSP) – i.e., 14 December 2000 – and is not retroactive.

When creating new names of genera, one therefore has to ensure that the proposed name is not already in use in the botanical or the zoological nomenclature. Centralized listing of names, as it exists for the prokaryotes, has not yet been introduced in the nomenclature of plants and animals. However, there are databases that can be consulted online, such as the Index Nominum Genericorum – a compilation of generic names of plants (http://botany.si.edu/

ing/) (accessed 30 May 2011), uBio (Universal Biological Indexer and Organizer) http://uio.mbl.edu/ (accessed 30 May 2011) and the lists of names in use in the zoological nomenclature found at the ZooBank – The Prototype Online Registry for Zoological Nomenclature (http://www.zoobank.org) (accessed 30 May 2011). Some older names of prokaryote genera exist in other nomenclatures as well. Thus, *Bacillus* is also the name of an insect, and *Proteus* is not only a genus of *Enterobacteriaceae* but also an amphibian. However, both these generic names were validly published before the acceptance of the new Principle 2 (i.e. they appeared in the 1980 Approved Lists [Skerman *et al.*, 1980, 1989]), and are therefore conserved in the prokaryote nomenclature.

Principle 3
The scientific names of all taxa are Latin or latinized words treated as Latin regardless of their origin. They are usually taken from Latin or Greek.

Principle 4
The primary purpose of giving a name to a taxon is to supply a means of referring to it rather than to indicate the characters or the history of the taxon.

'Names are labels, not descriptions' (Sneath, 2001). The type species of the genus *Natrialba* was non-pigmented, a feature unusual among the representatives of the family *Halobacteriaceae*, and this property was highlighted when naming the genus (Latin: *albus, -a, -um*=white). Later, red-pigmented species were added to the genus, but in spite of their colour the genus name *Natrialba* will remain attached to them. The rules of the Code do not prevent attaching the specific epithet *alkaliphilus* (-a, -um) to an isolate that that growth at low pH or *acidiphilus* (-a, -um) to a species that prefers alkaline environments. To avoid confusion, it is of course not recommended to name new organisms in this way.

Rule 6
The scientific names of all taxa must be treated as Latin; names of taxa above the rank of species are single words.

Recommendation 6 (as modified at the meetings of the ICSB/ICSP in Sydney in 1999) (De Vos and Trüper, 2000).

To form new bacterial names and epithets, authors are advised as follows.

1. Avoid names or epithets that are very long or difficult to pronounce.
2. Make names or epithets that have an agreeable form that is easy to pronounce when latinized.
3. Words from languages other than Latin or Greek should be avoided as long as equivalents exist in Latin or Greek or can be constructed by combining word elements from these two languages. *Exceptions*: names derived from typical local items such as foods, drinks or geographical localities for which no Latin or Greek names exist.

In spite of Recommendation 6 there are quite a few very long bacterial genus names and specific epithets around. The longest are *Thermoanaerobacterium thermosaccharolyticum* (42 letters), *Thermoanaerobacterium polysaccharolyticum* (40 letters), *Methanothermococcus thermolithotrophicus* (39 letters), *Thermoanaerobacter thermohydrosulfuricus* (39 letters) and *Thermoanaerobacterium thermosulfurigenes* (39 letters). Shorter names can be completely satisfactory, as shown by the examples *Dyella soli*, *Vibrio xuii* and *Yania flava*, being the shortest names (10 letters) currently in the lists. Whether names are difficult to pronounce depends to a large extent on the linguistic and cultural background of the person using the name. Specific epithets such as *nieuwersluisensis* (in the recently proposed but at the time of writing not yet validly published name *Ferrocurvibacter nieuwersluisensis*), *aotearoense* (in *Themoanaerobium aotearoense*) and *chahannaoensis* (in *Natrialba chahannaoensis*) may be easy to pronounce for native speakers of Dutch, Maori and Chinese, respectively, but may give some trouble to many others, in spite of the fact that indications how to pronounce the names are included with the etymology.

The main provisions for the formation of correct genus and species names for prokaryotes are given by Rules 10a and 12c of the Code:

Rule 10a

The name of a genus ... is a substantive, or an adjective used as a substantive, in the singular number and written with an initial capital letter. The name may be taken from any source and may even be composed in an arbitrary manner. It is treated as a Latin substantive.

Examples: Single Greek stem, *Clostridium;* two Greek stems, *Haemophilus;* single Latin stem, *Spirillum;* two Latin stems, *Lactobacillus;* hybrid name, Latin-Greek stems, *Flavobacterium;* latinized personal name, ...

A name may be composed of more elements; e.g. *Ectothiorhodospira* is composed of four Greek stems, with the following etymology: Gr. prep. *ektos*, outside; Gr. n. *theion* (Latin transliteration *thium*), sulfur; Gr. n. *rhodon*, rose; L. fem. n. *spira*, spiral; N.L. fem. n. *Ectothiorhodospira*, spiral rose with sulfur outside.

Recommendation 10a

The following Recommendations apply when forming new generic ... names.

1. Refrain from naming genera ... after persons quite unconnected with bacteriology or at least with natural science.
2. Give a feminine form to all personal generic ... names whether they commemorate a man or a woman.

Despite Recommendation 10a some genera are named after persons unconnected with natural science: *Afifella* (named after S. Afif, a British philosopher and painter), *Martelella* (named after E. Martel, a French explorer), *Serratia* (named after S. Serrati, an Italian physicist) etc.

Rule 12c

A specific epithet may be taken from any source and may even be composed arbitrarily.

A specific epithet must be treated in one of the three following ways.

1. As an adjective that must agree in gender with the generic name.

2. As a substantive (noun) in apposition in the nominative case.
3. As a substantive (noun) in the genitive case.

Recommendation 12c (as modified at the meetings of the ICSB/ICSP in Paris in 2002) (De Vos *et al.*, 2005)

Authors should attend to the following Recommendations, and those of Recommendation 6, when forming specific epithets.

1. Choose a specific epithet that, in general, gives some indication of a property or of the source of the species.
2. Avoid those that express a character common to all, or nearly-all, the species of a genus.
3. Ensure that, if taken from the name of a person, it recalls the name of one who discovered or described it, or was in some way connected with it, and possesses the appropriate gender.
4. Avoid in the same genus epithets which are very much alike, especially those that differ only in their last letters.
5. Avoid the use of the genitive and the adjectival forms of the same specific epithet to refer to two different species of the same genus.
6. If an ordinal adjective used for enumeration is chosen, then they may include numbers up to ten. *Example*: primus, secundus.

It is thus recommended to avoid formation of names such as *Halobacillus halophilus* and *Halobacillus salinus* [Recommendation 12c(2)], *Streptomyces cavourensis* (named after Cavour, a leading figure in the movement towards Italian unification) [Recommendation 12c(3)] and both *Pseudomonas palleroniana* and *Pseudomonas palleronii* [Recommendation 12c(5)].

Rule 27(2)
... The derivation (etymology) of a new name (and if necessary of a new combination) must be given.

The word 'etymology' is derived from the Greek etumos, true, and thus aims at the true, the literal sense of a word according to its origin. Etymology is a necessary element in biological nomenclature as it explains the existing (i.e. so far given) names and helps to form new names (Trüper, 2001). The source of the word elements using in the name should be stated (e.g. L. = Latin; N.L. = neo-Latin; M.L. = mediaeval Latin; Gr. = Greek), and the recommended ways to break the words should be given as well. Here is, as an example, the etymological information given in the recent description (November 2010) of *Natronoarchaeum mannanilyticum* gen. nov., sp. nov.:

- *Natronoarchaeum* (Na.tro.no.ar.chae'um. N.L. n. *natron* (arbitrarily derived from Arabic n. *natrun* or *natron*), soda, sodium carbonate; N.L. pref. *natrono-*, pertaining to soda; N.L. neut. n. *archaeum* archaeon (from Gr. adj. *archaios -ê -on*, ancient); N.L. neut. n. *Natronoarchaeum*, the soda archaeon).
- *Natronoarchaeum mannanilyticum* (man.na.ni.ly'ti.cum. N.L. neut. n. *mannanum*, mannan; N.L. neut. adj. *lyticum* (from Gr. neut. adj. *lutikon*), able to loosen, able to dissolve; N.L. neut. adj. *mannanilyticum*, mannan-dissolving).

Sometimes the information on the etymology and recommendations how the words may be broken is, indeed, important. A well-known case is the genus *Acidianus*, with the following etymology: L. adj. *acidus*, acid; L. masc. n. *Ianus*, a mythical Roman figure with two faces looking in opposite directions; N.L. masc. n. *acidianus*, acidic bifaced (bacterium), reflecting the growth conditions and the metabolism of the organisms. Based on this etymology, the proper pronunciation and breaking of the word would be A.cid.ia'nus and not A.ci.di.a'nus, 'suggesting a different meaning and causing suggestive jokes' (Trüper, 1999).

Rule 64
Diacritic signs are not used in names or epithets in bacteriology.
In names or epithets derived from words with such signs, the signs must be suppressed and the letters transcribed as follows: (1) ä, ö, and ü become *ae*, *oe*, and *ue*; (2) é, è, and ê become *e*; (3) ø, œ, and å become *oe*, *ae*, and *aa*, respectively.

◆◆◆◆◆◆ III. CREATION OF NAMES OF NEW GENERA AND SPECIES – A PRACTICAL GUIDE

Practical recommendations how to name a new prokaryote can be found in the Orthography appendix to the 'Bacteriological Code' (Trüper and Euzéby, 2009) and in review papers by Trüper (1996, 1999, 2001) and MacAdoo (1993). When using names of chemical compounds in new genus names and specific epithets, an article by Buchanan (1960) gives useful guidelines how to incorporate chemical terms into the nomenclature of prokaryotes.

The general guidelines given below provide only examples. It is always important to check the rules of the Code when proposing a new name, and it is always desirable to consult experts on prokaryote nomenclature (see below).

The following general guidelines apply when proposing new names:

1. Except for the last word element, only the stems are to be used. The stem to be used in making compounds is not always complete in the nominative. It is found by dropping the genitive ending.
2. Use of hyphens is not allowed.
3. The connecting vowel is -o- when the preceding element is of Greek, it is -i- when the preceding element is of Latin origin. There are some exceptions to this general rule. First of all, it applies only to new names. The rule was not followed in some of the old names such as *Lactobacillus* and *Bifidobacterium*, names that appeared in the Approved Lists of 1980 (Skerman *et al.*, 1980, 1989), and such names can no longer be changed based on the current version of the Code. The Judicial Commission in its meeting in Sydney in 1999 noted that the stabilization of nomenclature of well-established taxa deserves priority over orthographic correctness (De Vos and Trüper, 2000). Greek is more flexible than Latin about the connecting vowel, and other connecting vowels than -o- may be used if a precedent is found in Greek, as shown, e.g. in the

name *Corynebacterium* (Gr. n. *korune*, a club; L. neut. n. *bacterium*, a rod; N.L. neut. n. *Corynebacterium*, a club bacterium). Another important exception is when the -o- and the -i- designate the oxidation state of a chemical element. Iron in Latin is *ferrum*; ferro- is used in the chemical nomenclature to designate reduced iron Fe(II), and ferri- is the oxidized form Fe(III). Thus, *Ferroglobus* and *Ferroplasma* are organisms that oxidize Fe(II), and *Ferrimonas* reduces Fe(III). The specific epithets *ferrooxidans* and *ferrireducens* both are well formed.

4. A connecting vowel is dropped when the following word element starts with a vowel. This general guideline has some exceptions. For reasons of clarity or of previous usage, the connecting vowel -o- may be kept in word components like bio-, geo-, halo-, neo-, macro-, micro- etc., when a component follows that begins with a vowel. An interesting case is that of the genera *Thioalkalicoccus*, *Thioalkalimicrobium* and *Thioalkalivibrio*. After these genus names were proposed in 2000–2001, the original spelling was corrected to *Thialkalicoccus*, *Thialkalimicrobium* and *Thialkalivibrio* by the List Editor of the *International Journal of Systematic and Evolutionary Microbiology*. However, during the meeting of the Judicial Commission held in Paris in 2001 (De Vos *et al.*, 2005), it was decided that the recommendations governing the use of the connecting vowel should be overruled by the usage in chemistry and physics, and therefore the original spellings *Thioalkalicoccus*, *Thioalkalimicrobium* and *Thioalkalivibrio* were restored.
5. Words from languages other than Greek and Latin should be banned as long as equivalents exist in Greek or Latin or can be constructed by combining word elements from these two languages, and as far as they are not derived from names of geographical localities or local foods or drinks for which no equivalent Latin/Greek names exist.
6. To avoid confusion with names regulated by other Codes, the word elements *-myces*, *-phyces* and *-virus* should not be used in new names of prokaryotes.

Considerations of clarity and previous usage sometimes apply to the use of chemical nomenclature in names of prokaryotes. Formally, a genus name indicating the ability of a microorganism to reduce sulfate to sulfide should start with *Desulfati-*. Indeed, there are three genera, all created in the last decade, for which this form has been adopted: *Desulfatibacillum*, *Desulfatiferula* and *Desulfatirhabdium*. On the other hand, more than 40 genus names, including some proposed in recent years, use the less correct but more familiar form *Desulfo-*. Similarly, *Cellulosilyticum* and *Cellulosimicrobium* are well-formed names derived from N.L. n. *cellulosum* (cellulose); genitive: *cellulosi*; combinations based on the word elements *Cell-* (as in *Cellvibrio*) or *Cellulo-* (in *Cellulomonas* and *Cellulophaga*) (all names with standing in the nomenclature) should be avoided when creating new names.

A. New Genus Names

According to Rule 10a of the Code, the name of a genus ... is a substantive, or an adjective used as a substantive, in the singular number and written with an initial capital letter. The name may be taken from any source and may even be composed in an arbitrary manner. It is treated as a Latin substantive.

Based on this rule, there are only few limitations to the ways a prokaryote genus name can be formed, as long as the name is treated as a Latin substantive (noun) in the nominative case and in the singular number. However, the genus name *Faenia* is a Latin feminine plural name and it is therefore illegitimate.

Some genus names are derived from a single Latin stem (e.g. *Vibrio*, *Bacillus*) or a single Greek stem (e.g. *Clostridium*, *Thermus*). In other names, two or more Latin stems are combined (e.g. *Lactobacillus*, *Sulfolobus*), or two or more Greek stems (e.g. *Haemophilus*, *Halococcus*, *Ectothiorhodospira*), and still others are 'hybrid names' based on a combination of Latin and Greek stems: *Flavobacterium*, *Halorubrum*, *Halolactobacillus* etc. Looking at the new genus names added in recent years, one finds a number of name elements that are often used to indicate certain properties of the organism. Table 1 gives a (non-exhaustive) list. These word elements, in combination with others, will undoubtedly be used in the future as well to yield additional new names.

1. Genus names derived from personal names

Many generic names are latinized personal names to honour great microbiologists from the past and the present. Thus, names such as *Escherichia*, *Shigella*, *Pfennigia* and *Pasteuria* are named as a tribute to the contributions of Theodor Escherich, Kiyoshi Shiga, Norbert Pfennig and Louis Pasteur, respectively, to the science of microbiology. Not always are such names derived from the family name of the investigator: the genus name *Erwinia* honours Erwin F. Smith (1854–1929); both the first name and the last name were used in the name *Elizabethkingia* (in honour of Elizabeth O. King, who first described bacteria associated with infant meningitis) and in *Owenweeksia* (named after Owen B. Weeks, a pioneer of studies on *Flavobacterium*, *Cytophaga* and related organisms). The name *Simkania* was derived in an arbitrary way from the name Simona Kahane.

Based on Appendix 9 of the Code (Trüper and Euzéby, 2009), genus names can be derived from a personal name either directly or as a diminutive, both are always in the feminine gender. Table 2 gives examples and guidelines how such names should be formed. Note that when Henrique da Rocha Lima was honoured with a genus name – *Rochalimaea*, the particle 'da' was dropped and the two family names were combined. A third possibility to use personal names in genus names, not mentioned in Appendix 9, is the creation of compound, arbitrary names. Thus far there only appear to be four cases: the genera *Rummelliibacillus* named to honour astrobiologist John Rummel, *Kitasatospora* named to honour Japanese bacteriologist Kitasato, *Rathayibacter* to honour Australian plant pathologist E. Rathay and *Gordonibacter*, a tribute to the contributions of Jeffrey I. Gordon of Washington University School of Medicine, St. Louis.

According to Principle 2 of the Code as modified in 1999 (De Vos and Trüper, 2000), nomenclature of prokaryotes is no longer independent of botanical and zoological nomenclature. Therefore, when naming new taxa in the rank of

Table 1. Some word elements commonly used in prokaryote genus names and their etymology

Masculine gender		
	-arcus	L. masc. n. *arcus*, arch, bow
	-bacillus	L. dim. masc. n. *bacillus*, a small staff or rod
	-bacter	N.L. masc. n. *bacter*, a rod
	-coccus	N.L. masc. n. *coccus* (from Gr. n. *kokkos*), a berry
	-ferax	L. adj. *ferax*, fertile
	-fex	L. suff. *-fex* (from L. v. *facere*, to make)
	-globus	L. masc. n. *globus*, globe, sphere
	-oides	L. suff. *-oides* (from Gr. suff. *-eides*, from Gr. n. *eidos*, that which is seen, form, shape, figure), resembling, similar
	-philus	N.L. adj. *philus -a -um* (from Gr. adj. *philos -ê -on*), friend, loving
	-planus	L. masc. adj. *planus*, flat, plane
	-sinus	L. masc. n. *sinus*, a bent surface, curve
	-sipho	L. masc. n. *sipho*, little pipe to suck drinks through, a tube
	-vibrio	L. v. *vibro*, to set in tremulous motion, move to and fro, vibrate; N.L. n. *vibrio*, that which vibrates, and also a bacterial genus name of bacteria possessing a curved rod shape
	-vorax	L. adj. *vorax*, voracious
Feminine gender		
	-arcula	L. fem. n. *arcula*, small box
	-baca	L. fem. n. *baca*, berry
	-cystis	Gr. fem. n. *kustis*, bladder
	-monas	L. fem. n. *monas*, a unit, monad
	-musa	N.L. n. *Musa*, a scientific genus name, a banana (*Musa* sp.)
	-opsis	Gr. fem. n. *opsis*, aspect, appearance
	-phaga	Gr. v. *phagein*, to eat, to devour
	-pila	L. fem. n. *pila*, ball or sphere
	-rhabdus	Gr. fem. n. *rhabdos*, rod
	-saeta	L. fem. n. *saeta*, bristle
	-sarcina	L. fem. n. *sarcina*, a package, bundle, and also a generic name (*Sarcina*)
	-sphaera	L. fem. n. *sphaera*, a ball, globe, sphere
	-spina	L. fem. n. *spina*, spine
	-spira	L. fem. n. *spira*, coil, spire
	-spora	Gr. fem. n. *spora*, a seed, and in biology a spore
	-thrix	Gr. fem. n. *thrix*, thread
	-toga	L. fem. n. *toga*, Roman outer garment
Neuter gender		
	-bacterium	L. neut. n. *bacterium*, a small rod
	-bactrum	Gr. neut. n. *baktron*, a staff, stick, rod
	-bium	Gr. n. *bios*, life
	-filamentum	L. neut. n. *filamentum*, a spun thread
	-filum	L. neut. n. *filum*, thread, filament
	-genium	L. suff. *genius -a -um* (from L. v. *gigno*, to produce), producing
	-microbium	N.L. neut. n. *microbium* (from Gr. adj. *mikros*, small and Gr. n. *bios*, life), a microbe
	-nema	Gr. n. *nema*, thread
	-ovum	L. neut. n. *ovum*, egg
	-plasma	Gr. neut. n. *plasma*, anything formed or moulded, image, figure
	-spirillum	L. fem. n. *spira*, a spire; N.L. neut. n. *spirillum*, a spiral
	-sporangium	Gr. n. *spora*, seed; Gr. neut. n. *angeion*, a vessel
	-tomaculum	L. n. *tomaculum*, a kind of sausage

Table 2. Ways to form generic names from personal names

Personal name ending on	Add ending	Personal name	Latinized form	Diminutive ending	Personal name	Latinized form (diminutive ending)
-a	-ea	da Rocha Lima	*Rochalimaea*	Drop a, add -ella	Garcia	*Garciella*
					Labeda	*Labedella*
-e	-a	Benecke	*Beneckea*	-lla	Moore	*Moorella*
-e	-ia	Burke Lee	*Burkeia Leeia*	-lla	Jooste	*Joostella*
-i	-a	Nevski	*Nevskia*	-ella	Castellani	*Castellaniella*
					Henrici	*Henriciella*
-o	-a	Beggiato	*Beggiatoa*	-nella		Not found
		Soon-Woo Hong	*Soonwooa*[a]			
-o	-nia	Bizio	*Bizonia*	-nella	Cato	*Catonella*
		Costerton	*Costertonia*		Seino	*Seinonella*
-u	-ia	Simidu	*Simiduia*	-ella	Uruburu	*Uruburuella*
		Zhouia	*Zhou*		Zhi-Heng Liu	*Zhihengliuella*
-y	-a	Deley	*Deleya*	-ella	Lacey	*Laceyella*
					Winogradsky	*Winogradskyella*
-er	-a	Buchner	*Buchnera*	-ella	Koser	*Koserella*
		Lister	*Listeria*		Leadbetter	*Leadbetterella*
Any consonant	-ia	Kocur	*Kocuria*	-(i)ella	van Leeuwenhoek	*Leeuwenhoekiella*
		Escherich	*Escherichia*		Barnes	*Barnesiella*
		De Vos	*Devosia*		Salmon	*Salmonella*
		Erwin Smith	*Erwinia*[a]		Klug	*Klugiella*
		Wen-Xin Chen	*Wenxinia*[a]		Zobell	*Zobellella*

[a]Based on the first (given) name.

genus or higher, due consideration is to be given to avoiding names which are regulated by the Zoological Code and the International Code of Botanical Nomenclature. When in 2005 it was proposed to name a new genus in honour of Antonie van Leeuwenhoek, the discoverer of the prokaryote world, the name *Leeuwenhoekia* could not be used as it was already used for a mite. The name *Leeuwenhoekiella* was chosen instead.

It is recommended to refrain from naming genera after persons unconnected with bacteriology or at least natural sciences (recommendation 10a(1) of the Code). Further, it is a good practice to ask the person to be honoured by a scientific name for permission (as long as he/she is alive). Authors should refrain from naming bacteria after themselves or coauthors after each other in the same publication, as this may be considered immodest.

2. Arbitrary genus names

Rule 10a of the Code states that the name of a genus may be taken from any source and may even be composed in an arbitrary manner. In recent years, quite a number of such arbitrary names have been created, in many cases based on acronyms to designate research institutes and other institutions related to microbiology. A few examples:

- *Afipia*, from the abbreviation AFIP, for the Armed Forces Institute of Pathology.
- *Basfia*, derived from the chemical company BASF SE in Ludwigshafen, Germany.
- *Cedecea*, derived from the abbreviation CDC – the Centres for Disease Control and Prevention, Atlanta, Georgia.
- *Desemzia*, derived from the abbreviation DSMZ, the Deutsche Sammlung von Mikroorganismen und Zellkulturen (the German culture collection of microorganisms).
- *Emticicia*, formed from the acronym MTCC for Microbial Type Culture Collection and Gene Bank, Chandigarh, India.
- *Synergistes*, arbitrarily derived from English n. synergist.

B. New Specific Epithets

According to Rule 12c, a specific epithet must be treated in one of the three following ways: as an adjective that must agree in gender with the generic name, as a substantive (noun) in apposition in the nominative case or as a substantive (noun) in the genitive case. Below are examples of each of these options.

1. Adjectives as specific epithets

Many specific epithets are formed from Latin or latinized adjectives: *albus* (white), *aureus* (golden), *subtilis* (slender), *halophilus* (salt-loving) etc. In accordance to Rule 12(c)1, the gender should agree with the gender of the generic name. Table 3 provides examples of such adjectives used in prokaryote

Table 3. Examples of Latin adjectives in prokaryote nomenclature

	Masculine	Feminine	Neuter	English translation	Examples
First and second declensions	Parvus	Parva	Parvum	Small	*Limnohabitans parvus* *Turneriella parva* *Eperythrozoon parvum*
	Aureus	Aurea	Aureum	Golden	*Staphylococcus aureus* *Methylocapsa aurea* *Polyangium aureum*
	Piger	Pigra	Pigrum	Fat, lazy	*Desulfovibrio piger* *Undibacterium pigrum*
	Ruber	Rubra	Rubrum	Red	*Salinibacter ruber* *Rugamonas rubra* *Rhodospirillum rubrum*
Third declension	Celer	Celeris	Celere	Rapid	*Thermococcus celer* *Brevibacterium celere*
	Major (maior)	Major (maior)	Majus (maius)	More, larger	*Flectobacillus major* *Thiovulum majus*
	Agilis	Agilis	Agile	Quick	*Arthrobacter agilis* *Azomonas agilis* *Methylomicrobium agile*
	Simplex	Simplex	Simplex	Simple	*Pimelobacter simplex*
Infinitive ('present') participle	Natans	Natans	Natans	Swimming, floating	*Haloplanus natans*
	Aggregans	Aggregans	Agrregans	Aggregating	*Granulicella aggregans* *Methanocorpusculum aggregans*
	Flavescens	Flavescens	Flavescens	Becoming golden-yellow	*Muricauda flavescens* *Leptobacterium flavescens*
Perfect participle	Aggregatus	Aggregata	Aggregatum	Joined together	*Bryobacter aggregatus*
	Aerolatus	Aerolata	Aerolatum	Airborne	*Rhodocytophaga aerolata*
	Celatus	Celata	Celatum	Hidden	*Clostridium celatum*

nomenclature in the masculine, feminine and neuter gender. Participles are in the nomenclature treated as adjectives, thus enabling formation of specific epithets such as *ferrooxidans* (iron-oxidizing), *fermentans* (fermenting), *natans* (floating) etc. In most cases, these adjectives reflect some evident phenotypic property of the species. Infinitive (also named 'present') participles in the singular do not change with gender. According to the conjugations of Latin, they end on *-ans* (e.g. *natans* swimming, from natare), *-ens* (e.g. *salexigens* L. n. *sal*, *salis*, salt; L. v. *exigo*, to demand; N.L. part. adj. *salexigens*, salt-demanding) or *-iens* (e.g. *liquefaciens* L. part. adj. *liquefaciens* [from L. v. *liquefacio*], liquefying). Perfect participles change their ending with gender and are handled as

Table 4. Formation of specific epithets from personal names in the adjectival form

Ending of name	Example	Add the endings for gender			Examples
		Masculine	Feminine	Neuter	
Any consonant	Pasteur Soun Molisch	-ianus	-iana	-ianum	*Acetobacter pasteurianus* *Vulcanisaeta souniana* *Rhodospirillum molischianum*
-a	Migula Loya	-nus	-na	-num	*Aneurinibacillus migulanus* *Thalassomonas loyana*
-e		-anus	-ana	-anum	Not found
-i	Palleroni Li	-anus	-ana	-anum	*Pseudomonas palleroniana* *Cyclobacterium lianum*
-o		-anus	-ana	-anum	Not found [a]
-u		-anus	-ana	-anum	Not found
-y	Olley De Ley	-anus	-ana	-anum	*Shewanella olleyana* *Sulfurospirillum deleyianum*

[a] *Pseudoalteromonas (Alteromonas) espejiana* was named after the Chilean microbiologist Espejo; based on the current recommendations, *espejana* should be the preferred way to form the specific epithet.

adjectives of the first and second declension, e.g. *aggregatus* in *Bryobacter aggregatus* – L. masc. part. adj. *aggregatus*, joined together; *aerolata* in *Rhodocytophaga aerolata* – Gr. n. *aer*, air; L. fem. part. adj. *lata*, carried; N.L. fem. part. adj. *aerolata*, airborne; *celatum* in *Clostridium celatum* – L. neut. part. adj. *celatum*, hidden).

It is also possible to use the adjective to form specific epithets from personal names. The endings -nus, -anus or -ianus (masculine); -na, -ana or -iana (feminine); or -num, -anum or -ianum (neuter) are added, depending on the ending of the personal name (Table 4). This option to honour microbiologists from the past and the present has been surprisingly little used: the 1980 Approved Lists contain just four such names: *Acetobacter pasteurianus, Alteromonas espejiana, Clostridium pasteurianum* and *Rhodospirillum molischianum*. In the following 30 years, only 15 new ones were added: *Aeromonas sharmana, Aneurinibacillus migulanus, Cyclobacterium lianum, Hyphomonas hirschiana, Hyphomonas jannaschiana, Legionella gratiana, Phaeospirillum molischianum, Pseudomonas palleroniana, Shewanella colwelliana, Thalassomonas loyana, Thermus brockianus, Shewanella olleyana, Sulfurospirillum deleyianum* and *Vulcanisaeta souniana*.

Many specific epithets based on Latin or latinized adjectives are 'geographical' names reflecting the location where the species was first isolated. These adjectives typically carry the suffix -ensis (masculine, feminine) or -ense (neuter): *Paenibacillus huanensis* (of or belonging to Hunan, a province of China), *Mycobacterium marseillense* (pertaining to Marseille), *Muricauda lutaonensis* (pertaining to Lutao), *Halorubrum sodomense* (named for the biblical city of Sodom), *Thermoanaerobacterium aotearoense* (named after Aotearoa – 'the land of the long white cloud', being the Maori name for New Zealand) etc. It should be noted here that *Rhodothermus obamensis* is not in any way connected with the name of the president of the United States of America, but is a geographical name that refers to Obama, a hot spring in Japan.

It is good practice that in those cases when the location had a Latin name in antiquity, that name is used to make the specific epithet and not the modern name. Thus, we have *Haloferax lucentense* and not *'alicantense'*, as the Spanish city of Alicante was named Lucentum in Roman times; A *Mycobacterium* species from Nijmegen, the Netherlands, was named *Mycobacterium noviomagense* and not *'nijmegenense'*. In antiquity, Marseille was named Massilia, and therefore we have *Mycobacterium massiliense*, this in addition to *Mycobacterium marseillense*, a different species the type strain of which was also isolated from Marseille.

In recent years, the number of such 'geographical' names has strongly increased. In 2004, over a quarter of the newly proposed specific epithets were derived from local geographical names. This percentage has since decreased somewhat: it was almost 20% in the first 9 months of 2010. Trüper (2005) has called this phenomenon 'localimania', and mentioned the following possible reasons: (1) *ease*: everyone can coin such names, and no dictionary and no understanding of Latin is needed; (2) *less work*: it is not necessary to search for phenotypical properties that differentiate the new species; (3) local pride, national pride that could help funding; (4) geopolitical aims – to name organisms after areas of territorial dispute between two neighbouring countries, with the intention to show that this was done in order to support the claim of his/her own country. It is sad to note that, indeed, the creation of such 'geographical' names has been misused for geopolitical or other purposes not related to the science of microbiology. To clarify the meaning of such names, the following General Consideration was added to the Code in 2005 (Tindall *et al.*, 2008):

General Consideration 8
The International Code of Nomenclature of Bacteria (Prokaryotes) is an instrument of scientific communication. Names have meaning only in the context in which they were formed and used.

2. Substantives (nouns) in apposition in the nominative case

Rule 12c enables the formation of a specific epithet as a substantive (noun) in apposition in the nominative case. In apposition means the placing of a word or expression beside another so that the second explains and has the same grammatical construction as the first, i.e. the added nominative noun has an explanatory specifying function for the generic name (Trüper and de'Clari, 1997, 1998). For example, *Desulfovibrio gigas* may be understood as *Desulfovibrio* dictus *gigas*, *Desulfovibrio* called 'the giant'. Other similar cases are *Cupravidus necator* ('the killer'), *Ruminococcus obeum* ('the egg'), *Vibrio negripuchritudo* ('black beauty') and the recently named *Halorubrum chaoviator* ('traveller of the void', referring to the survival of the type strain following exposure to conditions in outer space during a space flight). *Bacteroides thetaiotaomicron* also belongs to this category ('a combination of the Greek letters theta, iota and omicron, relating to the morphology of vacuolated forms').

Also belonging to the group of specific epithets formed as nouns 'in apposition' are those that end with -cola (derived from incola, 'inhabitant', 'dweller') and

-cida ('the killer'). We thus have *Buchnera aphidicola* ('aphid dweller'), *Paenibacillus fonticola* ('inhabitant of a fountain'), *Salinivibrio costicola* ('rib dweller'), *Aliivibrio salmonicida* ('salmon killer') etc.

3. Substantives (nouns) in the genititve case

A far more common way to form specific epithets is the use of a noun in the genitive case. Most often the specific epithet is chosen to provide information on the environment from which the organism was isolated. The following examples, chosen randomly from the lists of recently added species, show how such epithets are formed:

- *Methanospirillum lacunae* (L. gen. n. *lacunae*, of a pond).
- *Moraxella porci* (L. gen. n. *porci*, of a pig).
- *Actinomadura scrupuli* (L. gen. n. *scrupuli*, of a small stone).
- *Paenibacillus residui* (L. gen. n. *residui*, of leavings, scraps, residues).
- *Nocardiopsis sinuspersici* (L. masc. n. *sinus persicus*, the Persian Gulf; L. gen. n. *sinuspersici*, of/from the Persian Gulf).
- *Siphonobacter aquaeclarae* (L. n. *aqua -ae*, water; L. adj. *clarus -a -um*, clear, bright; N.L. gen. n. *aquaeclarae*, of/from a clear water).
- *Cupravidus pampae* (N.L. n. *pampa* (from Quechuan noun *pampa*), pampa; N.L. gen. n. *pampae*, of pampa, the grassland plains of temperate South America).
- *Sphingopixis panaciterrulae* (N.L. n. *Panax -acis*, scientific name of ginseng; L. n. *terrula*, a field; N.L. gen. n. *panaciterrulae*, of a ginseng field).

Note the genitive form of nouns of the fourth declension in examples such as *Desulfovibrio portus* (L. gen. n. *portus*, of an estuary, of a river) and *Flectobacillus lacus* (L. gen. n. *lacus*, of a lake or pond).

The genitive noun form is also often used to form specific epithets derived from personal names, e.g. *Rhodobacter veldkampii* (L. gen. masc. n. *veldkampii*, of Veldkamp; named for H. Veldkamp, a Dutch microbiologist), *Vibrio shilonii* (N.L. gen. n. *shilonii*, of Shilo, named after the microbiologist Moshe Shilo) and *Syntrophomonas wolfei* (N.L. gen. n. *wolfei*, of Wolfe, to honour Ralph S. Wolfe). These names honour the achievements of three scientists/teachers to whom the author of this chapter owes much of his education in microbiology. When the genitive of a latinized personal name is used to form a specific epithet, the sex of the person to be honoured should be taken into consideration. Table 5 shows how specific epithets should be correctly formed from personal names as genitive nouns.

Based on the guidelines in Appendix 9 to the Code (Trüper and Euzéby, 2009), prefixes such as da, de, van, von, le, la, il etc. may be omitted or united to the name. An example in which the prefix was omitted is *Novispirillum itersonii* (N.L. gen. n. *itersonii*, of Iterson, named after the Dutch bacteriologist G. van Iterson). Examples in which the prefix was retained are *Rhodomicrobium vannielii* (N.L. gen. masc. n. *vannielii*, of van Niel; named for C.B. van Niel), *Flavobacterium degerlachei* (N.L. gen. masc. n. *degerlachei*, of de Gerlache, named in honour of Adrian de Gerlache) and *Tetrasphaera vanveenii* (N.L. gen. masc. n. *vanveenii*, of van Veen, referring to the late Dutch microbiologist W. L. van Veen).

Table 5. Formation of specific epithets from personal names as genitive nouns

Ending of name	Add for male person	Examples	Add for female person	Examples
-a	-e (classic)	Volta, *voltae* Komagata, *komagatae* Rogosa, *rogosae* Guinea, *guineae*	-e (first declension)	Renuka, *renukae* Johanna Döbereiner, *johannae*[b] Pamela Lee, *pamelaeae*[b]
	-i	Oshima, *oshimai*	-eae	Not found
	-ei	Mukohata, *mukohataei*	-iae	Not found
	-ii	Vora, *voraii*		
-e	-i	Hippe, *hippei* Hoyle, *hoylei*	-ae	Couble, *coubleae*
-é	-i	Doré, *dorei*[a]		
-i	-i	Modarski, *modarskii* Rossi, *rossii* Takai, *takaii* Hua-kui, *huakuii*[b]	-ae	Yabuuchi, *yabuuchiae*
-o	-nii	Shilo, *shilonii* Issachenko, *issachenkonii*	-niae	Cato, *catoniae*
	-nis	Not found		
-u	-ii	Simidu, *simiduii*	-iae	Joly-Guillou, *guillouiae*
-y	-i	De Ley, *deleyi* Walsby, *walsbyi* Euzéby, *euzebyi*[a]	-ae	Macy, *macyae* Olley, *olleyae*
-er	-i	Stutzer, *stutzeri* Stanier, *stanieri* Trüper, *trueperi*[a]	-ae	Wexler, *wexlerae* Lechevalier, *lechevalierae*
Any other letter	-ii	Pfennig, *pfennigii* Ørskov, *oerskovii*[a] Wynn-Williams, *wynnii* or *wynnwilliamsii* Sambhunath, *sambhunathii*[b]	-iae	Chen, *cheniae* Young, *youngiae* Doucet, *doucetiae* Roselyn Brown, *roselyniae*[b] Jia-Ling Wang, *jialingiae*[b] Isabel Spencer-Martins, *isabeliae*[b]

[a]é and è are transcribed as e; ü as ue, Ø as oe.
[b]Based on the first (given) name.

Note also the way special characters are transcribed: é and è are transcribed as e; ü as ue, ø as oe. Table 5 provides some examples.

C. Naming Candidate Taxa ('Candidatus')

Under the rules of the Code, valid publication of a species name is possible only for prokaryotes that can be cultured and maintained in pure culture, one condition being the deposition of a type strain in at least two publicly accessible service collections in different countries (De Vos and Trüper, 2000). To provide the opportunity to name species that have been relatively well characterized but cannot at present be maintained in pure culture, the category 'Candidatus' has been instituted, a provisional status that can be established for a putative taxon of an incompletely described prokaryote, awaiting its future cultivation and valid description of the name (Murray and Stackebrandt, 1995).

The category 'Candidatus' is not covered by rules of the Code, and therefore a name included in the category 'Candidatus' cannot be validly published. As of 10 May 2011, 98 such 'Candidati' have been described in the *International Journal of Systematic and Evolutionary Microbiology/International Journal of Systematic Bacteriology*, and additional ones have been documented in other journals.

Although the Code does not include provisions for the naming of 'Candidati', it is a good practice to apply the nomenclature rules and recommendations of the Code also when naming a new 'Candiatus', so that, when later the organism can be isolated in pure culture and fully characterized, the name can obtain standing in the prokaryote nomenclature. Some 'Candidati' are affiliated with genera that have standing in the nomenclature (e.g. '*Candidatus* Mycoplasma haemofelis', '*Candidatus* Streptomyces philanthi', '*Candidatus* Devosia euplotis'; for others new candidate genus names have been proposed, e.g. '*Candidatus* Neoehrlichia lotoris', '*Candidatus* Magnetoglobus multicellularis'. Note the use of quotation marks and the use of italic type for the word *Candidatus*, while the genus and species names in roman type.

D. Renaming Existing Taxa

Principle 1 of the Code reminds us that the essential points in nomenclature are: (1) Aim at stability of names, and (3) Avoid the useless creation of names. In spite of the aim for stability, changes of existing names are necessary from time to time because of our advanced insights in prokaryote taxonomy. When it becomes clear that two species formerly classified in the same genus are phylogenetically unrelated and/or their classification in different genera is warranted for other reasons, Rule 34 of the Code gives provisions for transferring species to another genus as *combinatio nova* (comb. nov.) or *nomen novum* (nom. nov.) if an author is obliged to substitute a new specific epithet as a result of homonymy. The original specific epithet is then retained in a new combination (for exceptions see Rule 41), but if the specific epithet is an adjective, its gender is adjusted if necessary so that it will agree with the gender of the genus name, as demanded by Rule 12c(1). A few examples:

- When the genus *Bacillus* was split into a number of genera, *Bacillus stearothermophilus* became *Geobacillus stearothermophilus* comb. nov.

- The reclassification of many species originally classified in the genus *Halobacterium* into a number of newly created species led to the renaming of *Halobacterium volcanii* as *Haloferax volcanii* comb. nov. and of *Halobacterium marismortui* as *Haloarcula marismortui* comb. nov.
- Renaming *Rhodospirillum sodomense* as *Rhodovibrio sodomensis* comb. nov. required a change of the gender of the specific epithet from neuter to masculine.
- When *Alcaligenes eutrophus* was renamed *Ralstonia eutropha* comb. nov., and later *Wautersia eutropha* comb. nov., the specific epithet was modified because of the female gender of the new genus names.

The original specific epithet is changed in a *nomen novum.*

- When *Cytophaga aquatilis* was reclassified in the genus *Flavobacterium,* it was necessary to substitute a new specific epithet because the name *Flavobacterium aquatile* already existed. The authors chose the epithet *hydatis* and they proposed *Flavobacterium hydatis* nom. nov.

◆◆◆◆◆◆ IV. THE SPECIAL CASE OF THE CYANOBACTERIA

Among the prokaryotes, the cyanobacteria present special nomenclature problems, as the group is traditionally treated under the rules of the International Code of Botanical Nomenclature (McNeill *et al.*, 2007) (as cyanophyta or blue-green algae) (Compère, 2005). In 1962, Roger Stanier and Cornelis van Niel, in their classic paper on 'The concept of a bacterium' wrote: 'The distinctive property of bacteria and blue-green algae is the prokaryotic nature of their cells.' Soon after, bacteriologists proposed to include the group under the rules of the Bacteriological Code. A formal statement that included the cyanobacteria among the organisms covered by the Bacteriological Code was made in General Consideration 5, as modified in 1999 (De Vos and Trüper, 2000):

General Consideration 5
This *Code of Nomenclature* applies to all Prokaryotes. . . .
Explanatory Note to General Consideration 5: 'Prokaryotes' covers those organisms that are variously recognized as, e.g. *Schizomycetes, Bacteria, Eubacteria, Archaebacteria, Archaea, Schizophycetes, Cyanophyceae* and *Cyanobacteria.*

Problems arose, as the two Codes are very different and are incompatible in a number of essential aspects:

1. One of the key provisions under the 'Bacteriological Code' is central listing of names validly published, centralized in the *International Journal of Systematic and Evolutionary Microbiology.* Under the International Code of Botanical Nomenclature, few restrictions exist on the journal in which new names may be validly published, and as a result, the existing botanical nomenclature information is widely scattered. Central listing does not yet

exist. In most cases, priority of publication dates back to 1753, although for two groups of Cyanophyta later dates are in use: 1886 and 1892.

2. Under the rules of the Bacteriological Code, the nomenclatural type of a species is a living type strain, maintained in pure culture. In contrast, type specimens of species under the Botanical Code may not be living plants or cultures but herbarium specimens, illustrations, or in some cases cultures preserved in a metabolically inactive state.
3. The Bacteriological Code is not independent of the Botanical Code. Therefore, it is currently impossible to describe under the Bacteriological Code a new species of a genus that was earlier named under the provisions of the Botanical Code. Such genus names have no valid status under the rules of the Bacteriological Code, making the new combination illegitimate.

The 1980 approved list of bacterial names (Skerman *et al.*, 1980, 1989) did not contain any names of cyanobacteria, and only very few species names have since been validly published under the 'Bacteriological Code'. Attempts are underway to solve the nomenclature problems of this group and to find a solution that will satisfy bacteriologists and botanists alike (Oren and Tindall, 2005).

♦♦♦♦♦♦ V. FINAL COMMENTS

The guidelines and examples provided above should give sufficient information to allow an author to construct and assign correctly formed names to new genera and species being described. However, it should be emphasized that only the essentials are given here. The 'Bacteriological Code' and the orthography appendix (Trüper and Euzéby, 2009), as well as more detailed treatises on prokaryote nomenclature (Trüper, 1999, 2001), should be consulted for additional information.

New names proposed in papers submitted to the *International Journal of Systematic and Evolutionary Microbiology* are subjected to proper 'quality control' by nomenclature experts. This is also true for names earlier effectively published elsewhere and submitted to be included in the Validation Lists (Tindall *et al.*, 2006). Also some other journals have qualified editorial staff to ensure that new names proposed comply with the rules of the Code. However, not all editors of scientific journals are equally aware of the demands of the nomenclature, and this had led to the appearance of some malformed names in recent years. The name of the widespread marine bacterium now known as '*Candidatus* Pelagibacter ubique' (Rappé *et al.*, 2002) (a name even often used in the literature as simply *Pelagibacter ubique*!) will never be validated in this form, as *ubique* – everywhere is an adverb, and based on Rule 12c an adverb cannot be used as a specific epithet. Another intriguing name is 'Ferroplasma acidarmanus' (Edwards *et al.*, 2000). Here the authors gave no clue about the etymology of the proposed specific epithet. It is not even clear whether acidarmanus is meant to be a Neo-Latin adjective (and if so, the gender is incorrect as the genus name has the neuter gender) or a noun 'in apposition'. I have been unable to find the word 'armanus' in any Latin dictionary.

Table 6. Neo-Latinists (in alphabetical order), willing to help microbiologists with formation of correct names of new taxa

Name	Address	E-mail address
Microbiologists with a good knowledge of Latin		
Prof. Dr Jean Euzéby	Ecole Nationale Veterinaire, 23, chemin des Capelles, B.P. 87614, F-30176 Toulouse cedex 3, France	Can be found at http://www.bacterio.net
Prof. Dr Bernhard Schink	Fachbereich Biologie, Universität Konstanz, Universitätsstr. 10, D-78457 Konstanz, Germany	bernhard.schink@uni-konstanz.dea
Other expert Neo-Latinists with understanding of prokaryote nomenclature		
Prof. Dr Reuben P. Bell	Director of Medical Humanities, University of New England College of Osteopathic Medicine, 11 Hills Beach, Biddleford, ME 04005, USA	reubenpbell@gmail.com
Prof. Dr Jean-Louis Charlet	Université de Provence, Résidence Beaumanoir, bâtiment 3, Allée des Lilas, F-13100 Aix-en-Provence, France	charlet@mmsh.univ-aix.fr
Prof. Dr Olga Anna Duhl	Lafayette College, Department of Foreign Languages and Literatures, Easton, PA 18042, USA	duhlo@lafayette.edu
Dr Nikolaus Groß	Verlag Leo Latinus.Hauptstraße 63 D-89250 Senden, Germany	spqr@leolatinus.com
Hubert W. Hawkins	105 Hill Drive, Manquin, VA 23106, USA	uppinghamhouse@aol.com
Dr Justo Hernandez	University of Laguna, Teneriffe; Home address: Calle Doctor González Coviella 11, E-38004 Santa Cruz de Teneriffe, Canary Islands, Spain	justoh79@hotmail.com
Dr Heinrich Kuhn	Seminar für Geistesgeschichte und Philosophie der Renaissance, Ludwig-Maximilians-Universität/ Hauptpost 44, Geschwister-Scholl-Platz 1, D-80539 München, Germany	hck@lrz.uni-muenchen.de
Prof. Dr Lopez-Munos	Departamento de Filologia: Filologia Latina, Universidad de Almeria, E-04120 Almeria, Spain	manuel.lopez@ual.es; manuel.lopezmunoz@gmail.com
Dr Kurt August Neuhausen	Institut für Griechische und Lateinische Philologie, Universität Bonn, Am Hof 1, D-53115 Bonn, Germany	k.a.neuhausen@uni-bonn.de
Dr Florentina Nicolae	Ovidius University Constanta, Faculty of Letters, Aleea Universitatii 1, Campus 1, 900472, Constanta, Romania	nicolae_flori@yahoo.com

(Continued)

Table 6. (Continued)

Name	Address	E-mail address
Dr Mark T. Riley	Department of Foreign Languages, California State University, Sacramento, CA 95819, USA	triley@csus.edu
Prof. Dr Joaquín José Sánchez-Gázquez	Área de Filología Latina, Departamento de Filología, Facultad de Humanidades, Universidad de Almería, Ctra. de Sacramento, s/n. E-04120 Almería, Spain	jjsanche@ual.es, jojosanche@gmail.com
Prof. Dr Wilfried Stroh	Abteilung für Griechische und Lateinische Philologie, Universität München, Geschwister-Scholl-Platz 1, D-80539 München, Germany	stroh@klassphil.uni-muenchen.de
Dr Marietta Szendrényi	Csiki u. 9, H-6753 Szeged, Hungary	sz.marietta@invitel.hu, szundne2@invitel.hu
Dr Adrie H. van der Laan	Erasmus Center for Early Modern Studies, PO Box 22140, NL-3003 DC Rotterdam, The Netherlands	vanderlaan@erasmus.org

Based on Trüper (2007) and updated for March 2011.

It is always preferable to consult experts on nomenclature before submitting proposals for new names of prokaryote taxa to scientific journals. Fortunately, there are quite a few microbiologists with an in-depth knowledge of Latin, as well as Latin experts with an interest in biological nomenclature, who are willing to help. Table 6 presents a list, based on the list compiled by Trüper in 2007 and updated in 2011, of experts willing to help authors getting the nomenclature right.

◆◆◆◆◆◆ VI. INTERNET RESOURCES

The International Code of Bacterial Nomenclature (the Bacteriological Code), 1990 version (Lapage *et al.*, 1992), is available online at http://www.ncbi.nlm.nih.gov/books/NBK8817/ (accessed 30 May 2011).

Appendix 9 (Orthography) of the International Code of Nomenclature of Prokaryotes, the updated version of the Bacteriological Code currently in preparation (Trüper and Euzéby, 2009), is available for open access at http://ijs.sgmjournals.org/cgi/content/full/59/8/2107 and at http://www.ifmb-a.uni-bonn.de/icnp-appendix-9.pdf (accessed 30 May 2011).

The International Committee on Systematics of Prokaryotes (ICSP) and its taxonomic subcommittees – http://www.the-icsp.org (accessed 30 May 2011).

The *International Journal of Systematic and Evolutionary Microbiology* (http://ijs.sgmjournals.org) (accessed 30 May 2011), the official journal of the ICSP, and the only journal in which names of new taxa of prokaryotes can be validly published.

The List of Prokaryotic Names with Standing in Nomenclature, compiled and maintained by Prof. Jean Euzéby of the National Veterinary School of Toulouse, France (http://www.bacterio.net) (accessed 30 May 2011). This website contains a wealth of information on prokaryote nomenclature, and it is updated monthly with the publication of the latest issue of the *International Journal of Systematic and Evolutionary Microbiology*. Much useful information presented in this chapter was derived from this website.

Index Nominum Genericorum – A compilation of generic names of plants, published for organisms covered by the International Code of Botanical Nomenclature (http://botany.si.edu/ing/) (accessed 30 May 2011).

uBio (Universal Biological Indexer and Organizer) (http://uio.mbl.edu/) (accessed 30 May 2011).

A list of over 250 names of genera of cyanobacteria (a database of cyanobacterial genera), prepared by Prof. Jiří Komárek and Dr Tomáš Hauer (University of South Bohemia České Budějovice, and the Academy of Sciences of the Czech Republic, Institute of Botany, Třeboň, Czech Republic), is found at http://www.cyanodb.cz/main (accessed 30 May 2011).

The International Code of Botanical Nomenclature (the Vienna Code; 2006) (McNeill *et al.*, 2007) is available online at http://ibot.sav.sk/icbn/main.htm (accessed 30 May 2011).

Lists of names in use in the zoological nomenclature can be found at the ZooBank The Prototype Online Registry for Zoological Nomenclature (http://www.zoobank.org) (accessed 30 May 2011).

Bergey's Manual of Systematic Bacteriology, the major handbook with descriptions of species and higher taxa of prokaryotes, providing a systematic framework for their classification – The Taxonomic Outline of Bacteria and Archaea http://www.bergeys.org/outlines.html (accessed 30 May 2011).

NamesforLife (http://services.namesforlife.com/home) (accessed 30 May 2011), set up by Prof. George Garrity (Technology Innovation Center, East Lansing, Michigan, USA), contains an extensive database of names in use in the nomenclature of microorganisms.

The website of the Deutsche Sammlung von Mikroorganismen und Zellkulturen (DSMZ, Braunschweig, Germany) contains a section named 'Bacterial Nomenclature Up-to-Date' (http://www.dsmz.de/microorganisms/main.php?contentleft_id=14) (accessed 30 May 2011).

♦♦♦♦♦♦ ACKNOWLEDGEMENT

I thank Prof. Jean Euzéby of the National Veterinary School of Toulouse, France, for his valuable comments.

References

Buchanan, R. E. (1960). Chemical terminology and microbiological nomenclature. *Int. Bull. Bacteriol. Nomencl. Taxon.* **10**, 16–22. Reprinted from: Buchanan, R. E. (1994).

Taxonomic notes: an aid to formation of bacterial names. Chemical terminology and microbiological nomenclature. *Int. J. Syst. Bacteriol.* 44, 588-590.

Buchanan, R. E., St., John-Brooks, R. and Breed, R. S. (1948). International Bacteriological Code of Nomenclature. *J. Bacteriol.* **55**, 287–306.

Buchanan, R. E., Holt, J. G. and Lessel, E. F. (eds), (1966). *Index Bergeyana* Williams & Wilkins, Baltimore, MD.

Colwell, R. R., Clayton, R. A., Ortiz-Conde, B. A., Jacobs, D. and Russek-Cohen, E. (1995). The microbial species concept and biodiversity. In *Microbial Diversity and Ecosystem Function* (D. Allsopp, R. R. Colwell and D. L. Hawksworth, Eds), pp. 3–15. CAB International, Wallingford, CT.

Compère, P. (2005). The nomenclature of the Cyanophyta under the Botanical Code. *Algol. Stud.* **117**, 31–37.

De Vos, P. and Trüper, H. G. (2000). Judicial Commission of the International Committee on Systematic Bacteriology. IXth International (IUMS) Congress of Bacteriology and Applied Microbiology. Minutes of the meetings, 14, 15 and 18 August 1999, Sydney, Australia. *Int. J. Syst. Evol. Microbiol.* **50**, 2239–2244.

De Vos, P., Trüper, H. G. and Tindall, B. J. (2005). Judicial Commission of the International Committee on Systematics of Prokaryotes. Xth International (IUMS) Congress of Bacteriology and Applied Microbiology. Minutes of the meetings, 28, 29 and 31 July and 1 August 2002, Paris, France. *Int. J. Syst. Evol. Microbiol.* **55**, 525–532.

Edwards, K. J., Bond, P. L., Gihring, T. M. and Banfield, J. F. (2000). An archaeal iron-oxidizing extreme acidophile important in acid mine drainage. *Science* **287**, 1796–1799.

Euzéby, J. (1997). List of bacterial names with standing in nomenclature: a folder available on the internet. *Int. J. Syst. Bacteriol.* **47**, 590–592.

Garrity, G. M. and Holt, J. G. (2001). The road map to the *Manual*. In *Bergey's Manual of Systematic Bacteriology, 2nd ed., Vol. 1. The Archaea and the Deeply Branching and Phototrophic Bacteria* (D. R. Boone and R. W. Castenholz, Eds), pp. 119–166. Springer, New York.

Lapage, S. P., Sneath, P. H. A., Lessel, E. F., Skerman, V. B. D., Seeliger, H. P. R. and Clark, W. A. (eds), (1992). *International Code of Nomenclature of Bacteria (1990 Revision). Bacteriological Code* American Society for Microbiology, Washington, D.C.

Locke, J. (1689). *An Essay Concerning Humane Understanding Printed for Tho.* Basset, London.

MacAdoo, T. O. (1993). Nomenclatural literacy. In *Handbook of New Bacterial Systematics* (M. Goodfellow and A. G. O'Donnell, Eds), pp. 339–358. Academic Press, London.

McNeill, J., Barrie, F. R., Burdet, H. M., Demoulin, V., Hawksworth, D. L., Marhold, K., Nicolson, D. H., Prado, J., Silva, P. C., Skog, J. E., Wiersema, J. H. and Turland, N. J. (Eds.) (2007). International Code of Botanical Nomenclature (Vienna Code) adopted by the Seventeenth International Botanical Congress Vienna, Austria, July 2005, Gantner, Ruggell.

Murray, R. G. E. and Stackebrandt, E. (1995). Taxonomic Note: implementation of the provisional status Candidatus for incompletely described procaryotes. *Int. J. Syst. Bacteriol.* **45**, 186–187.

Oren, A. and Tindall, B. J. (2005). Nomenclature of the cyanophyta/ cyanobacteria/ cyanoprokaryotes under the International Code of Nomenclature of Prokaryotes. *Algol. Stud.* **117**, 39–52.

Rappé, M. S., Connon, S. A., Vergin, K. L. and Giovannoni, S. J. (2002). Cultivation of the ubiquitous SAR11 marine bacterioplankton clade. *Nature* **418**, 630–633.

Rosselló-Mora, R. and Amann, R. (2001). The species concept for prokaryotes. *FEMS Microbiol. Rev.* **25**, 39–67.

Skerman, V. B. D., McGowan, V. and Sneath, P. H. A. (1980). Approved lists of bacterial names. *Int. J. Syst. Bacteriol.* **30**, 225–420.

Skerman, V. B. D., McGowan, V. and Sneath, P. H. A. (eds) (1989). *Approved Lists of Bacterial Names (Amended Edition).* American Society of Microbiology, Washington, D.C.

Sneath, P. H. A. (2001). Bacterial nomenclature. In *Bergey's Manual of Systematic Bacteriology, 2nd ed., Vol. 1. The Archaea and the Deeply Branching and Phototrophic Bacteria* (D. R. Boone and R. W. Castenholz, Eds), pp. 83–88. Springer-Verlag, New York.

Stackebrandt, E., Frederiksen, W., Garrity, G. M., Grimont, P. A. D., Kämpfer, P., Maiden, M. C. J., Nesme, X., Rosseló-Mora, R., Swings, J., Trüper, H. G., Vauterin, L., Ward, A. C. and Whitman, W. B. (2002). Report of the ad hoc committee for the re-evaluation of the species definition in bacteriology. *Int. J. Syst. Evol. Microbiol.* **52**, 1045–1047.

Stanier, R. Y. and van Niel, C. B. (1962). The concept of a bacterium. *Arch. Mikrobiol.* **42**, 17–35.

Tindall, B. J., Kämpfer, P., Euzéby, J. and Oren, A. (2006). Valid publication of names of prokaryotes according to the rules of nomenclature: past history and current practice. *Int. J. Syst. Evol. Microbiol.* **56**, 2715–2720.

Tindall, B. J., De Vos, P. and Trüper, H. G. (2008). Judicial Commission of the International Committee on Systematics of Prokaryotes. XIth International (IUMS) Congress of Bacteriology and Applied Microbiology. Minutes of the meetings, 23, 24 and 27 July, San Francisco, CA, USA. *Int. J. Syst. Evol. Microbiol.* **58**, 1737–1745.

Trüper, H. G. (1996). Help! Latin! How to avoid the most common mistakes while giving Latin names to newly discovered prokaryotes. *Microbiología SEM* **12**, 473–475.

Trüper, H. G. (1999). How to name a prokaryote? Etymological considerations, proposals and practical advice in prokaryote nomenclature. *FEMS Microbiol. Rev.* **23**, 231–249.

Trüper, H. G. (2001). Etymology in nomenclature of prokaryotes. In "Bergey's Manual of Systematic Bacteriology, 2nd ed., Vol. 1. The Archaea and the Deeply Branching and Phototrophic Bacteria" (D. R. Boone and R. W. Castenholz, Eds), pp. 89–99. Springer-Verlag, New York.

Trüper, H. G. (2005). Is 'localimania' becoming a fashion for prokaryote taxonomists? *Int. J. Syst. Evol. Microbiol.* **55**, 1753.

Trüper, H. G. (2007). Neo-Latinists worldwide willing to help microbiologists. *Int. J. Syst. Evol. Microbiol.* **57**, 1164–1166.

Trüper, H. G. and de'Clari, L. (1997). Taxonomic note: necessary correction of specific epithets formed as substantives (nouns) 'in apposition'. *Int. J. Syst. Bacteriol.* **47**, 908–909.

Trüper, H. G. and de'Clari, L. (1998). Taxonomic note: necessary correction of specific epithets formed as substantives (nouns) 'in apposition'. *Int. J. Syst. Bacteriol.* **48**, 615.

Trüper, H. G. and Euzéby, J. P. (2009). International Code of Nomenclature of Prokaryotes. Appendix 9: orthography. *Int. J. Syst. Bacteriol.* **59**, 2107–2113.

Wayne, L. G., Brenner, D. J., Colwell, R. R., Grimont, P. A. D., Kandler, O., Krichevsky, M. I., Moore, L. H., Moore, W. E. C., Murray, R. G. E., Stackebrandt, E., Starr, M. P. and Trüper, H. G. (1987). Report of the ad hoc committee on reconciliation of approaches to bacterial systematics. *Int. J. Syst. Bacteriol.* **37**, 463–464.

Index

Index

C

D

Index

N

O

P

Q

R

S

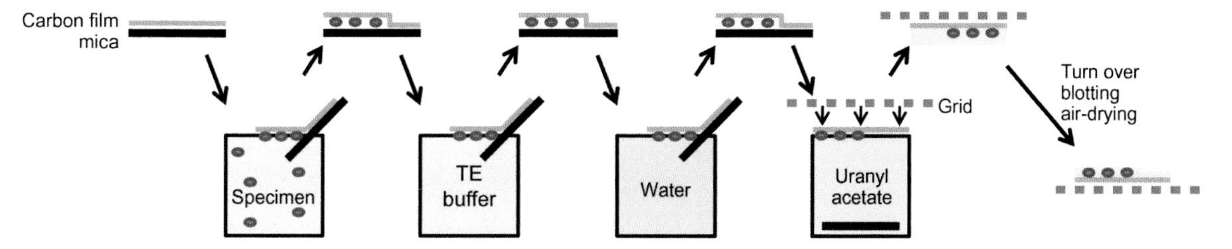

Chapter 4, Figure 2. Scheme for negative-staining applying carbon film on mica. (see page 76 of this book)

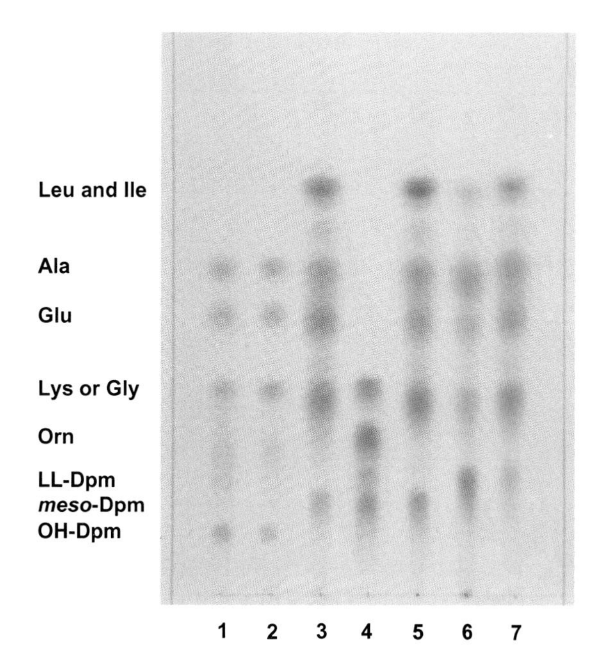

Chapter 5, Figure 3. Thin-layer chromatogram of hydrolysates (4.0 N HCl, 100°C, 16 h) of peptidoglycan preparations (lanes 1 and 2) and whole cells (lanes 3, 5–7) and of a standard (lane 4) consisting of *meso*-Dpm, LL-Dpm, Orn and Lys (each 5 mg/ml). Chromatographic conditions see Protocol 1.
Lane 1: *Rugosimonospora acidiphila* DSM 45227^T (containing OH-Dpm)
Lane 2: *Rugosimonospora africana* DSM 45228^T (containing OH-Dpm)
Lane 3: *Brevibacterium frigoritolerans* DSM 8801^T (containing *meso*-Dpm)
Lane 5: *Bacillus subtilis* subsp. *subtilis* DSM 10^T (containing *meso*-Dpm)
Lane 6 : *Kribbella antibiotica* DSM 15501^T (containing LL-Dpm)
Lane 7: *Kribbella flavida* DSM 17836^T (containing LL-Dpm)
Abbreviations: OH-Dpm, 2,6-diamino-3-hydroxypimelic acid; Dpm, 2,6-diaminopimelic acid. (see page 108 of this book)

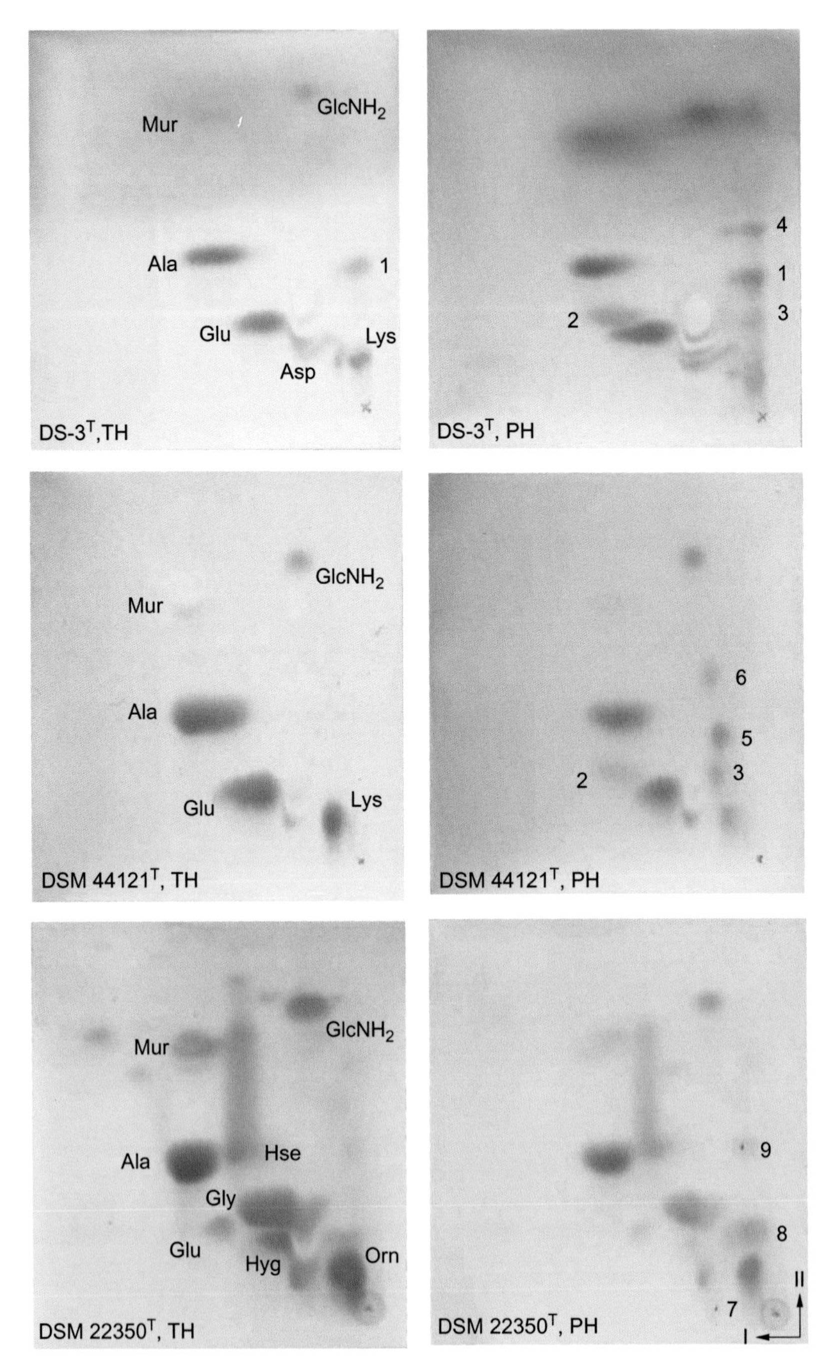

Chapter 5, Figure 4. Two-dimensional cellulose thin-layer chromatograms of total (TH; 4.0 N HCl, 100°C, 16 h) and partial (PH; 4.0 N HCl, 100°C, 0.75 h) hydrolysates of the peptidoglycan of *Isoptericola dokdonensis* DS-3^T (A11.31; A4α L-Lys-D-Asp; Yoon *et al.*, 2006), *Promicromonospora sukumoe* DSM 44121^T (A11.59; A4α L-Lys-L-Ala-D-Glu; Stackebrandt and Schumann, 2004) and

Chapter 5, Figure 4. (*Continued*) *Marisediminicola antarctica* DSM 22350^T (B6; B2β, position 1: Gly, position 3: L-Hse, interpeptide bridge: D-Glu(Hyg)-Gly-D-Orn; Li *et al.* (2010); see also Schleifer, 1970). Chromatographic conditions see Protocol 9.
Abbreviations: Mur, muramic acid; $GlcNH_2$, glucosamine; Hse, homoserine; Hyg, *threo*-3-hydroxy-glutamic acid; I and II, indicating the direction of development with solvent systems I and II.
Peptides:

1. D-Asp-L-Lys
2. L-Ala – D-Glu
3. L-Lys – D-Ala
4. D-Asp – L-Lys – D-Ala
5. L-Ala-L-Lys
6. L-Ala – L-Lys – D-Ala
7. Gly–Hyg, Gly – Hyg – Gly or Gly – Hyg – Hse
8. D-Orn – D-Ala
9. Gly–D-Orn–D-Ala

Peptide 1 is stable under conditions of total hydrolysis. Peptides 7 and 9 appear yellow shortly after spraying with ninhydrin reagent and heating before they turn to bluish. (see page 118 of this book)

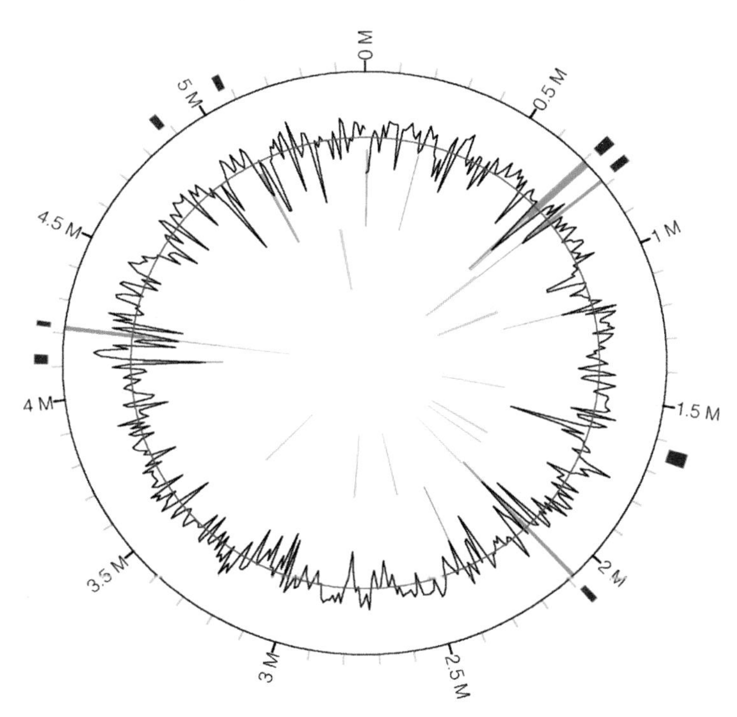

Chapter 18, Figure 5. Identification of genomic islands in *Acidovorax avenae* ATCC 19860 by different prediction tools: SWGIS (red), IslandPath (blue), SIGI-HMM (orange) and IslandPick (green boxes). Black line histogram shows GC% variations. (see page 428 of this book)